HIGHLIGHTS IN CONDENSED MATTER PHYSICS

Related Titles from AIP Conference Proceedings

685   Molecular Nanostructures: XVII International Winterschool/Euroconference on Electronic Properties of Novel Materials
Edited by Hans Kuzmany, Jörg Fink, Michael Mehring, and Siegmar Roth, October 2003, 0-7354-0154-3

678   Lectures on the Physics of Highly Correlated Electron Systems VII: Seventh Training Course in the Physics of Correlated Electron Systems and High-Tc Superconductors
Edited by A. Avella and F. Mancini, August 2003, 0-7354-0147-0

629   Lectures on the Physics of Highly Correlated Electron Systems VI: Sixth Training Course in the Physics of Correlated Electron Systems and High-Tc Superconductors
Edited by F. Mancini, October 2002, 0-7354-0083-0

580   Lectures on the Physics of Highly Correlated Electron Systems V: Fifth Training Course in the Physics of Correlated Electron Systems and High-Tc Superconductors
Edited by Ferdinando Mancini, July 2001, 0-7354-0019-9

527   Lectures on the Physics of Highly Correlated Electron Systems IV: Fourth Training Course in the Physics of Correlated Electron Systems and High-Tc Superconductors
Edited by Ferdinando Mancini, July 2000, 1-56396-950-5

500   The Physics of Electronic and Atomic Collisions: XXI International Conference
Edited by Yukikazu Itikawa, Kazuhiko Okuno, Hiroshi Tanaka, Akira Yagishita, and Michio Matsuzawa, February 2000, 1-56396-777-4

438   Lectures on the Physics of Highly Correlated Electron Systems
Edited by Ferdinando Mancini, July 1998, 1-56396-789-8

To learn more about these titles, or the AIP Conference Proceedings Series, please visit the webpage **http://proceedings.aip.org/proceedings**

HIGHLIGHTS IN CONDENSED MATTER PHYSICS

Salerno, Italy   9-11 May 2003

EDITORS
Adolfo Avella
Roberta Citro
Canio Noce
Mario Salerno
Università degli Studi di Salerno
Salerno, Italy

SPONSORING ORGANIZATIONS
Università degli Studi di Salerno
Istituto Italiano per gli Studi Filosofici
INFM
IIASS

Melville, New York, 2003
AIP CONFERENCE PROCEEDINGS ■ VOLUME 695

Editors:

Adolfo Avella
Roberta Citro
Canio Noce
Mario Salerno

Dipartimento di Fisica "E. R. Caianiello"
Università degli Studi di Salerno
Via S. Allende
84081 Baronissi (SA)
ITALY

E-mail: avella@sa.infn.it
citro@sa.infn.it
canio@sa.infn.it
salerno@sa.infn.it

Authorization to photocopy items for internal or personal use, beyond the free copying permitted under the 1978 U.S. Copyright Law (see statement below), is granted by the American Institute of Physics for users registered with the Copyright Clearance Center (CCC) Transactional Reporting Service, provided that the base fee of $20.00 per copy is paid directly to CCC, 222 Rosewood Drive, Danvers, MA 01923. For those organizations that have been granted a photocopy license by CCC, a separate system of payment has been arranged. The fee code for users of the Transactional Reporting Service is: 0-7354-0167-5/03/$20.00.

© 2003 American Institute of Physics

Individual readers of this volume and nonprofit libraries, acting for them, are permitted to make fair use of the material in it, such as copying an article for use in teaching or research. Permission is granted to quote from this volume in scientific work with the customary acknowledgment of the source. To reprint a figure, table, or other excerpt requires the consent of one of the original authors and notification to AIP. Republication or systematic or multiple reproduction of any material in this volume is permitted only under license from AIP. Address inquiries to Office of Rights and Permissions, Suite 1NO1, 2 Huntington Quadrangle, Melville, N.Y. 11747-4502; phone: 516-576-2268; fax: 516-576-2450; e-mail: rights@aip.org.

L.C. Catalog Card No. 2003114750
ISBN 0-7354-0167-5
ISSN 0094-243X
Printed in the United States of America

Contents

Group Photo of the Participants vii
Preface ix
Signatures x
A Homage to Ferdinando Mancini: A Friend, a Colleague, a Scientist xi
    M. Marinaro and G. Scarpetta
The Genesis of the Boson Transformation xiii
    V. Srinivasan

Phonon-Assisted Strong Cooper Pair Interaction in Highly Correlated Cuprate Systems 1
    M. Tachiki
Aspects of Unconventional Density Waves 10
    K. Maki, B. Dóra, and A. Virosztek
Superconductors as Giant Atoms: Qualitative Aspects 21
    J. E. Hirsh
High Temperature Superconductors: Experimental Implications of a Variational Theory of the Superconducting State 34
    M. Randeria, A. Paramekanti, and N. Trivedi
Critical Temperature of High-T_c Superconductors in the Bipolaron Model 42
    A. S. Alexandrov
Polarons: Recent Developments 47
    J. T. Devreese
Charge Dynamics of t-J Model and Anomalous Bond-Stretching Phonons in Cuprates 65
    P. Horsh and G. Khaliullin
Tools for Studying Quantum Emergence near Phase Transitions 75
    M. Imada, S. Onoda, T. Mizusaki, and S. Watanabe
Antiferromagnetic Exchange and Spin-Fluctuation Pairing Mechanisms in Cuprates 92
    N. M. Plakida
Magnetic Fluctuations and Resonant Peak in Doped Antiferromagnets 101
    P. Prelovek, I. Sega, and J. Bona
A Criterion for the Transition of a Three Dimensional Bravais Lattice from Bulk to Molecular Behaviour 108
    M. Ghanashyam Krishna and V. Srinivasan
Bose-Einstein Condensation in Thermo Field Dynamics 114
    H. Matsumoto
Coherent Structures of Bose-Einstein Condensates in Optical Lattices 126
    B. B. Baizakov, V. V. Konotop, and M. Salerno
Multifractal Analysis of Various PDF in Turbulence Based on Generalized Statistics: A Way to Tangle in Superfluid He 135
    T. Arimitsu and N. Arimitsu
Combinatorial Aspects of Exclusion and Parastatistics 145
    S. Chaturvedi

Charge and Phase Dynamics in a Stack of Intrinsic Josephson Junctions 152
    T. Koyama

Self-Consistent Mean-Field Theory for Frustrated Josephson Junction Arrays 164
    F. P. Mancini, P. Sodano, and A. Trombettoni

Magnetic Interactions in Transition Metal Oxides with Orbital Degrees of Freedom 176
    A. M. Oleś

Orbital Physics Versus Spin Physics 188
    L. F. Feiner and A. M. Oleś

Local Moment Systems: Magnetism and Electronic Correlations 196
    W. Nolting, W. Müller, C. Santos, and P. Sinjukow

Superconductivity and Ferromagnetism: How to Find a Compromise 215
    M. Cuoco, P. Gentile, and C. Noce

Superfluid Properties of the Boson-Fermion Model 230
    R. Micnas, S. Robaszkiewicz, and A. Bussmann-Holder

Composite Operators and Algebra Constraints: A Formalism for Highly Interacting Systems 240
    F. Mancini

Self-Energy Corrections within the Composite Operator Method 258
    A. Avella

The Cumulant Expansion Approach for Strongly Correlated Electron Models 268
    R. Citro and M. Marinaro

Pseudospin-Electron Model for Strongly Correlated Electron Systems (Thermodynamics and Dynamics) 281
    I. V. Stasyuk

Electron Correlations at Nanoscale 291
    J. Spałek, A. Rycerz, E. M. Görlich, and R. Zahorbeński

Spin Ordering in the One-Dimensional Kondo Lattice and Double-Exchange Models 304
    D. J. Garcia, K. Hallberg, M. Avignon, and B. Alascio

Low Temperature Transport Properties of Strongly Interacting Systems — Thermal Conductivity of Spin-1/2 Chains 315
    N. Andrei, E. Shimshoni, and A. Rosch

Decoherence in Dissipative Systems 327
    F. Guinea

Structural Biology of Viruses: A Case of Synergy 333
    E. J. Mancini

Compact Dark Objects and Gravitational Microlensing Towards the Large Magellanic Cloud 339
    L. Mancini, P. Jetzer, and G. Scarpetta

Summary of the Conference "Highlights in Condensed Matter Physics" 348
    A. M. Oleś

Our Own Roots 353
    F. Mancini

Author Index 357

Preface

This book contains the invited lectures delivered at the International Conference *Highlights in Condensed Matter Physics 2003* (HCMP03), held in Salerno at Arechi Castle from 9 to 11 May 2003 to celebrate the 60th birthday of Professor Ferdinando Mancini. It also contains his sons' dedicated contributions as well as ours.

When we had decided how to honor Professor Ferdinando Mancini, we contacted many outstanding scientists involved in the fields which Professor Mancini mostly contributed to during his long career in condensed matter physics (CMP). We received from them all an enthusiastic consensus and to all of these eminent scientists we express our gratitude for promptly and friendly accepting our invitation to deliver lectures at HCMP03.

The articles contained in this volume give a status report on our understanding of the complicated and fascinating field of CMP. Many different topics are discussed and presented in this book; they have in common the quantum complexity of many body problems and the lack of exact solutions in the majority of the cases. In spite of these difficulties, important achievements have been made in recent years and some of them are summarized here. Modern developments are presented and explained in tutorial style, emphasizing the decisive ideas and the hot topics of current and future research in this field. The authors, working on different aspects of solid state physics and employing a variety of techniques, mainly treat the following themes: strongly correlated electron systems and high temperature superconductivity. Special emphasis has also been given to low dimensional systems and Bose-Einstein condensation. In this respect, we are confident that this book will attract the interest of researchers with different expertise in condensed matter physics.

We deeply thank our colleagues, Professors Gaetano Scarpetta and Mario Fusco-Girard, who joined us in the Local Organizing Committee. We wish to express our deepest and sincere gratitude to them.

We wish to acknowledge the help received, at various stages of this venture, from many people in our department. Their contributions are extremely appreciated: Renzo Rubele, Madalina Ciobanu, Domenico Borrelli, Vincenzo Di Marino and Bernardo Amoruso.

We would also like to acknowledge the financial support from Istituto Italiano per gli Studi Filosofici (Naples) in the person of Avv. Gerardo Marotta, Università degli Studi di Salerno and its Facoltà di Scienze MM.FF.NN. and Dipartimento di Fisica "E. R. Caianiello", Provincia di Salerno, Istituto Nazionale di Fisica della Materia (INFM) and Istituto Internazionale per gli Alti Studi Scientifici (IIASS, Vietri sul Mare (SA), Italy). The conference was also sponsored by Regione Campania and Ministero dell'Istruzione, dell'Università e della Ricerca (MIUR).

We are pleased to conclude in wishing Nando to remain so varied in his activity and so deeply involved in research for many more years to come.

Adolfo Avella, Roberta Citro, Canio Noce, and Mario Salerno
Baronissi (SA), October 2, 2003

Hideki Matsumoto

I. V. Stasyuk

G. Japaridze

真木和美

Masashi Tachiki 立木昌

IMR. Tohoku Univ.

Happy birthday! Many more!

11/5/03

Jozef Spałek

Happy Birthday and many years active life!

Sasha

Natasha Perkins

Mohit Randeria

A Homage to Ferdinando Mancini: a Friend, a Colleague, a Scientist

The friendship between Nando and I began many years ago. It was in 1965 when we met at the Department of Theoretical Physics of the University of Naples, where Nando was studying to earn his degree in Physics and I was at the very beginning of my academic career. One day, Nando came into my office and asked me to assist him in the preparation of his thesis. I accepted. Remarkably, this was my first experience as a supervisor, even though I am not quite sure that Nando was aware of this circumstance at the time. After that we continued our collaboration and have had many opportunities to work together on topics such as superconductivity theory, field theory at finite temperature, and more recently, strongly correlated systems. Besides the scientific work, our collaboration has covered other activities, I mean academic duties and the promotion of science in Salerno. These activities started in 1972 when the Faculty of Science was established. He was among the pioneers, sharing with some of us the hard job of constructing and developing a scientific faculty in a place where scientific institutions of *Alta Cultura* were completely absent. The perseverance and firmness that Nando has always shown in overcoming difficulties has been an example to all of us. In conclusion, I find it appropriate on the occasion of his sixtieth birthday, to stress on his large contribution to the development of physics in Salerno, and express to him my congratulations for his important anniversary and best wishes for many active and happy years.

Maria Marinaro

Let me express my affection and deep esteem to my old friend Ferdinando Mancini, on the occasion of this conference, celebrating his sixtieth birthday; our friendship goes back to more than 37 years ago, when I was still a student at the University of Naples. The friendship between our two families strengthened after his return from Milwaukee, where he obtained the Ph.D. in Physics working in the group of Prof. H. Umezawa. We were young and animated by a pioneer spirit and in 1972 we both decided, together with a little group of colleagues, guided by prof. Eduardo R. Caianiello, to move from the University of Naples to Salerno to take part in the difficult enterprise of founding the Science Faculty in the new Salerno University. Ferdinando Mancini has contributed on different levels to the development of physics research in our department and to the reinforcement of the science faculty: first of all with his scientific activity in theoretical physics of condensates states; then his teaching in university courses, through which he molded several young generations of physicists; training young researchers, some of which now belong to the staff of the Physics Department; organizing, since 1996, the "Training Course in the Physics of Correlated Electron Systems and High-T_c Superconductors", held in Vietri sul Mare (Salerno), a fascinating and fast-developing research area of considerable relevance and interest to the physics community; taking on the responsibility for the University management, participating for more than ten years in the University Board of Directors, as elected member; and last but not least, accepting three times to be designed as Director of the Physics Department: under his last Direction he had success in the process of reunification of the two previously separated Physics Departments. In conclusion, my best wishes to you, Ferdinando, to your wife Fernanda and to your three marvellous children, which have followed you in your steps, all heading for the difficult and exciting road of research in physics.

Gaetano Scarpetta

The Genesis of the Boson Transformation

V. Srinivasan

School of Physics, University of Hyderabad - 500 046 (India)

Abstract. The boson transformation is a powerful method of handling space time dependent ground states. It bypasses the Gorkov equations and makes calculations simple. The method is described and Mancini's contribution to the development of this method is discussed.

It gives me great pleasure in writing this Festschrift article about Prof. Manicini's early research life in Milwaukee. Prof. Mancini was a student of Hiroomi Umezawa at the University of Wisconsin, Milwaukee From (1968-1972). He started his research on superconductivity in 1968. At that time Prof. Umezawa continued his work on spontaneous symmetry breaking, which he started at Naples, under the title "Dynamical rearrangement of symmetries". He had three papers; the first on the Nambu model, the second on liquid helium-4 with Sen and the third on superconductivity with Leplae. Umezawa's approach was to expand the Heisenberg field in terms of a complete set of free fields whose masses were unknown and had to be computed self consistently. He termed it the dynamical map and around the same time Haag also advocated a similar approach and it is known as the Haag expansion. Whenever spontaneous symmetry breaking occurs, there occurs in the set of asymptotic fields, a massless particle, or in solid state parlance, a phonon-like dispersion mode named the Nambu-Goldstone mode appears. In the first paper in the series "Dynamical rearrangement of symmetries" on the Nabmu model Umezawa observed that when the Heisenberg field, namely here the fermion field is expanded in terms of a complete set of in-fields namely the free massive fermion field and the massless Bose field, there is an additional degree of freedom. If in the expansion the Bose field $B(x)$ is replaced by $B(x) + f(x)$ where $f(x)$ is a c number field satisfying the same equation as $B(x)$, then the field equation for the Heisenberg field satisfies the same equation as the original one. Since this calculation was done using the chain approximation, it was not a theorem but had to be verified case by case. Much later in 1973 it was proved exactly. It is at this juncture Mancini started his research career. He took up the task of verifying this in the context of superconductivity. This transformation was named the boson transformation; what this transformation does is to take the space independent ground state into a space time dependent ground state. The Bose condensation is controlled by $f(x)$. When $B(x)$ is replaced by $B(x) + f(x)$, the observables like energy, current etc., acquire c number terms. Taking vacuum expectation values, one is led to a set of linear equations in terms of $f(x)$. In the context of superconductivity $2f(x)$ is the phase of the order parameter. The conventional approach for treating these problems is the Gorkov approach, which close to T_c or zero temperature leads to the usual Landau-Ginsburg equations, which are nonlinear equations. The first physical example to be investigated using the boson transformation has the ground state properties of type-II superconductors. This program was initiated by Prof. Mancini.

In this case the equations for the ground state are

$$\nabla^2 f(x) = 0 \qquad (1)$$

$$\vec{J} = \frac{c}{4\pi\lambda_L^2 e} \int d^3 y\, c(x-y) \left[\nabla f(y,t) - \frac{e}{c\hbar}\vec{A}(y,t)\right] \qquad (2)$$

and the usual Maxwell equations

$$\operatorname{curl} \vec{H} = \frac{4\pi \vec{J}}{c}, \quad \vec{H} = \operatorname{curl} \vec{A} \qquad (3)$$

Here λ_L is the London penetration depth. These equations when solved give the property of the ground state. For a single vortex ground state when the external field is along the z-direction

$$f(x) = \tan^{-1}\left(\frac{X}{Y}\right) \qquad (4)$$

This when substituted in (2) and (3) and solved gives

$$H(x) = \frac{\nu\phi}{2\pi\lambda_L^2} \int dk \frac{kc(k)}{k^2 + \frac{c(k)}{\lambda_L^2}} J_0(kr)$$

Here $c(k)$ is the Fourier transform of $c(x-y)$ and $[B(x,t),B(y,t)] = c(\vec{x}-\vec{y})$.
At that time $c(k)$ was calculated upto k^2 order and was found to be

$$c(k) = 1 - k^2 \xi^2$$

with $\alpha = \frac{2\pi^2}{45}$ and ξ is the coherence length. Since $f(x)$ satisfies a linear equation, multivortex solutions are easy to obtain. For if f_1 and f_2 are solutions of the Laplace equation so is $f_1 + f_2$. This way the magnetization for an equilateral triangular lattice, was computed. Also the field due to a single vortex and its current computed. These had surprises. For a single vortex the magnetic field reversed sign when $\kappa = \frac{\lambda}{\xi} < 1.2$, showing the existence of an attractive vortex interaction. The magnetization curves around Hc_1 showed a Van der Vaal behavior when $\kappa < 1.2$. At that time while some experiments on Niobium and Vanadium supported Abrikosov's work while some did not. At this time Mancini, along with the others in the group gave the correct picture and called these superconductors whose $\kappa > \frac{1}{\sqrt{2}} < 1.2$ type-II/I superconductors. It was shown that the phase transition from the Meissner state to the mixed state is not second order but first order and this is due to the attractive vortex interactions. About the same time using L.G. equations Jacobs and Eileenberger got to the same conclusions. This was a very nice timely piece of research initiated and completed by Mancini along with the group members.

Since the boson method was an operator formalism, the usual imaginary time formalism, which was created by Umezawa was not applicable. The success of this method to describe magnetic properties of superconductors created an urgency for the need of an operator formulation of statistical mechanics. This was done immediately by Umezawa and later Nando, Marinaro et al showed its relationship with the Mills method. Umezawa in a private conversation had acknowledged the contribution of Nando and me in the genesis of TFD during his 60th Birthday. Mancini applied the Boson method to the time dependent Josephson effect, with success. It should be recalled that it is difficult to solve time dependent L.G. equations.

In the 70's when Umezawa was away on a sabbatical to Alberta, Mancini worked on the Coulomb effects on superconductivity along with me. We used the Bethe-Salpeter equation and solved it using the chain approximation. Instead of only one bound state, the plasmon, we found two; when the Coulomb field was removed the other bound state persisted. The fermion mass was Δ and this was (2Δ). We wrote a paper with this prediction fitted *Coulomb effects and collective modes in superconductivity* Preprint No. 4867-7172. It was rejected by JMP on the grounds that this new quasi-particle does not exist. Ten years later it was experimentally discovered by Suryakumar et al. These results are also summarized in the Physics Reports Vol. 10C No. 4, 1974, titled *Derivation and application of the boson method in superconductivity*. See for instance Eq.(6.4.2) of this reference.

Later in 1973 just before the conference in Capri which launched the Weinberg-Salam model, Umezawa gave an exact proof of the boson transformation; a simpler proof can be found in the book *Thermo field dynamics and condensed states*. A very elegant description of crystals, defects etc., is given here. The language changed a little, but very much the technology started by Prof. Mancini. Also that there is a quasi-susy because of the amplitude mode and Higgs could be a $t\bar{t}$ bound state are discussed now. But it was Prof. Mancini who initiated these things. The classification of superconductors, the attractive vortex interaction responsible for the type II/I behavior of Vanadium and Niobium, the prediction of the amplitude mode, and the development of a new formalism of TFD are the topics Nando contributed. it shows both his powerful intuition and his computational abilities.

I wish many more happy years of research and happiness to Nando.

The publications and research work of Prof. Mancini during this period is contained in the review article *Derivation and application of the Boson method in superconductivity* by L. Leplae, H. Umezawa and F. Mancini, Vol. 10C, No. 4 (1974).

Phonon-assisted strong Cooper pair interaction in highly correlated cuprate systems

Masashi Tachiki

National Institute for Materials Science 1-2-1 Sengen, Tsukuba 305-0047, Japan

Abstract. The dispersion of the in-plane CuO bond stretching LO phonon mode in the high-T_C cuprate superconductors shows strong softening with the optimum doped samples near the zone boundary, but the overdoped samples with zero T_C does not show the anomalous softening. This fact suggests that the phonon is closely related to the mechanism of the high-T_C superconductivity. The softening can be described with a negative electronic dielectric function that results in overscreening of the inter-site Coulomb interaction in the highly correlated cuprate systems. We propose that such a strong electron-phonon coupling of specific modes can form basis for phonon mechanism of high-temperature superconductivity. On the basis of the model, with the Eliashberg theory using the experimentally determined electron dispersion and dielectric function, we demonstrate the possibility of superconductivity with the order parameter of the symmetry and the transition temperature well in excess of 100K.

INTRODUCTION

The cuprate high T_C superconductors are originally Hubbard or charge transfer insulators. When carriers are doped, the cuprates becomes highly correlated metals. In the metal, the temperature dependence of many physical quantities such as the Hall [1] and Seebeck coefficients [2] are very anomalous. In the insulating state, the cuprates are antiferromagnets. In the metallic state the antiferromagnetic spin fluctuations have been observed neutron scattering, nuclear magnetic resonance, etc. The spin fluctuation-assisted pairing interaction induces the d-wave superconductivity. Therefore, many peoples considered that the spin fluctuations are the origin of the high T_C superconductivity. However, there are some contradictions for the spin mechanism. Kambe et al. observed the spin magnetic fluctuation using nuclear magnetic resonance in $Tl_2Ba_2CuO_8$ and obtained the same spin fluctuations both for the sample with T_C =85K and the over-doped sample with T_C =0 [3].

Naito et al. fabricated $La_{2-x}Ba_xCuO_4$ by epitaxial growth and observed that the sample has the transition temperature of 47K which is higher than T_C =30 of bulk sample [4]. The spin fluctuation of the former sample is lower than that of the latter sample. Recently observed two superconductors, an electron-doped infinite layer superconductor with T_C =43 [5] and MgB_2 with T_C =39 [6] have weak spin fluctuations.

Just after the discovery of high T_C superconductor, isotope effect was observed in $La_{2-x}Ba_xCuO_4$ and $La_{2-x}Sr_xCuO_4$. Anomalous softening was observed in the dispersion of the in-plane Cu-O bond stretching LO phonon of $La_{2-x}Sr_xCuO_4$ [7, 8, 9] and $YBa_2Cu_3O_{7-\delta}$ [10, 11] by neutron inelastic scattering. Quite recently, Mizuki and Yamada using the synchrotron radiation facility-SPring 8, made the experiments of x-ray inelastic scattering of $La_{2-x}Sr_xCuO_4$, and obtained the following result [12]. The dispersions of the ab-plane bond stretching LO phonon mode in $La_{2-x}Sr_xCuO_4$ with a near optimum-doped carrier concentration show the strong softening of the dispersion same as those measured by neutron inelastic scattering. However, in the over-doped samples with x=0.3 and thus with almost T_C =0, the softening of the LO phonon is almost disappears and recovers to the normal phonon dispersion. This fact suggests that the phonon is closely related to the mechanism of high T_C superconductivity.

SOFTENING OF THE DISPERSION OF THE IN-PLANE BOND STRETCHING LO PHONON

The spectral intensity of the longitudinal charge oscillation is given by

$$\rho(\mathbf{q},\omega) = -\frac{1}{\pi} Im \frac{1}{\varepsilon(\mathbf{q},\omega)}, \quad (1)$$

The dynamical dielectric function is given by the sum of the electric dielectric function $\varepsilon_{el}(\mathbf{q},\omega)$ and the ionic dielectric function $\varepsilon_{ion}(\mathbf{q},\omega)$ as

$$\varepsilon(\mathbf{q},\omega) = \varepsilon_{el}(\mathbf{q},\omega) + \varepsilon_{ion}(\mathbf{q},\omega) - 1 \quad (2)$$

A minus unity in the right hand side of Eq. (2) comes from the fact that all the dielectric functions should be unity at the high frequency limit. We express the ionic dielectric function on a conventional form,

$$\varepsilon_{ion}(\mathbf{q},\omega) = \frac{\omega_{LO}^2 - \omega^2}{\omega_{TO}^2 - \omega^2}, \quad (3)$$

$\omega_{LO}(\mathbf{q},\omega)$ and $\omega_{TO}(\mathbf{q},\omega)$ being respectively the frequencies of the bare longitudinal and transverse optical phonons in the insulating state. We express the electronic dielectric constant as

$$\varepsilon_{el}(\mathbf{q},\omega) = \varepsilon_1(\mathbf{q},\omega) + i\varepsilon_2(\mathbf{q},\omega), \quad (4)$$

where $\varepsilon_1(\mathbf{q},\omega)$ and $\varepsilon_2(\mathbf{q},\omega)$ are the real and imaginary part of the electronic dielectric constant, respectively. If we insert the expressions Eqs. (3) and (4) into Eq. (1), we have

$$\rho(\mathbf{q},\omega) = \frac{1}{[\varepsilon_1(\mathbf{q},\omega^*)]^2} \frac{[\omega_{LO}^2(\mathbf{q}) - \omega_{TO}^2(\mathbf{q})]}{2\pi\omega^*(\mathbf{q})} \frac{\Delta(\mathbf{q})}{[\omega - \omega^*(\mathbf{q})]^2 + \Delta^2(\mathbf{q})}, \quad (5)$$

where

$$\omega^*(\mathbf{q})^2 = \omega_{TO}^2(\mathbf{q}) + \frac{[\omega_{LO}^2(\mathbf{q}) - \omega_{TO}^2(\mathbf{q})]}{\varepsilon_1[\mathbf{q},\omega^*(\mathbf{q})]} \quad (6)$$

and

$$\Delta(\mathbf{q}) = \frac{\varepsilon_2(\mathbf{q},\omega^*)}{\varepsilon_1^2(\mathbf{q},\omega^*)} \frac{\omega_{LO}^2(\mathbf{q}) - \omega_{TO}^2(\mathbf{q})}{2\omega^*(\mathbf{q})} \quad (7)$$

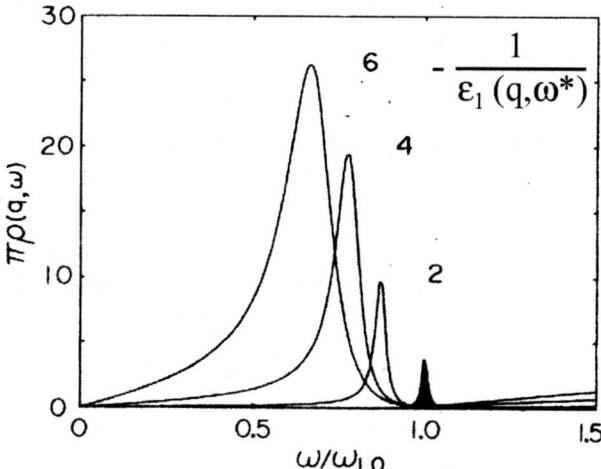

FIGURE 1. Spectral intensities of the charge-transfer oscillations associated with the LO phonon as functions of the normalized frequency ω/ω_{LO} for several values of $-1/\varepsilon_1(\mathbf{q},\omega^*)$

The frequency $\omega^*(\mathbf{q})$ is the frequency of the LO phonon softened by the mixing with charge oscillations. For that the measured LO frequency is softened lower than the bare TO phonon, the real part of the dielectric constant should be negative in the wave number and the frequency range as seen in Eq. (6). An example of the spectral density calculated from Eq. (5) is shown in Fig. 1. In the figure, the black peak corresponds to the spectral intensity of the bare LO phonon. We assumed a small damping constant for the LO and TO phonons. As seen in Fig. 1, as the absolute value of $1/\varepsilon_1(\mathbf{q},\omega^*)$ increases, the peak of the spectral function shifts to a lower frequency and the total intensity increases. This behavior is important to increase the phonon-mediated attractive interaction as shown later. In the next section, we briefly explain the meaning of the negative electronic dielectric constant or the over-screening effect.

MEANING OF OVER-SCREENING

For simplicity, we consider a static case. We write a staggered external electric fields as $D(\mathbf{q})$ and the induced electric charge polarization as $P_{el}(\mathbf{q})$, and define the electronic charge susceptibility $\chi_{el}(\mathbf{q})$ by

$$4\pi P_{el}(\mathbf{q}) = \chi_{el}(\mathbf{q})D(\mathbf{q}) \tag{8}$$

then, combining an electromagnetic relations $D(\mathbf{q}) = E(\mathbf{q}) + 4\pi P_{el}(\mathbf{q}) = \varepsilon_{el}(\mathbf{q},0)E(\mathbf{q})$ with Eq. (8), we have

$$\frac{1}{\varepsilon_{el}(\mathbf{q},0)} = 1 - \chi_{el}(\mathbf{q},0). \tag{9}$$

According to the Kramers-Krönig relation, we have

$$\chi_{el}(\mathbf{q},0) = 2\int_0^\infty d\omega \frac{1}{\omega} \rho_{el}(\mathbf{q},\omega), \tag{10}$$

where $\rho_{el}(\mathbf{q},\omega)$ is the spectral intensity of the electronic charge fluctuations given by

$$\rho_{el}(\mathbf{q},\omega) = -\frac{1}{\pi} Im[\frac{1}{\varepsilon_{el}(\mathbf{q},\omega)}]. \tag{11}$$

The spectral intensity is always positive and therefore we have an inequality from Eqs. (9), (10), and (11),

$$\frac{1}{\varepsilon_{el}(\mathbf{q},0)} \leq 1. \tag{12}$$

The inequality Eq. (12) gives two regions $\varepsilon_{el}(\mathbf{q},0) \geq 1$ and $\varepsilon_{el}(\mathbf{q},0) \leq 0$. The former case is commonly seen for most materials, and the latter case may appear in highly correlated systems like cuprate superconductors. In the cuprates, doped carriers are moving by interacting with antiferromagnetically fluctuating spins, and the kinetic energy of the carriers are much decreases by the friction. At that time, as seen from the Kramers-Krönig relation Eq. (10), $\chi_{el}(\mathbf{q},0)$ possibly becomes larger than unity, causing a negative electronic dielectric constant. At present we are using some model and studying this possibility.

SUPERCONDUCTING GAP FUNCTION AND TRANSITION TEMPERATURE

The quasi-particles are renormalized due to the strong on-site Coulomb interaction. The interaction increases the effective mass of the quasi-partcles, but does not change e of the particles. Therefore, the effective potential acting between quasi-particles $\mathbf{k}$ and $\mathbf{k}'$ is written as

$$V_{eff}(\mathbf{q},\omega) = \frac{V(\mathbf{q})}{\varepsilon(\mathbf{q},\omega)}, \tag{13}$$

where $V(\mathbf{q})$ is the bare Coulomb interaction and $\mathbf{q}$ is $\mathbf{k} - \mathbf{k}'$. With the normal screening the effective potential is always smaller than the bare potential. However, in the case of over-screening $\varepsilon_1(\mathbf{q},\omega)$ can be negative, and thus $\varepsilon_1(\mathbf{q},\omega)$ contributes to the effective interaction and works to enhance the phonon-mediated attractive interaction as seen later. Since the over-screening effect comes from various kinds of the correlation effect, $\varepsilon(\mathbf{q},\omega)$ is a complicate function of degrees of freedoms of charge, spin, and lattice. This effect is the core of the present mechanism. Consequently, the attractive interaction should exist between the quasi-particles with $\mathbf{k}$ and $\mathbf{k}'$, when $\mathbf{q} = \mathbf{k} - \mathbf{k}'$ is in the $\mathbf{q}$ regions where the phonon softening occurs as shown later. The spectral intensity function of the total charge fluctuations is expressed by using Eqs. (1), (2), and (3) as

$$\rho(\mathbf{q},\omega) = -\frac{1}{\pi}Im[\frac{1}{\varepsilon(\mathbf{q},\omega)}]$$
$$= -\frac{1}{\pi}Im[\frac{1}{\varepsilon_{el}(\mathbf{q},\omega)}] + \frac{\omega_{LO}^2(\mathbf{q}) - \omega_{TO}^2}{\varepsilon_{el}(\mathbf{q},\omega_{LO}^*(\mathbf{q}))^2}\delta(\omega^2 - \omega_{LO}^{*2}(\mathbf{q})), \quad (14)$$

where the first term in the right hand side of Eq.(14) is the electronic spectral intensity $\rho_{el}(\mathbf{q},\omega)$, and $\omega_{LO}^*(\mathbf{q})$ is the LO phonon frequency renormalized by charge fluctuations. If we use the spectral representation for $1/\varepsilon(\mathbf{q},\omega)$, the effective interaction Eq.(13) is written as

$$V_{eff}(\mathbf{q},\omega) = V(\mathbf{q})/\varepsilon(\mathbf{q},\omega) = V(\mathbf{q})[1 - 2\int_0^\infty d\Omega \frac{\Omega\rho(\mathbf{q},\Omega)}{\Omega^2 - (\omega + i\delta)^2}]. \quad (15)$$

Using Eq.(15) we set up the Eliashberg equation linearized with respect to the gap function $\Delta(\mathbf{k},i\omega_n)$ as

$$\Delta(\mathbf{k},i\omega_n) = -T\sum_\ell \sum_{\mathbf{k}'} V_{eff}(\mathbf{k} - \mathbf{k}', i\omega_n - i\omega_\ell) \frac{\Delta(\mathbf{k}',\omega_\ell)}{\xi_{\mathbf{k}'}^2 + \omega_\ell^2}, \quad (16)$$

where $\omega_n = (2n+1)\pi T$ with n being integers, and ξ_k is the quasi-particle energy measured from the Fermi level. We use an approximation that the damping of the quasi-particles is neglected. However, the modification of the band structure due to the correlation effect is taken into account by using the model band structure determined by the experimental results of angle-resolved photoemmision. From Eq. (16) we obtain the superconducting order parameter $\Phi(\mathbf{k}) = Re\Delta(\mathbf{k},\xi_\mathbf{k})$ to a good approximation as

$$\Phi(\mathbf{k}) = -\sum_\mathbf{k} K(\mathbf{k},\mathbf{k}') \frac{\tanh \xi_{\mathbf{k}'}/2T}{2\xi_{\mathbf{k}'}} \Phi(\mathbf{k}'). \quad (17)$$

We abbreviated the detailed calculation and an approximation to obtain Eq. (17). The kernel in Eq. (17) is written as

$$K(\mathbf{k},\mathbf{k}') = v(\mathbf{k}-\mathbf{k}')[\frac{1}{\varepsilon_{el}(\mathbf{q},0)} + 2\int_0^\infty d\Omega \frac{|\xi_\mathbf{k}| + |\xi_{\mathbf{k}'}|}{\Omega(\Omega + |\xi_\mathbf{k}| + |\xi_{\mathbf{k}'}|)} \rho_{el}(\mathbf{q},\Omega)$$
$$-\frac{1}{\varepsilon_{el}(\mathbf{q},\omega_{LO}^*(\mathbf{q}))} \frac{\omega_{LO}^{*2}(\mathbf{q}) - \omega_{TO}^2}{\omega_{LO}^*(\mathbf{q})[\omega_{LO}^*(\mathbf{q}) + |\xi_\mathbf{k}| + |\xi_{\mathbf{k}'}|]}]. \quad (18)$$

Roughly speaking, in the right hand side of Eq. (18) the first two terms are the electronic contribution and the third term is the phonon contribution. However the electronic contribution is mixed in the third term as seen in Eq. (18). Therefore, the kernel has a vibronic nature. In conventional system with $\varepsilon_1(\mathbf{q},\omega) \geq 1$ the third term in Eq. (18) cannot be enhanced. However, when $\varepsilon_1(\mathbf{q},\omega) < 0$ and its absolute value is small, the phonon contribution can be strongly enhanced. It is physically understood that this enhancement comes from the increase of the spectral intensity when $\varepsilon_1(\mathbf{q},\omega)$ is negative and the frequency of the LO phonon is softened as discussed in section II.

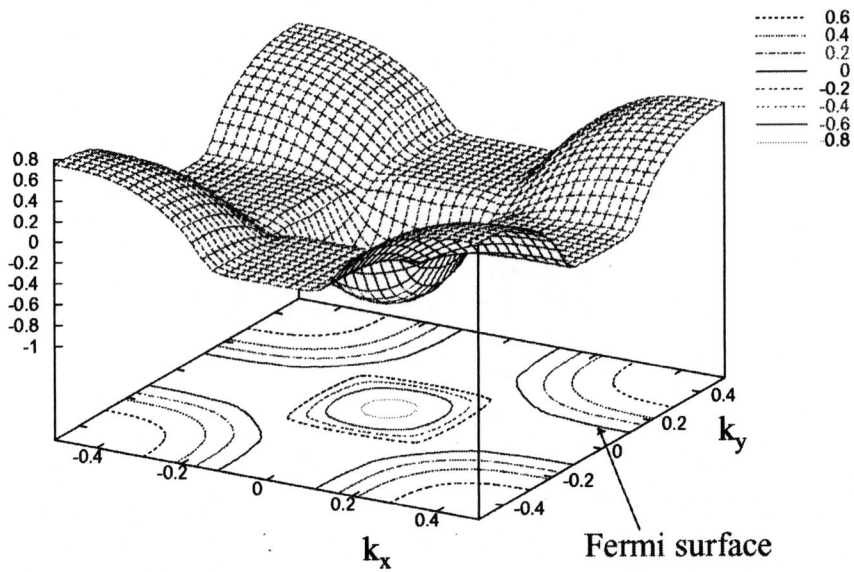

FIGURE 2. Electron energy band structure measured by angle-resolved photoemission in optimum doped cuprates [13, 14]

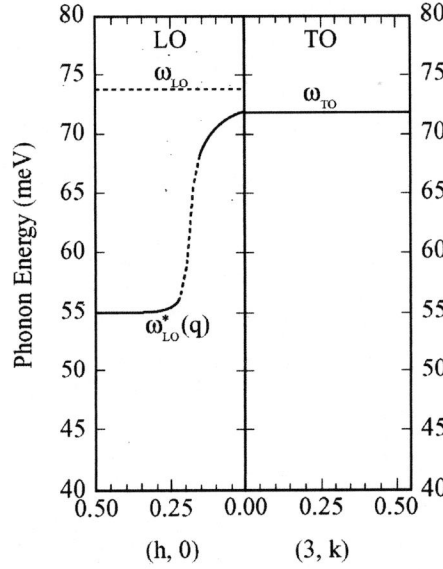

FIGURE 3. LO and TO phonon dispersions of the B_{2g} mode in YBCO by neutron inelastic scattering [15]

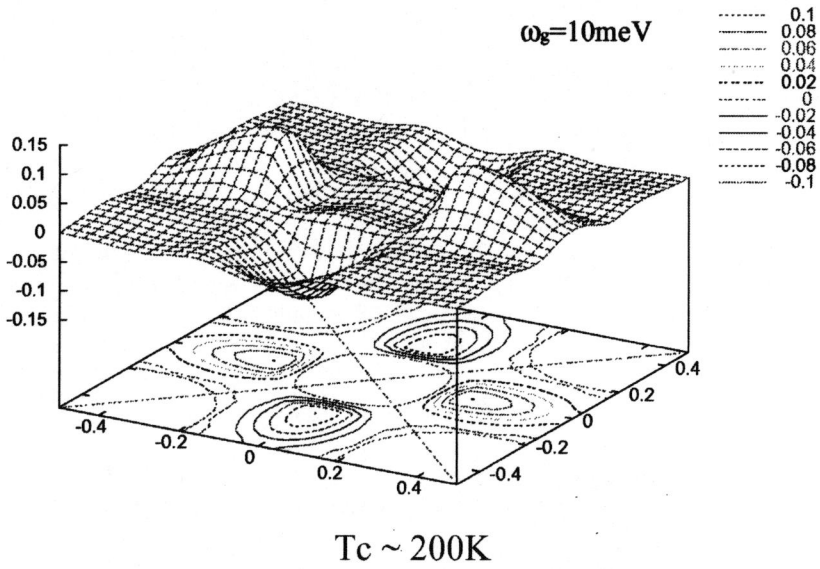

Tc ~ 200K

FIGURE 4. Superconducting order parameter in arbitrary units

SUPERCONDUCTING GAP SYMMETRY AND T_C

In Eq. (18), the quasi-particle energy ξ_k is determined by utilizing the experimental data of the angle-resolved photoemission. The band structure is almost universal for all the cuprate superconductor with optimum dope. The characteristic of the band is that it has flat regions along the k_x and k_y axes which originates from the strong correlation effect. Hereafter, we express the wave number in units of $2\pi/a$. The figure 2 shows the angle-resolved photoemission data in the optimum doped cuprates. The Figure 3 schematically shows the dispersions of LO and TO phonon measured by Chung et al. using optimum doped YBCO [15]. The anomalous softening occurs narrow regions around the q_x and q_y axes. We used the experimental values for $\omega_{LO}^*(q)$ in the third term in the bracket of Eq. (18). The dielectric function in Eq. (18) is approximated by its real part and its value is estimated from the experimental values in Fig. 3 using Eq. (6). Then, solving Eq. (17), we obtained the infinitesimal superconducting order parameter at T_c shown in Fig. 4 and the value of T_c. As seen in Fig. 4, the order parameter has a d-wave symmetry. The transition temperature is of order of 100K.

DISCUSSIONS AND SUMMARY

The strong softening in the LO phonon of the in-plane Cu-O stretching mode in cuprates was observed by neutron inelastic scattering and x-ray inelastic scattering of synchrotron radiation. This phenomenon is explained by the fact that the phonons strongly mix with charge fluctuations of low frequencies. The low frequency charge fluctuations may occur from the fact that the doped carriers are moving around interacting with spin fluctuations. This effect may give a large friction for the charge motion. For this prediction we need a theoretical confirmation using a model calculation.

In the state of the phonon softened, the spectral intensity of the longitudinal charge oscillation becomes enormously large. By this effect the phonon-mediated attractive interaction is strongly enhanced and the superconducting transition is in excess of 100K. Since the phonon anomaly is localized in narrow regions, the superconducting order parameter is of $d(k_x^2 - k_y^2)$ symmetry.

The phenomena mentioned above occur on doped-cuprates that are highly correlated systems of charge, spin, and crystal lattice. In these systems, the frequencies of the charge oscillation may be strongly softened and the Coulomb interaction is over-screen.

ACKNOWLEDGMENTS

This research has been done by collaborations with Dr. Masahiko Machida at Japan Atomic Energy Institute and Prof. Takeshi Egami University of Pennsylvania. I would like to express sincere thanks to them.

REFERENCES

1. Chien, T. R., Wang, Z. Z., and Ong, N. P., *Phys. Rev. Lett.* **67**, 2088 (1991); Kendoriza, C., Mandrus, D., Mihaly, L., and Forro, L., *Phys. Rev. B* **46**, 14293 (1992); Carington, A., Machenzie, A. P., Lin, C. T., and Cooper, J. R., *Phys. Rev. Lett.* **69**, 2855 (1992); Ginzberg, D. M., Lee, W. C., and Stupp, S. E., *Phys. Rev. B* **47**, 12167 (1993); Nishikawa, T., Takeda, J., and Sato, M., *J. Phys. Soc. Jpn.* **63**, 1441 (1994); Takeda, J., Nishikawa, T., and Sato, M., *Physica C* **231**, 293 (1994).
2. Uher, C., Kaiser, A. B., Gmelin, E., and Waltz, I., *Phys. Rev. B* **36**, 5676 (1987); Sera, M., and Sato, M., *Physica C* **185-189**, 1339 (1991); Cohn, J. L., Skelton, E. F., and Wolf, S. A., *Phys. Rev. B* **45**, 13140 (1992); Xin, Y., Wang, K. W., Fan, C. X., Sheng, Z. Z., and Chang, F. T., *Phys. Rev. B* **48**, 557 (1993).
3. Kambe, S., Yasuoka, H., Hayashi, A., and Ueda, Y., *Phys. Rev. B* **47**, 2825 (1993).
4. Sato, H., Tsukada. A., Naito, M., and Matsuda, A., *Phys. Rev. B* **62**, R7999 (2000).
5. Chen, C.-T., Seneor, P., Yeh, N.-C., Vasquez, R. P., Bell, L. D., Jung, C. U., Kim, J. Y., Park, M.-S., Kim, H.-J., and Lee, S.-I., *Phys. Rev. Lett.* **88**, 227002 (2002).
6. Nagamatsu, J., Nakagawa, N., Muranaka, T., Zenitani, Y., and Akimitsu, J., *Nature* **410**, 63 (2001).
7. Pintschovius, L., and Reichardt, W., *in Physical Properties of High Temperature Superconductors iV*, ed. Gingberg, D., (Singapore, World Science) 295 (1994).
8. Phintschovius, L., Pyka, N., Reichardt, W., Rumiantsev, A. Y., Mitrofanov, N. L., Ivanov, A. S., Collin, G., and Bourges, P., *Physica C* **185-189**, 156 (1991).
9. McQueeney, R. J., Petrov, Y., Egami, T., Yethiraj, M., Shirane, G., and Endoh, Y., *Phys, Rev. Lett.* **82**, 628 (1999).
10. Egami, T., McQueeney, R. J., Chung, J.-H., Yethiraj, M., Arai, M., Inamura, Y., Endoh, Y., Tajima, S., Frost, C., and Dogan, F., *Applied Physics B*, in press.
11. Egami, T., Chung, J.-H., McQueeney, R. J., Yethiraj, M., Mook, H. A., Frost, C., Petrov, Y., Dogan, F., Inamura, Y., Arai, M., Tajima, S., and Endoh, Y., *Physica B* **316-317**, 62 (2000).
12. Yamada, K., and Mizuki, J., (Private communication)
13. Ino, A., Kim, C., Nakamura, M., Yoshida, T., Mizokawa, T., Fujimori, A., Shen, Z.-X., Kakeshita, T., Eisaki, H., and Uchida, S., *Pjhys. Rev. B***65**, 094504 (2002); Lu, D. H., Feng, D. L., Armitage, N. P., Shen, K. M., Damascelli, A., Kim, C., Ronning, F., Shen, Z.-X., Bonn, D. A., Liang, R., Hardy, W. N., Rykov, A. I., and Tajima, S., *Phys. Rev. lett.* **86**, 4370 (2001); Norman, M. R., Ding, H., Randeria, M.,

Campuzano, J. C., Yokoya, T., Takeuchi, T., Takahashi, T., Mochiku, T., Kadowaki, K., Guptasarma, P., and Hinks, D. G., *Nature* **392**, 157 (1998).
14. Lu, D. H., Feng, D. L., Armitage, N. P., Shen, K. M., Damascelli, A., Kim, C., Ronnig, F., Shen, Z.-X., Bonn, D. A., Liang, R., Hardy, W. N., Rykov, A. I., and Tajima, S., *Phys. Rev. Lett.* **86**, 4370 (2001).
15. Chung, J.-H., Egami, T., McQueeney, R. J., Yethiraj, M., Arai, M., Yokoo, T., Petrov, Y., Mook. H. A., Endoh, Y., Tajima, S., Frost, C., and Dogan, F., *Phys. Rev. B***67**, 014517 (2003).

Aspects of unconventional density waves

Kazumi Maki*, Balázs Dóra† and Attila Virosztek**‡

*Department of Physics and Astronomy, University of Southern California, Los Angeles
CA 90089-0484, USA
†The Abdus Salam ICTP, Strada Costiera 11, I-34014, Trieste, Italy
**Department of Physics, Budapest University of Technology and Economics, H-1521 Budapest, Hungary
‡Research Institute for Solid State Physics and Optics, P.O.Box 49, H-1525 Budapest, Hungary

Abstract. Recently many people discuss unconventional density waves (i.e. unconventional charge density waves (UCDW) and unconventional spin density waves (USDW)). Unlike in conventional density waves, the quasiparticle spectrum in these systems is gapless. Also these systems remain metallic. Indeed it appears that there are many candidates for UDW. The low temperature phase of α-(BEDT-TTF)$_2$KHg(SCN)$_4$, the antiferromagnetic phase in URu$_2$Si$_2$, the CDW in transition metal dichalcogenite NbSe$_2$, the pseudogap phase in high T_c cuprate superconductors, the glassy phase in organic superconductor κ-(BEDT-TTF)$_2$Cu[N(CN)$_2$]Br. After a brief introduction on UCDW and USDW, we shall discuss some of the above systems, where we believe we have evidence for unconventional density waves.

INTRODUCTION

As is well known quasi-one dimensional electron systems have four canonical ground states: s-wave (spin-singlet) superconductor, p-wave (spin-triplet) superconductor, charge density wave (CDW) and spin density wave (SDW) [1, 2, 3, 4]. All of these states have quasiparticle (QP) energy gap Δ and their QP density decreases exponentially at low temperatures ($T \ll \Delta$). Also the thermodynamics of these states is practically described by the BCS theory of s-wave superconductors[5]. Indeed except for p-wave superconductors these ground states have been found and their properties are actively persued even today. As to p-wave superconductors it is most likely realized in quasi-one dimensional superconductor Bechgaard salts or (TMTSF)$_2$X with X=PF$_6$ and ClO$_4$[6]. The thermal conductivity measurement show the presence of energy gap, and most recent NMR study indicates the triplet pairing[6].

However, since 1979 a new class of superconductors have appeared on the scene: heavy fermion superconductors (1979), organic superconductors (1980), high T_c cuprate superconductors (1986), Sr$_2$RuO$_4$ (1994) and rare earth transition metal borocarbides (1994). Now most of these new superconductors look like unconventional and/or nodal[7, 8]. However only recently d$_{x^2-y^2}$-symmetry of both hole and electron doped superconductors have been established[9]. Also the superconductivity in Sr$_2$RuO$_4$ is f-wave[10, 11, 12] and the one in YNi$_2$B$_2$C and LuNi$_2$B$_2$C is s+g-wave[13, 14]. Therefore unconventional superconductivity has taken center stage in the 21st century physics.

Parallel to these developments many people consider unconventional and/or nodal density waves (UDW)[15, 16, 17, 18]. We believe now that the low-temperature phase (LTP) of α-(BEDT-TTF)$_2$KHg(SCN)$_4$[19, 20, 21, 22], the antiferromagnetic phase in URu$_2$Si$_2$[23, 24], the CDW in 2H-NbSe$_2$[25], the pseudogap phase in high T_c cuprates[16, 18, 26] and the glassy phase in κ-(ET)$_2$ salts[27] belong to UDW.

PHYSICAL PROPERTIES OF UCDW AND USDW

First of all the thermodynamic properties of UDW are very well described in terms of mean field theory like the BCS one. In fact the thermodynamics of most of UDW is described in terms of the BCS theory for d-wave superconductors[8, 28]. Qualitatively the thermodynamics of d-wave superconductors is not much different from the one for s-wave superconductors. In particular a clear jump in the specific heat at $T = T_c$ (transition temperature) is observable in both cases. On the other hand at low temperatures, unlike in conventional DW, there are nodal excitations, giving rise to the power law specific heat like $C \sim T^2$. Also due to the nodal excitations UCDW and USDW are metallic down to $T = 0K$. Further unlike conventional density wave there is no clear x-ray or spin signal indicating the phase transition, since $\langle \Delta(\mathbf{k}) \rangle = 0$. Here $\langle \ldots \rangle$ means average over the Fermi surface. For this reason UDW is an important candidate for states with hidden order parameter.

For the existence of UDW we need higher dimensionality and competing interactions. Therefore we can see here clearly the paradigm shift from quasi-one dimensional systems to quasi-two dimensional and three dimensional systems. Also in order to study UDW experimentally we need more subtle and delicate technique. In this context the angular dependent magnetoresistance provides a unique window to study UDW.

ANGULAR DEPENDENT MAGNETORESISTANCE IN α-(BEDT-TTF)$_2$KHG(SCN)$_4$

The LTP in α-(BEDT-TTF)$_2$MHg(SCN)$_4$ with M=K, Tl, Rb is still controversial. This compound is quasi-two dimensional system with 1D like and 2D like Fermi surfaces as shown in Fig. 1[29]. From the magnetic phase diagram in a magnetic field $\mathbf{H} \parallel b^*$, it is believed that the LTP is not SDW but a kind of CDW[30]. We have proposed recently that UCDW can account for a number of features in LTP of α-(BEDT-TTF)$_2$KHg(SCN)$_4$ including the threshold electric field[19, 20, 31, 32]. More recently we have discovered that the angular dependent magnetoresistance (ADMR) observed in LTP can be interpreted in terms of Landau quantization of the quasiparticle spectrum in UCDW[15, 22, 33]. First let us assume that the QP spectrum in UCDW is given by[22, 33]

$$E(\mathbf{k}) = \sqrt{\xi^2 + \Delta^2(\mathbf{k})} - \varepsilon_0 \cos(2\mathbf{b}'\mathbf{k}), \qquad (1)$$

where $\xi \approx v_a(k_a - k_F)$, v_a is the Fermi velocity, $\Delta(\mathbf{k}) = \Delta\cos(ck_z)$, $\mathbf{b}'$ is the vector lying outside of the $a-c$ plane and ε_0 is the parameter describing the imperfect nesting[34, 35, 36, 21]. In fitting the experimental data we discovered that 1. Eq. (1) gives only one

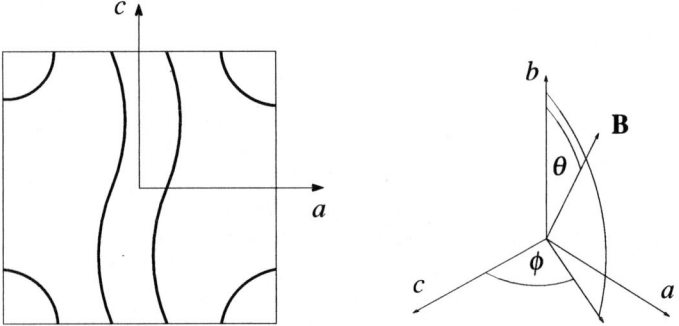

FIGURE 1. The Fermi surface of α-(BEDT-TTF)$_2$KHg(SCN)$_4$ is shown in the left panel. In the right one the geometrical configuration of the magnetic field with respect to the conducting plane is plotted.

single dip in ADMR, 2. therefore the imperfect nesting term has to be generalized as

$$\varepsilon_0 \cos(2\mathbf{b}'\mathbf{k}) \longrightarrow \sum_n \varepsilon_n \cos(2\mathbf{b}'_n \mathbf{k}), \tag{2}$$

where $\mathbf{b}'_n = b'[\hat{\mathbf{r}}_b + \tan(\theta_n)(\hat{\mathbf{r}}_a \cos\phi_0 + \hat{\mathbf{r}}_c \sin\phi_0)]$, $\varepsilon_n = \varepsilon_0 2^{-|n|}$, $\tan(\theta_n) = \tan(\theta_0) + nd_0$. Indeed ADMR has a broad peak at $\mathbf{H} \parallel b^*$ (or $\theta = 0$) and exhibits a number of dips at $\theta = \theta_n$ (see Fig. 3!)

$$\tan(\theta_n)\cos(\phi - \phi_0) = \tan(\theta_0) + nd_0, \tag{3}$$

where $\tan\theta_0 \simeq 0.5$, $d_0 \simeq 1.25$, $\phi_0 \simeq 27°$ and $n = 0, \pm 1, \pm 2\ldots$ [37, 38]. Now in the presence of magnetic field $\mathbf{H}$ with the orientation described by θ and ϕ (see Fig. 1), the QP spectrum changes to

$$E_n = \pm\sqrt{2nv_a\Delta ce|B\cos\theta|}, \tag{4}$$

where $n = 0, 1, 2\ldots$. This is readily obtained following Ref. [15]. The contribution from the imperfect nesting term is considered as a perturbation and the lowest order corrections to the energy spectrum are given by:

$$E_0^1 = E_1^1 = -\sum_m \varepsilon_m \exp(-y_m), \tag{5}$$

$$E_1^2 = -\sum_m \varepsilon_m(1 - 2y_m)\exp(-y_m), \tag{6}$$

where $y_m = v_a b'^2 e|B\cos(\theta)|[\tan(\theta)\cos(\phi - \phi_o) - \tan(\theta_m)]^2/\Delta c$. The $n = 1$ level was twofold degenerate, but the imperfect nesting term splits the degeneracy by E_1^1 and E_1^2. Also the imperfect nesting term breaks the particle-hole symmetry. When $\beta E_1 \gg 1$ ($\beta = (k_BT)^{-1}$), the quasiparticle transport in the quasi-one dimensional Fermi surface is dominated by the quasiparticles at $n = 0$ and $n = 1$ Landau levels. Considering that there are 2 conducting channels and only the quasi-one dimensional one is affected by the appearance of UCDW, the ADMR is written as

$$R(B,\theta,\phi)^{-1} = 2\sigma_1\left(\frac{\exp(-\beta E_1) + \cosh(\beta E_1^1)}{\cosh(\beta E_1) + \cosh(\beta E_1^1)} + \frac{\exp(-\beta E_1) + \cosh(\beta E_1^2)}{\cosh(\beta E_1) + \cosh(\beta E_1^2)}\right) + \sigma_2 \tag{7}$$

Here σ_1 and σ_2 are the conductivities of the $n = 1$ Landau level and quasi-two dimensional channels, in which the contribution of the $n = 0$ Landau level was melted, respectively. The same expressions were found for $\Delta(\mathbf{k}) = \Delta \sin(ck_z)$.

Eq. (7) is compared to the ADMR data taken from a single crystal of α-(BEDT-TTF)$_2$KHg(SCN)$_4$ for the temperature interval 1.4-20 K under magnetic field up to 15 T[33]. The ADMR data are consistent with the previous reports[38, 39, 40]. In Fig. 2 we compare the B dependence of the magnetoresistance at $T = 1.4$ K and $T = 4.14$ K and the T dependence of the magnetoresistance for $B = 15$ T, for $\theta = 0°$. In fitting the temperature dependence of the resistivity, we assumed $\Delta(T)/\Delta(0) = \sqrt{1 - (T/T_c)^3}$, which was found to be very close to the exact solution of $\Delta(T)$[17]. The influence of imperfect nesting terms in these cases is negligible, since they contribute only close to $\theta = \theta_n$.

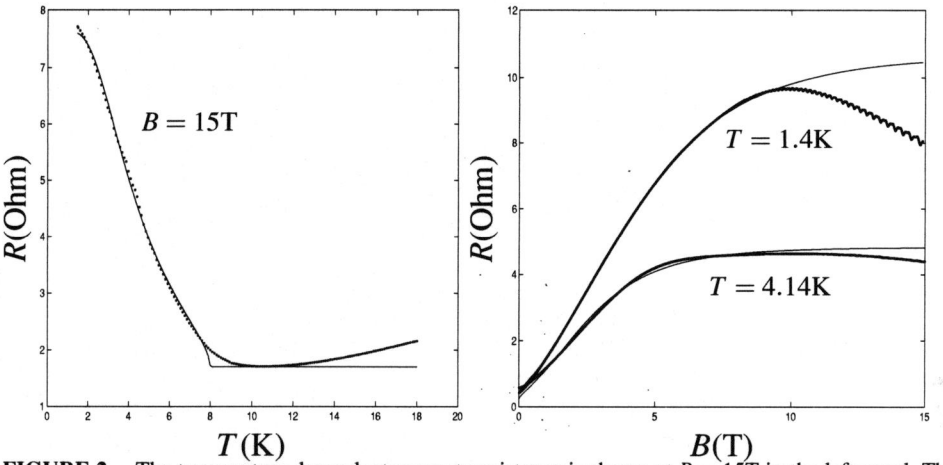

FIGURE 2. The temperature dependent magnetoresistance is shown at $B = 15$T in the left panel. The dots are the experimental data, the solid line is our fit. In the right panel, the magnetoresistance is plotted for $T = 1.4$K and 4.14K as a function of magnetic field. The thick solid line is the experimental data, the thin one denotes our fit based on Eq. (7).

Clearly the fitting becomes better as T decreases and/or B increases. Also for $T = 1.4$ K Shubnikov-de Haas oscillation becomes visible around $B = 10$ T, then the fitting starts breaking away. Clearly in this high field region the quantization of Fermi surface itself starts interfering with the quantization described above. Also the deviation of the theoretical curve from the experimental one above T_c in Fig. 2 is due to the fact that the higher Landau levels contribute in this high temperature regime. From these fittings we can deduce $\sigma_2/\sigma_1 \sim 0.1$ and 0.3, and by assuming the mean field value of Δ (17 K), we get $v_a \sim 6 \times 10^6$ cm/s. In Fig. 3 we show the experimental data of ADMR as a function of θ for current parallel and perpendicular to the conducting plane for $T = 1.4$ K, $B = 15$ T and $\phi = 45°$. As is readily seen the fittings are excellent. From this we deduce $\sigma_2/\sigma_1 \sim 0.1$, $b' \sim 30$ Å, $\varepsilon_0 \sim 3$ K. This b' is comparable to the lattice constant $b = 20.56$ Å. Finally we show in Fig. 4 R versus θ for different ϕ and compare with the experimental data side by side. Perhaps there are still differences in some details but the overall agreement is very striking. The present model can describe a similar figure found in Ref. [40] rather well.

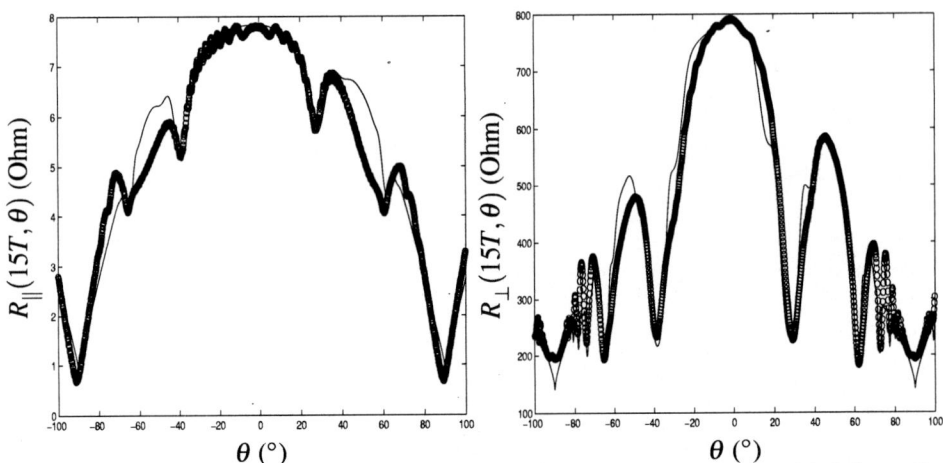

FIGURE 3. The angular dependent magnetoresistance is shown for current parallel (left panel) and perpendicular (right panel) to the $a-c$ plane at $T = 1.4$K, $B = 15$T, $\phi = 45°$. The open circles belong to the experimental data, the solid line is our fit based on Eq. (7).

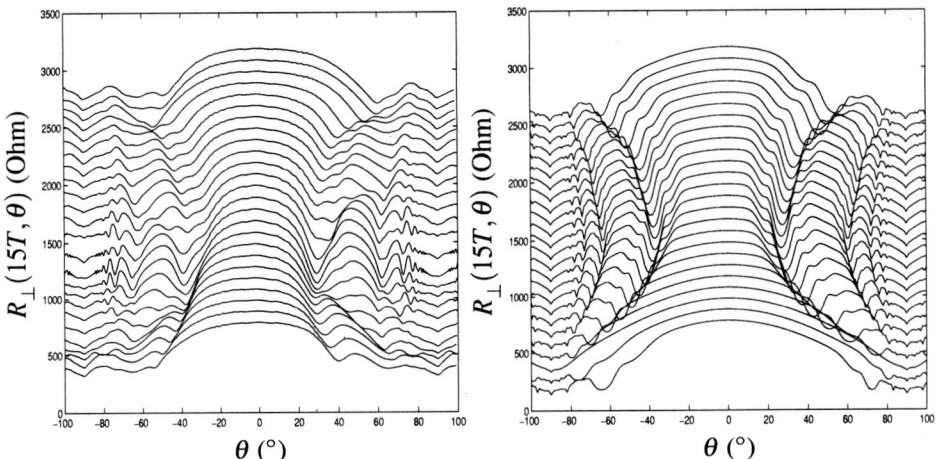

FIGURE 4. ADMR is shown for current perpendicular to the $a-c$ plane at $T = 1.4$K and $B = 15T$ for $\phi = -77°$, $-70°$, $-62.5°$, $-55°$, $-47°$, $-39°$, $-30.5°$, $-22°$, $-14°$, $-6°$, $2°$, $10°$, $23°$, $33°$, $41°$, $48.5°$, $56°$, $61°$, $64°$, $67°$, $73°$, $80°$, $88.5°$, $92°$ and $96°$ from bottom to top. The left (right) panel shows experimental (theoretical) curves, which are shifted from their original position along the vertical axis by $n \times 100$Ohm, $n = 0$ for $\phi = -77°$, $n = 1$ for $\phi = -70°$, ...

In summary the Landau quantization of the QP spectrum of UDW as proposed by Nersesyan et al.[15] can account for the striking ADMR found in LTP of α-(BEDT-TTF)$_2$KHg(SCN)$_4$. Very similar ADMR have been seen also in M=Rb and Tl compounds. Therefore we conclude that LTP in α-(BEDT-TTF)$_2$MHg(SCN)$_4$ salts should be UCDW. Also we believe that ADMR provides clear signature for the presence of

UCDW and USDW. Therefore this technique can be exploited for other possible candidates of UDW.

PSEUDOGAP PHASE IN HIGH T_C CUPRATES

We believe that the most important legacy of high T_c cuprates is that the mean field theory like the Landau theory of Fermi liquid[41, 42, 43] and the BCS theory of superconductivity[5] works in the quasi-two dimensional system with strong electron correlations[44, 45]. Of course the Fermi liquid theory as formulated by Landau for a spherical Fermi surface cannot be applied directly to high T_c cuprates. In particular the quasi-two dimensionality and the resulting nesting feature of the Fermi surface has to be considered. But this feature is readily handled in terms of the renormalization group theory of two dimensional Fermi liquid[46, 47, 48]. Also as to the superconductivity the one band Hubbard model will give the simplest starting point. Then as discussed by Scalapino and others, $d_{x^2-y^2}$ superconductivity follows immediately[49, 50]. Further if one limits oneself to single crystals of optimally doped high T_c cuprates, one can do quantitative test of the BCS theory of d-wave superconductor in the weak coupling limit[28, 45, 51]. Unfortunately until now only three kinds of single crystals of high T_c cuprates are available: LSCO, YBCO and Bi2212. If you compare $\Delta(0)/k_B T_c = 2.14$, the weak coupling theory prediction[28] for d-wave superconductor ($\Delta(0)$ is the maximum of the energy gap) to the one obtained for the optimally doped single crystals, we obtain 2.14, 2.8, 5 for LSCO, YBCO and Bi2212 respectively. This means that the superconductivity in LSCO is very close to the weak coupling limit, the one in YBCO is moderately in the strong coupling limit, while Bi2212 is definitely in the strong coupling limit[8]. In Fig. 5 a generic phase diagram of the hole doped high T_c cuprates is shown. It is still controversial where T^* line hits the superconducting transition temperature curve T_c. But from the validity of the mean field theory at optimal doping we assume that it hits somewhat in the underdoped side. Then it is possible that the extension of this line continues to $T = 0$ K at $x = 0.15$ at the quantum critical point. On this point we may refer to an earlier resistivity measurement in high magnetic field though it is limited unfortunately to only LSCO system[52]. Therefore the d-wave superconductivity in high T_c cuprates is well understood in terms of two dimensional one band Hubbard model except one caveat: what means T^*? Earlier it was believed that T^* is a crossover temperature where either superconducting or antiferromagnetic fluctuations becomes important[53]. More recently possible phase transition to d-wave density wave at $T = T^*$ has been proposed[16, 18, 26]. The most serious objection to this model is that no jump in the specific heat at $T = T^*$ has been observed until now, though many physical quantities like nuclear spin lattice relaxation rate T_1^{-1}, magnetic susceptibility, electric conductivity exhibit kinks at T^*[54]. The d-wave nature of density wave has been established by angular dependent photo-electron spectrum study[55]. Another less indirect signature of d-wave is the surprising relation $\Delta(0)/T^* = 2.14$ (the weak coupling result for d-wave density wave as well as for d-wave superconductor[17, 28]) established by STM study of $\Delta(0)$ (the energy gap in the density of states at $T = 0$ K) in LSCO, YBCO and Bi2212[56]. Therefore the only remaining question is if it is UCDW

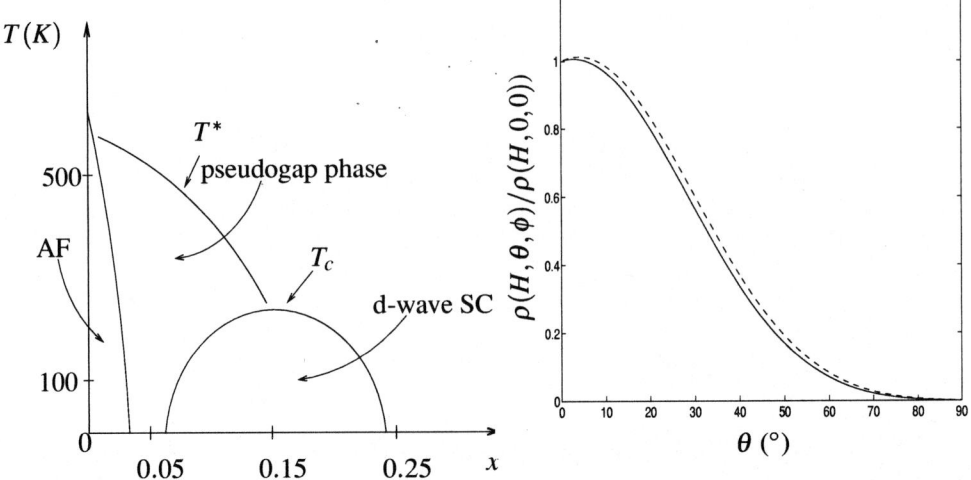

FIGURE 5. Left panel: The schematic phase diagram of high T_c cuprates. Right panel: The angular dependent magnetoresistance is shown as a function of θ for current parallel to the a-axis for $\phi = 0°$ (dashed line) and $\phi = 45°$ (solid line)

or USDW. We have proposed recently that USDW can interpret very readily two crucial experiments observed in the pseudogap phase in high T_c cuprates YBCO and Bi2212: the weak antiferromagnetism[57] and the optical dishroism in ARPES[58]. Sidis et al. observed the appearance of the weak antiferromagnetism at $T = T^*$. This feature is qualitatively very similar to the weak antiferromagnetism observed in URu$_2$Si$_2$[24]. Unfortunately the temperature dependence of the intensity of the AF amplitude is rather different from the one in URu$_2$Si$_2$. But there are a few more possible contributions what we have neglected. In this picture the spin configuration of USDW is given by $S^{\pm} = S_x \pm i S_y$ lying in the $a-b$ plane. There are many attempts to describe this feature in terms of orbital angular momentum, but these models look too artificial. Perhaps the optical dishroism observed in the pseudogap phase in Bi2212 is still more controversial[58]. Indeed this is predicted by Chandra Varma based on a three band Hubbard model with a complicated order parameter[59]. There are many works trying to reinterpret this feature in terms of orbital currents associated with d-wave density wave[60].

One of the natural consequence of d-wave SDW with spin component $S^{\pm}$ is the optical dichroism as observed by Kaminski et al.[58]. The fact that the spin component lies in the $a-b$ plane is consistent with neutron scattering experiment[57]. Making use of the standard procedure to calculate ARPES, we find

$$I_{\pm} \sim 1 \pm \frac{\Delta(\mathbf{k})}{E(\mathbf{k})} \tag{8}$$

or

$$P = \frac{I_+ - I_-}{I_+ + I_-} = \frac{\Delta(\mathbf{k})}{E(\mathbf{k})}, \tag{9}$$

where $\Delta(\mathbf{k}) = \Delta\cos(2\phi)$ and $E(\mathbf{k})$ is the QP energy. Eq. (8) tells that the optical dichroism is proportional to $\cos(2\phi)$. In particular $P = 0$ for $\mathbf{k}$ in the nodal directions while P takes the maximum value at the antinodal directions. These facts are consistent with experiment[58]. We expect also in a uniform ground state the 100 % dichroism. But small dichroism is mostly due to the nonuniform ground state. We further expect the spin polarization of the outcoming electrons parallel to the photon polarization.

Also we propose that the angular dependent magnetoresistance will be a powerful method to investigate the d-wave density wave in high T_c cuprates[61]. In a magnetic field $\mathbf{H}$ applied as shown in Fig. 1 (after replacing $a \to b, b \to c, c \to a$), the QP spectrum in d-wave density wave changes to

$$E_\pm = \sqrt{2\sqrt{2}e|H|\Delta(v_F a|\cos(\theta)| + v_c c\sin(\theta)|\sin(\phi \pm \frac{\pi}{4})|)}, \qquad (10)$$

where v_F and v_c are the Fermi (in-plane) and perpendicular velocity, respectively. Again we followed Ref. [15] and neglected the imperfect nesting terms for simplicity. Therefore the magnetoresistance is given by

$$\frac{\rho(H,\theta,\phi) - \rho(0,\theta,\phi)}{\rho(0,\theta,\phi)} = \frac{e^{\beta(E_+ + E_-)} - 1}{e^{\beta E_+} + e^{\beta E_-} + 2} \qquad (11)$$

A typical θ dependence is shown in Fig 5.

Although this cannot distinguish between d-wave CDW and SDW, at least this will provide a unique test of the UDW proposed in high T_c cuprates.

Very recently d-wave symmetry of the superconductivity in heavy fermion layered compound CeCoIn$_5$ and in the organic superconductor κ-(BEDT-TTF)$_2$Cu(NCS)$_2$ have been established[62, 63]. Further both of these superconductors lie in the vicinity of a kind of antiferromagnetic state (most likely a kind of SDW)[27, 64]. Perhaps the most surprising phenomenon is the dependence of superconductivity in κ-(BEDT-TTF)$_2$Cu[N(CN)$_2$]Br on the cooling rate[27]. For clarity we consider two extreme cases: the well annealed crystals are kept at liquid N$_2$ temperature for three days before final slow cooling to the temperature region around 10 K, while the quenched crystals were cooled down to the liquid He temperature within one hour. Surprisingly the superconducting transition temperature is little affected by the different cooling procedure. But from the diamagnetic response it is shown that the superfluid density in the quenched sample is less than 1% of the annealed sample. Further the temperature dependence of the superfluid density of the annealed sample is consistent with the one in d-wave superconductor, while the one for the quenched sample can be interpreted in terms of the one in s-wave superconductor. Therefore we suspect that the origin of the controversy over d-wave versus s-wave superconductivity lies in the question of the cooling rate. It is well understood that disorder in the ethylene groups attached to the BEDT-TTF molecule is destructive to superconductivity, though we do not know how. The slow cooling through the glassy transition temperature (100 K-70 K) where the ethylene group disorder sets in, helps to form more ordered ethylene groups[64]. Also it is very likely that disorder in the ethylene group is more disastrous to superconductivity than to SDW. Then a natural question is if this kind of SDW is USDW or not. Unfortunately, there is no experimental data on the characterization of this antiferromagnetic order parameter. Therefore we are sure that ADMR will be very useful to clarify this question.

Also can the weak superconductivity or gossamer superconductivity[65] found in the quenched sample be described in terms of coexisting d-wave superconductivity and d-wave SDW? We believe this is one of the most interesting questions in organic superconductors.

CONCLUDING REMARKS

We have seen that UCDW and USDW are very likely realized in organic conductors, in heavy fermion systems and in the pseudogap phase in high T_c cuprates. Also we have proposed that the angular dependent magnetoresistance will provide a unique probe to discover UDW. In particular we have identified successfully UCDW in α-$(BEDT-TTF)_2KHg(SCN)_4$. Also we have pointed out that there are many similarities among the pseudogap phase in high T_c cuprates, the glassy phase in organic superconductor κ-$(BEDT-TTF)_2X$ and the 115 compounds in heavy fermion systems including $CeCoIn_5$ and $PuCoGa_5$[66]. The latter system with superconducting transition temperature $T_c = 18$ K is of great interest. Also as unconventional superconductivity becomes the superconductivity of the 21st century, we are confident that UCDW and USDW will be the density wave in this new century.

ACKNOWLEDGMENTS

We are very much pleased to dedicate this work for the 60th birthday of Professor Mancini, our friend and our colleague. We wish for Nando a number of coming fruitful years. Also we thank Fernando for founding the training school in true Platonic tradition, which provides us a small civilized corner in the present turbulent universe. We thank Mario Basletić, Bojana Korin-Hamzić, Amir Hamzić, M. V. Kartsovnik, Marko Pinterić, Silvia Tomić and Peter Thalmeier for discussions and collaborations on related subjects. This work was supported by the Hungarian National Research Fund under grant numbers OTKA T032162 and TS040878.

REFERENCES

1. Sólyom, J., *Adv. Phys.*, **28**, 201 (1979).
2. Grüner, G., *Density waves in solids*, Addison-Wesley, Reading, 1994.
3. Lang, M., *Superconducting Review*, **2**, 1 (1996).
4. Ishiguro, T., Yamaji, K., and Saito, G., Springer, Berlin, 1999.
5. Bardeen, J., Cooper, L. N., and Schrieffer, J. R., *Phys. Rev.*, **108**, 1175 (1957).
6. Lee, I. J., Brown, S. E., Clark, W. G., Strouse, M. J., Naughton, M. J., Kang, W., , and Chaikin, P. M., *Phys. Rev. Lett.*, **88**, 017004 (2002).
7. Sigrist, M., and Ueda, K., *Rev. Mod. Phys.*, **63**, 239 (1991).
8. Maki, K., "Introduction to d-wave superconductivity," in *Lectures in the Physics of Highly Correlated Electron Systems*, edited by F. Mancini, AIP Conference Proceedings 438, Woodbury, 1998, p. 83.

9. Tsuei, C. C., and Kirtley, J. R., *Rev. Mod.Phys.*, **72**, 969 (2000).
10. Izawa, K., Takahashi, H., Yamaguchi, H., Matsuda, Y., Suzuki, M., Sasaki, T., Fukase, T., Yoshida, Y., Settai, R., and Onuki, Y., *Phys. Rev. Lett.*, **86**, 2653 (2001).
11. Dóra, B., Maki, K., and Virosztek, A., *Europhys. Lett.*, **62**, 426 (2003).
12. Kee, H.-Y., Maki, K., and Chung, C. H., Phys. Rev. B (in press).
13. Maki, K., Thalmeier, P., and Won, H., *Phys. Rev. B*, **65**, 140502(R) (2002).
14. Izawa, K., Kamata, K., Nakajima, Y., Matsuda, Y., Watanabe, T., Nohara, M., Takagi, H., Thalmeier, P., and Maki, K., *Phys. Rev. Lett.*, **89**, 137006 (2002).
15. Nersesyan, A. A., and Vachnadze, G. E., *J. Low T. Phys.*, **77**, 293 (1989).
16. Benfatto, L., Caprara, S., and Di Castro, C., *Eur. Phys. J. B*, **17**, 95 (2000).
17. Dóra, B., and Virosztek, A., *Eur. Phys. J. B*, **22**, 167 (2001).
18. Chakravarty, S., Laughlin, R. B., Morr, D. K., and Nayak, C., *Phys. Rev. B*, **63**, 094503 (2001).
19. Basletić, M., Korin-Hamzić, B., Kartsovnik, M. V., and Müller, H., *Synth. Met.*, **120**, 1021 (2001).
20. Dóra, B., Virosztek, A., and Maki, K., *Phys. Rev. B*, **64**, 041101(R) (2001).
21. Dóra, B., Virosztek, A., and Maki, K., *Physica B*, **312-313**, 571 (2002).
22. Dóra, B., Maki, K., Korin-Hamzić, B., Basletić, M., Virosztek, A., Kartsovnik, M. V., and Müller, H., *Europhys. Lett.*, **60**, 737 (2002).
23. Ikeda, H., and Ohashi, Y., *Phys. Rev. Lett.*, **81**, 3723 (1998).
24. Virosztek, A., Maki, K., and Dóra, B., *Int. J. Mod. Phys. B*, **16**, 1667 (2002).
25. Castro-Neto, A. H., *Phys. Rev. Lett.*, **86**, 4382 (2001).
26. Dóra, B., Virosztek, A., and Maki, K., *Acta Physica Polonica B*, **34**, 571 (2003).
27. Pinterić, M., Tomić, S., Prester, M., Drobac, D., and Maki, K., *Phys. Rev. B*, **66**, 174521 (2002).
28. Won, H., and Maki, K., *Phys. Rev. B*, **49**, 1397 (1994).
29. Singleton, J., *Rep. Prog. Phys.*, **63**, 1161 (2000).
30. Andres, D., Kartsovnik, M. V., Biberacher, W., Weiss, H., Balthes, E., Müller, H., and Kushch, N., *Phys. Rev. B*, **64**, 161104(R) (2001).
31. Dóra, B., Virosztek, A., and Maki, K., *Phys. Rev. B*, **65**, 155119 (2002).
32. Dóra, B., Maki, K., and Virosztek, A., *Phys. Rev. B*, **66**, 165116 (2002).
33. Maki, K., Dóra, B., Kartsovnik, M., Virosztek, A., Korin-Hamzić, B., and Basletić, M., *Phys. Rev. Lett.*, **90**, 256402 (2003).
34. Yamaji, K., *J. Phys. Soc. Japan*, **51**, 2787 (1982).
35. Yamaji, K., *J. Phys. Soc. Japan*, **52**, 1361 (1983).
36. Huang, X. Z., and Maki, K., *Phys. Rev. B*, **40**, 2725 (1989).
37. Sasaki, T., and Toyota, N., *Phys. Rev. B*, **49**, 10120 (1994).
38. Kovalev, A. E., Kartsovnik, M. V., Shibaeva, R. P., Rozenberg, L. P., Schegolev, I. F., and Kushch, N. D., *Solid State Commun.*, **89**, 575 (1994).
39. Caulfield, J., Singleton, J., Hendriks, P. T. J., Perenboom, J. A. A. J., Pratt, F. L., Doporto, M., Hayes, W., Kurmoo, M., and Day, P., *J. Phys. Cond. Mat.*, **6**, L155 (1994).
40. Hanasaki, N., Kagoshima, S., Miura, N., and Saito, G., *J. Phys. Soc. Japan*, **65**, 1010 (1996).
41. Landau, L. D., *Soviet Phys. JETP*, **3**, 920 (1957).
42. Landau, L. D., *Soviet Phys. JETP*, **5**, 101 (1957).
43. Landau, L. D., *Soviet Phys. JETP*, **8**, 104 (1958).
44. Maki, K., and Won, H., "From Superfluid ^{3}He to Triplet Superconductor $Sr_2 2RuO_4$," in *Fluctuating paths and fields*, edited by Janke, W. et al., World Scientific, Singapore, 2001.
45. Hussey, N. E., *Advances in Physics*, **51**, 1685 (2002).
46. Shankar, R., *Rev. Mod. Phys.*, **66**, 129 (1994).
47. Metzner, W., Castellani, C., and Di Castro, C., *Advances in Physics*, **47**, 317 (1998).
48. Houghton, A., Kwon, H. J., and Marston, J. B., *Advances in Physics*, **49**, 141 (2000).
49. Scalapino, D. J., *Phys. Rep.*, **250**, 329 (1995).
50. Pao, C. H., and Bickers, N. E., *Phys. Rev. Lett.*, **72**, 1870 (1994).
51. Sun, Y., and Maki, K., *Europhys. Lett.*, **32**, 355 (1995).
52. Ando, Y., Boebinger, G. S., Passnet, A., Kimura, T., and Kishio, K., *Phys. Rev. Lett.*, **75**, 4462 (1995).
53. Won, H., and Maki, K., *Physica C*, **282**, 1837 (1997).
54. Tallon, J. L., and Loram, J. W., *Physica C*, **349**, 53 (2000).
55. Timusk, T., and Statt, B., *Rep. Prog. Phys.*, **62**, 61 (1999).
56. Kugler, M., Fischer, O., Renner, Ch., Ono, S., and Ando, Y., *Phys. Rev. Lett.*, **86**, 4911 (2001).

57. Sidis, Y., Ulrich, C., Bourges, P., Bernhard, C., Niedermayer, C., Regnault, L. P., Andersen, N. H., and Keimer, B., *Phys. Rev. Lett.*, **86**, 4100 (2001).
58. Kaminski, A., S. Rosenkranz, H. M. F., Campuzano, J. C., Li, Z., Raffy, H., Cullen, W. G., You, H., Olson, C. G., Varma, C. M., and Höchst, H., *Nature*, **416**, 610 (2002).
59. Varma, C. M., *Phys. Rev. B*, **61**, R3804 (2000).
60. Nguyen, H. K., and Chakravarty, S., *Phys. Rev. B*, **65**, 180519 (2002).
61. Maki, K., Dóra, B., and Virosztek, A., *J. Phys. IV France*, **12**, Pr9–45 (2002).
62. Izawa, K., Yamaguchi, H., Matsuda, Y., Shishido, H., Settai, R., and Onuki, Y., *Phys. Rev. Lett.*, **87**, 057002 (2001).
63. Izawa, K., Yamaguchi, H., Sasaki, T., and Matsuda, Y., *Phys. Rev. Lett.*, **88**, 027002 (2002).
64. Müller, J., Lang, M., Stèglich, F., Schlueter, J. A., Kini, A. M., and Sasaki, T., *Phys. Rev. B*, **65**, 144521 (2002).
65. Laughlin, R. B., cond-mat/0209269.
66. Sarrao, J. L., Morales, L. A., Thompson, J. D., Scott, B. L., Stewart, G. R., Wastin, F., Boulet, J. R. P., Colineau, E., and Lander, G. H., *Nature*, **420**, 297 (2002).

Superconductors as giant atoms: Qualitative aspects

J.E. Hirsch

Department of Physics, University of California, San Diego
La Jolla, CA 92093-0319

Abstract. When the Fermi level is near the top of a band the carriers (holes) are maximally dressed by electron-ion and electron-electron interactions. The theory of hole superconductivity predicts that only in that case can superconductivity occur, and that it is driven by *undressing* of the carriers at the Fermi energy upon pairing. Indeed, experiments show that dressed hole carriers in the normal state become undressed electron carriers in the superconducting state. This leads to a description of superconductors as giant atoms, where undressed time-reversed electrons are paired and propagate freely in a uniform positive background. The pairing gap provides rigidity to the wavefunction, and electrons in the giant atom respond to magnetic fields the same way as electrons in diamagnetic atoms. We predict that there is an electric field in the interior of superconductors and that the charge distribution is inhomogeneous, with higher concentration of negative charge near the surface; that the ground state of superconductors has broken parity and possesses macroscopic spin currents, and that negative charge spills out when a body becomes superconducting.

I. INTRODUCTION

The Bloch-Landau description of metals is appropriate to describe the normal state. It describes dressed quasiparticles (electrons or holes) whose charge sign depends on the location of the Fermi level in the band, and whose effective mass is different from the free electron mass due to electron-ion and electron-electron interactions. Instead, we propose that in superconductors the carriers of the supercurrent are *bare electrons* and that superconductors should be understood as giant atoms.

Unlike the conventional BCS-Eliashberg theory, the theory of hole superconductivity discussed here is proposed to describe *all* superconductors[1]. It proposes that superconductivity arises only when the Fermi level is near the top of an electronic energy band, and is driven by *undressing*[2] of the carriers at the Fermi energy upon pairing. It requires the existence of electron-ion and electron-electron interaction but *not* of electron-phonon interaction, so that it predicts that superconductivity can occur in a solid even if the ionic mass is infinite. On the other hand it predicts that superconductivity cannot occur if there are no *antibonding electrons* at the Fermi energy, i.e. if the Fermi level is close to the bottom of the band[3].

When carriers 'undress' they become more free-electron-like. We argue that certain experiments in superconductors show that the superfluid carriers are in fact completely free-electron like. Hence they have completely 'undressed' from both electron-ion and electron-electron interactions. We argue that carriers at the Fermi energy are most highly dressed by both the electron-electron and the electron-ion interaction when the Fermi level is close to the top of the band. Furthermore, the formalism shows that the dressing is highest when the ions are negatively charged[4]. Naturally, the most favorable situation for superconductivity is when the carriers are most heavily dressed in the normal state, because that is when the most is gained by 'undressing'. This is the case for high T_c cuprates, where the 'dressing' in the normal state is most apparent. However the same physics applies to all superconductors, whether high, medium or low T_c[5]. The superfluid electrons are undressed, paired, and otherwise completely free, so that they behave very much like electrons in diamagnetic atoms. The difference between microscopic atoms and 'giant atom' superconductors is the same as that between Rutherford atoms and 'Thomson atoms'[6].

In our previous work we have stressed the dressing of carriers due to the electron-electron interaction. However, by changing the mass of the electron from its bare free electron value, the electron-ion interaction *also* dresses the bare electron, increasingly so as the Fermi level moves up in the band. Realizing that 'undressing' involves undressing from *both* the electron-electron *and* the electron-ion interaction leads to the giant atom scenario[7].

II. DRESSING FROM THE ELECTRON-ION INTERACTION

When an electronic energy band is filled from bottom to top, the electrons that come in first are free to choose the state that best suits them. Also, when only few electrons exist in the band they will adjust their state to take maximum advantage of the crystal potential, without being much affected by interaction with each other. These happy 'bonding electrons' have a low energy, a smooth wave function with large amplitude between ions to take maximum advantage of the electron-ion potential while minimizing their kinetic energy, and an effective mass not very different from the free electron mass. Each of them contributes to the electrical conductivity of the metal, to the low frequency optical conductivity, and to the cohesion of the solid by giving rise to an effective attraction between ions (hence the name 'bonding').

The situation is very different for the unhappy electrons that come in near the end of the band-filling process, the 'antibonding electrons'. Because of the Pauli principle these electrons cannot take full advantage of the electron-ion potential, since they have to be orthogonal to the pre-existing bonding electrons. Hence their wavefunction has to oscillate greatly, having maximum amplitude at the ionic sites and vanishing in-beween, as shown schematically in Figure 1. With such constraints in fact they would rather reside on isolated ions, which is why they give rise to a repulsive force between the ions (hence the name 'antibonding'), and sometimes succeed in breaking up the solid or at least driving it into another more stable configuration (which is why lattice instabilities are associated with the existence of these electrons at the Fermi energy). The large oscillations in the wave function give rise to a large kinetic energy.

Furthermore the antibonding electrons do not contribute to the electrical conductivity of the solid nor to the low frequency optical conductivity, as a matter of fact they 'anticontribute'. When an external force is applied to the electron (eg by applying an electric potential difference to the solid) both the electron and the ionic lattice will pick

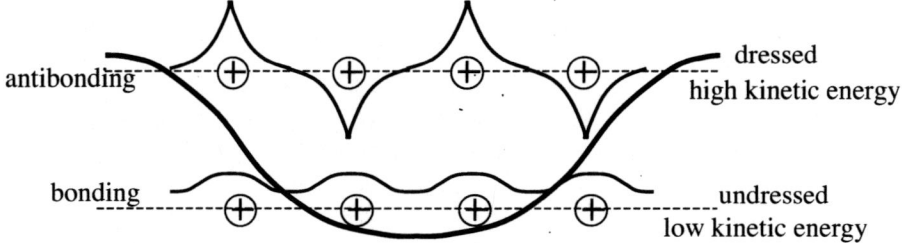

Electronic energy band

FIGURE 1. Electronic states in a band. The states at the bottom (bonding) have a high density of charge in-between the ions, and a smooth wave function. The states at the top (antibonding) have a node in the charge density between the ions and a spiky wave function.

up the added mechanical momentum:

$$F\Delta t = \Delta p_{el} + \Delta p_{latt} \quad (1)$$

According to semiclassical transport theory,

$$F = \hbar \dot{k} \quad (2)$$

$$\Delta p_{el} = m_e \Delta v_{el} = \frac{m_e}{\hbar} \Delta(\frac{\partial \varepsilon_k}{\partial k}) = m_e \, (\frac{1}{\hbar^2} \frac{\partial^2 \varepsilon_k}{\partial^2 k}) \hbar k \Delta t = \frac{m_e}{m^*} F \Delta t \quad (3)$$

with m^* the effective mass, which is *negative* for antibonding electrons, and m_e the free electron mass. Hence the electron acquires a momentum that is *opposite* to the applied force, and the ionic lattice picks up the difference:

$$\Delta p_{latt} = (1 - \frac{m_e}{m^*}) F \Delta t \quad (4)$$

If we quantify the 'dressing' of the free electron by the momentum transfer to the lattice Eq. (4), it is clear that the electron-ion interaction increasingly dresses the electron as the Fermi level goes up in the band. This is clearly seen in 'nearly free electron' theory, i.e. perturbation theory, where the band energy is given by

$$\varepsilon_k = \varepsilon_k^0 + \sum_K \frac{|U_K|^2}{\varepsilon_k^0 - \varepsilon_{k+K}^0} \quad (5)$$

with $\varepsilon_k^0 = \hbar^2 k^2 / 2m_e$ the free electron energy and U_K the electron-ion Fourier component for reciprocal lattice vector K. The effective mass is close to the free electron mass near the bottom of the band ($k = 0$) where the energy denominator in Eq. (5) is largest. As k increases the effective mass increases, then diverges and becomes negative and decreases in magnitude. The momentum transfer to the lattice Eq. (4) increases monotonically throughout this process. Beyond weak coupling perturbation theory, it is clear that this will be qualitatively similar for all energy bands since m^* is positive near the bottom and negative near the top.

Similarly the electrical conductivity *per electron* as well as the Drude weight per electron decrease monotonically as the Fermi level rises. The Drude weight is given by

$$\frac{2}{\pi e^2} \int_{intraband} d\omega \sigma_1(\omega) = \frac{n_e}{m^*} \qquad (6)$$

The equality holds for the Fermi level close to the bottom of the band, with n_e the number of electrons. Near the top of the band, n_e is replaced by the number of holes, $n_h = 2 - n_e$, and m^* in Eq. (6) by its absolute value. The difference between the Drude weight Eq. (6) and n_e/m_e, the corresponding value for free electrons, also quantifies the amount of 'dressing' and this difference increases as the Fermi level rises from the bottom to the top of the band. It also represents the optical spectral weight that is transfered from low intra-band frequencies to higher inter-band frequencies due to the electron-ion interaction.

Superconductivity can be understood as originating in the desire of antibonding electrons at the Fermi energy to contribute rather than 'anticontribute' to the electrical conductivity. To do so they have to become like the bonding electrons at the bottom of the band, getting around the fact that those single particle states are already occupied. By pairing the antibonding electrons will be able to avoid the Pauli principle, however to do so, it is necessary to recourse to the electron-electron interaction. This will be easiest for the antibonding electrons when the influence of the electron-ion interaction force is small, which is the case when the ions are negatively charged[4].

III. DRESSING FROM THE ELECTRON-ELECTRON INTERACTION

Another essential aspect of 'dressing' arises from the effect of the electron-electron interaction. We have dealt with this aspect extensively in our published work on hole superconductivity and hence will only discuss it briefly here. Dressing from the electron-electron interaction parallels the dressing from the electron-ion interaction discussed above. To separate the two aspects, consider a tight binding band structure in a hypercubic lattice with nearest neighbor hopping only, so that the band effective mass has the same magnitude for carriers at the bottom and the top of the band. We ignore here the aspect of dressing associated with the *sign* of the effective mass discussed above. When including the electron-electron interaction we also find that the dressing increases as the Fermi level goes up in the band, and results in a situation where holes (carriers at the top of the band) are highly dressed and electrons (carriers at the bottom of the band) are undressed[8]. Here the 'dressing' manifests itself in both an increase in the effective mass m^* and a decrease in the quasiparticle weight z, which are approximately related by $m^* = 1/z$. The many-body Hamiltonian that describes this physics is a dynamic Hubbard model[9], and its projection to a low energy effective Hamiltonian yields a Hubbard model with correlated hopping[10], with the hopping amplitude decreasing as the Fermi level rises in the band.

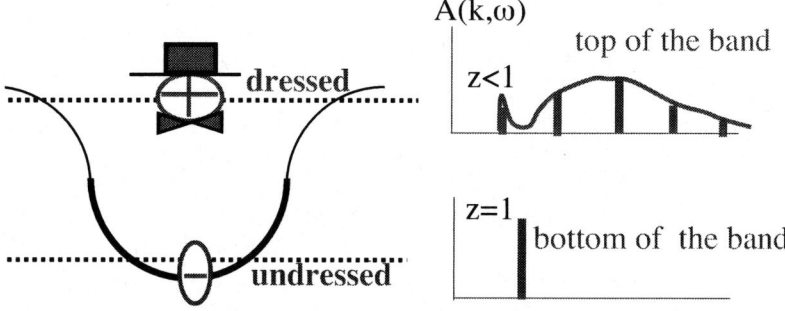

FIGURE 2. Electronic states in a band. The states at the bottom have a spectral function $A(k, \omega)$ with a single quasiparticle peak of weight $z = 1$. The states at the top of the band have a spectral function with a small quasiparticle peak ($z < 1$) and a broad incoherent spectrum at higher energies.

IV. SUPERCONDUCTIVITY FROM UNDRESSING: EXPERIMENTAL EVIDENCE

We have argued in the previous sections that as the Fermi level goes up in the band the carriers become increasingly dressed by *both* the electron-ion and the electron-electron interaction. Experiments show that in superconductors carriers are dressed in the normal state, that the superconducting transition involves 'undressing', and that carriers are completely undressed in the superconducting state, as follows:

A. Evidence that carriers in the normal state are dressed

A look at the periodic table shows that essentially all superconducting elements have positive Hall coefficient, indicating hole transport, or equivalently large 'dressing' of the bare electron by the electron-ion interaction. The same is true for compounds[11]. In the highest T_c materials, the cuprates, carriers in the normal state are so heavily dressed that the validity of the quasiparticle concept has been called into doubt, especially in the underdoped regime. This is seen from photoemission and optical conductivity measurements, that show small amplitude for low frequency coherent response and large amplitude for high frequency incoherent response in the normal state, indicating small quasiparticle weight and large effective mass of carriers in the normal state, i.e. large 'dressing'[12, 13]. Upon hole doping the normal state carriers become less dressed, consistent with the fact that the Fermi level goes down in the band and dressing due to both electron-electron and electron-ion interaction decreases as discussed above. The fact that superconductors are often close to lattice instabilities indicates the presence of 'antibonding' electrons, i.e. electrons 'dressed' by the electron-ion interaction, and the fact that superconductors are usually poor conductors of electricity in the normal state indicates 'dressing' of normal state carriers.

B. Evidence that the superconducting transition involves undressing

This is most clearly seen in systems where dressing is highest in the normal state. Photoemission experiments in cuprates show the emergence of a quasiparticle peak as the system goes superconducting (increase in z)[12]. The effect is greatest in the underdoped regime where the dressing in the normal state is highest, as expected. This indicates that dressing is suppressed by the transition to superconductivity and indirectly that the effective mass decreases through the relation $m^* = 1/z$. Optical experiments show directly that the effective mass decreases, as they show a transfer of spectral weight from high to low frequencies as a system goes superconducting[14, 15, 16]. Concerning undressing from the electron-ion interaction, we argue that this is evidenced by the observed sign reversal of the Hall coefficient below T_c, that turns from positive to negative before it goes to zero[17]. This suggests that the dressing from the electron-ion interaction that caused the Hall coefficient to be positive in the normal state is eliminated when going superconducting.

C. Evidence that carriers in the superconducting state are undressed

The 'Bernoulli potential', the electric potential that develops between regions in the superconductor where the superfluid is moving with different velocities reveals the sign of the charge carriers. All experiments performed indicate that its sign is consistent with negative electrons being the charge carriers[18], even though conventional theory predicts its sign should be the one corresponding to the sign of the charge carriers in the normal state[19]. The measured magnitude is also consistent with carriers in the superconducting state being free electrons. The sign and magnitude of the magnetic field that exists in the interior of a rotating superconductor (London field)[20] indicate that the carriers of the superfluid current are undressed electrons[21]. This is also indicated by the sign and magnitude of the gyromagnetic effect, the change in angular momentum of a superconducting body that occurs when an external magnetic field is applied[22].

V. HOW IS UNDRESSING RELATED TO PAIRING AND SUPERCONDUCTIVITY?

We know that dressing increases when the Fermi level rises in the band, and is largest with the band almost full when the carriers are holes. When holes pair, the band becomes locally less full, hence undressing should occur since locally the Fermi level has moved down to where the carriers are less dressed. And of course we know from BCS theory that pairing is associated with superconductivity. The dynamic Hubbard model predicts that pairing and superconductivity should occur at low T because undressing causes a lowering of kinetic energy, hence a decrease of free energy at low T compared to the normal state[9]. The correlated hopping term in the low energy effective Hamiltonian shows clearly how the effective mass and the kinetic energy decrease when the system goes superconducting[10]. The dynamic Hubbard model also shows that the quasiparticle weight increases and that transfer of spectral weight from high to low frequency occurs upon pairing which is a signature of undressing[23].

VI. HISTORICAL PRECEDENTS

Our theory involves pairing as BCS theory does, and in addition it incorporates many ideas that were discussed earlier but were completely abandoned after BCS theory became established.

The idea that superconductors exhibit quantum mechanics at a macroscopic scale became generally accepted around the mid 1930's after the Meissner effect was discovered and London's phenomenological theory proposed. With it came the idea that electronic wavefunctions in superconductors extend coherently over the entire macroscopic body. This naturally follows from considering the expression for the atomic diamagnetic susceptibility

$$\chi_0 = -\frac{e^2}{6m_e c^2} <r^2> \qquad (7)$$

which is small when $<r^2>$ is of atomic dimensions but grows without bounds for macroscopic samples. When the sample becomes big the effect of the magnetic field generated by the electrons themselves becomes important and the susceptibility is

$$\chi = \frac{\chi_0}{1-4\pi\chi_0} \to -\frac{1}{4\pi} \qquad (8)$$

The idea that the electronic wavefunction extends coherently over the entire superconducting body is an essential ingredient of the 'giant atom' concept. In fact, after Meissner's discovery and London's theory it was quite common to refer to superconductors as being analogous to 'big atoms' or 'giant atoms'[24]. This was by reference to Eq. (7) and diamagnetism. However other possible consequences of the 'giant atom' concept like inhomogeneous distribution of positive and negative charge, as in atoms, or the possibility of spin-orbit coupling, were not considered at that time.

The idea of a 'non-viscous electronic fluid' describing the superconducting electrons was popular at that time[25], implying no momentum transfer but rather a complete detachment between electrons and lattice. Specifically, Kronig[26] proposed to ignore the discrete ionic potential altogether and replace it by 'jellium'. This picture is inconsistent with antibonding states at the top of the band which transfer large momentum to the ionic lattice, and suggests instead that the superfluid carriers behave more like electrons at the bottom of the band. At the same time the fact that superconductors in the normal state exhibit hole rather than electron transport was well known and discussed at the time[27], so the connection between transition to superconductivity and 'undressing' from the electron-ion interaction was somehow implicit in the discussions back then. Of course the concept of pairing was not being discussed at the time. Also the idea that superconductivity may involve a reduction of the carrier's effective mass was discussed (by no other than Bardeen!) in the pre-BCS era[28].

As another precedent worth citing, Meissner wondered whether the electrons that carry the supercurrent are the same electrons that carry the normal current, or whether new carriers become available in the superconducting state[29]. He favored the latter, based on the observation that superconducting elements have more than one electron outside closed shells. This is in qualitative agreement with the principle discussed here: antibonding electrons 'anticontribute' to electrical conductivity in the normal state, and they carry the supercurrent after they undress in the superconducting state.

VII. THE ROTATING SUPERCONDUCTOR PUZZLE

A simply connected superconducting body rotating with angular velocity $\vec{\omega}$ has a magnetic field

$$\vec{B} = -\frac{2m_e c}{e} \vec{\omega} \qquad (9)$$

throughout its interior[20], with m_e the *free* electron mass. Its origin when a superconducting body is put into rotation can be understood as follows[25]: the rotating positive ions generate an electric field through Faraday's law:

$$\oint \vec{E} \cdot \vec{dl} = -\frac{1}{c}\frac{d}{dt}\int \vec{B} \cdot \vec{ds} \qquad (10)$$

hence for a uniform magnetic field the electric field at distance r from the axis of rotation is

$$E = -\frac{r}{2c}\frac{dB}{dt} \qquad (11)$$

and the superfluid changes its velocity according to

$$m_e \frac{dv_s}{dt} = eE = -\frac{er}{2c}\frac{dB}{dt} \qquad (12)$$

For the superfluid initially at rest the final velocity is then

$$v_s = -\frac{er}{2m_e c}B \qquad (13)$$

In the interior the superfluid rotates together with the lattice, so that if the body is rotating with angular velocity ω, $v_s = \omega r$ and

$$\omega = -\frac{e}{2m_e c}B \qquad (14)$$

from which Eq. (9) follows. Alternatively, Eq. (9) can also be directly derived from the London equation

$$\vec{\nabla} \times \vec{v}_s = -\frac{e}{m_e c}\vec{B} \qquad (15)$$

for $\vec{v}_s = \vec{\omega} \times \vec{r}$ as appropriate for rigid rotation, since $\vec{\nabla} \times (\vec{\omega} \times \vec{r}) = 2\vec{\omega}$.

There is a problem with this however. The centripetal force required for an electron to rotate with angular velocity ω at radius r is only one-half the one provided by the Lorentz force due to the magnetic field Eq. (9). In other words, an electron in a magnetic field B rotates at the cyclotron frequency $\omega = eB/m_e c$ rather than at the Larmor frequency Eq. (14). The giant atom scenario resolves this difficulty as explained below[30].

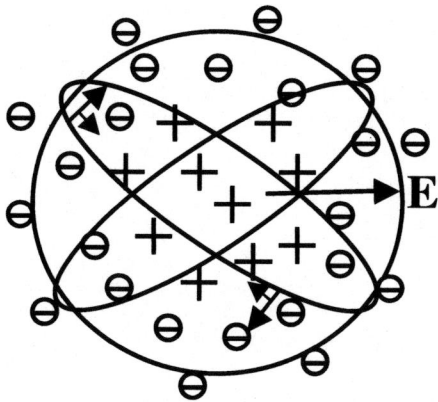

FIGURE 3. Superconductor as a giant atom (schematic). Electrons are pushed out of the interior towards the surface. An electric field pointing out exists in the interior of the superconductor. Some electrons will 'spill out' outside the surface of the superconducting body[7].

VIII. THE GIANT ATOM SCENARIO

If we accept that the wavefunction of the superconducting electron extends coherently over macroscopic distances it is difficult to imagine how it could 'know' about the microscopic variation of the ionic potential over the scale of Angstrom and adjust accordingly, as antibonding electrons have to do. Rather, it is natural to conclude that the antibonding electrons at the Fermi energy manage to detach themselves completely from the lattice ('undress' from the electron-ion interaction) and adopt a 'long wavelength' wavefunction that does not 'see' microscopic details. In that case each of these electrons will see a smeared positive charge distribution, screened self-consistently by all the other electrons in the system. We call this the 'giant atom' scenario[7] (Figure 3).

In microscopic atoms, electrons do interact with each other, yet the response to a magnetic field involves the free electron mass. Similarly we argue that superfluid electrons in the giant atom superconductor respond with their free mass, as evidenced by Eq. (9), even though they interact with other electrons. This follows from galilean invariance. It is only if it can see the discrete non-translationally invariant ionic potential that the electron can respond with an effective mass different from the bare mass, since only in that case can the lattice pick up some momentum.

For electrons in filled shells of atoms, the change in velocity upon application of a magnetic field is[31]

$$\Delta v = -\frac{e}{m_e c} A \qquad (16)$$

with A the magnetic vector potential. This follows from microscopic quantum mechanics, as well as from simple classical arguments: in an atom, the centripetal acceleration

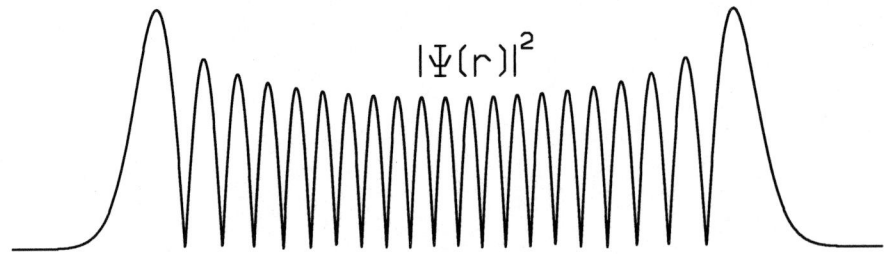

FIGURE 4. Wave function of harmonic oscillator for large quantum number. The wave function amplitude is largest near the region of maximum classical elongation. As a consequence the electronic density is not uniform, in contrast to the plane-wave free-electron model.

for the electron's orbit is provided by the electric field from the ionic charge, E_{ion}

$$\frac{m_e v^2}{r} = eE_{ion} \qquad (17)$$

and the change in centripetal acceleration upon application of a magnetic field B is provided by the magnetic Lorentz force

$$\frac{2m_e v \Delta v}{r} = \frac{ev}{c} B \qquad (18)$$

from which Eq. (16) follows for $\vec{A} = \vec{B} \times \vec{r}/2$. For a rotating superconductor, Eq. (9) for the magnetic field in its interior is consistent with the relation Eq. (18) for $\Delta v = \omega r$. The derivation of Eq. (18) from Eq. (17) is only valid if $\Delta v \ll v$ and indicates how to resolve the puzzle raised by the magnetic field in the rotating superconductor: *the electrons need to be rotating already (as in the ordinary atom) at a speed much larger than ωr before the superconductor is set into rotation.*

However, this can only be possible if there is an electric field inside the superconductor that provides their centripetal acceleration in the absence of magnetic field, as in the case of the atom Eq. (17). This implies that the electronic charge distribution in the superconductor, just as in the atom, cannot be homogeneous, rather there has to be more negative charge near the surface and more positive charge in the interior. But that is precisely what the model of hole superconductivity predicts, that the superconductor expels negative charge from its interior[32].

These considerations together with the 'undressing' phenomenology lead us to the model of a macroscopic 'Thomson atom', where a positive uniform charge distribution exists over the volume of the superconductor[7], except within a penetration length of the surface where the compensating negative charge resides. Consider a spherical or cylindrical geometry for simplicity. The electric field from a uniform charge distribution is linear in r, and the potential acting on the electron is a harmonic oscillator potential. A wavefunction for such a potential for large quantum number is shown in Figure 4. It is not uniform but has larger amplitude for large r. This is of course obvious from the fact that an oscillator spends most of its time near the region of maximum elongation, and justifies self-consistently the assumption that the electron sees a positive charge distribution in the interior.

Of course the giant atom scenario with an electric field in the interior is not enough. Electrons in such a potential, even in the absence of imperfections, will not exhibit the Meissner effect but rather the much weaker Landau diamagnetism. This is because there will be a rearrangement when a magnetic field is applied[33]: electrons that are orbiting with their angular momentum parallel to the applied field will increase their velocity, and would lower their energy by turning around and joining the electrons orbiting with angular momentum antiparallel to the applied magnetic field which are slowed down by it. This is where the pairing condition comes in: if for every electron with $k \uparrow$ there is a partner with $-k \downarrow$ and it costs a finite energy Δ to break up the pair, it will prevent (unless the magnetic field is strong enough) the electron from 'turning over' because that would break the pair. So the BCS energy gap provides the necessary 'rigidity' for the pair to respond like London rather than Landau electrons.

The pairing requires that for every $k \uparrow$ electron there is a partner $-k \downarrow$ electron, but it *does not* require that there be a corresponding $-k \uparrow, k \downarrow$ pair. In fact we argue that this will quite generally not be the case, due to spin-orbit coupling. The energy of an electron in a macroscopic orbit will be lower if its orbital angular momentum is antiparallel to its spin, because in its restframe it sees a magnetic field from the positive background that is parallel to its spin magnetic moment. Thus we argue that spin up electrons will be orbiting predominantly in one direction, and spin down electrons in the opposite direction, as shown schematically in Fig. 3, giving rise to a macroscopic spin current which, as explained above, is fully compatible with the pairing concept.

From a purely classical point of view, such a spin current is inescapable if we accept the concept that an outward pointing electric field exists in the interior of superconductors and that the negative charge density is higher near the surface. In order for a superfluid electron at radius r to be in mechanical equilibrium it needs to be rotating so that $m_e v^2/r$ equals the force from the electric field pulling it in. In the absence of magnetic field and rotation of the superconductor there is no charge current field $v_s(r)$ of the superfluid in the interior of the superconductor. But there can and will be a spin current field $v_{s\sigma}(r)$ such that $v_{s\sigma}(r) = -v_{s,-\sigma}(r)$.

Some experimental consequences of the giant atom scenario were discussed in Ref.[7]. We predict that the spin current should give rise to a quadrupolar electric field around superconducting rings, that should be experimentally detectable, and to forces between superconducting rings due to these electric fields. Furthermore we predict that just as in regular atoms the negative charge will not be confined to the region of the positive charge but rather will leak out of the superconducting body, of course in much smaller proportion than for a regular atom. This is likely to play a role in the proximity effect[34]. Finally, since in the normal state the negative charge distribution is homogeneous we predict that a radially outward electron current is generated when the system goes into the superconducting state, to generate the nonhomogeneous charge distribution characteristic of the superconductor. The interaction of this radial current with the ionic background will deflect up- and down-spin electrons tangentially in opposite directions to give rise to a spin current of precisely the sign discussed above through the mechanism discussed in Ref.[35]. Furthermore, the radially outgoing electron current in the presence of a magnetic field will give rise to screening currents that will cancel the magnetic field in the interior of the superconductor as required by the Meissner effect.

The concept of an electric field in the interior of the superconductor however raises a question: why isn't it screened by mobile charges, as electric fields are screened in the interior of normal metals? For the superfluid electrons we have argued that the electric field produces a force that is the centripetal force that sustains the superfluid spin current. However at finite temperature there will also be thermally excited quasiparticles which can move . Why is it that they do not screen the electric field in the interior and nullify it?

This question in fact also arises in the conventional framework when the superconductor is rotating, since an electric field needs to exist to balance the forces on the rotating superfluid electrons, and has never been addressed in that framework. For our case there appears to be a simple answer: the theory of hole superconductivity predicts that quasiparticles are *positively charged*[36], with charge given by

$$Q_k = e(u_k^2 - v_k^2) \tag{19}$$

which is positive on the average due to electron-hole asymmetry. Positive quasiparticles will be pushed *out* by the electric field inside superconductors and not screen the field. In equilibrium a density gradient of quasiparticles will be established so that the electrochemical potential for quasiparticles will be constant throughout the volume of the superconductor.

IX. CONCLUSION

The theory considered here proposes that superconductors in the normal state have carriers at the Fermi energy that are highly dressed by both electron-ion and electron-electron interactions, and that when a metal becomes superconducting these carriers completely undress from both electron-ion and electron-electron interactions and become free-electron-like, except for the pairing correlations. Superconductivity is driven by electron undressing, and the resulting system is a macroscopic 'Thomson atom' with paired time-reversed electrons. The facts that the theory may lead to an understanding of a variety of puzzling phenomena in superconductors[30], that it provides a unified scheme to explain superconductivity in all superconductors[37], that it has a plausible microscopic foundation[3], and that it can be understood in terms of a single underlying physical principle, undressing, argue for its validity. Ultimately its validation or refutation will come from experiments testing its predictions. Various of its experimental predictions are consistent with observations as discussed in the references.

ACKNOWLEDGMENTS

The author is grateful to the organizers and participants of the meeting "Highlights in Condensed Matter Physics", Salerno, Italy, May 9-12, 2003, for the opportunity to present and discuss these concepts in a stimulating environment.

REFERENCES

1. See www.physics.ucsd.edu/~jorge/hole.html for a complete list of references.
2. J.E. Hirsch, Phys.Rev.B **62**, 14487 (2000); Phys.Rev.B **62**, 14498 (2000).
3. J.E. Hirsch, Phys.Rev.B **65**, 184502 (2002).
4. J.E. Hirsch, Phys.Rev. B**48**, 3327 (1993).
5. J.E. Hirsch, Physica C **158**, 326 (1989).
6. J.J. Thomson, 'The corpuscular theory of matter', Archibald Constable and Co., London, 1907.
7. J.E. Hirsch, Phys.Lett. A **309**, 457 (2003).
8. J.E. Hirsch, Phys.Rev. B**67**, 035103 (2003).
9. J.E. Hirsch, Phys. Rev. Lett. **87**, 206402 (2001).
10. J.E. Hirsch and F. Marsiglio, Phys. Rev. B **39**, 11515 (1989); Phys.Rev. B**45**, 4807 (1992); Phys. Rev. B **62**, 15131 (2000).
11. I.M. Chapnik, Sov,Phys. Doklady **6**, 988 (1962).
12. H. Ding, J. R. Engelbrecht, Z. Wang, J. C. Campuzano, S.C. Wang, H.B. Yang, R. Rogan, T. Takahashi, K. Kadowaki, and D. G. Hinks, Phys. Rev. Lett. **87**, 227001 (2001).
13. S.Uchida, T.Ido, H.Takagi, T. Arima, Y. Tokura and S. Tajima, Phys. Rev. B **43**, 7942 (1991).
14. H. J. A. Molegraaf, C. Presura, D. van der Marel, P. H. Kes, and M. Li Science **295**, 2239 (2002).
15. A.F. Santander-Syro et al, cond-mat/0111539 (2001), Europhys.Lett. **62**, 568 (2003).
16. M.V. Klein and G. Blumberg, Science **283**, 42 (1999).
17. S.J. Hagen et al, Phys.Rev. B**41**, 11630 (1990).
18. J. Bok and J. Klein, Phys.Rev.Lett. **20**, 660 (1968); T.D. Morris and J.B. Brown, Physica **55**, 760 (1971); Y.N. Chiang and O.G. Shevchenko, Low Temp. Phys. **22**, 513 (1966).
19. C.J. Adkins and J.R. Waldram, Phys.Rev.Lett. **21**, 76 (1968).
20. A.F. Hildebrand , Phys.Rev.Lett. **8**, 190 (1964); A.A. Verheijen et al, Physica B **165-166**, 1181 (1990).
21. J.E. Hirsch, cond-mat/0211643.
22. I.K. Kikoin and S.W. Gubar, J.Phys. USSR **3**, 333 (1940).
23. J.E. Hirsch, Phys.Rev. B**66**, 064507 (2002).
24. F. London and H. London, Physica **2**, 341 (1935)
25. R. Becker, F. Sauter and C. Heller, Z. Phys. **85**, 772 (1933).
26. R. de L. Kronig, Z. Phys. **78**, 744 (1932); **80**, 203 (1933).
27. I.Kikoin and B. Lasarev, Physik.Zeits. d. Sowjetunion **3**, 351 (1933); A. Papapetrou, Z. F. Physik **92**, 513 (1934); M. Born and K.C. Cheng, Nature **161**, 1017 (1948).
28. J. Bardeen, Phys.Rev. **81**, 829 (1951).
29. W. Meissner, Ergeb.Exakte Naturw. **11**, 219 (1932).
30. J.E. Hirsch, cond-mat/0305542.
31. J.C. Slater, 'Quantum Theory of Atomic Structure', Vol. II, McGraw-Hill, New York, 1960, Chpt. 23.
32. J.E. Hirsch, Phys.Lett.A **281**, 44 (2001).
33. H. Welker, Z. Phys. **114**, 525 (1939).
34. H. Meissner, Phys.Rev. **117**, 672 (1960).
35. J.E. Hirsch, Phys.Rev.B **60**, 14787 (1999).
36. J.E. Hirsch, Phys. Rev. Lett. **72**, 558 (1994).
37. J.E. Hirsch, in "Studies of High Temperature Superconductors", ed. by A. Narlikar, Nova Sci. Pub., New York, 2002, Vol. 38, p. 49.

High temperature superconductors: Experimental implications of a variational theory of the superconducting state

Mohit Randeria*[†], Arun Paramekanti** and Nandini Trivedi*[†]

*Tata Institute of Fundamental Research, Mumbai 400 005, India
[†]Department of Physics, University of Illinois at Urbana-Champaign, Urbana, IL 61801
**Department of Physics and Kavli Institute for Theoretical Physics, University of California, Santa Barbara, California 93106-4030

Abstract. We review recent work on the properties of a d-wave superconducting state in which strong Coulomb interactions suppress double occupancy. We find that pairing and phase coherence show qualitatively different trends as a function of doping: the pairing scale decreases monotonically with hole doping x while the order parameter shows a non-monotonic "dome". We obtain detailed results for the doping-dependences of various quantities including the momentum distribution, nodal quasiparticle weights and dispersions, optical spectral weight, and superfluid density. Our results are in remarkable agreement with existing data on the high T_c cuprates and some of our predictions have been recently verified.

1. WAVE FUNCTION FOR HIGH T_C CUPRATES

Since the discovery of the cuprates in 1987, it has become increasingly clear that these materials cannot be described by simple BCS theory, and the effect of strong Coulomb correlations in a doped Mott insulator must be taken into account. A key problem is how to accurately treat strong correlations in a two-dimensional Fermi system.

The variational method has often been highly successful, especially in problems without an obvious small parameter, and very early on P. W. Anderson [1] proposed that the cuprates should be thought of in terms of Gutzwiller projected variational wavefunctions. Despite some early progress [2, 3, 4, 5], which focussed primarily on the ground state energy, this approach was not fully explored until recently.

In this article, we briefly review our recent work [6, 7, 8] on the properties of superconducting wavefunctions in which double occupancy is strongly suppressed by short-range Coulomb interactions. Our focus has not been on the ground state energy, nor on whether a certain wave function is the lowest energy state for particular Hamiltonian. Rather we have focussed on experimentally observable properties related to the SC ground state and low-lying excitations. The advantage of working with variational wavefunctions of the kind explored here, is that, in contrast to slave boson theories, one deals directly with the physical electron degree of freedom.

We have found that this approach leads to remarkable insights into the magnitudes and doping dependences of various observables as the system evolves from a Fermi liquid for hole doping $x \gtrsim 0.35$ to a Mott insulator at $x = 0$.

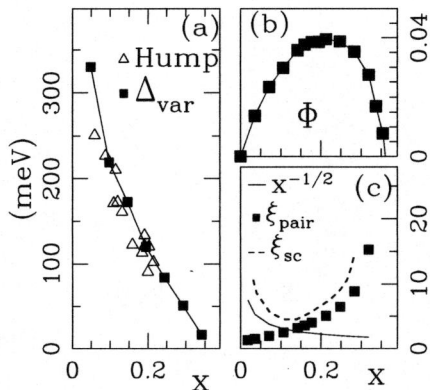

FIGURE 1. (a):Doping dependence of the variational parameter Δ_{var} (filled squares) that characterizes the strength of pairing, compared with the ARPES "hump" scale (open triangles) from Ref. [9]; (see Fig. 2). Note that there are no adjustable parameters. (b): d-wave SC order parameter $\Phi(x)$ showing a non-monotonic x-dependence similar to the experimental $T_c(x)$. (c): Plot of the "pair size" $\xi_{pair} = v_F/\Delta_{var}$, interhole separation $x^{-1/2}$, and the SC coherence length defined through $\xi_{SC} \geq min(x^{-1/2}, \xi_{pair})$

Here we will keep technical details to a minimum, referring the reader to the original references [6, 7, 8], and emphasize the final results and their comparison with experiments. We use

$$|\Psi_0\rangle = \exp(-iS)P|\Psi_{BCS}\rangle \qquad (1)$$

to describe the superconducting (SC) ground state of the cuprates. Ψ_0 has three pieces: (1) the d-wave BCS wave function, (2) the Gutzwiller projection operator P which eliminates all configurations with double occupancy in view of the large Coulomb U, and (3) a unitary transformation [3] $\exp(-iS)$, calculated perturbatively in t/U, which is well known in the derivation of the t-J model. Ψ_{BCS} is completely specified by the internal pair wave function $\varphi(\mathbf{k}) = v_\mathbf{k}/u_\mathbf{k} = \Delta_\mathbf{k}/[\xi_\mathbf{k} + \sqrt{\xi_\mathbf{k}^2 + \Delta_\mathbf{k}^2}]$ which is chosen to have a d-wave form with $\xi_\mathbf{k} = \varepsilon(\mathbf{k}) - \mu_{var}$ and $\Delta_\mathbf{k} = \Delta_{var}(\cos k_x - \cos k_y)/2$. The d-wave state is chosen because it is known to be the lowest energy solution over a wide doping range [4, 5] and also well established experimentally.

We choose the dispersion $\varepsilon(\mathbf{k})$ with hopping parameters $t = 300$ meV, $t' = -t/4$, consistent with band structure calculations and ARPES experiments and the Coulomb $U = 12t$ corresponding to $J = 100$ meV, which is consistent with neutron and Raman data. At each hole doping x, the optimal values of the two variational parameters μ_{var} and Δ_{var} are determined by minimizing the ground state energy of the 2D Hubbard model, which is a minimal one-band model which captures the strong correlation physics. The energy, and all other equal-time correlations, are calculated using the variational Monte Carlo method which is the only known scheme for computing expectation values treating the projection operator exactly.

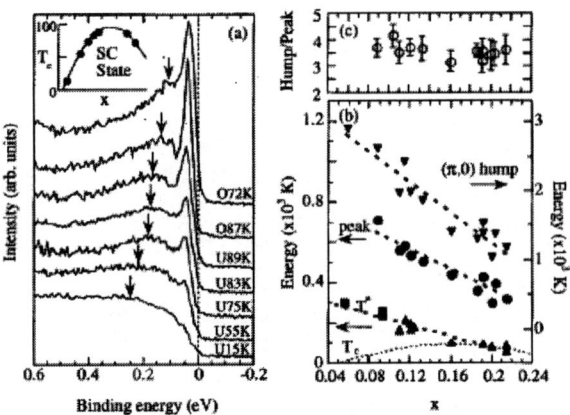

FIGURE 2. (a):SC state ARPES spectra [9] at $\mathbf{k} = (\pi, 0)$ showing the incoherent hump scale for Bi2212 samples with different dopings (corresponding T_c's are shown in inset). (b): Doping dependence of T_c, pseudogap T^*, peak (gap) and hump energies in the SC state. (c): Hump to peak energy ratio as a function of hole doping x. (From Ref. [9]).

2. PAIRING VS. SC COHERENCE: TWO ENERGY SCALES

One of our most striking results [6, 7] is that upon inclusion of strong correlations, pairing and SC long range order show qualitatively different doping dependences as shown in Fig. 1. The pairing scale, related to the variational parameter Δ_{var}, decreases monotonically with hole doping x. Through an analysis of spectral function moments, we show that Δ_{var} can be related to a characteristic energy scale for incoherent excitations at $(\pi, 0)$. This motivates us to compare its doping dependence with that of the "hump scale" observed [9] in the SC state ARPES lineshape at $\mathbf{k} = (\pi, 0)$ (see Fig. 2). We find remarkable agreement (Fig. 1(a)) between the value and the doping dependence of the calculated Δ_{var} and the observed ARPES "hump scale" [9]. The doping dependence of Δ_{var} resembles that of the pseudogap temperature scale T^*.

In contrast, the SC order parameter Φ, defined in terms of the of-diagonal long range order in the pair-pair correlation function, shows a non-monotonic doping dependence (Fig. 1(b)) like that of $T_c(x)$. For $x \geq x_c \simeq 0.35$, the system is a Fermi liquid, and Δ_{var} vanishes due to loss of pairing and the SC order parameter Φ vanishes. With decreasing x the pairing builds up and so does the SC order parameter at least down to $x \simeq 0.2$. However, for smaller x the pairing in the wave function as characterized by Δ_{var} keeps on decreasing with underdoping but the Coulomb correlations kill the SC order. Precisely at half-filling, $x = 0$, projection fixes the number to be $n = 1$ at each site, leading to large quantum phase fluctuations which destroy SC order. One can show that $\Phi(x) \sim x \Delta_{var}(x)/J$.

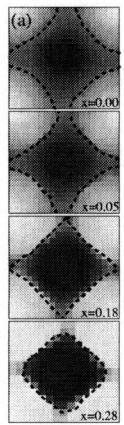

FIGURE 3. Grayscale plots of the momentum distribution $n(k)$ at various dopings (black=1, white=0). Dashed curve denotes the "Fermi surface" (FS) contour on which $n(k) = 1/2$, which is essentially the same as the noninteracting FS.

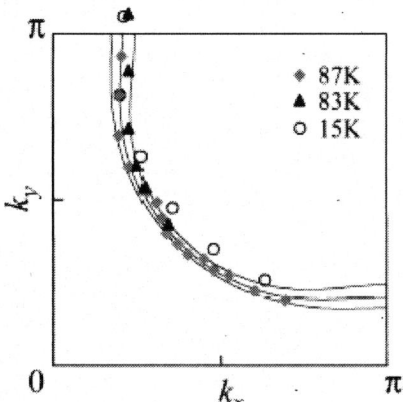

FIGURE 4. ARPES Fermi surfaces for samples with $T_c = 87K, 83K, 15K$ showing a large volume. The solid lines are fits are tight binding estimates of the Fermi surface at 18%, 13% and 6% assuming rigid band behavior. (From Ref. [10]).

3. NODAL QUASIPARTICLES:

We focus on the low lying excitations in the SC state and find that with decreasing x one approaches the Mott insulator at $x = 0$ as follows:

(a) We find a large underlying Fermi surface at each x as probed by the momentum distribution shown in Fig. 3. This is in good agreement with ARPES measurements [10] in Fig. 4.

(b) We find sharp gapless nodal quasiparticles in the SC state (Fig. 5), with nodal k_F

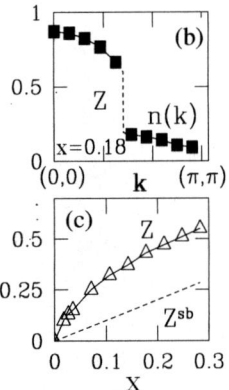

FIGURE 5. Top panel: The momentum distribution $n(k)$ along $(0,0) \to (\pi,\pi)$ for $x = 0.18$. The discontinuity in $n(k)$ signals gapless quasiparticles with a weight Z. Bottom panel: Doping dependence of Z. Also shown is the slave boson mean field theory result $Z^{sb} = x$.

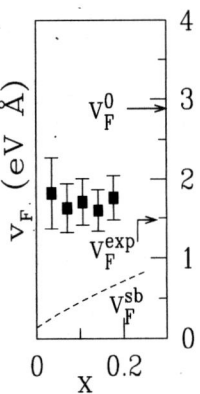

FIGURE 6. Nodal quasiparticle velocity v_F estimated from spectral function moments compared with the bare v_F^0, ARPES estimates (v_F^{exp}) [12], and slave boson (sb) estimate. The doping independence of v_F together with $Z(x)$ from Fig. 5 constrains the structure of the self energy [6, 7, 8].

weakly dependent on x, but quasiparticle weight Z strongly x-dependent and vanishing as $x \to 0$ (Fig. 5). Our prediction for nodal $Z(x)$ has recently been verified by ARPES experiments [11].

(c) The nodal quasiparticle velocity v_F is very weakly doping dependent (Fig. 6) despite the strong x dependence of Z. ARPES [12] finds strong evidence for a weakly x-dependent v_F.

Our results for the nodal quasiparticles were *not* obtained by working with variational single-particle excited states, e.g., Gutzwiller projected Bogoliubov quasiparti-

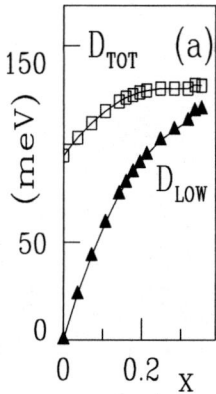

FIGURE 7. Integrated optical spectral weight D_{tot} and the low energy integrated D_{low} (Drude weight) as a function of x.

cles. Given the very broad linewidths of the spectral functions, it is difficult to use simple variational excited states and obtain useful results on the coherent piece. To cirumvent this problem, we made novel use of the singular structure of spectral function moments [6, 7] to focus on quantities like Z and v_F which characterize the low energy properties of the nodal quasiparicles.

4. DRUDE WEIGHT:

The x-dependence of the total optical spectral weight D_{tot} and the low frequency Drude weight D_{low} shown in Fig. 7 are calculated using sum rules which relate moments of the optical conductivity to appropriately defined equal time correlators [6, 7]. In marked contrast to D_{tot}, which includes transitions to the upper Hubbard band, D_{low} vanishes because of the projection and gives an insulator at $x = 0$. At low doping, we find that D_{tot} is a weak function of x while D_{low} increases more rapidly, reflecting a rapid transfer of spectral weight from the upper to the lower Hubbard band with doping.

Both the magnitude and doping dependence of D_{low} are in agreement with optical data on the cuprates [14, 13] (see Fig. 8 for data on YBCO). Using our calculated $D_{low} \approx 90$ meV (at optimality) and a lattice spacing $a = 3.85\text{Å}$, we find the plasma frequency $\omega_p^* = \sqrt{4\pi D_{low} e^2/a} \sim 2.0 eV$. Further, the experimental ω_p^{*2} also vanishes linearly in the low doping regime. We also predict that $D_{low} \sim Z$ over the entire range $0 \le x \le x_c$.

5. SUPERFLUID STIFFNESS:

We show that the superfluid stiffness $D_s \le D_{low}$. The vanishing of D_{low} at small x implies that D_s vanishes as $x \to 0$, consistent with the Uemura plot [10]. Further,

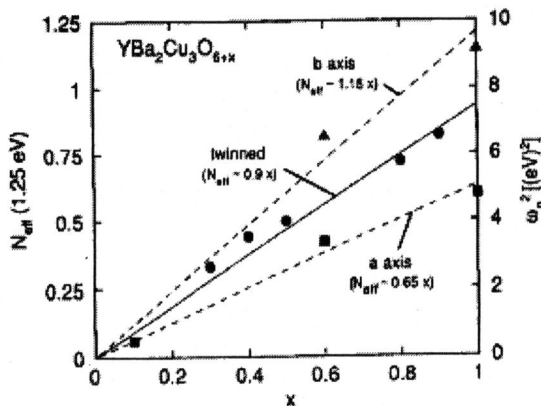

FIGURE 8. Experimental variation of the Drude weight plasma frequency ω_p^2 with doping (From Ref. [13]).

we can put a lower bound on the penetration depth $\lambda_L^{-2} = 4\pi e^2 D_s/(\hbar^2 c^2 d_c)$, where d_c is the mean-interlayer spacing. Using $d_c = 7.5$Å for BSCCO and our calculated $D_{low} \approx 90 meV$ at optimality, we find $\lambda_L \geq 1350$Å. The measured value in optimally doped BSCCO is $\lambda \simeq 2100$Å[10]. This agreement is quite satisfactory, given that at $T = 0$ the experimentally measured superfluid density is reduced by impurity effects and by quantum phase fluctuations both of which are not included in our calculation.

6. CONCLUSION:

The simplest strongly correlated wave function in Eq. (1) is thus found to be extremely successful in describing the SC state properties of high T_c cuprates and the evolution of the ground state from a Fermi liquid at large doping, to a d-wave SC down to a Mott insulator at half filling. The SC dome does not require introducing, by hand, any competing order, but is rather a natural consequence of Mott physics at half filling. The dichotomy of a large pairing energy scale and a small superfluid stiffness is also naturally explained leading to a pairing induced pseudogap in the underdoped region.

ACKNOWLEDGMENTS

MR would like to take this opportunity to acknowledge the hospitality of Professor F. Mancini when he lectured at the Vietri Sul Mare Summer School several years ago, and also the invitation to visit this most beautiful part of Italy once again for this Conference. MR and NT gratefully acknowledge the hospitality of the Physics Department at University of Illinois and support through DOE grant DEFG02-91ER45439 and DARPA grant N0014-01-1-1062. AP was supported by NSF DMR-9985255 and PHY-07949 and the

Sloan and Packard foundations. We acknowledge the use of computational facilities at TIFR, including those provided by the DST Swarnajayanti Fellowship.

REFERENCES

1. Anderson, P. W., *Science*, **235**, 1196 (1987).
2. Zhang, F.-C., Joynt, R., Rice, T. M., and Shiba, H., *Supercond. Sci. and Tech.*, **68**, 425 (1987).
3. C. Gros, R. J., and Rice, T. M., *Z. Phys. B*, **68**, 425 (1987).
4. Yokoyama, H., and Shiba, H., *J. Phys. Soc. Jpn.*, **57**, 2482 (1988).
5. Gros, C., *Phys. Rev. B*, **38**, 931 (1988).
6. Paramekanti, A., Randeria, M., and Trivedi, N., *Phys. Rev. Lett*, **87**, 217002 (2001).
7. Paramekanti, A., Randeria, M., and Trivedi, N., *cond-mat/0305611* (2003).
8. Randeria, M., Paramekanti, A., and Trivedi, N., *cond-mat/0307217* (2003).
9. Campuzano, J.-C., Ding, H., Norman, M. R., Fretwell, H. M., Randeria, M., Kaminski, A., Mesot, J., Takeuchi, T., Sato, T., Yokoya, T., Takahashi, T., Mochiku, T., Kadowaki, K., Guptasarma, P., Hinks, D. G., Konstantinovic, Z., Li, Z. Z., and Raffy, H., *Phys. Rev. Lett.*, **83**, 3709 (1999).
10. Ding, H., Norman, M. R., Yokoya, T., Takeuchi, T., Randeria, M., Campuzano, J.-C., Takahashi, T., Mochiku, T., and Kadowaki, K., *Phys. Rev. Lett.*, **78**, 2628 (1997).
11. Yoshida, T., Zhou, X. J., Sasagawa, T., Yang, W. L., Bogdanov, P. V., Lanzara, A., Hussain, Z., Fujimori, A., Eisaki, H., Shen, Z.-X., Kakeshita, T., and Uchida, S., *Phys. Rev. Lett.*, **91**, 027001 (2003).
12. Zhou, X. J., Yoshida, T., Lanzara, A., Bogdanov, P. V., Keller, S. A., Shen, K. M., Yang, W. L., Ronning, F., Sasagawa, T., Kakeshita, T., Noda, T., Eisaki, H., Uchida, S., Lin, C. T., Zhou, F., Xiong, J. W., Ti, W. X., Zhao, Z. X., Fujimori, A., Hussain, Z., and Shen, Z.-X., *Nature*, **423**, 398 (2003).
13. Cooper, S. L., Reznik, D., Kotz, A., Karlow, M. A., Liu, R., Klein, M. V., Lee, W. C., Giapintzakis, J., Ginsberg, D. M., Veal, B. W., and Paulikas, A. P., *Phys. Rev. B*, **47**, 8233 (1993).
14. Orenstein, J., Thomas, G. A., Millis, A. J., Cooper, S. L., Rapkine, D. H., Timusk, T., Schneemeyer, L. F., and Waszczak, J. V., *Phys. Rev. B*, **42**, 6342 (1990).

Critical temperature of high-T_c superconductors in the bipolaron model*

A.S. Alexandrov

Department of Physics, Loughborough University, Loughborough LE11 3TU, UK

Abstract. The extension of the BCS theory to the strong-coupling regime with small bipolarons numerically accounts for high superconducting critical temperatures and isotope effects, of many high-T_c superconductors.

I. PARAMETER-FREE DESCRIPTION OF T_c

An ultimate goal of the theory of superconductivity is to provide an expression for T_c as a function of some well-defined parameters characterizing the material. In the framework of the BCS theory the Eliashberg equation for the gap function properly takes into account a realistic phonon spectrum and retardation of the electron-phonon interaction. But applying a theory of this kind to high-T_c cuprates is problematic. Since bare electron bands are narrow, strong correlations result in the Mott insulating state of undoped parent compounds. As a result, the Coulomb pseudopotential, μ^*, and the electron-phonon coupling constant λ are ill defined in doped cuprates, and polaronic effects are important as in many doped semiconductors [1]. Hence, an estimate of T_c in cuprates within the BCS theory appears to be an exercise in calculating μ^* rather than T_c itself. Also one cannot increase λ without accounting for a polaron collapse of the band. This appears at $\lambda \approx 1$ [2].

On the other hand, the bipolaron theory [2] provides a parameter-free expression for T_c [3], which fits the experimentally measured T_c in many cuprates for any level of doping. T_c is calculated using the density sum rule as the Bose-Einstein condensation (BEC) temperature of $2e$ charged bosons on a lattice. Just before the discovery [4] we predicted T_c as high as $\approx 100K$ using an estimate of the bipolaron effective mass [5]. Uemura [6] established a correlation of T_c with the in-plane magnetic field penetration depth measured by μsR technique in many cuprates as $T_c \propto 1/\lambda_{ab}^2$. The technique is based on the implantation of spin polarized muons. It monitors the time evolution of the muon spin polarization. He concluded that cuprates are neither BCS nor BEC superfluids but they are in a crossover region from one to the other, because the experimental T_c was found about 3 or more times below the BEC temperature.

Here we calculate T_c of a bipolaronic superconductor taking properly into account the microscopic band structure of bipolarons in layered cuprates [2]. We arrive at a parameter-free expression for T_c [3], which in conrast to Ref. [6] involves not only the in-plane, λ_{ab} but also the out-of-plane, λ_c, magnetic field penetration depth, and a normal state Hall ratio R_H just above the transition. It describes the experimental data for a few dozen different samples clearly indicating that many cuprates are in the BEC rather than in the crossover regime.

The energy spectrum of bipolarons is two-fold degenerate in cuprates [2]. One can

TABLE I: Experimental data on $T_c(K)$, ab and c penetration depth(nm), Hall coefficient $(10^{-3}(cm^3/C))$, and calculated values of T_c respectively for $La_{2-x}Sr_xCuO_4$ (La), $YBaCuO(x\%Zn)$ (Zn), $YBa_2Cu_3O_{7-x}$ (Y) and $HgBa_2CuO_{4+x}$ (Hg) compounds

Compound	T_c^{exp}	λ_{ab}	λ_c	R_H	$T_c(3D)$	T_c
La(0.2)	36.2	200	2540	0.8	38	41
La(0.22)	27.5	198	2620	0.62	35	36
La(0.24)	20.0	205	2590	0.55	32	32
La(0.15)	37.0	240	3220	1.7	33	39
La(0.1)	30.0	320	4160	4.0	25	31
La(0.25)	24.0	280	3640	0.52	17	19
Zn(0)	92.5	140	1260	1.2	111	114
Zn(2)	68.2	260	1420	1.2	45	46
Zn(3)	55.0	300	1550	1.2	35	36
Zn(5)	46.4	370	1640	1.2	26	26
Y(0.3)	66.0	210	4530	1.75	31	51
Y(0.43)	56.0	290	7170	1.45	14	28
Y(0.08)	91.5	186	1240	1.7	87	88
Y(0.12)	87.9	186	1565	1.8	75	82
Y(0.16)	83.7	177	1557	1.9	83	89
Y(0.21)	73.4	216	2559	2.1	47	59
Y(0.23)	67.9	215	2630	2.3	46	58
Y(0.26)	63.8	202	2740	2.0	48	60
Y(0.3)	60.0	210	2880	1.75	43	54
Y(0.35)	58.0	204	3890	1.6	35	50
Y(0.4)	56.0	229	4320	1.5	28	42
Hg(0.049)	70.0	216	16200	9.2	23	60
Hg(0.055)	78.2	161	10300	8.2	43	92
Hg(0.055)	78.5	200	12600	8.2	28	69
Hg(0.066)	88.5	153	7040	6.85	56	105
Hg(0.096)	95.6	145	3920	4.7	79	120
Hg(0.097)	95.3	165	4390	4.66	61	99
Hg(0.1)	94.1	158	4220	4.5	66	105
Hg(0.101)	93.4	156	3980	4.48	70	107
Hg(0.101)	92.5	139	3480	4.4	88	127
Hg(0.105)	90.9	156	3920	4.3	69	106
Hg(0.108)	89.1	177	3980	4.2	58	90

apply the effective mass approximation at $T \simeq T_c$, because T_c should be less than the bipolaron bandwidth. Also a three-dimensional correction to the spectrum is important for the Bose-Einstein condensation which is well described by the tight-binding approximation as

$$E_{\mathbf{K}}^{x,y} = \frac{\hbar^2 K_{x,y}^2}{2m_x^{**}} + \frac{\hbar^2 K_{y,x}^2}{2m_y^{**}} + 2t_\perp[1 - \cos(K_z d)], \tag{1}$$

where d is the interplane distance and $t_\perp$ is the inter-plane bipolaron hopping integral. Substituting the spectrum, Eq.(1) into the density sum rule,

$$\sum_{\mathbf{K}, i=(x,y)} [\exp(E_{\mathbf{K}}^i/T_c) - 1]^{-1} = n_b \tag{2}$$

one readily obtains T_c as (in ordinary units)

$$k_B T_c = f\left(\frac{t_\perp}{k_B T_c}\right) \times \frac{3.31\hbar^2 (n_B/2)^{2/3}}{(m_x^{**} m_y^{**} m_c^{**})^{1/3}}, \tag{3}$$

where the coefficient $f(x) \approx 1$ is a function of the anisotropy, $t_\perp/(k_B T_c)$, and $m_c^{**} = \hbar^2/(2|t_\perp|d^2)$.

This expression is rather ambiguous because the effective mass tensor as well as the bipolaron density n_b are not well known. Fortunately, we can express the band-structure parameters via in-plane,

$$\lambda_{ab} = \left[\frac{m_x^{**} m_y^{**}}{8\pi n_B e^2 (m_x^{**} + m_y^{**})}\right]^{1/2}$$

and out-of-plane penetration depths,

$$\lambda_c = \left[\frac{m_c^{**}}{16\pi n_b e^2}\right]^{1/2}$$

(we use $c = 1$). The bipolaron density is expressed through the in-plane Hall ratio (above the transition) as

$$R_H = \frac{1}{2en_b} \times \frac{4m_x^{**} m_y^{**}}{(m_x^{**} + m_y^{**})^2}, \tag{4}$$

which leads to

$$T_c = 1.64 f\left(\frac{t_\perp}{k_B T_c}\right)\left(\frac{eR_H}{\lambda_{ab}^4 \lambda_c^2}\right)^{1/3}. \tag{5}$$

Here T_c is measured in Kelvin, eR_H in cm^3 and λ in cm. The coefficient f is about unity in a very wide range of $t_c/(k_B T_c) \geq 0.01$. Hence, the bipolaron theory yields a parameter-free expression, which unambiguously tells us how near cuprates are to the BEC regime,

$$T_c \approx T_c(3D) = 1.64 \left(\frac{eR_H}{\lambda_{ab}^4 \lambda_c^2}\right)^{1/3}. \tag{6}$$

We compare two last expressions with the experimental T_c of more than 30 different cuprates, for which both λ_{ab} and λ_c are measured along with $R_H(T_c + 0)$ in Table 1. The Hall ratio has a strong temperature dependence above T_c. Therefore, we use the experimental Hall ratio just above the transition. In a few cases (mercury compounds), where $R_H(T_c + 0)$ is unknown, we take the inverse chemical density of carriers (divided by e) as R_H. For almost all samples the theoretical T_c fits experimental values within an experimental error bar for the penetration depth (about $\pm 10\%$). There are a few Zn doped YBCO samples (Table), whose critical temperature is higher than the theoretical one. If we assume that the degeneracy of the bipolaron spectrum is removed by the random potential of Zn, then the theoretical T_c would be almost the same as the experimental values for these samples as well.

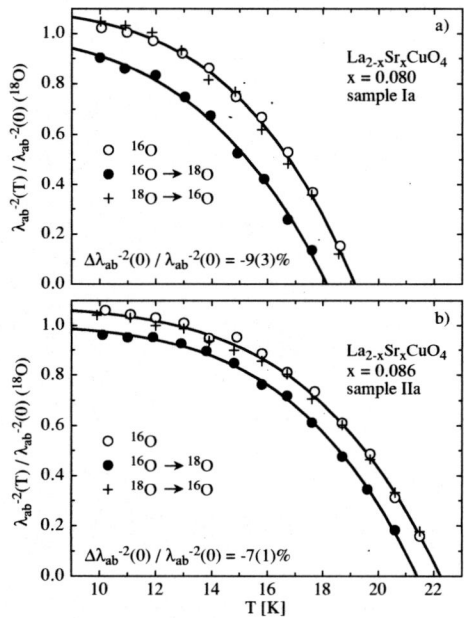

FIG. 1: The isotope effect on the magnetic field penetration depth in two samples of $La_{2-x}Sr_xCuO_4$ [8] (courtesy of J. Hofer).

II. ISOTOPE EFFECT ON T_c AND ON SUPERCARRIER MASS

The advances in the fabrication of the isotope substituted samples made it possible to measure a sizable isotope effect, $\alpha = -d\ln T_c / d\ln M$ in many high-T_c oxides. This led to a general conclusion that phonons are relevant for high T_c. Moreover the isotope effect in cuprates was found to be quite different from the BCS prediction, $\alpha = 0.5$ (or less). Several compounds showed $\alpha > 0.5$ and a small negative value of α was found in $Bi - 2223$.

These features of the isotope effect, in particular large values in low T_c cuprates, an overal trend to lower value as T_c increases, and a small or even negative α in some high T_c cuprates were understood using the isotope exponents of (bi)polaronic superconductors [7]. With increasing ion mass the bipolaron mass increases and the Bose-Einstein condensation temperature T_c decreases in the bipolaronic superconductor. On the contrary an increase of the ion mass leads to a band narrowing and to an enhancement of the polaron density of states, and to an inrease of T_c in polaronic superconductors. Hence the isotope exponent in T_c can distinguish the BCS like polaronic superconductivity with $\alpha < 0$, and the Bose-Einstein condensation of small bipolarons with $\alpha > 0$. Moreover, underdoped cuprates, which are definitely in the BEC regime, could have $\alpha > 0.5$, as observed.

Another prediction of the bipolaron theory is an isotope effect on the carrier mass,

$$\alpha_{m^*} = 0.5 \ln \frac{m^*}{m}, \tag{7}$$

which is linked with the isotope effect on T_c.

Remarkably, this prediction was experimentally confirmed by Zhao et al.[8] providing a compelling evidence for the polaronic carriers in doped cuprates. The effect was observed in the London penetration depth of isotope-substituted cuprates, Fig.1. The carrier density is unchanged with the isotope substitution of O^{16} by O^{18}, so that the isotope effect on λ_{ab} measures directly the isotope effect on the carrier mass. In particular, the carrier mass isotope exponent $\alpha_{m^*} = d\ln m^*/d\ln M$ was found as large as $\alpha_{m^*} = 0.8$ in $La_{1.895}Sr_{0.105}CuO_4$. Then, according to Eq.(7) the polaron mass enhancement should be $m^*/m \approx 5$ in this material. This corresponds to the in-plane bipolaron mass as large as $m^{**} \approx 10 m_e$. The in-plane magnetic field penetration depth, calculated with this mass is $\lambda_{ab} = [m^{**}/8\pi n e^2]^{1/2} \approx 316$nm, where n is the hole density. It agrees well with the experimental one, $\lambda_{ab} \simeq 320$nm. Using the measured values of $\lambda_{ab} = 320$ nm, $\lambda_c = 4160$ nm, and of $R_H = 4 \times 10^{-3}$ cm^3/C (just above T_c) we obtain $T_c = 31K$ in astonishing agreement with the experimental value $T_c = 30$ K in this compound.

These simple numerical results as well as a great number of other experimental facts (see, for review, [9]) present a piece of strong evidence for a novel state of electronic matter in the layered cuprates; this is the charged Bose-liquid of bipolarons.

REFERENCES

[1] Alexandrov, A. S., and Mott, N. F., Rep. Prog. Phys. **57**, 1197 (1994).
[2] Alexandrov, A. S., in 'Models and Phenomenology for Conventional and High-temperature Superconductivity' (Course CXXXVI of the Intenational School of Physics 'Enrico Fermi'), eds. G. Iadonisi, J.R. Schrieffer and M.L. Chiofalo, (IOS Press, Amsterdam), p. 309 (1998).
[3] Alexandrov, A. S., and Kabanov V. V., Phys. Rev. B **59** (1999).
[4] Bednorz, J. G., and Müller, K. A., Z. Phys. B**64**, 189 (1986).
[5] Alexandrov, A. S., and Kabanov V. V., Sov. Phys. Solid State, **28**, 631 (1986).
[6] Uemura, Y. J., in 'Polarons and Bipolarons in High-Tc Superconductors and Related Materials', E.K.H. Salje, A.S. Alexandrov and W.Y. Liang (Cambridge University Press, Cambridge), p.453 (1995).
[7] Alexandrov, A. S., Phys. Rev. B**46**, 14932. (1992).
[8] Zhao, G., Hunt, M. B., Keller, H., and Müller, K. A., Nature **385**, 236 (1997).
[9] Alexandrov, A. S., and Edwards, P. P., Physica C**331**, 2000; Devreese, J. T., in *Encyclopedia of Applied Physics*, vol. 14, p. 383 (VCH Publishers, 1996).

Dedicated to Professor Ferdinando Mancini on the occasion of his 60th birthday

Polarons: Recent developments

J. T. Devreese

Theoretische Fysica van de Vaste Stoffen (TFVS),
Universiteit Antwerpen, Universiteitsplein 1, B-2610 Antwerpen, Belgium;
also at: Technische Universiteit Eindhoven, P. O. Box 513, 5600 MB Eindhoven, The Netherlands.

Abstract.
In this presentation three recent contributions to the theory of continuum (or "Fröhlich"-) polarons are discussed. (i) Using a generalization of the Jensen-Feynman variational principle within the path-integral formalism for identical particles, the ground-state energy of a confined N-polaron system is studied as a function of N and of the electron-phonon coupling strength. (ii) Cyclotron-resonance (CR) spectra of a gas of interacting polarons in a GaAs/AlAs quantum well are theoretically investigated taking into account the magnetoplasmon-phonon mixing and the band non-parabolicity. The calculated CR spectra are in a good agreement with experimental data. The theory explains that, for a high-density polaron gas, anticrossing of the CR spectra occurs near the GaAs TO-phonon frequency rather than near the GaAs LO-frequency. (iii) A theoretical investigation of the optical properties of stacked quantum dots is presented, which is based on the non-adiabatic approach.

INTRODUCTION

Three recent contributions to the theory of continuum polarons are considered.

1. INTERACTING POLARONS IN A QUANTUM DOT

1.1. Ground-state properties of interacting polarons in a quantum dot

Many-electron states in quantum dots have been investigated theoretically by various approaches, e. g., the Hartree-Fock method [1, 2, 3], the density-functional theory [4, 5, 6, 7, 8, 9, 10], the quantum Monte Carlo simulation [11], the variational Monte Carlo method and Padé approximation [12, 13], numerical diagonalization of the Hamiltonian in a finite-dimensional basis [14]. The electron-phonon interaction was not taken into account in these investigations.

In this section, a system of N electrons with mutual Coulomb repulsion and interacting with the lattice vibrations is considered. A parabolic confinement potential, characterised by the frequency parameter Ω_0, is assumed. The total number of electrons is $N = \sum_\sigma N_\sigma$, where N_σ is the number of electrons with spin projection $\sigma = \pm 1/2$. A canonical ensemble is treated, where the number of electrons N_σ for each σ is fixed. The bulk phonons (characterized by wave vectors $\mathbf{q}$ and frequencies $\omega_\mathbf{q}$) are described by the complex coordinates Q_q. The full set of electron and phonon coordinates is denoted by $\bar{x} \equiv \{x_{j,\sigma}\}$ and $\bar{Q} \equiv \{Q_q\}$.

The partition function $Z(\{N_\sigma\},\beta)$ of the system can be expressed as a path integral over all electron and phonon coordinates:

$$Z(\{N_\sigma\},\beta) = \sum_P \frac{(-1)^{\xi_P}}{N_{1/2}!N_{-1/2}!} \int d\bar{x} \int_{\bar{x}}^{P\bar{x}} D\bar{x}(\tau) \int d\bar{Q} \int_{\bar{Q}}^{\bar{Q}} D\bar{Q}(\tau) e^{-S[\bar{x}(\tau),\bar{Q}(\tau)]}, \quad (1)$$

where $S[\bar{x}(\tau),\bar{Q}(\tau)]$ is the "action" functional:

$$S[\bar{x}(\tau),\bar{Q}(\tau)] = -\frac{1}{\hbar}\int_0^{\hbar\beta} L(\dot{\bar{x}},\dot{\bar{Q}};\bar{x},\bar{Q}) d\tau \quad (2)$$

with the Lagrangian of the electron-phonon system $L(\dot{\bar{x}},\dot{\bar{Q}};\bar{x},\bar{Q})$. The parameter $\beta \equiv 1/(k_B T)$ is inversely proportional to temperature T. In order to take the Fermi-Dirac statistics into account, the integral over the electron paths $\{\bar{x}(\tau)\}$ in (1) contains a sum over all permutations P of the electrons with equal spin projections, with ξ_P denoting the parity of a permutation P.

The path integral over the phonon variables in (2) can be calculated analytically [15]. As a result, the partition function of the electron-phonon system (1) factorises into a product of a free-phonon partition function with a partition function $Z_P(N_{1/2},N_{-1/2}|\beta)$ of interacting polarons, which is a path integral over the electron coordinates only:

$$Z(\{N_\sigma\},\beta) = Z_P(\{N_\sigma\},\beta) \prod_q \frac{1}{2\sinh(\beta\hbar\omega_{LO}/2)}. \quad (3)$$

$$Z_P(\{N_\sigma\},\beta) = \sum_P \frac{(-1)^{\xi_P}}{N_{1/2}!N_{-1/2}!} \int d\bar{x} \int_{\bar{x}}^{P\bar{x}} D\bar{x}(\tau) e^{-S_P[\bar{x}(\tau)]}. \quad (4)$$

where $S_P[\bar{x}(\tau)]$ results from the elimination of the phonon variables and contains the "influence phase" of the phonons. It describes the phonon-induced retarded interaction between the electrons, including the retarded self-interaction of each electron. The free energy of a system of interacting polarons $F_P(\{N_\sigma\},\beta)$ is related to the partition function (4) by the relation:

$$F_P(\{N_\sigma\},\beta) = -\frac{1}{\beta}\ln Z_P(\{N_\sigma\},\beta). \quad (5)$$

At present no method is known to calculate the non-Gaussian path integral (4) analytically. For *distinguishable* particles, the Jensen-Feynman variational principle [15] provides a convenient approximation technique. It yields a lower bound to the partition function, and consequently an upper bound to the free energy.

The formulation of a variational principle for the free energy for a system of *identical particles* is a non-trivial problem. However, it can be shown [16] that the path-integral approach to the many-body problem for a fixed number of identical particles can be formulated as a Feynman-Kac functional defined on a state space for N indistinguishable particles, by imposing an ordering on the configuration space and by the introduction of a set of boundary conditions at the boundaries of this state space. The path integral (in the imaginary-time variable) for identical particles was shown to be *positive* within this state space. This implies that a many-body extension of the Jensen-Feynman inequality was found, which can be used for interacting identical particles (Ref. [16], p. 4476). A more detailed analysis of this variational principle for both local and retarded interactions can be found in Ref. [17]. It is required that the potentials are symmetric with respect to all permutations of the particle positions, and that both the exact propagator and the model propagator are antisymmetric (for fermions) with respect to permutations of any two electrons at any point in time. This means that those propagators must be defined on the same configuration space. Keeping in mind these requirements, the variational inequality for identical particles has the same form as the standard Jensen-Feynman variational principle:

$$F_P \leq F_0 + \frac{1}{\beta} \langle S_P - S_0 \rangle_{S_0}, \tag{6}$$

where S_0 is a model action with corresponding free energy F_0. The angular brackets mean a weighted average over the paths

$$\langle (\bullet) \rangle_{S_0} = \frac{\sum_P \frac{(-1)^{\xi_P}}{N_{1/2}! N_{-1/2}!} \int d\bar{x} \int_{\bar{x}}^{P\bar{x}} D\bar{x}(\tau) (\bullet) e^{-S_0[\bar{x}(\tau)]}}{\sum_P \frac{(-1)^{\xi_P}}{N_{1/2}! N_{-1/2}!} \int d\bar{x} \int_{\bar{x}}^{P\bar{x}} D\bar{x}(\tau) e^{-S_0[\bar{x}(\tau)]}}. \tag{7}$$

It should be recalled that S_0 now must fulfill the properties, which were mentioned above.

In Ref. [18], we have chosen a model system consisting of N electrons with coordinates $\bar{x} \equiv \{x_{j,\sigma}\}$ coupled to N_f "fictitious" particles with coordinates $\bar{y} \equiv \{y_j\}$ in a harmonic confinement potential with elastic interparticle interactions as studied in Ref. [19]. The Lagrangian of this model system takes the form

$$L_M(\dot{\bar{x}}, \dot{\bar{y}}; \bar{x}, \bar{y}) = -\frac{m}{2} \sum_\sigma \sum_{j=1}^{N_\sigma} (\dot{x}_{j,\sigma}^2 + \Omega^2 x_{j,\sigma}^2) + \frac{m\omega^2}{4} \sum_{\sigma,\sigma'} \sum_{j=1}^{N_\sigma} \sum_{l=1}^{N_{\sigma'}} (x_{j,\sigma} - x_{l,\sigma'})^2$$

$$-\frac{m_f}{2} \sum_{j=1}^{N_f} (\dot{y}_j^2 + \Omega_f^2 y_j^2) - \frac{k}{2} \sum_\sigma \sum_{j=1}^{N_\sigma} \sum_{l=1}^{N_f} (x_{j,\sigma} - y_l)^2. \tag{8}$$

The frequencies Ω, ω, Ω_f, the mass of the fictitious particle m_f, and the force constant k are variational parameters. Clearly, this Lagrangian is symmetric with respect to electron permutations. Using this model Lagrangian we obtain an upper bound to the free energy F_{var}. Both the free energy and the correlation functions of the model system can be calculated analytically using the generation-function technique [20]. Further on, the zero-temperature case is considered.

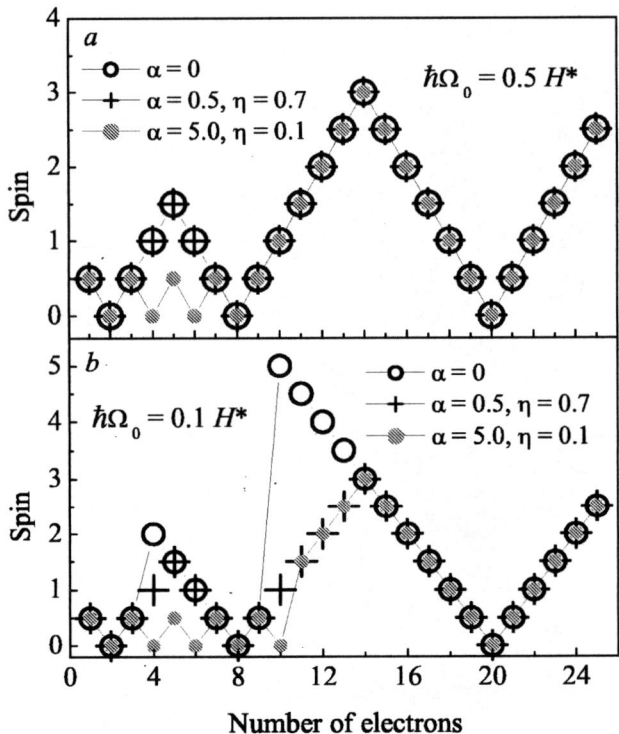

FIGURE 1. Total spin of the system of interacting polarons (for different coupling strengths) in a parabolic quantum dot as a function of the number of electrons for $\hbar\Omega_0 = 0.5\ H^*$ (a) and for $\hbar\Omega_0 = 0.1\ H^*$ (b). (From Ref. [18].)

In Fig. 1, the total spin S of a system of interacting polarons in their ground state is plotted as a function of the number of electrons in a quantum dot for different values of the confinement frequency Ω_0, of the electron-phonon coupling constant α and of the parameter $\eta = \varepsilon_\infty/\varepsilon_0$. The parameter $\hbar\Omega_0$ is measured in effective Hartrees ($H^* = [m/(m_0 \varepsilon_\infty^2)] \times 1$ Hartree). For closed-shell systems $S = 0$. For open-shell systems, where all shells except the upper one are filled, the electrons in the upper (partly filled) shell are distributed in such a way that the total spin S takes its maximal possible value (Hund's rule [21]). As seen from Fig. 1 (a), for a quantum dot with $\hbar\Omega_0 = 0.5\ H^*$ at $\alpha = 0$ and at $\alpha = 0.5$, the shell filling does obey Hund's rule.

For sufficiently small Ω_0, a spin-polarized state for a system of interacting electrons in a quantum dot can become energetically more favourable than a state satisfying Hund's rule. For a quantum dot with $\hbar\Omega_0 = 0.1\ H^*$, the spin-polarized state at $\alpha = 0$ appears to be energetically most favourable for $N = 4$ and $N = 10$ (i. e. for spin-polarized systems where all shells are half-filled), as seen from Fig. 1 (b).

In the strong-coupling case ($\alpha \gg 1$ and $\eta \ll 1$), the total spin of an open-shell system for the ground state can take its minimal possible value, as seen from Fig. 1(a) for $\alpha = 5, \eta = 0.1$ at $N = 4$ to 6. This trend to minimize the total spin is a consequence of the electron-phonon interaction, presumably due to the fact that the phonon-mediated electron-electron attraction can overcome the Coulomb repulsion.

In summary, confined few-electron systems without the electron-phonon interaction can exist in one of two phases: the spin-polarized state and a state obeying Hund's rule, depending on the confinement frequency (see, e. g., Ref. [14]). For interacting few-polaron systems, besides the above two phases, there may occur also a third phase — the state with minimal spin — in quantum dots of polar substances with sufficiently strong electron-phonon coupling $\alpha \geq 3$ (for instance, of high-T_c superconductors [22]).

1.2. Optical properties of interacting polarons in a quantum dot

To investigate the optical properties of the many-polaron system, in Ref. [18], the memory-function formalism of Ref. [23] is extended to the case of interacting polarons in a quantum dot. Within this technique, the optical conductivity for a system of interacting polarons in a parabolic confinement potential is given in terms of the memory function $\chi(\omega)$,

$$\text{Re}\sigma(\omega) = -\frac{e^2}{m} \frac{\omega \text{Im}\chi(\omega)}{[\omega^2 - \Omega_0^2 - \text{Re}\chi(\omega)]^2 + [\text{Im}\chi(\omega)]^2}, \qquad (9)$$

where $\chi(\omega)$ is

$$\chi(\omega) = \sum_q \frac{2|V_q|^2 q^2}{3N\hbar\omega} \int_0^\infty (e^{i\omega t} - 1) \text{Im}\left[T^*_{\omega_{LO}}(t) \langle \rho_q(t) \rho_{-q}(0) \rangle_M\right] dt. \qquad (10)$$

Here, $T_\omega(t) = \cos[\omega(t - i\hbar\beta/2)]/\sinh(\beta\hbar\omega/2)$ is the phonon Green's function, while $\langle \rho_q(t) \rho_{-q}(0) \rangle_M$ is the density-density correlation function calculated using the model system of electrons harmonically interacting with fictitious particles [with the Lagrangian (8)]. It should be noted that the optical conductivity (9) differs from that for a translationally invariant polaron system both by the explicit form of $\chi(\omega)$ and by the presence of Ω_0^2 in the denominator. For $\alpha \to 0$, the optical conductivity (9) tends to a δ-peak at $\omega = \Omega_0$,

$$\lim_{\alpha \to 0} \text{Re}\sigma(\omega) = \frac{\pi e^2}{2m} \delta(\omega - \Omega_0). \qquad (11)$$

For a translationally invariant system $\Omega_0 \to 0$ this small-coupling limit (11) reproduces the "central peak" of the polaron optical conductivity (see Eq. (5) of Ref. [24]).

In the zero-temperature limit, the memory function (10) becomes

$$\chi(\omega) = \lim_{\varepsilon \to +0} \frac{2\alpha}{3\pi N \omega} \left(\frac{\omega_{LO}}{A}\right)^{3/2}$$

$$\times \sum_{p_1=0}^{\infty} \sum_{p_2=0}^{\infty} \sum_{p_3=0}^{\infty} \frac{(-1)^{p_3}}{p_1! p_2! p_3!} \left(\frac{a_1^2}{N\Omega_1 A}\right)^{p_1} \left(\frac{a_2^2}{N\Omega_2 A}\right)^{p_2} \left(\frac{1}{NwA}\right)^{p_3}$$

$$\times \left\{ \begin{bmatrix} \sum_{m=0}^{\infty} \sum_{n=0}^{\infty} \sum_{\sigma} [f_1(n,\sigma|\{N_\sigma\},\beta) - f_2(n,\sigma;m,\sigma|\{N_\sigma\},\beta)]|_{\beta \to \infty} \\ \times \left(\frac{1}{\omega - \omega_{LO} - [p_1\Omega_1 + p_2\Omega_2 + (p_3 - m + n)w] + i\varepsilon} - \frac{1}{\omega + \omega_{LO} + p_1\Omega_1 + p_2\Omega_2 + (p_3 - m + n)w + i\varepsilon} \right. \\ \left. + \mathscr{P}\left(\frac{2}{\omega_{LO} + p_1\Omega_1 + p_2\Omega_2 + (p_3 - m + n)w}\right) \right) \\ \times \sum_{l=0}^{m} \sum_{k=n-m+l}^{n} \frac{(-1)^{n-m+l+k}\Gamma(p_1+p_2+p_3+k+l+\frac{3}{2})}{k!l!} \left(\frac{1}{wA}\right)^{l+k} \binom{n+2}{n-k}\binom{2k}{k-l-n+m} \end{bmatrix} \right.$$

$$+ \begin{bmatrix} \left(\frac{1}{\omega - \omega_{LO} - (p_1\Omega_1 + p_2\Omega_2 + p_3 w) + i\varepsilon} - \frac{1}{\omega + \omega_{LO} + p_1\Omega_1 + p_2\Omega_2 + p_3 w + i\varepsilon} \right. \\ \left. + \mathscr{P}\left(\frac{2}{\omega_{LO} + p_1\Omega_1 + p_2\Omega_2 + p_3 w}\right) \right) \\ \times \sum_{m=0}^{\infty} \sum_{n=0}^{\infty} \sum_{\sigma,\sigma'} f_2(n,\sigma;m,\sigma'|\{N_\sigma\},\beta)|_{\beta \to \infty} \\ \times \sum_{k=0}^{n} \sum_{l=0}^{m} \frac{(-1)^{k+l}\Gamma(p_1+p_2+p_3+k+l+\frac{3}{2})}{k!l!} \left(\frac{1}{wA}\right)^{k+l} \binom{n+2}{n-k}\binom{m+2}{m-l} \end{bmatrix} \right\}, \quad (12)$$

where $\mathscr{P}$ denotes the principal value, A is defined as $A \equiv \left[\sum_{i=1}^{2} a_i^2/\Omega_i + (N-1)/w\right]/N$, Ω_1, Ω_2, and w are the eigenfrequencies of the model system, a_1 and a_2 are the coefficients of the canonical transformation which diagonalizes the model Lagrangian (8). $f_1(n,\sigma|\{N_\sigma\},\beta)$ and $f_2(n,\sigma;m,\sigma'|\{N_\sigma\},\beta)$ are, respectively, the one-electron and the two-electron distribution functions in the canonical ensemble.

The changes of the shell filling schemes, which occur when varying the confinement frequency, also manifest themselves in the spectra of the optical conductivity. In Fig. 2, optical conductivity spectra for $N = 20$ polarons are presented for a quantum dot with the parameters of CdSe: $\alpha = 0.46$, $\eta = 0.656$ [25] and with different values of the confinement energy $\hbar\Omega_0$. In this case, the spin-polarized ground state changes to the ground state satisfying Hund's rule with increasing $\hbar\Omega_0$ in the interval $0.0421 H^* < \hbar\Omega_0 < 0.0422 H^*$.

In Fig. 3, optical conductivity spectra are represented for several values of the confinement energy $\hbar\Omega_0$ for $N = 14$ polarons in a quantum dot with $\alpha = 5$, $\eta = 0.1$. (The results of the all-coupling theory for large values of α are shown here for the sake of enhanced visualization of the phonon 'sidebands'. Future technological progress might allow one to fabricate quantum dots of materials with relatively large values of α.) For relatively weak confinement, the ground state is spin-polarized, like for $\hbar\Omega_0^{(1)} = 0.00883\, H^*$ (panel a). With increasing confinement, the transition from a spin-polarized state (with total spin $S = 7$) to the state obeying Hund's rule (with $S = 3$) occurs between $\hbar\Omega_0^{(1)} = 0.00883\, H^*$ (panel b) and $\hbar\Omega_0^{(2)} = 0.00884\, H^*$ (panel c) of the confinement parameter.

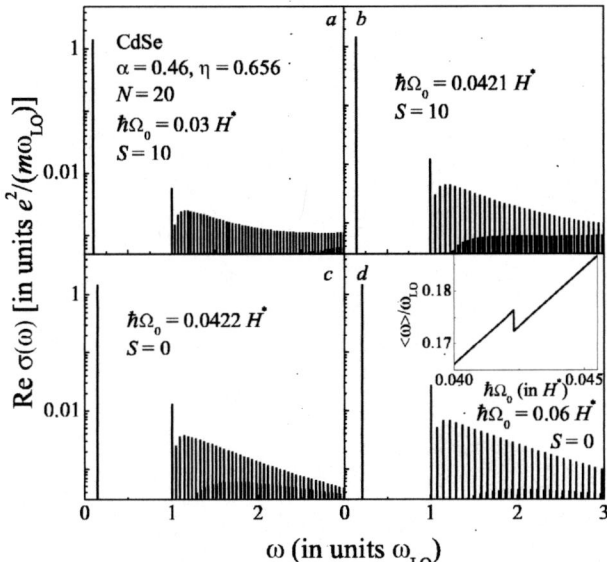

FIGURE 2. Optical conductivity spectra of $N = 20$ interacting polarons in quantum dots with $\alpha = 0.46$, $\eta = 0.656$ for different confinement energies close to the transition from a spin-polarized ground state to a ground state obeying Hund's rule. *Inset*: the first frequency moment $\langle \omega \rangle$ of the optical conductivity as a function of the confinement energy. (From Ref. [18].)

In the insets to Figs. 2, 3, the first frequency moment of the optical conductivity

$$\langle \omega \rangle \equiv \frac{\int_0^\infty \omega \text{Re}\sigma(\omega) d\omega}{\int_0^\infty \text{Re}\sigma(\omega) d\omega}, \tag{13}$$

as a function of $\hbar\Omega_0$ shows a *discontinuity*, at the value of the confinement energy corresponding to the change of the shell filling schemes from the spin-polarized ground state to the ground state obeying Hund's rule. This discontinuity should be observable in optical measurements.

The shell structure for a system of interacting polarons in a quantum dot is clearly revealed when analysing both the addition energy and the first frequency moment of the optical conductivity. The addition energy $\Delta(N)$ needed to put an extra electron into a quantum dot containing N electrons is defined as

$$\Delta(N) = E^0(N+1) - 2E^0(N) + E^0(N-1), \tag{14}$$

where $E^0(N)$ is the ground-state energy. In Figs. 4, 5, we show both the function

$$\Theta(N) \equiv \langle \omega \rangle|_{N+1} - 2\langle \omega \rangle|_N + \langle \omega \rangle|_{N-1}, \tag{15}$$

and the addition energy $\Delta(N)$.

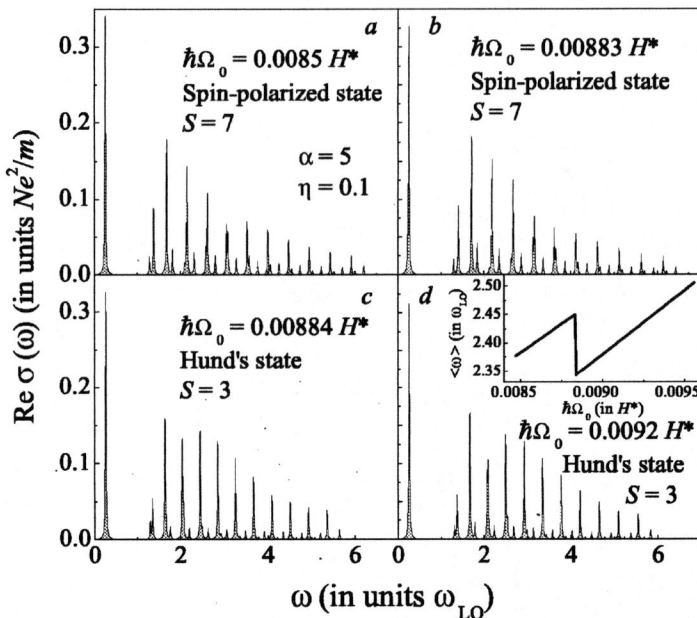

FIGURE 3. Optical conductivity spectra of $N = 14$ interacting polarons in quantum dots with $\alpha = 5$, $\eta = 0.1$, for different confinement frequencies close to the transition from a spin-polarized ground state to a ground state obeying Hund's rule. *Inset*: the first frequency moment $\langle \omega \rangle$ of the optical conductivity as a function of the confinement energy. (From Ref. [18].)

As seen from Figs. 4, 5, distinct peaks appear in $\Theta(N)$ and $\Delta(N)$ at the "magic numbers" corresponding to closed-shell configurations at $N = 8, 20$ for the states obeying Hund's rule in panels a,b and to half-filled-shell configurations at $N = 10, 20$ for the spin-polarized states in panels c,d of Fig. 5. In the case when the shell filling scheme is one and the same for different N (see panels a,b in Figs. 4, 5, where the filling obeys Hund's rule), each of the peaks of $\Theta(N)$ corresponds to a peak of the addition energy. In the case when the shell filling scheme changes with varying N (panels c,d), the function $\Theta(N)$ exhibits pronounced minima for N corresponding to the change of the filling scheme from the states, obeying Hund's rule, to the spin-polarized states.

It follows that measurements of the addition energy and the first frequency moment of the optical absorption as a function of the number of polarons in a quantum dot can reflect the difference between open-shell and closed-shell configurations. In particular, the closed-shell configurations may be revealed through peaks in the function $\Theta(N)$. The filling patterns for a many-polaron system in a quantum dot can be determined from the analysis of the first moment of the optical absorption for different numbers of polarons. The appearance of minima in the function $\Theta(N)$ will then indicate a transition from the states which are filled according to Hund's rule to the spin-polarized states.

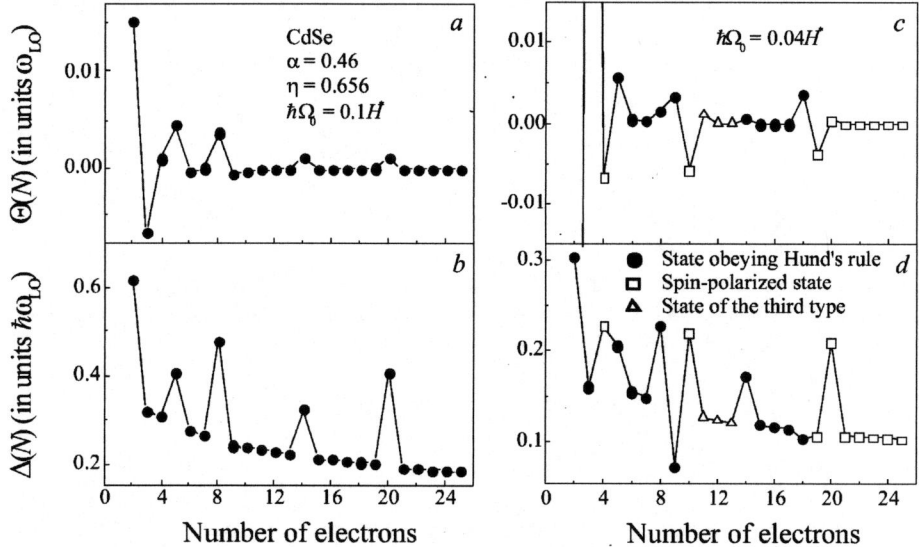

FIGURE 4. The function $\Theta(N)$ and the addition energy $\Delta(N)$ for systems of interacting polarons in CdSe quantum dots with $\alpha = 0.46$, $\eta = 0.656$ for $\Omega_0 = 0.358\omega_{LO}$ (panels a,b) and for $\Omega_0 = 0.143\omega_{LO}$ (panels c,d). Open squares denote the spin-polarized ground state; full dots denote the ground state, obeying Hund's rule; open triangles denote the ground state of the third type, with more than one partly filled shells, which is not totally spin-polarized. (From Ref. [18].)

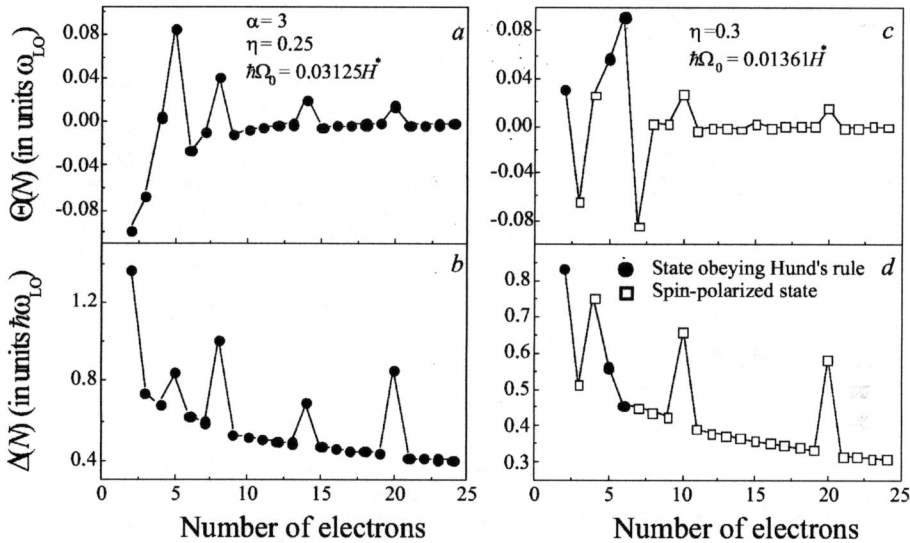

FIGURE 5. The function $\Theta(N)$ and the addition energy $\Delta(N)$ for systems of interacting polarons in quantum dots with $\alpha = 3$, $\eta = 0.25$ and $\Omega_0 = \omega_{LO}$ (panels a,b) and with $\alpha = 3$, $\eta = 0.3$ and $\Omega_0 = 0.5\omega_{LO}$ (panels c,d). (From Ref. [18].)

2. MANY-POLARON CYCLOTRON RESONANCE IN QUANTUM WELLS

In the presence of a magnetic field, resonant magnetopolaron coupling (i. e., the anticrossing of zero-phonon and one-phonon states of the polaron system) can occur. Numerous experiments on the cyclotron resonance give clear evidence of this resonant magnetopolaron coupling. The resonant magnetopolaron coupling manifests itself near the LO-phonon frequency for low electron densities (see, e. g., Refs. [26, 27, 28, 29, 30, 31]) and also for higher electron densities [32].

Cyclotron-resonance measurements performed on semiconductor quantum wells with high electron density [33, 34] reveal anticrossing near the TO-phonon frequency rather than near the LO-phonon frequency. In Ref. [34], this effect is interpreted by invoking mixing between magnetoplasmons and phonons and in terms of a resonant coupling of the electrons with the mixed magneto-plasmon-phonon modes.

In Ref. [35], CR spectra for a polaron gas in a GaAs/AlAs quantum well are theoretically investigated, taking into account (i) the electron-electron interaction and the screening of the electron-phonon interaction, (ii) the magnetoplasmon-phonon mixing, (iii) the electron-phonon interaction with all the phonon modes specific for the quantum well under investigation. As a result of this mixing, different magnetoplasmon-phonon modes appear in the quantum well, which give contributions to the CR spectra.

In Ref. [35], a finite-barrier quantum well of width d is considered. The quantum well (medium 1) with high-frequency dielectric constant ε_1 is placed into a matrix (medium 2) with high-frequency dielectric constant ε_2. Both these media are supposed to be polar. The external magnetic field $\mathbf{B}$ is applied parallel to the z-axis. The symmetric gauge is chosen for the vector potential of the magnetic field.

The Hamiltonian of the system of electrons interacting with the phonons and with each other in the quantum well is

$$H = \sum_{nlm\sigma} E_{nl\sigma} a^+_{nlm\sigma} a_{nlm\sigma} + \sum_{\lambda,\mathbf{q}} \hbar\omega_{\lambda,\mathbf{q}} b^+_{\lambda,\mathbf{q}} b_{\lambda,\mathbf{q}}$$
$$+ \frac{1}{\sqrt{S}} \sum_{\lambda,\mathbf{q}} \sum_{n,k} \left(\gamma_{\lambda,\mathbf{q}}\right)_{nk} \rho_{nk}(\mathbf{q}) \left(b_{\lambda,\mathbf{q}} + b^+_{\lambda,-\mathbf{q}}\right)$$
$$+ \frac{1}{2S} \sum_{\mathbf{q}} \sum_{n_1 k_1 n_2 k_2} V_C(k_1, n_1; n_2, k_2|\mathbf{q}) \mathcal{N}\left[\rho^+_{k_1 n_1}(\mathbf{q}) \rho_{n_2 k_2}(\mathbf{q})\right], \qquad (16)$$

where S is the area of the quantum well in the xy-plane, $\mathcal{N}[\ldots]$ is the normal product symbol, $a^+_{nlm\sigma}$ ($a_{nlm\sigma}$) is a creation (annihilation) operator for an electron in the one-particle state

$$\Psi_{nlm}(\rho,\varphi,z) = \psi_n(z) \phi_{lm}(\rho,\varphi),$$
$$\phi_{lm}(\rho,\varphi) = \frac{1}{\sqrt{2\pi}} e^{im\varphi} \Phi_{lm}(\rho). \qquad (17)$$

The wave function $\psi_n(z)$ corresponds to the n-th size-quantized subband for motion along the z-axis, while $\phi_{lm}(\rho,\varphi)$ is the wave function characterizing the "in-plane" motion of an electron with definite z-projection m of its angular momentum, l is a Landau level quantum number. $\rho_{nk}(\mathbf{q})$ is the electron density operator

$$\rho_{nk}(\mathbf{q}) = \sum_{lml'm'\sigma} \left(e^{i\mathbf{q}\cdot\mathbf{r}_\parallel}\right)_{lm,l'm'} \hat{a}^+_{nlm\sigma} \hat{a}_{kl'm'\sigma}, \tag{18}$$

with matrix element

$$\left(e^{i\mathbf{q}\cdot\mathbf{r}_\parallel}\right)_{lm,l'm'} = \int_0^\infty \rho d\rho \int_0^{2\pi} d\varphi\, e^{i\mathbf{q}\cdot\mathbf{r}_\parallel} \phi^*_{lm}(\rho,\varphi) \phi_{l'm'}(\rho,\varphi). \tag{19}$$

Within the memory-function technique, the real part of the optical conductivity (in the Faraday configuration) in the local-parabolic-band approximation can be written as [36, 37]

$$\mathrm{Re}\,\sigma(\omega) = -\frac{n_s e^2}{m_b} \mathrm{Im}\, \frac{1}{\omega - \omega_c - \chi(\omega,\omega_c)/\omega + i\varepsilon} \quad (\varepsilon \to +0), \tag{20}$$

where n_s is the 2D electron density and $\chi(\omega)$ is the memory function. In order to take into account the splitting of the CR peaks due to the nonparabolicity of the conduction band, Eq. (20) is generalized in the following way,

$$\mathrm{Re}\,\sigma(\omega) = \sum_{n,l,\sigma} \varkappa_{nl\sigma} \mathrm{Re}\,\sigma_{nl\sigma}(\omega), \tag{21}$$

where each contribution $\mathrm{Re}\,\sigma_{nl\sigma}(\omega)$ due to the transitions $(l \to l+1)$ is calculated within the local-parabolic-band approximation

$$\mathrm{Re}\,\sigma_{nl\sigma}(\omega) = -\frac{n_s e^2}{m_b} \mathrm{Im}\, \frac{1}{\omega - \omega_c^{(nl\sigma)} - \chi\left(\omega,\omega_c^{(nl\sigma)}\right)/\omega + i\varepsilon} \quad (\varepsilon \to +0). \tag{22}$$

The transition frequency

$$\omega_c^{(nl\sigma)} \equiv \frac{E_{n,l+1,\sigma} - E_{nl\sigma}}{\hbar} \tag{23}$$

is used in Eq. (22) instead of the cyclotron frequency ω_c in Eq. (20). The weight $\varkappa_{nl\sigma}$ is proportional to the number of open channels for the transitions $(l \to l+1)$. The normalized weights $\varkappa_{nl\sigma}$ are

$$\varkappa_{nl\sigma} = \frac{f_{nl\sigma}(1 - f_{n,l+1,\sigma})}{\sum_{n',l',\sigma'} f_{n'l'\sigma'}(1 - f_{n',l'+1,\sigma'})}, \tag{24}$$

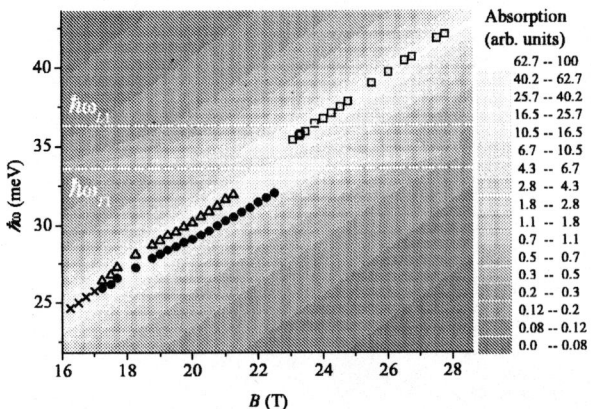

FIGURE 6. Density map of the magnetoabsorption spectra for a 10-nm GaAs/AlAs quantum well as calculated in Ref. [35]. Symbols indicate peak positions of the experimental spectra (which are taken from Fig. 3 of Ref. [34]). Dashed lines show LO- and TO-phonon energies in GaAs. (After Ref. [35].)

where $f_{nl\sigma}$ is the Fermi occupation number. The memory function $\chi(\omega,\omega_c)$ has the form (cf. Refs. [36, 38])

$$\chi(\omega,\omega_c) = -\sum_{n,k}\sum_\lambda \int \frac{d^2\mathbf{q}}{(2\pi)^2} \frac{\left|(\gamma_{\lambda,\mathbf{q}})_{nk}\right|^2 q^2}{n_s \hbar m_b \varepsilon^2(q)}$$

$$\times \int_{-\infty}^{\infty} dt\, e^{-\varepsilon t}\left(e^{i\omega t}-1\right) \operatorname{Im}\left[\mathscr{D}_\lambda(\mathbf{q},t)G_{nk}(\mathbf{q},t)\right] \quad (\varepsilon \to +0), \quad (25)$$

where $\varepsilon(q)$ is the static screening factor [36], $\mathscr{D}_\lambda(\mathbf{q},t)$ is the phonon Green's function

$$\mathscr{D}_\lambda(\mathbf{q},t) \equiv -i\Theta(t)\left[\left\langle b_{\lambda,\mathbf{q}}(t)b_{\lambda,\mathbf{q}}^+(0)\right\rangle + \left\langle b_{\lambda,-\mathbf{q}}^+(t)b_{\lambda,-\mathbf{q}}(0)\right\rangle\right], \quad (26)$$

and $G_{nk}(\mathbf{q},t)$ is the electron density-density Green's function,

$$G_{nk}(\mathbf{q},t) = -i\Theta(t)\frac{1}{S}\left\langle \rho_{nk}(\mathbf{q},t)\rho_{nk}^+(\mathbf{q},0)\right\rangle, \quad (27)$$

with the Heaviside step function $\Theta(t)$. The averaging in Eqs. (26) and (27) is performed on the equilibrium statistical operator of the electron-phonon system.

In Fig. 6, a set of the calculated magnetoabsorption spectra for a 10-nm GaAs/AlAs quantum well is plotted as a density map. This map shows the magnetoabsorption intensity as a function of both the magnetic field and the frequency. For comparison of these spectra with the experimental data of Ref. [34], the peak positions from Fig. 3 of

Ref. [34] are shown in the same graph. Splitting of the peaks occurs because the electron Landau levels in a non-parabolic conduction band are non-equidistant. In Fig. 6, the higher-energy part of each split peak corresponds to the transitions between the Landau levels $(0 \to 1)$, while the lower-energy part corresponds to the transitions $(1 \to 2)$. With increasing magnetic-field strength, the upper filled level with $l = 1$ becomes less populated, so that the relative intensity of the lower-energy peak diminishes. For the series of peaks indicated by crosses the splitting due to the band nonparabolicity is not experimentally resolved.

It is clearly seen from Fig. 6, that for the high-density polaron gas, anticrossing of the CR spectra occurs near the GaAs TO-phonon frequency ω_{T1} rather than near the GaAs LO-phonon frequency ω_{L1} for both the experimental and the calculated spectra. This effect is in contrast with the cyclotron resonance of a low-density polaron gas in a quantum well, where anticrossing occurs near the LO-phonon frequency. The appearance of the anticrossing frequency close to ω_{T1} instead of ω_{L1} is due to the screening of the electron-phonon interaction by the plasma vibrations. A similar effect appears also for the magnetophonon resonance: as shown in Ref. [39], the magnetoplasmon-phonon mixing leads to the shift of the resonant frequency of the magnetophonon resonance in quantum wells from $\omega \approx \omega_{L1}$ to $\omega \approx \omega_{T1}$.

3. OPTICAL PROPERTIES OF POLARONIC EXCITONS IN QUANTUM DOTS: ROLE OF NON-ADIABATICITY

In this section, I present a theoretical investigation of the optical properties of quantum dots, which is based on the non-adiabatic approach developed in Refs. [40, 41, 42]. Non-adiabaticity is an inherent property of exciton-phonon systems in various quantum-dot structures. Non-adiabaticity *drastically enhances the efficiency of the exciton-phonon interaction*, especially when the exciton levels are separated with energies close to the phonon energies. Also note that "intrinsic" excitonic degeneracy can lead to enhanced efficiency of the exciton-phonon interaction. The effects of non-adiabaticity are important to interpret the surprisingly high intensities of the phonon 'sidebands' observed in the optical absorption, the photoluminescence and the Raman spectra of quantum dots. Deviations of the phonon-peak sidebands, observed in some experimental optical spectra, from the Franck-Condon progression, which is prescribed by the commonly used adiabatic approximation, find a natural explanation within our non-adiabatic approach [40, 41, 42, 43].

Here, the non-adiabatic approach is applied to the particular case of stacked self-assembled quantum dots. Recently, stacked quantum dots have gained an increasing attention (see, e. g., Refs. [10, 44, 45, 46, 47, 48]) due to a possibility to finely control their energy spectra, what makes stacked quantum dots very promising for future nanodevices [45, 46, 48]. A parallelepiped-shaped quantum dot is considered as a model for a self-assembled quantum dot in a stack. The exciton-phonon interaction is taken into account for all phonon modes specific for these quantum dots (bulk-like, half-space and interface phonons).

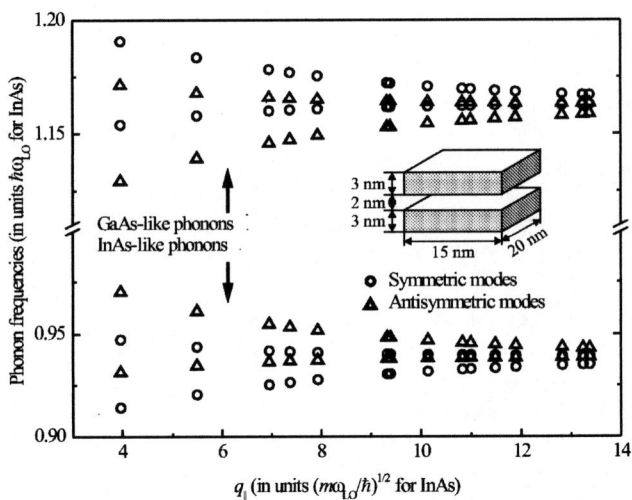

FIGURE 7. Interface-phonon frequencies for two stacked parallelepiped-shaped InAs/GaAs quantum dots with sizes $\ell_x = 20$ nm, $\ell_y = 15$ nm, and $\ell_z = 3$ nm. The inter-dot distance is $\Delta\ell_z = 2$ nm. (From Ref. [51].)

In Fig. 7, interface-phonon frequencies are represented for stacked InAs/GaAs quantum dots formed by two InAs parallelepipeds with sizes $\ell_x = 20$ nm, $\ell_y = 15$ nm, and $\ell_z = 3$ nm, and inter-dot distance along the z-axis $\Delta\ell_z = 2$ nm. The frequencies are plotted as a function of the in-plane wave number $q_\parallel$, which takes discrete values due to the quantization of the phonons in the xy-plane. Owing to the stacking, the optical-phonon spectra in the system of stacked self-assembled quantum dots acquire a more complicated structure than those in a single quantum dot. In particular, in a stack of N quantum dots, each interface-phonon frequency of a single quantum dot splits into N branches. The splitting of the interface-phonon frequencies is due to the electrostatic interaction between the optical polar vibrations of the different quantum dots.

These features of the optical-phonon spectrum of stacked quantum dots are manifested in their optical properties. We calculate the optical absorption spectrum of stacked quantum dots using the Kubo formula. Within our non-adiabatic approach the following expression results for the linear coefficient of the optical absorption by the exciton-phonon system in a quantum-dot structure:

$$\alpha(\Omega) \propto \operatorname{Re} \sum_{\beta,\beta'} d_\beta^* d_{\beta'} \int_0^\infty dt\, e^{i(\Omega - \Omega_\beta + i\varepsilon)t} \langle \beta | \bar{U}(t) | \beta' \rangle, \qquad (28)$$

where Ω is the frequency of the incident light, while d_β and Ω_β are, respectively, the electric dipole matrix element and the Franck-Condon frequency of a transition between the exciton vacuum state and the one-exciton state $|\beta\rangle$. The evolution operator averaged over the phonon ensemble, $\bar{U}(t)$, is

$$\bar{U}(t) = \mathrm{T}\exp\left\{-\frac{1}{\hbar^2}\sum_\lambda \int_0^t dt_1 \int_0^{t_1} dt_2\right.$$
$$\left.\times\left[(\bar{n}_\lambda+1)e^{-i\omega_\lambda(t_1-t_2)}\gamma_\lambda(t_1)\gamma_\lambda^\dagger(t_2) + \bar{n}_\lambda e^{i\omega_\lambda(t_1-t_2)}\gamma_\lambda^\dagger(t_1)\gamma_\lambda(t_2)\right]\right\}. \quad (29)$$

In Eq. (29), T is the time ordering operator, the index λ labels the phonon modes specific for the quantum-dot structure under consideration, ω_λ are phonon frequencies, $\gamma_\lambda(t)$ are the exciton-phonon interaction amplitudes in the interaction representation, and $\bar{n}_\lambda = \left[\exp(\hbar\omega_\lambda/k_B T) - 1\right]^{-1}$.

Within the adiabatic approximation, which has been widely used to calculate the optical spectra of quantum dots, non-diagonal matrix elements of the exciton-phonon interaction are neglected when calculating $\alpha(\Omega)$ as given by Eq. (28) with Eq. (29). In the adiabatic approach [49, 50] one supposes that (i) both the initial and the final states of a quantum transition are non-degenerate, (ii) the energy differences between the exciton states are much larger than the phonon energies. It has been shown in Refs. [40, 41, 42, 43] that these conditions are often violated for optical transitions in quantum dots. In other words, the exciton-phonon system in a quantum dot can be essentially *non-adiabatic*. The polaron interaction for an exciton in a degenerate state results in *internal non-adiabaticity* ("the proper Jahn–Teller effect"), while the existence of exciton levels separated by an energy comparable with the LO-phonon energy leads to *external non-adiabaticity* ("the pseudo Jahn–Teller effect").

In Ref. [40], a method was proposed to calculate the absorption spectrum given by Eqs. (28) and (29) for a spherical quantum dot taking into account non-adiabaticity of the exciton-phonon system. This approach has been further refined in Ref. [51]: for the matrix elements of the evolution operator a closed set of equations has been obtained using a diagrammatic technique. This set describes the effect of non-adiabaticity both on the intensities and on the positions of the absorption peaks.

In Figs. 8 and 9 the calculated optical absorption spectra are shown for a single quantum dot and for a system of two stacked quantum dots, respectively. From the comparison of the spectra obtained in the adiabatic approximation with those resulting from the non-adiabatic approach, the following effects of non-adiabaticity are revealed.

First, *polaron shift* of the zero-phonon lines with respect to the bare-exciton levels is larger in the non-adiabatic approach than in the adiabatic approximation. Second, there is a strong *increase of the intensities of the phonon satellites* as compared to those given by the adiabatic approximation. This increase can be by more than two orders of magnitude. Third, in the optical absorption spectra found within the non-adiabatic approach, there appear phonon satellites related to *non-active bare exciton states*.

Fourth, the optical-absorption spectra demonstrate the crucial role of *non-adiabatic mixing* of different exciton and phonon states in quantum dots. This results in a rich structure of the absorption spectrum of the exciton-phonon system [52, 41, 42]. For the stacked quantum dots, this effect is enhanced in the (quasi-) resonant case, when

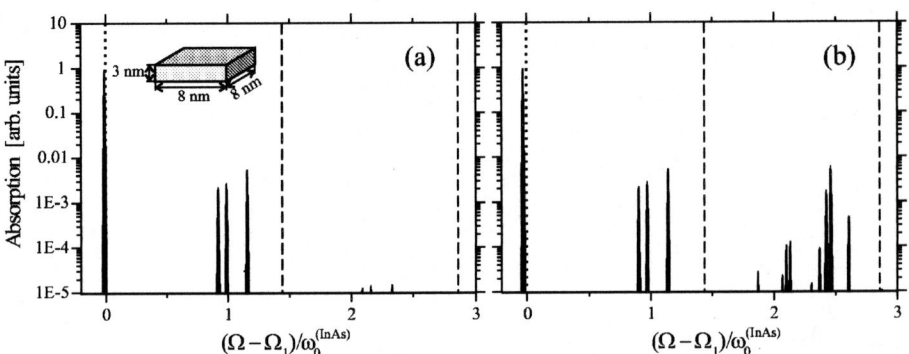

FIGURE 8. Absorption spectra, calculated for a single quantum dot with the adiabatic approximation [panel (a)] and with the non-adiabatic approach [panel (b)]. Optically active and non-active energy levels of a bare exciton are shown as dotted and dashed lines, respectively. Ω_1 is the transition frequency for the lowest state of a bare exciton. (From Ref. [51].)

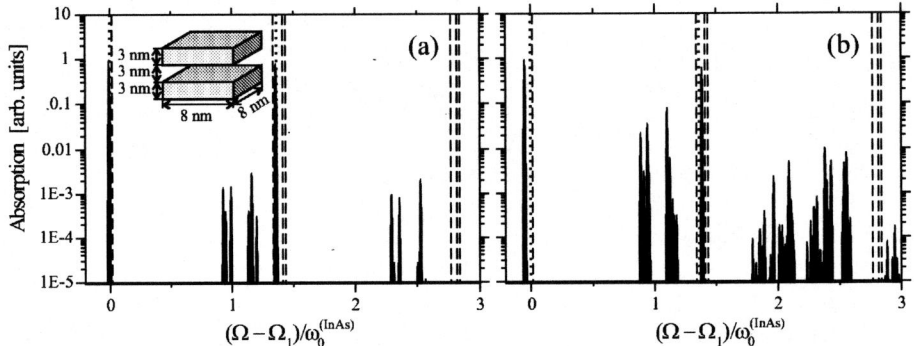

FIGURE 9. Absorption spectra, calculated for a system of two stacked quantum dots with the adiabatic approximation [panel (a)] and with the non-adiabatic approach [panel (b)]. Optically active and non-active energy levels of a bare exciton are shown as dotted and dashed lines, respectively. Ω_1 is the transition frequency for the lowest state of a bare exciton. (From Ref. [51].)

the exciton-level splitting, caused by the coupling between quantum dots, is close to a LO phonon energy [see panel (b) in Fig. 9]. Similar conclusions about the pronounced influence of the exciton-phonon interaction on the optical spectra of quantum dots have been recently formulated in Ref. [53] in terms of a strong coupling regime for excitons and LO phonons. Such a strong coupling regime is a particular case of the non-adiabatic mixing related to a (quasi-) resonance which arises when the spacing between exciton levels is close to the LO phonon energy.

Due to non-adiabaticity, multiple absorption peaks appear in spectral ranges characteristic for phonon satellites. From the states, which correspond to these peaks, the system can rapidly relax to the lowest emitting state. Therefore, in the photoluminescence excitation (PLE) spectra of quantum dots, pronounced peaks can be expected in spectral ranges characteristic for phonon satellites. New experimental evidence of the enhanced phonon-assisted absorption due to effects of non-adiabaticity has been recently provided by PLE measurements on single self-assembled InAs/GaAs [54] and InGaAs/GaAs [55] quantum dots.

ACKNOWLEDGMENTS

As is manifest from the list of references, I had the pleasure of collaborations on the physics of polarons with several colleagues, post-doctoral scientists and students. I like to thank them all here. I thank V. Fomin, F. Brosens, V. Gladilin, S. Klimin and J. Tempere for discussions during the preparation of this manuscript. This work has been supported by the BOF NOI (UA-UIA), GOA BOF UA 2000, IUAP, FWO-V projects G.0287.95, G.0071.98, G.0274.01, G.0435.03, the W.O.G. WO.025.99N (Belgium) and the European Commission GROWTH Programme, NANOMAT project, contract No. G5RD-CT-2001-00545.

REFERENCES

1. Yannouleas, C. and Landman, U., *Phys. Rev. Lett.* **82**, 5325-5328 (1999).
2. Bednarek, S., Szafran, B., and Adamowski, J., *J. Phys. Rev. B* **59**, 13036-13042 (1999).
3. Szafran, B., Adamowski, J., and Bednarek, S., *Phys. Rev. B* **61**, 1971-1977 (2000).
4. Lee, I.-H., Rao, V., Martin, R. M. and Leburton, J.-P., *Phys. Rev. B* **57**, 9035-9042 (1998).
5. Hirose, K. and Wingreen, N. S., *Phys. Rev. B* **59**, 4604-4607 (1999).
6. Reimann, S. M., Koskinen, M., Kolehmainen, J., Manninen, M., Austing, D. G., and Tarucha, S., *Eur. Phys. J. D* **9**, 105-110 (1999).
7. Kim, Y.-H., Lee, I.-H., Nagaraja, S., Leburton, J.-P., Hood, R. Q., and Martin, R. M., *Phys. Rev. B* **61**, 5202-5211 (2000).
8. Jiang, T. F., Tong, X.-M., and Chu, S.-I., *Phys. Rev. B* **63**, 045317, 9 pages (2001).
9. Partoens, B., and Peeters, F. M., *Physica E* **6**, 577-580 (2000).
10. Partoens, B., and Peeters, F. M., *Phys. Rev. Lett.* **84**, 4433-4436 (2000).
11. Egger, R., Häusler, W., Mak, C. H., and Grabert, H., *Phys. Rev. Lett.* **82**, 3320-3323 (1999).
12. Gonzalez, A., Partoens, B., and Peeters, F. M., *Phys. Rev. B* **56**, 15740-15743 (1997).
13. Gonzalez, A., Partoens, B., Matulis, A., and Peeters, F. M., *Phys. Rev. B* **59**, 1653-1656 (1999).
14. Reimann, S. M., Koskinen, M., and Manninen, M., *Phys. Rev. B* **62**, 8108-8113 (2000).
15. Feynman, R. P., *Statistical Mechanics*, Benjamin, Massachusetts, 1972.
16. Lemmens, L. F., Brosens, F., and Devreese, J. T., *Phys. Rev. E* **53**, 4467 - 4476 (1996).
17. Devreese, J. T., "Note of the path-integral variational approach in the many-body theory", in *Fluctuating Paths and Fields*, World Scientific, Singapore, 2001, pp. 289 - 304.
18. Klimin, S. N., Fomin, V. M., Brosens, F., and Devreese, J. T., to be published.
19. Devreese, J. T., Klimin, S. N., Fomin, V. M., and Brosens, F., *Solid State Commun.* **114**, 305 - 310 (2000).
20. Brosens, F., Devreese, J. T., and Lemmens, L. F., *Phys. Rev. E* **55**, 227 - 236 (1997); **55**, 6795 - 6802 (1997); **58**, 1634 - 1643 (1998).
21. See e. g. Messiah, A., *Quantum Mechanics*, North-Holland, Amsterdam, 1966, vol. 2, p. 702.
22. Devreese, J. T., and Tempere, J., *Solid State Commun.* **106**, 309-313 (1998).
23. Devreese, J. T., De Sitter, J., and Goovaerts., M., *Phys. Rev. B*, **5**, 2367 - 2381 (1972).
24. Devreese, J. T., Lemmens, L. F., and Van Royen, J., *Phys. Rev. B* **15**, 1212 - 1214 (1977).
25. Kartheuser, E., in *Polarons in Ionic Crystals and Polar Semiconductors*, North-Holland, Amsterdam, 1972, pp. 717-733.

26. McCombe, B. D., and Kaplan, R., *Phys. Rev. Lett.* **21**, 756 - 759 (1968).
27. Chang, Y.-H., McCombe, B. D., Mercy, J.-M., Reeder, A. A., Ralston, J., and Wicks, G. A., *Phys. Rev. Lett.* **61** 1408 - 1411 (1988).
28. Brummel, M.A., Nicholas, R.J., Hopkins, M.A., Harris, J.J., and Foxon, C.T., *Phys. Rev. Lett.* **58**, 77-80 (1987).
29. Hu, C.M., Batke, E., Köhler, K., Ganser, P. *Phys. Rev. Lett.* **76**, 1904-1907 (1996).
30. Vaughan, T.A., Nicholas, R.J., Langerak, C.J.G.M., Murdin, B.N., Pidgeon, C.R., Mason, N.J., and Walker, P.J., *Phys. Rev. B* **53**, 16481-16484 (1996).
31. Wang, Y.J., Nickel, H.A., McCombe, B.D., Peeters, F.M., Shi, J.M., Hai, G.Q., Wu, X.-G., Eustis, T.J., and Schaff, W., *Phys. Rev. Lett.* **79**, 3226-3229 (1997).
32. Świerkowski, L., Szymański, J., Simmonds, P.E., Fisher T.A., Skolnick, M.S., *Phys. Rev. B* **51**, 9830-9835 (1995).
33. Ziesmann, M., Heitmann, D., and Chang, L. L., *Phys. Rev. B* **35** 4541 - 4544 (1987).
34. Poulter, A. J. L., Zeman, J., Maude, D. K., Potemski, M., Martinez, G., Riedel, A., Hey, R., and Friedland, K. J., *Phys. Rev. Lett.* **86**, 336 - 339 (2001).
35. Klimin, S. N. and Devreese, J. T., to be published.
36. Wu, X.-G., Peeters, F. M., and Devreese, J. T., *Phys. Rev. B* **36**, 9760 - 9764 (1987).
37. Wu, X., Peeters, F. M., and Devreese, J. T., *Phys. Rev. B* **40**, 4090 - 4094 (1989).
38. Hai, G. Q., Peeters, F. M., and Devreese, F. M., *Phys. Rev. B* **47** 10358 - 10374 (1993).
39. Afonin, V. V., Gurevich, V. L.. and Laiho, R., *Phys. Rev. B* **62**, 15913 - 15924 (2000).
40. Fomin, V. M., Gladilin, V. N., Devreese, J. T., Pokatilov, E. P., Bałaban, S. N., and Klimin, S. N., *Phys. Rev. B* **57**, 2415 - 2425 (1998).
41. Devreese, J. T., Fomin, V. M., Gladilin, V. N., Pokatilov, E. P., and Klimin, S. N., *Nanotechnology* **13**, 163 - 168 (2002).
42. Devreese, J. T., Fomin, V. M., Pokatilov, E. P., Gladilin, V. N., and Klimin, S. N., *Phys. Stat. Sol. (c)* **0**, 1189-1192 (2003).
43. Pokatilov, E. P., Klimin, S. N., Fomin, V. M., Devreese, J. T., and Wise, F. W., *Phys. Rev. B* **65**, 075316, 8 pages (2002).
44. Anisimovas, E., and Peeters, F. M., *Phys. Rev. B* **65**, 233302 (2002) (4 pages).
45. Holleitner, A. W., Blick, R. H., and Eberl, K., *Appl. Phys. Letters* **82**, 1887-1889 (2003).
46. Rebohle, L. Schrey, F. F., Hofer, S., Strasser, G., and Unterrainer, K., *Appl. Phys. Letters* **81**, 2079-2081 (2002).
47. Bednarek, S., Chwiej, T., Adamowski, J., and Szafran, B., *Phys. Rev. B* **67**, 205316, 6 pages (2003).
48. Shi, B., and Xie, Y. H., *Appl. Phys. Letters* **82**, 4788-4790 (2003).
49. Pekar, S. I., *Zh. Eksp. Teor. Fiz.* **20**, 267 - 270 (1950).
50. Huang, K., and Rhys, A., *Proc. R. Soc. London, Ser. A* **204**, 406 - 423 (1950).
51. Gladilin, V. N., Klimin, S. N., Fomin, V. M., and Devreese, J. T., to be published.
52. Gladilin, V. N., Balaban, S. N., Fomin, V. M., and Devreese, J. T., in: *Proc. 25th Int. Conf. on the Physics of Semiconductors, Osaka, Japan, 2000*, Springer, Berlin, 2001, Part II, pp. 1243-1244.
53. Verzelen, O., Ferreira, R., and Bastard, G., *Phys. Rev. Lett.* **88**, 146803, 4 pages (2002).
54. Lemaître, A., Ashmore, A. D., Finley, J. J., Mowbray, D. J., Skolnick, M. S., Hopkinson, M., and Krauss, T. F., *Phys. Rev. B* **63**, 161309(R), 4 pages (2001).
55. Zrenner, A., Findeis, F., Baier, M., Bichler, M., and Abstreiter, G., *Physica B* **298**, 239-245 (2001).

Charge dynamics of t-J model and anomalous bond-stretching phonons in cuprates

P. Horsch* and G. Khaliullin*

Max-Planck-Institut für Festkörperforschung, D-70569 Stuttgart, Germany

Abstract. The density response of a doped Mott-Hubbard insulator is discussed starting from the t-J model in a slave boson $1/N$ representation. In leading order $O(1)$ the density fluctuation spectra $N(\mathbf{q},\omega)$ are determined by an undamped collective mode at large momentum transfer, in striking disagreement with results obtained by exact diagonalization, which reveal a very broad dispersive peak, reminiscent of strong spin-charge coupling. The $1/N$ corrections introduce the polaron character of the bosonic holes moving in a uniform RVB background. The resulting $N(\mathbf{q},\omega)$ captures all features observed in diagonalization studies, fulfills the appropriate sum rules, and apart from the broadening of the collective mode shows a new low energy feature at the energy $\chi J + \delta t$ related to the polaron motion in the spinon background. It is further shown that the low energy structure, which is particularly pronounced in $(\pi,0)$ direction, describes the strong renormalization and anomalous damping of the highest bond-stretching phonons in $La_{2-x}Sr_xCuO_4$.

Dedicated to Ferdinando Mancini on the occasion of his 60th birthday

INTRODUCTION

High-temperature superconductors are doped Mott-Hubbard insulators, therefore the density response is expected to be very different from that of weakly correlated Fermi systems. The low-energy density response, i.e. in the frequency range inside the Mott-Hubbard or charge transfer gap, is proportional to the doping, i.e., described by transitions within the lower Hubbard band. For a complete understanding of the physics ruling these systems both the spin and the charge response must be considered in the full momentum $\mathbf{q}$ and frequency ω domain. The spin response has been and still is extensively explored, mainly stimulated by a wealth of experimental data provided by neutron scattering. A similarly powerful experimental tool is missing for the charge response, with the exception of optical conductivity, which however provides information only in the limit $\mathbf{q} \to 0$. In the absence of solid experimental information about the detailed structure of $N(\mathbf{q},\omega)$, the best we can do to test a theory is the computer experiment. This provides for small clusters an unbiased answer how $N(\mathbf{q},\omega)$ looks like, e.g., for the t-J model[1, 2], and how this quantity depends on the key parameters, namely the magnetic exchange interaction J and the doping δ.

In one dimension the Hubbard physics is basically simple (although not mathematically), as it is characterized by charge and spin separation, which implies that the density response is basically that of non-interacting spinless fermions, i.e., showing vanishing excitation energy at $q = 0$ and $4k_F$. This is even true for the 1D t-J model away from

$J = 0$ and $2t$, where the model is not exactly solvable[1]. As we shall see the 2D model relevant for the cuprates shows a very different behavior compared to the 1D case.

Exact diagonalization studies for the t-J model[1, 2] have revealed that the dynamical density response $N(\mathbf{q}, \omega)$ for the 2D model is characterized by *strong spin-charge coupling*. These calculations show several features unexpected from the point of view of weakly correlated fermion systems which have to be explained by theory: (i) the strong suppression of low energy $2k_F$ scattering in the density response, (ii) a broad incoherent peak at high energy (several t) whose shape is rather insensitve to changes of hole concentration and exchange interaction J, (iii) the very different form of $N(\mathbf{q}, \omega)$ compared to the spin response function $S(\mathbf{q}, \omega)$, which share common features in weakly correlated fermionic systems, and finally (iv) there are at low energy ($\sim J$) some features that do depend on doping and J. It is worthwhile to note that finite temperature diagonalization studies[3] show only weak temperature dependence for $T < 0.3t$ even at low energy.

While considerable analytical work has been done to explain the spin response of the t-J model only few authors analysed $N(\mathbf{q}, \omega)$. Wang et al.[4] studied collective excitations in the density channel and found sharp peaks at large momenta corresponding to free bosons. Similar results were obtained by Gehlhoff and Zeyher[5] using the X-operator formalism and by Foussats and Greco[6] using a path integral representation for X-operators. Lee et al.[7] obtained a broad incoherent density fluctuation spectrum by considering a model of bosons coupled to a quasistatic disordered gauge field, while Jackeli and Plakida[8] employed the memory function approach.

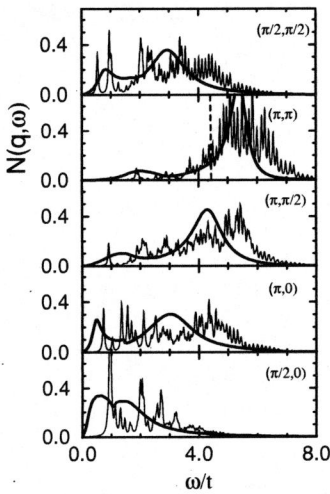

FIGURE 1. Comparison of $N(\mathbf{q}, \omega)$ obtained by the slave-boson theory[9] (solid lines) with diagonalization data[1] for a periodic 4×4 cluster with $J/t = 0.4$ and doping $\delta = 0.25$. The dashed line in the (π, π) spectrum indicates the δ-function collective peak obtained when polaron effects are neglected.

Starting from a slave-boson representation we show that the essential features observed in the numerical studies can be obtained in the framework of the Fermi-liquid phase of the t-J model at zero temperature[9]. Our main findings are: (i) at low momenta the main effect of strong correlations is to transfer spectral weight from particle-hole excitations into a pronounced collective mode. Because of the strong damping of this mode (linear in q) due to the coupling to the spinon particle-hole continuum, this collective excitation is qualitatively different from a sound mode. (ii) At large momenta we find a strict similarity of $N(\mathbf{q},\omega)$ with the spectral function of a single hole moving in a uniform RVB spinon background. In this regime $N(\mathbf{q},\omega)$ consists of a broad peak at high energy whose origin is the fast, incoherent motion of bare holes. (iii) The polaronic nature of dressed holes leads to the formation of a second peak at lower energy, which is more pronounced in $(\pi,0)$ direction in agreement with diagonalization studies[1, 2].

We show that the anomalous renormalization of certain phonon modes as observed in inelastic neutron scattering provides a sensitive test of the pecularities of the low-energy density response. In this contribution we shall analyse the strong, doping-dependent renormalization of the highest breathing phonon modes which is a generic feature in the high-T_c compounds.

SLAVE BOSON THEORY OF DENSITY RESPONSE

We start from the N-component generalization of the slave-boson t-J Hamiltonian[10, 4] $H_{tJ} = H_t + H_J$, which is obtained by replacing the constrained electron creation operators $\tilde{c}^+_{i,\sigma} = c^+_{i,\sigma}(1-n_{i,-\sigma}) \to f^+_{i,\sigma}h_i$:

$$H_t = -\frac{2t}{N} \sum_{<i,j>\sigma} (f^+_{i\sigma}h^+_j h_i f_{j\sigma} + h.c.), \qquad (1)$$

$$H_J = \frac{J}{N} \sum_{<i,j>\sigma\sigma'} f^+_{i\sigma}f_{i\sigma'}f^+_{j\sigma'}f_{j\sigma}(1-h^+_i h_i)(1-h^+_j h_j), \qquad (2)$$

where $f^+_{i,\sigma}$ is a fermionic (spinon) operator, $\sigma = 1,\cdots,N$ is the fermionic flavor index, and h_i denotes the bosonic holes. These operators obey standard commutation rules, yet the number of these auxiliary particles must obey the constraint $\sum_\sigma f^+_{i\sigma}f_{i\sigma} + h^+_i h_i = N/2$. The original t-J model is recovered for $N = 2$.

The slave boson parametrization provides a straightforward description of the strong suppression of density fluctuations of constrained electrons through the representation of the density response in terms of a dilute gas of bosons. A common treatment of model (1) is the density-phase representation ("radial" gauge[11]) of the bosonic operator $h_i = r_i \exp(i\theta_i)$ with the subsequent $1/N$-expansion around the Fermi-liquid saddle point. While this gauge is particularly useful to study the low energy and momentum properties, it is not very convenient for the study of the density response in the full ω and $\mathbf{q}$ space. Formally the latter follows in the radial gauge from the fluctuations of r_i^2. However, if one considers for example convolution type bubble diagrams, one realizes that their contribution to the static structure factor is correctly of order $1/N$, but is not proportional to the density of holes δ as it should be. According to Arrigoni et al[12] such unphysical results originate from a large negative pole contribution in the $\langle r_{-\mathbf{q}} r_{\mathbf{q}} \rangle_\omega$ Green's function of the real field r, which is hard to control by a perturbative treatment of phase fluctuations.

We follow therefore Popov[13] using the density-phase treatment only for small momenta $q < q_0$, while keeping the original particle-hole representation of the density operator, h^+h, at large momenta. More precisely $h_i = r_i \exp(i\theta_i) + b_i$, where $b_i = \sum_{|\mathbf{q}|>q_0} h_\mathbf{q} \exp(i\mathbf{q}\mathbf{R}_i)$. The cutoff q_0 is introduced dividing "slow" (collective) variables represented by r and θ from "fast" (single-particle) degrees of freedom b_i and b_i^+. As discussed by Popov[13] this "mixed" gauge is particularly useful for finite temperature studies to control infrared divergences. We start formally with the "mixed" gauge and keep only terms of order δ and $1/N$ in the bosonic self energies. In this approximation our zero temperature calculations become quite straightforward: The cutoff $q_0 < \delta$ actually does not enter in the results and we arrive finally at the Bogoliubov theory for a dilute gas of bosons moving in a fluctuating spinon background.

The Lagrangian corresponding to the model (1) is then given by (the summation over σ is implied)

$$L = \sum_i \left(f_{i\sigma}^+ (\frac{\partial}{\partial \tau} - \mu_f) f_{i\sigma} + b_i^+ (\frac{\partial}{\partial \tau} - \mu_b) b_i \right) + H_t + H_J$$
$$+ \frac{i}{\sqrt{N}} \sum_i \lambda_i \left(f_{i\sigma}^+ f_{i\sigma} + (r_i + b_i^+)(r_i + b_i) - \frac{N}{2} \right), \quad (3)$$

$$H_t = -\frac{2t}{N} \sum_{<ij>} f_{i\sigma}^+ f_{j\sigma} (b_j^+ b_i + r_i r_j + r_j b_i + b_j^+ r_i) + h.c. \quad (4)$$

Here the λ field is introduced to enforce the constraint, and μ_f, μ_b are fixed by the particle number equations $\langle n_f \rangle = \frac{N}{2}(1-\delta)$ and $\langle r_i^2 + b_i^+ b_i \rangle = \frac{N}{2}\delta$, respectively. The uniform mean field solution $r_i = r_0\sqrt{N/2}$ leads in the large N limit to the renormalized narrow fermionic spectrum $\xi_\mathbf{k} = -z\tilde{t}\gamma_\mathbf{k} - \mu_f$, with $\tilde{t} = J\chi + t\delta$, $\gamma_\mathbf{k} = \frac{1}{2}(\cos k_x + \cos k_y)$, $\chi = \sum_\sigma \langle f_{i\sigma}^+ f_{j\sigma} \rangle/N$, and $z = 4$ the number of nearest neighbors. In the $N = \infty$ limit $\chi_\infty \simeq 2/\pi^2$ is given by that of free fermions, while for the original t-J model its value should be larger[14] due to Gutzwiller projection. In the following $\chi = \frac{3}{2}\chi_\infty$ will be used. Distinct from the finite-temperature gauge-field theory of Nagaosa and Lee[15] the bond-order phase fluctuations acquire a characteristic energy scale in this approach[4], and the fermionic ("spinon") excitations can be identified with Fermi-liquid quasiparticles. The mean field spectrum of bosons is $\omega_\mathbf{q} = 2z\chi t(1-\gamma_\mathbf{q})$. Thus the effective mass of holes $m_h^0 \propto 1/t$ is much smaller than that of the spinons.

Due to the diluteness of the bosonic subsystem, $\delta \ll 1$, the density correlation function $\chi_{\mathbf{q},\omega} = \langle \delta n^h \delta n^h \rangle_{\mathbf{q}\omega}$ is mainly given by the condensate induced part which is represented by the Green's function $\langle (b_\mathbf{q}^+ + b_{-\mathbf{q}})(b_\mathbf{q} + b_{-\mathbf{q}}^+) \rangle_\omega$ for $q > q_0$, and $2\langle r_{-\mathbf{q}} r_\mathbf{q} \rangle_\omega$ for $q < q_0$, respectively:

$$\chi_{\mathbf{q}\omega} \simeq \frac{N}{2} r_0^2 \left(\langle (b_\mathbf{q}^+ + b_{-\mathbf{q}})(b_\mathbf{q} + b_{-\mathbf{q}}^+) \rangle_{q>q_0} + 2\langle r_{-\mathbf{q}} r_\mathbf{q} \rangle_{q<q_0} \right). \quad (5)$$

The $1/N$ self-energy corrections to these functions are calculated in a conventional way[11, 10] expanding $r_i = (r_0\sqrt{N} + (\delta r)_i)/\sqrt{2}$ and considering Gaussian fluctuations around the mean field solution. Neglecting all terms of order δ/N and q_0^2/N, only one relevant $1/N$ contribution remains which corresponds to the dressing of the slave-boson

Green's function by spinon particle-hole excitations. Within this approximation and at zero temperature no divergences occur at low momenta, thus one can take the limit $q_0 \to 0$. The final result for the dynamic stucture factor (*normalized by the hole density*) is:

$$N_{\mathbf{q},\omega} = \frac{2}{\pi} Im\left((\omega_{\mathbf{q}} a + S_{\mathbf{q},\omega}^{(1/N)} - \mu_b)/D_{\mathbf{q},\omega}\right), \quad (6)$$

$$D_{\mathbf{q},\omega} = (\omega_{\mathbf{q}} a + S_{\mathbf{q},\omega}^{(1/N)} - \mu_b)(\omega_{\mathbf{q}} + S_{\mathbf{q},\omega}^{(1)} + S_{\mathbf{q},\omega}^{(1/N)} - \mu_b) - (\omega a - A_{\mathbf{q},\omega}^{(1/N)})^2. \quad (7)$$

The origin of the contribution

$$S_{\mathbf{q},\omega}^{(1)} = z t r_0^2 \left(\frac{(1+\Pi_2)^2}{\Pi_1} - \Pi_3\right)_{\mathbf{q},\omega}, \quad (8)$$

$$\Pi_m = z t \sum_{\mathbf{k}} \frac{n(\xi_{\mathbf{k}}) - n(\xi_{\mathbf{k+q}})}{\xi_{\mathbf{k+q}} - \xi_{\mathbf{k}} - \omega - i0^+} (\gamma_{\mathbf{k}} + \gamma_{\mathbf{k+q}})^{m-1}, \quad (9)$$

is the indirect interaction of bosons via the spinon band due to the hopping term (which gives Π_3 in (4)) and due to the coupling to spinons via the constraint field λ. The latter channel provides a repulsion between bosons, making $S^{(1)}(\omega=0)$ positive and therefore ensuring the stability of the uniform mean-field solution.

The $1/N$ self energies $S^{(1/N)}$ and $A^{(1/N)}$ are essentially a single boson property. They are given by the symmetric and antisymmetric combinations (with respect to $\omega + i0^+ \to -\omega - i0^+$) of the self energy

$$\Sigma_{\mathbf{q},\omega}^{(1/N)} = \frac{4}{N} \sum_{|\mathbf{k}|<k_F<|\mathbf{k'}|} (z t \gamma_{\mathbf{k'-q}})^2 G_{\mathbf{q+k-k'}}^0 (\omega + \xi_{\mathbf{k}} - \xi_{\mathbf{k'}}). \quad (10)$$

Here $G_{\mathbf{q}}^0(\omega) = (\omega - \omega_{\mathbf{q}} - \Sigma_{\mathbf{q},\omega}^{(1/N)} + \mu_b)^{-1}$ is the Green's function for a single boson moving in a uniform RVB background. Although in the context of $1/N$ theory the G^0 function in (5) should be considered as a free propagator, we shall use here the selfconsistent polaron picture for a single hole[16]. This is crucial when comparing the theory for $N=2$ with diagonalization studies. Finally, the constants a and μ_b in (3) are given by $(1 - t r_0^2/\tilde{t})$ and $S^{(1/N)}(\omega = \mathbf{q} = 0)$, respectively. The parameter r_0^2 in Eq.8, which formally corresponds to the condensate fraction in our theory, is determined selfconsistently from $r_0^2 = \delta - \sum_{\mathbf{q}\neq 0} \tilde{n}_{\mathbf{q}}$. The momentum distribution $\tilde{n}_{\mathbf{q}} = \langle b_{\mathbf{q}}^+ b_{\mathbf{q}} \rangle$ is calculated from the corresponding bosonic Green's function for finite hole-density. The interactions implied in the polaron formation lead to a significant reduction of the number of bosons in the condensate[9], and affects the balance of the selfenergies $S^{(1)}$ and $S^{(1/N)}$ in Eq. (7).

The structure of $N(\mathbf{q},\omega)$ (6) in the small $\omega, \mathbf{q}$ limit is mainly controlled by the interaction of bosons represented by the $S^{(1)}$ term ($\propto r_0^2$), while the internal polaron structure of the boson determined by $S^{(1/N)}$ is less important. $N(\mathbf{q},\omega)$ consists of a weak spinon particle-hole continuum with cutoff $\propto v_F q$, and a very pronounced linear collective mode which nearly exhausts the sum rule. The velocity of this mode is always

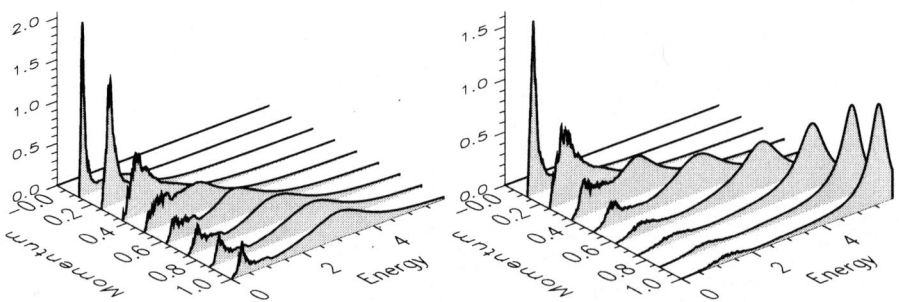

FIGURE 2. Density fluctuation spectra $N(\mathbf{q},\omega)$ for $J/t = 0.4$ and $\delta = 0.15$ along $(\pi,0)$ (left) and (π,π) directions (right). Energy in units of t.

somewhat smaller than the spinon Fermi velocity, $v_s \leq v_F \simeq z\tilde{t}$, which implies a strong damping $\propto \omega$ (or q) of this mode (Fig.2).

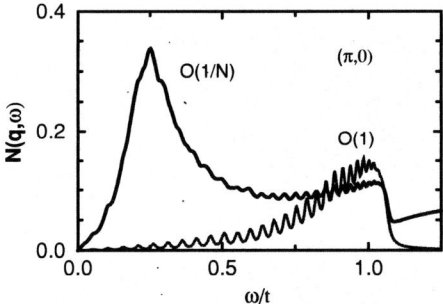

FIGURE 3. Low energy density response at $(\pi,0)$ for $J = 0.3$ and $\delta = 0.18$. The peak centered at $\sim (\chi J + \delta t)$ due to the polaron motion appears in the $1/N$ order on top of the spinon particle-hole continuum with edge at energy $z(\chi J + \delta t)$. Small oscillations in the data result from the discretization of the energy scale in the numerical solution and have no physical meaning.

The density response $N(\mathbf{q},\omega)$ (Fig.2) at large momenta, $q > \delta$, which we can compare with diagonalization results (Fig.1), is dominated by the properties of the single boson selfenergy $S^{(1/N)}$. The calculated density response of the t-J model has three characteristic features on different energy scales: (i) The main spectral weight of the excitations at large momenta is located in an energy region of order of several t. This high energy peak is very broad and incoherent as a result of the strong coupling of bosons to low-energy spin excitations. The position of this peak and its shape are rather insensitive to the ratio $J/t \leq 1$ in agreement with conclusions of [1, 2]. This is simply due to the fact that the high-energy properties of the t-J model are controlled by t. (ii) The theory predicts also a second peak at lower energy which is more pronounced in the direction $(\pi,0)$ (see also Fig.3), while its weight is strongly suppressed for $\mathbf{q}$ near (π,π). The origin of this excitation is due to the formation of a polaron-like band of dressed bosons. The relative weight of this contribution increases with J as a result of the increasing spinon bandwidth. (iii) In addition there is the spinon particle-hole continuum which is generated by $S^{(1)}$ with relatively small weight ($\propto \delta^2$). At $(\pi,0)$ the high energy cutoff of the spinon

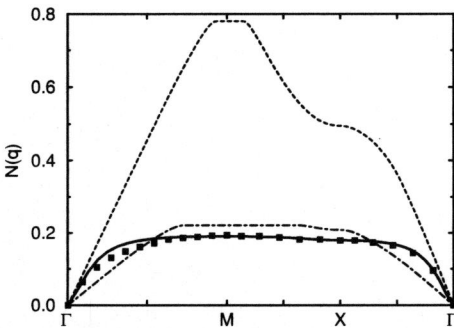

FIGURE 4. Comparison of static structure factor $N(\mathbf{q})$ (solid line) calculated for $J = 0.4$ with the result for a Gutzwiller projected wave function[17] (squares) and spinless fermions (dash-dotted) for $\delta = 0.213$. The comparison with the result obtained for non-interacting fermions with spin (dashed line) shows the large reduction of the density response due to the constraint.

continuum is at $z(\chi J + \delta t)$ (Fig.3), while the polaron peak is at about $(\chi J + \delta t)$. We note that the polaron peak in $N(\mathbf{q}, \omega)$ at optimal doping is in the same energy range as the high energy longitudinal optical phonons in cuprates, and as we shall discuss below leads to a strong damping and anomalous renormalization of these modes for certain wave numbers.

Figure 4 provides a comparison of the static charge structure factor $N(\mathbf{q})$ with that of a Gutzwiller projected wave function and with that for uncorrelated electrons. The latter comparison shows the strong suppression of the density response due to correlations. Finally we note that our calculated structure $N(\mathbf{q})$ factor satisfies the sum rule for constrained fermions $\sum_q N(\mathbf{q}) = \delta(1-\delta)$ within 1% [9].

RENORMALIZATION OF BREATHING PHONONS

Inelastic neutron scattering experiments on high-T_c superconductors have shown that in particular the highest energy longitudinal optic phonon branch near $(\pi, 0)$ softens and broadens strongly as holes are doped into the insulating parent compound. Whereas the corresponding breathing mode at (π, π) shows much smaller softening and no anomalous broadening (Fig.5). This effect appears to be generic for cuprates and detailed neutron scattering studies have been reported for $La_{2-x}Sr_xCuO_4$ [18, 19, 20, 21], $YBa_2Cu_3O_{6+x}$[19, 22, 23] and also for Bi-based cuprates[24]. This certainly shows that there are phonons which couple rather strongly to the charge carriers that contribute to superconductivity, and one may ask what is special about the interaction of strongly correlated electrons and phonons.

The renormalization of these phonons can be calculated in the framework of the t-J model, since the breathing oxygen motion of these modes (see Fig.5) modulates the energy of a hole in a Zhang-Rice singlet state, and therefore couples directly to the density of doped holes. Expanding the Zhang-Rice energy $E_{ZR} = 8\frac{t_{pd}^2}{\Delta\varepsilon}$ with respect to the

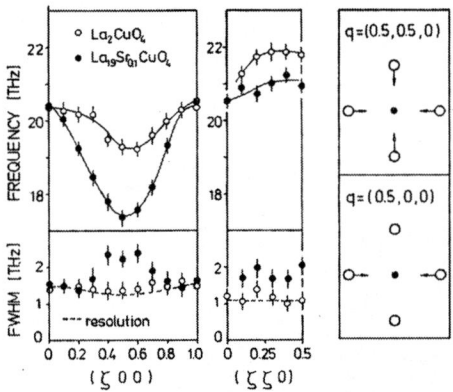

FIGURE 5. Doping dependence of the high energy bond-stretching phonons along $(\zeta,0,0)$ and $(\zeta,\zeta,0)$ directions for undoped La_2CuO_4 (open circles) and $La_{1.85}Sr_{0.15}CuO_4$ (filled circles) as obtained by neutron scattering. The lower panel gives the line width at half maximum. The displacements of oxygen ions (a) for the $\mathbf{q} = (\pi,\pi)$ breathing phonon distortion and (b) for the $(\pi,0)$ half-breathing mode are indicated by arrows in the right panel. Here $(\pi,0)$ corresponds to $(0.5,0,0)$ reduced lattice units. (reprinted from Pintschovius und Reichardt[19] with permission from World Scientific)

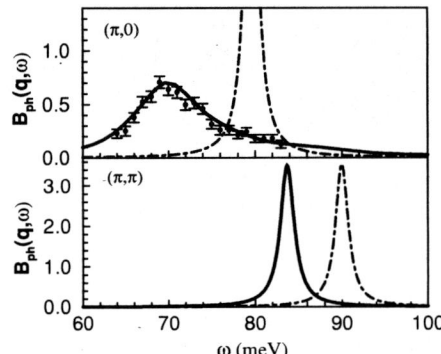

FIGURE 6. Calculated phonon spectral function B_{ph} for $(\pi,0)$ and (π,π) breathing phonons for $\delta = 0.15$, $t = 0.4$ eV, $J = 0.12$ eV and $\xi = 0.25$ (solid lines) compared to undoped system (dash-dotted lines) and the inelastic neutron scattering data[20] for $La_{1.85}Sr_{0.15}CuO_4$. (reprinted from Ref.[25] with permission from Elsevier)

oxygen displacements $u^i_\alpha = u_\alpha(R_i + \delta^O_\alpha)$, $\alpha = x,y$, of the four O-neighbors at $R_i + \delta^O_\alpha$ around the Cu-hole at R_i yields the linear electron-phonon coupling [25]

$$H_{e-ph} = g \sum_i (u^i_x - u^i_{-x} + u^i_y - u^i_{-y}) h^+_i h_i. \quad (11)$$

We assume that the resonance integral obeys the Harrison relation $t_{pd} \propto r_0^{-7/2}$, where r_0 is the Cu-O distance, and obtain $g = 7E_{ZR}/4r_0$, i.e., $g \approx 2\text{eV}/\text{Å}$. The lattice part of the Hamiltonian is determined by the force constant $K \approx 25\text{eV}/\text{Å}^2$ for the longitudinal O-motion. Due to the structure of H_{e-ph} the breathing modes couple directly to $\chi_{\mathbf{q},\omega}$.

We have studied the renormalization of the phonon Green's functions along $(\pi,0)$ and (π,π) directions

$$D^{ph}_{\mathbf{q},\omega} = \frac{\omega_{\mathbf{q},0}}{\omega^2 - \omega^2_{\mathbf{q},0}(1 - \alpha_{\mathbf{q}}\chi_{\mathbf{q},\omega})}, \quad (12)$$

where $\omega_{\mathbf{q},0}$ is the bare phonon frequency, i.e. measured in the undoped parent compound, and $\alpha_{\mathbf{q}} = \frac{4g^2}{K}(\sin^2 q_x/2 + \sin^2 q_y/2)$. Based on the parameters of the pd-model we estimate for the dimensionless coupling constant $\xi = g^2/tK \sim 0.3 - 0.5$. The coupling constant $\alpha_{\mathbf{q}}^2$ vanishes at the Γ point and becomes maximal at the zone edges. This explains the marginal changes of renormalized phonon frequency $\omega_{\mathbf{q}}$ at $\mathbf{q} = (0,0)$. The particularly strong increase of the renormalization for $\mathbf{q}$ along $(\pi,0)$ is a combined effect of the strong increase of $\alpha_{\mathbf{q}}^2$ and the low energy polaron structure in the density response.

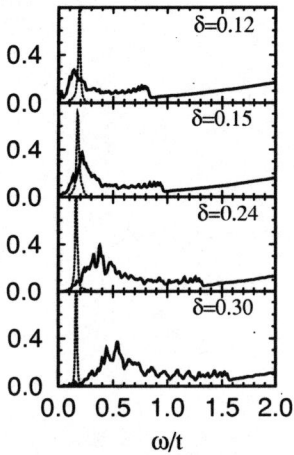

FIGURE 7. Doping dependence of low-energy density response at $(\pi,0)$ (solid lines). As a consequence of the scaling of the polaron structure $\propto (\chi J + \delta t)$ there is a strong change in the renormalization and damping of the $(\pi,0)$ optical phonon (dashed lines), which is at $\omega_0 = 0.2t$ in the undoped system.

Figure 6, calculated for typical parameters for cuprate superconductors, shows the strong renormalization of the $(\pi,0)$ half-breathing mode for $La_{1.85}Sr_{0.15}CuO_4$ with a twice as large shift as for the (π,π) breathing phonon. The large damping of the $(\pi,0)$ phonon results from the hybridization with the large polaron peak in $N(\mathbf{q},\omega)$ at this momentum and is consistent with the experimental data[20]. The phonon energies of the undoped parent compound $\omega_{\mathbf{q},0} = 80(90)$ meV for $(\pi,0)$ and (π,π), respectively, are taken from Ref.[19].

The strong doping dependence of this effect, shown in Fig.7, is due to the scaling $\propto (\chi J + \delta t)$ of the polaron peak position in $N(\mathbf{q},\omega)$. In particular our results imply that the large damping of the $(\pi,0)$ optical phonon should disappear at larger doping concentrations.

SUMMARY

We have outlined a 1/N slave-boson theory for the density response of the t-J model, which explains the data obtained by exact diagonalization. We demonstrated that the predicted low energy polaron structure in the density response, which is particularly pronounced along $(\pi,0)$, explains the anomalous doping induced line width and shift of the longitudinal planar $(\pi,0)$ phonon. The energy of the polaron peak is determined by the spinon energy scale, therefore we predict a nontrivial doping dependence for the phonon renormalization. In that respect further neutron scattering studies of the doping dependence of phonons would provide a sensitive test for the low energy density response as well as for the spin structure in the different doping regimes.

ACKNOWLEDGMENTS

We are grateful to L. Pintschovius for several illuminating discussions about the interpretation of neutron scattering data and for permission to reprint Figure 5.

REFERENCES

1. T. Tohyama, P. Horsch, and S. Maekawa, Phys. Rev. Lett. **74** 980 (1995).
2. R. Eder, Y.Ohta, and S. Maekawa, Phys. Rev. Lett. **74** 5124 (1995).
3. J. Jaklic and P. Prelovsek, Adv. Phys. **49**, 1 (2000).
4. Z. Wang, Y. Bang, and G. Kotliar, Phys. Rev. Lett. **67**, 2733 (1991).
5. L. Gehlhoff and R. Zeyher, Phys. Rev. **52**, 4635 (1995).
6. A. Foussats and A. Greco, Phys. Rev. B **65**, 195107 (2002).
7. D.K.K. Lee, D.H. Kim and P.A. Lee, Phys. Rev. Lett. **76**, 4801 (1996).
8. G. Jackeli and N.M. Plakida, Phys. Rev. B **60**, 5266 (1999).
9. G. Khaliullin and P. Horsch, Phys. Rev. B **54**, R9600 (1996).
10. G. Kotliar and J. Liu, Phys. Rev. B **38**, 5142 (1988).
11. N. Read and D.M. Newns, J. Phys. C **16**, 3273 (1983).
12. E. Arrigoni *et al.*, Physics Reports **241**, 291 (1994).
13. V.N. Popov, Functional Integrals in Quantum Field Theory and Statistical Physics, (D. Reidel, Dordrecht, 1983).
14. F.C. Zhang, C. Gros, T.M. Rice, and H. Shiba, Supercond. Sci. Technol. **1**, 36 (1988); R.B. Laughlin, J. Low Temp. Phys. **90**, 443 (1995).
15. N. Nagaosa and P. Lee, Phys. Rev. Lett. **64**, 2450 (1990).
16. C.L. Kane, P.A. Lee and N. Read, Phys. Rev. B **39**, 6880 (1989).
17. C. Gros and R. Valenti, Phys. Rev. B **50**, 11313 (1994).
18. L. Pintschovius *et al.*, Physica C **185-189**, 156 (1991).
19. L. Pintschovius and W. Reichardt, *Physical Properties of High Temperature Superconductors IV*, edited by D. Ginsberg (World Scientific, Singapore,1994), p. 295.
20. R. J. McQueeney, T. Egami, G. Shirane, and Y. Endoh, Phys. Rev. B **54**, R9689 (1996); R. J. McQueeney *et al.*, Phys. Rev. Lett. **82**, 628 (1999).
21. L. Pintschovius and M. Braden, Phys. Rev. B **60**, R15039 (1999).
22. W. Reichardt, J. Low Temp. Phys. **105**, 807 (1996).
23. L. Pintschovius, W. Reichardt, M. Kläser, T. Wolf, and H. v. Löhneisen, Phys. Rev. Lett. **89**, 037001 (2002).
24. B. Renker *et al.*, Physica C **162-164**, 462 (1989).
25. G. Khaliullin and P. Horsch, Physica C **282-287**, 1751 (1997).

Tools for Studying Quantum Emergence near Phase Transitions

Masatoshi Imada*[†], Shigeki Onoda**, Takahiro Mizusaki[‡]* and Shinji Watanabe*

*Institute for Solid State Physics, University of Tokyo, Kashiwanoha, Kashiwa, Chiba, 277-8581, Japan
[†]PRESTO, Japan Science and Technology Corporation
**Tokura Spin Superstructure Project, ERATO, Japan Science and Technology Corporation, Department of Applied Physics, University of Tokyo, Hongo 7-3-1, Tokyo 113-8656, Japan
[‡]Institute of Natural Sciences, Senshu University, Higashimita, Tama, Kawasaki, 214-8580, Japan

Abstract. We review recent studies on developing tools for quantum complex phenomena. The tools have been applied for clarifying the perspective of the Mott transitions and the phase diagram of metals, Mott insulators and magnetically ordered phases in the two-dimensional Hubbard model. The path-integral renormalization-group (PIRG) method has made it possible to numerically study correlated electrons even with geometrical frustration effects without biases. It has numerically clarified the phase diagram at zero temperature, $T = 0$, in the parameter space of the onsite Coulomb repulsion, the geometrical frustration amplitude and the chemical potential. When the bandwidth is controlled at half filling, the first-order transition between insulating and metallic phases is evidenced. In contrast, the filling-control transition shows diverging critical fluctuations for spin and charge responses with decreasing doping concentration. Near the Mott transition, a nonmagnetic spin-liquid phase appears in a region with large frustration effects. The phase is characterized remarkably by gapless spin excitations and the vanishing dispersion of spin excitations. Magnetic orders quantum mechanically melt through diverging magnon mass. The correlator projection method (CPM) is formulated as an extension of the operator projection theory. This method also allows an extension of the dynamical mean-field theory (DMFT) with systematic inclusion of the momentum dependence in the self-energy. It has enabled determining the phase diagram at $T > 0$, where the boundary surface of the first-order metal-insulator transition at half filling terminates on the critical end curve at $T = T_c$. The critical end curve is characterized by the diverging compressibility. The single particle spectra show strong renormalization of low-energy spectra, generating largely momentum dependent and flat dispersion. The results of two tools consistently suggest that the strong competitions of various phases with underlying diverging compressibility can induce an emergence near a new type of quantum criticality distinguished from the conventional and simple quantum critical phenomena.

1. INTRODUCTION

A striking feature in the properties of the two-dimensional Hubbard model is the enhancement of the charge compressibility, when the filling is controlled near half filling. In fact, although the compressibility κ is strictly vanishing in the Mott insulating state, it was reported a decade ago in a quantum Monte Carlo study that the compressibility appears to diverge as $\kappa \propto \delta^{-1}$ with the decreasing doping concentration δ in the metallic phase toward the Mott insulator [1, 2]. This is in sharp contrast with the transition to the band insulator in two dimensions. If this criticality simply characterizes the Mott transition, though the universality class would be nontrivial, it would be possible to discuss the quantum critical phenomena within the conventional concept of the quantum phase transition. In fact, a scaling description described by the dynamical exponent $z = 4$ has been proposed by one of the authors [3]. In an idealistic limit without any further instabilities, this scaling description was supported by a number of analyses including the localization length, Drude weight and Hall coefficient [4, 5, 6, 7]. The consistency of various quantities by this scaling description also implies that the Fermi surface instability to the Mott insulator emerges quite inhomogeneously in the momentum space with a positive feedback effect leading to a singular collapse of the Fermi surface [8, 9].

However, it turns out that the divergence of the compressibility, the most characteristic critical feature, drives the collapse of this quantum criticality itself by a tiny perturbation or by an additional parameter in the model [8, 9]. The reason is that when the doping is decreased, in contrast with the one-dimensional systems, the diverging charge compressibility implying the diverging density of states near the Fermi level necessarily drives various types of instability for symmetry breakings if the Fermi degeneracy is accompanied. Hence before the real critical point is reached, the originally rather simple quantum criticality itself becomes fragile and can be truncated by other symmetry breakings if a tiny additional perturbation exists. The candidates for the symmetry breakings include magnetic orders, charge orders (or stripes), phase separation and superconductivity. In fact, the divergence of the charge susceptibility at zero wave number directly results in the phase separation if the long-ranged Coulomb force is ignored as in the Hubbard model. At least this divergence signals that the effective interaction of charged excitations is transformed from repulsive to attractive and simultaneously the dressing of particles diverges. The resultant competitions of orders and its remaining fluctuations may introduce a large residual entropy even at low temperatures. As far as the authors know, this type of cascading quantum criticality originated from one quantum phase transition has never been considered before and offers a new type of physics for the quantum critical phenomena. The critical region may not be characterized by simple boring critical exponents but by possible cascading and hierarchical structure formations.

Recently, the bandwidth-control transition between the Mott insulator and metals, a different route from the filling-control transition discussed above, has been studied in detail by two new methods. One is the path-integral renormalization-group (PIRG) method [12, 13], which is a numerical scheme of wavefunction renormalization group. The other is the correlator projection method (CPM) [14], which is an analytical way for the systematic inclusion of low-energy excitations by improving the operator projection method [15]. This method improves the deficiency of the dynamical mean field theory [16] by taking account of spatial fluctuations. The PIRG calculation [17, 18] and the CPM [14] both have given consistent results. Namely, the bandwidth-control transition of the Hubbard model is of the

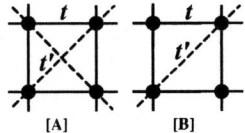

FIGURE 1. Lattice structure of geometrically frustrated lattices [A] on a square lattice and [B] on an anisotropic triangular lattice. The nearest- and next-nearest-neighbor transfers are denoted by t (solid bonds) and t' (dashed bonds), respectively.

first-order with a jump in n_D defined by the average of double occupation of up- an down-spin electrons on the same site. This first-order character is in contrast with the above mentioned filling-control case, where diverging fluctuations are observed as we mentioned above. The first-order transition indicates that there exists a strong attractive interaction of charged excitations. A sharp contrast between the bandwidth-control and filling-control transitions requires more detailed studies for the complete understanding of the Mott transition. We review recent theoretical studies to clarify the whole mechanism and phase diagram of the Mott transition.

In strongly correlated electron systems as in transition metal oxides and organic compounds, particularly near their Mott insulating phases, various competing orders and their fluctuations are universally observed and the subject of intensive studies in recent two decades [10]. It is apparent that such severe competitions at least partially originate from the tendency toward the underlying compressibility divergence mentioned above. To understand the striking complexity of these competing orders and their fluctuations as compared to weakly correlated systems, it is important to understand the underlying mechanism and physics of the criticality leading to the compressibility divergence and the subsequent collapse into a cascade-type fluctuations and competing orders.

In addition to enormous number of studies in transition metal compounds, we note two interesting recent experimental studies, one on the organic compounds and the other on the adsorbed monolayer ^{3}He on graphite [19]. Fournier et al. [20] have shown that the first-order transition between the Mott insulator and a metal (at low temperatures, a superconductor) takes place in an organic material, $\kappa-$ET compound, and its first-order boundary starting from zero temperature extends up to the critical end point at around 34K in the plane of pressure and temperature. In this study, the pressure controls the relative bandwidth to the interaction strength while the electron filling is fixed, and therefore, bandwidth-control transition is realized. In fact, according to the extended Hückel calculation [21] for κ-type ET compound, a minimal model may be the single-band Hubbard model at half filling on an anisotropic triangular lattice defined later in Eq.(1) on the lattice structure given in Fig. 1[B]. As we already discussed, this first-order transition with the critical end point was predicted before this experiments [18, 14].

On the other hand, the ^{3}He monolayer adsorbed on a HD layer and graphite substrate shows a typical filling-control transition to the Mott insulating state by changing the ^{3}He density. The Mott insulating state is realized at the commensurate density with the periodic potential formed by the underlying HD layer [19]. When the density approaches the Mott insulating state in the liquid phase, it shows striking increase and critical divergence of the effective mass probed by the specific heat coefficient and the magnetic susceptibility. The liquid phase nicely obeys the Fermi liquid criterion. Although a very weak first-order character (namely phase coexistence or phase separation) may appear, this critical divergence implies the basic continuous character of the filling-control transition in contrast with the bandwidth-control transition observed in the organic compounds.

These two different experiments performed on relatively clean and simple systems as compared to transition metal oxides show a similar contrast between bandwidth-control and filling-control transitions to theoretical studies which we mentioned above. We will discuss below the state of the art of the theoretical understanding and the perspectives of the Mott transition in more detail.

We study the Hubbard model with nearest and next-nearest neighbor transfers, t and t', respectively on a two dimensional lattice defined by

$$\mathcal{H} = \mathcal{H}_t + \sum_i H_{Ui} - \mu M \tag{1}$$

$$\mathcal{H}_t = -t \sum_{\langle ij \rangle} (c_{i\sigma}^\dagger c_{j\sigma} + h.c.) \tag{2}$$

$$\mathcal{H}_t' = -t' \sum_{\langle kl \rangle} (c_{k\sigma}^\dagger c_{l\sigma} + h.c.) \tag{3}$$

$$\mathcal{H}_{Ui} = U(n_{i\uparrow} - \frac{1}{2})(n_{i\downarrow} - \frac{1}{2}), \tag{4}$$

where $M \equiv \sum_{i\sigma} n_{i\sigma}$ and $n_{i\sigma} = c_{i\sigma}^\dagger c_{i\sigma}$ with the creation (annihilation) operator $c_{i\sigma}^\dagger (c_{i\sigma})$ of an electron at the site i with the spin σ. Here μ is the chemical potential and U is the onsite Coulomb repulsion. The t' term introduces the geometrical frustration. We first review the two methods, PIRG and CPM and then summarize the results on bandwidth-control and filling-control transitions. The implications of various results and relationship are also discussed.

2. THEORETICAL METHODS

2.1. Path-Integral Renormalization-Group (PIRG) Method

The PIRG method is a numerical scheme for obtaining the ground state of interacting many-fermion systems. In the PIRG method [12, 13], the optimized ground-state wavefunction $|\Phi\rangle$ is obtained as a linear combination of states as $|\Phi\rangle = \sum_{l=1}^{L} c_l |\varphi_l\rangle$ with a given L, by optimizing a numerically chosen non-orthogonal basis $\{|\varphi_l\rangle\}$. The ground state is approached by successive renormalization processes in the path integral. Our method optimizes both the basis $|\varphi_l\rangle$ and the coefficients c_l. It is apparent that this method does not have the sign problem, because the explicit wavefunction is constructed. With increasing L, $|\Phi\rangle$ is systematically improved from the starting variational state at $L=1$, such as the Hartree-Fock state. This method is not categorized to the Monte Carlo-type method. This may be rather viewed as a numerical procedure to find the true ground state by the extrapolation from the best variational ground-state wavefunction within the allowed dimension of the Hilbert space, L. This is also viewed as a numerical procedure of the wavefunction renormalization-group scheme in the imaginary time direction.

The renormalization to lower and lower energy state is achieved by successively operating the projection operator $\exp[-\tau H]$ with a finite τ to the initial trial wavefunction $|\Phi_0\rangle$. After operating $\exp[-\tau H]$, the dimension of the obtained state in our chosen basis functions expands through the off-diagonal element of H. In the process of successive operations, the dimension increases exponentially. Then we seek for the best truncation of the Hilbert space within the allowed memory and computation time. Under the constraint that the number of the stored states, L, is kept constant, we iterate the process of projection and truncation to lower the energy of the resultant wavefunction until the convergence.

For the basis functions we take Slater determinants. We note that in the PIRG method, the best Hartree-Fock result should be reproduced at $L=1$. The wavefunctions are systematically improved toward the true ground state with increasing L.

To operate $\exp[-\tau H]$ to a Slater determinant, we take the path integral formalism and $\exp[-\tau H]$ is approximated by $\exp[-\tau(H_t - \mu M)]\exp[-\tau\sum_i H_{Ui}]$ for sufficiently small τ. Then we use the Stratonovich-Hubbard transformation for the interaction part. Because of the summation over the Stratonovich variables, the number of states after the operations of $\exp[-\tau H]$ exponentially increases with increasing number of operations. Then, truncation of the states is performed so as to select the truncated state which has lower and lower energy. This projection and truncation process is repeated until the convergence. For a better convergence, we extrapolate to the zero-energy variance by a linear function of the energy variance defined by $\Delta_E = (\langle E^2 \rangle - \langle E \rangle^2)/\langle E \rangle^2$ if Δ_E is small, where $\langle E^2 \rangle = \langle \Phi|H^2|\Phi \rangle / \langle \Phi|\Phi \rangle$ and $\langle E \rangle = \langle \Phi|H|\Phi \rangle / \langle \Phi|\Phi \rangle$. This works because the variance disappears if the exact ground state is obtained. Readers are referred for the details of the algorithm to Ref. [13]. This method has been applied to a number of different systems successfully [17, 18, 22, 23].

We have also extended this algorithm for studies on the excitation spectra [24]. We have improved the original PIRG algorithm to obtain the lowest energy state with specific quantum numbers such as the total spin S and total momentum $\mathbf{k}$ when they commute with the Hamiltonian. Such quantum number projection can be performed by constructing lower symmetry states. For example, the state with a specified total spin is obtained from the rotation of a state $|\Psi\rangle$ in the spin space using the rotation operator $\mathcal{R}(\phi)$ and then an integration over the rotation angle ϕ. For a state with a specified total momentum, it is obtained by a spatial translation with a shift of $\mathbf{r}$ by the translation operator $\mathcal{L}(\mathbf{r})$. A state with specific quantum numbers S and $\mathbf{k}$ is obtained from the weighted integration as $|\Phi(S,\mathbf{k})\rangle = \int d\phi W_R(S,\phi) R(\phi) \sum_\mathbf{r} W_L(\mathbf{k},\mathbf{r})\mathcal{L}(\mathbf{r})|\Psi\rangle$. Here the weights W_R and W_L are chosen to specify the quantum numbers S and $\mathbf{k}$. For example, $W_L(\mathbf{k},\mathbf{r}) = \exp[i\mathbf{k}\cdot\mathbf{r}]$. After this projection, the lowest energy state with a specified total spin and the total momentum can be obtained. The quantum number projection can

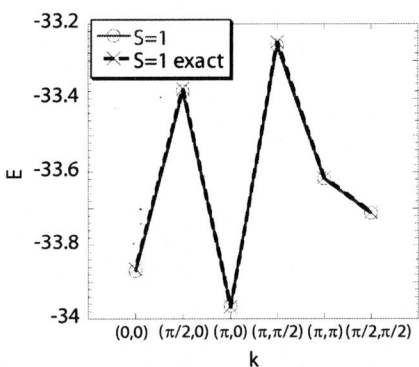

FIGURE 2. Comparison of dispersion for S=1 lowest energy states between PIRG (circles) and exact results (crosses) for the two-dimensional Hubbard model at $U = 5.7, t = 0, t' = 0.5$ on 4×4 lattice with the periodic boundary condition. The momenta are denoted as (k_x, k_y) with k_x and k_y being scaled by $1/L_x$ and $L_x = 4$.

also be performed for other type of symmetry operations such as spatial inversions, and rotations if the Hamiltonian retains these symmetries. The computational time, however, increases with the increase of the symmetrization projections. The physical quantity A can be calculated from $\langle A \rangle = \langle \Psi|A|\Phi\rangle / \langle \Psi|\Phi\rangle$.

The quantum-number projection procedure can be taken to the wavefunctions already obtained in the above PIRG method. This improves the wavefunctions of the ground state and enables extracting the excitation spectra if a fraction of excited states remain even after the PIRG procedure. This estimate becomes worse if we calculate excited states with a higher excitation energy, because such component is already projected out by the PIRG process and almost missing. A better accuracy particularly for the excited states is obtained by taking the renormalization and optimization procedure of the PIRG by lowering the energy with the states after the quantum number projection in each iteration step. We call this algorithm as quantum-number projected PIRG (QP-PIRG) method. This quantum number projection procedure further improves the accuracy of the energy estimate substantially even for the ground state.

The accuracy of this method was carefully examined and confirmed in many examples. For example, all the ground states and excitation spectra studied on 4×4 lattice show roughly 4 digit accuracy in comparison with the available exact results. Figure 2 shows the comparison of the $S = 1$ excitation spectra calculated by the QP-PIRG method with the exact ones for the Hubbard model on 4 by 4 lattice at $U = 5.7, t = 1.0$ and $t' = 0.5$. Here L is taken only up to around 100, while the dimension of the whole Hilbert space is beyond 10^8.

We have also extended our algorithm to treat the system in the grand canonical ensemble [25]. For this purpose, the particle-hole transformation for only up-spin electrons is applied to the Hamiltonian and the basis states. To reach the ground state at the fixed chemical potential μ, in this notation, the Stratnovich-Hubbard transformation which hybridizes up-spin and down-spin electrons is introduced.

The PIRG for the grand canonical ensemble (GPIRG) method is efficient for direct calculations of the chemical-potential dependence in physical quantities.

2.2. Correlator Projection Method (CPM)

The operator projection theory or memory function formalism [26] is an analytical framework to obtain the dynamics of interacting systems in an unperturbative fashion. It was improved to take account of the low-energy excitation in a self-consistent manner [15]. Based on a series of Dyson equations or continued-fraction expansion of the Green's function, the method systematically improves the mean-field, conserving [27, 28, 29] and Hubbard-type [30, 31, 32] approximations. By taking account of spin correlations as well as the Hubbard-band splitting, this gives a unified scheme of the filling-control metal-insulator transition beyond the early pictures [33, 30, 34, 35]. However, the self-consistent approximation [15] fails in reproducing a Mott insulator in the particle-hole asymmetric cases at the finite couplings [36] because of insufficient estimates of local dynamics. To improve this deficiency, an alternative method to Refs. [15, 37, 38] has been proposed. [14] to take account of momentum dependences ignored in the dynamical mean-field approximation (DMFA) [16]. We call this scheme CPM.

The formalism of the CPM is based on the formal and hierarchical Dyson equations for the electron Green function $G(\omega, \mathbf{k})$ described in the upper left panel of Fig. 3. Here, ω and $\mathbf{k}$ describe the frequency and momentum, respectively. They are obtained by the projection in the equation of motion for $c_{\mathbf{i}\sigma}$ [15]. Coefficients are given in Refs. [15, 31]: With the electron transfer $t_{\mathbf{i}\sigma,\mathbf{j}\sigma'}$, $\varepsilon_{\mathbf{k}}^{(1,1)}$ and $\varepsilon_{\mathbf{k}}^{(2,2)}$ are the Fourier transforms of $-t_{\mathbf{i}\sigma,\mathbf{j}\sigma} - \mu + U\langle n\rangle/2$ and $-2\tilde{t}_{\mathbf{i}\sigma,\mathbf{j}\sigma} - \mu + U(1 - \langle n\rangle/2) - \varepsilon_{\mathrm{cor}}/\mathcal{M}$, respectively. We also define $\varepsilon^{(2,1)} = U^2\mathcal{M}$ with $\mathcal{M} = \frac{\langle n\rangle}{2}(1 - \frac{\langle n\rangle}{2})$. In these definitions, $\varepsilon_{\mathrm{cor}} = -\sum_{\mathbf{j}\sigma} t_{\mathbf{i}\sigma,\mathbf{j}\sigma}\langle(1/2 - n_{\mathbf{i}\sigma})(c_{\mathbf{i}\sigma}^\dagger c_{\mathbf{j}\sigma} + h.c.)\rangle/2$ and $\tilde{t}_{\mathbf{i}\sigma,\mathbf{j}\sigma} = t_{\mathbf{i}\sigma,\mathbf{j}\sigma}(\langle n_{\mathbf{i}\sigma} n_{\mathbf{j}\sigma}\rangle/4 + \langle \vec{S}_{\mathbf{i}\sigma} \cdot \vec{S}_{\mathbf{j}\sigma}\rangle - \langle \Delta_{\mathbf{i}\sigma}^\dagger \Delta_{\mathbf{j}\sigma}\rangle)/\mathcal{M}$ are introduced with the charge, spin and local-pair operators $n_{\mathbf{i}\sigma}$, $\vec{S}_{\mathbf{i}\sigma}$ and $\Delta_{\mathbf{i}} \equiv \langle c_{i\uparrow} c_{i\downarrow}\rangle$, respectively. Physically, $\varepsilon^{(2,1)}$ generates the Hubbard band splitting. $\tilde{t}_{\mathbf{i}\sigma,\mathbf{j}\sigma}$ introduces a $\mathbf{k}$ dependence of $\Sigma_1(\omega, \mathbf{k})$ mainly through the superexchange interaction. $\langle n_{\mathbf{i}\sigma} n_{\mathbf{j}\sigma}\rangle$, $\langle \vec{S}_{\mathbf{i}\sigma} \cdot \vec{S}_{\mathbf{j}\sigma}\rangle$ and $\langle \Delta_{\mathbf{i}\sigma}^\dagger \Delta_{\mathbf{j}\sigma}\rangle$ in $\tilde{t}_{\mathbf{i}\sigma,\mathbf{j}\sigma}$ may be determined self-consistently by solving the additional coupled correlator projection equations for *two-particle* operators. However, this has not been performed yet. An alternative way is to independently determine them from the two-particle self-consistent method [29].

An approximation has to be introduced in the highest-order self-energy part $\Sigma_n(\omega, \mathbf{k})$ to obtain the explicit solution. Here, we show the results of the second-order projection, where we take up to $n = 2$. In earlier studies, Σ_2 has been calculated by a two-site method [39] and a self-consistent decoupling approximation [15]. However, with these methods, it was difficult to discuss the metal-insulator transition when a particle-hole asymmetry exists at half filling because of insufficient treatment of the local dynamics. In the extension to CPM [14], DMFA was employed in the highest order self-energy. This enables more accurate calculation of

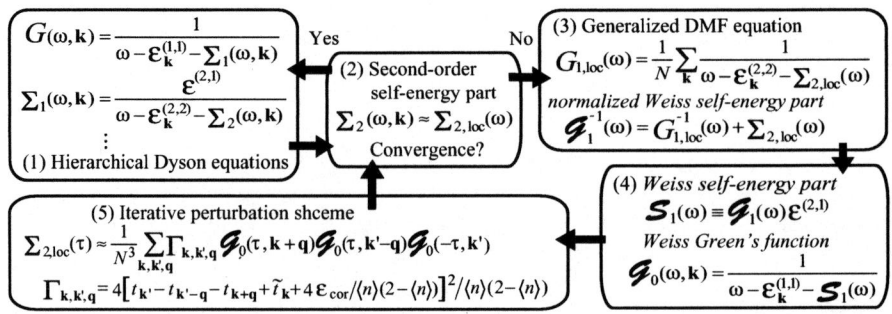

FIGURE 3. Schematic self-consistent loop of the currently proposed formalism of CPM. N is the number of the sites. The CPM uses the generalized DMFA to calculate the local dynamics of $\Sigma_n(\omega, \mathbf{k})$, instead of the self-consistent decoupling approximation in the OPM [15]. The first-order CPM is reduced to a conventional DMFA [16].

the local dynamics of $\Sigma_2(\omega, \mathbf{k})$ by ignoring its $\mathbf{k}$ dependence. This is a systematic extension of the conventional DMFA. In fact, in contrast with the original DMFA, $\Sigma_1(\omega, \mathbf{k})$ contains $\mathbf{k}$ dependences.

The generalized DMFA scheme in the case of $n = 2$ is illustrated in the flow lines in Fig. 3: First, by taking an arbitrary $\Sigma_{2,\text{loc}}$ in the procedure (2), the local *normalized* self-energy part $G_{1,\text{loc}}(\omega)$ is calculated in the DMFA procedure (3). Then, the *normalized Weiss self-energy part* $\mathcal{G}_1$ is obtained from $G_{1,\text{loc}}$ and $\Sigma_{2,\text{loc}}$. By following the procedure (4), the *Weiss self-energy part* $\mathcal{S}_1$ obtained from $\mathcal{G}_1$ generates the *Weiss Green's function* $\mathcal{G}_0$ by replacing Σ_1 with $\mathcal{S}_1$ in the Dyson equation. To calculate $\Sigma_{2,\text{loc}}(\omega)$, we employ the iterative perturbation scheme [16] using $\mathcal{G}_0(\omega, \mathbf{k})$ in the procedure (5). Then, the loop continues to (2) by replacing the original $\Sigma_{2,\text{loc}}$ with $\Sigma_{2,\text{loc}}$ obtained in the procedure (5). This iteration procedure is continued until the convergence of $\Sigma_{2,\text{loc}}$. A self-consistent solution is obtained after the iteration of this loop. The converged $\Sigma_2(\omega)$ is used to calculate $G(\omega, \mathbf{k})$ by substituting into the Dyson equation in the upper left panel. In this loop, conservation laws for $G(\omega, \mathbf{k})$ and $G(\omega, \mathbf{k})\Sigma_1(\omega, \mathbf{k})$ are satisfied. For details, readers are referred to Refs. [15, 14].

When we take the lower level approximation, namely $n = 1$, CPM is reduced to the original DMFA [16]. With higher-order projections, spatial correlations are analytically restored in a systematic fashion.

3. BANDWIDTH-CONTROL TRANSITION

3.1. Results from PIRG

Recently, the PIRG method was applied to the two-dimensional Hubbard model (1) on the square lattice [17, 18]. In Fig. 4, the obtained phase diagram is illustrated in the parameter space of U/t and t'/t. The stabilized phases are a consequence of severe competitions of various possible phases. The phase diagram contains the antiferromagnetic insulator (AFI), paramagnetic metal (PM) and the nonmagnetic insulator (NMI). The other phases such as superconducting and stripe phases are not stabilized. These phases are in fact destabilized with increasing L (namely, increasing quantum fluctuations) even when we input these mean-field solutions at $L = 1$ as the initial state. However, it does not strictly exclude a possibility of extremely small order parameter. We will discuss this issue further below. A remarkable result is that the metal-insulator transition is of the first-order type with a jump of the averaged double occupation n_D. Furthermore, the phase diagram contains a nonmagnetic insulator (NMI) near the metal-insulator transition boundaries.

Figure 5 shows the jump of the double occupation. Since the control parameter for the bandwidth-control transition is U/t, the conjugate quantity to U should be a relevant quantity to determine the character of the transition. This conjugate quantity is nothing but the averaged double occupation n_D defined by $n_D = \langle n_{i\uparrow} n_{i\downarrow} \rangle$. The jump in n_D indicates there occurs the level crossing of the insulating and metallic states. The jump tends to decrease with increasing the frustration parameter t'/t, indicating that the transition would become more continuous type with increasing the frustration.

In the present NMI phase, the absence of various symmetry breakings including the antiferromagnetic (AF) order has been shown [17, 18]. This NMI phase is stabilized under a severe competition with metallic and AF insulating (AFI) phases. The phase diagrams show quantum melting of spin orders at higher U than the Mott transition. This quantum melting of the spin order is ascribed to enhanced charge fluctuations and increasing double occupation near the Mott transition, which cannot be studied in the Heisenberg models. The appearance of the NMI near the Mott transition is a natural consequence since, with decreasing U, the spin solid may quantum mechanically melt before the melting of the Mott insulator itself as we know larger fragility of the spin long-ranged order than the Mott insulator itself in low-dimensional systems. The existence of the nonmagnetic insulating phase was recently identified experimentally in organic compounds [40, 41]

The nature of the NMI phase has been further explored together with the quantum number projection to clarify the excitation spectra [42]. In the NMI phase, typical system size dependences of the spin excitation gap ΔE between the singlet ground state and the lowest triplet state are shown in Fig. 6. The data points in Fig. 6 indicate that the triplet excitations become gapless in the thermodynamic limit. The data also imply that the uniform susceptibility becomes a nonzero constant. Except in 1D systems, this result is the first numerical evidence by unbiased calculations for the existence of gapless excitations without apparent long-ranged order in the Mott insulator.

The dispersion of the $S = 1$ excitation in the NMI phase has strong and monotonic system size dependence and for large system sizes, it surprisingly becomes vanishingly small. The size dependence shows very quick collapse of the dispersion with increasing system size and may not be fitted by a power of the inverse system size as in the electron-hole excitations in metals. The collapse implies that the triplet excitations cannot propagate as a collective mode. In addition to $S = 1$

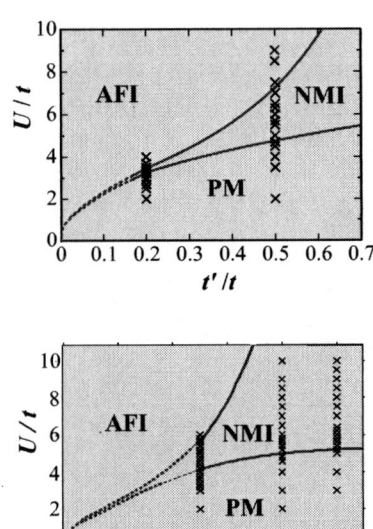

FIGURE 4. Phase diagrams of the Hubbard model in the parameter space of U scaled by t, and the frustration parameter t'/t. AFI, PM, and NMI represent the AF insulating, paramagnetic metallic and nonmagnetic insulating phases, respectively. The upper panel shows the result for the model [A] in Fig. 1 and the lower panel, for the model [B]. Calculations were performed at the cross points.

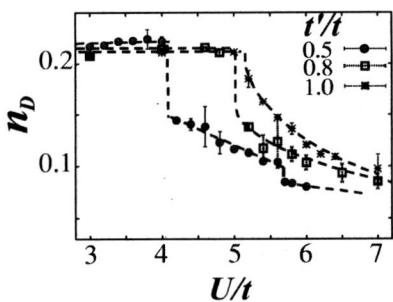

FIGURE 5. Averaged double occupation n_D of the two-dimensional Hubbard model as a function of U/t for $t'/t = 0.5, 0.8$ and 1.0 on the anisotropic triangular lattice given in Fig. 1.

excitations, the total singlet state $(S = 0)$ at any total momentum $\mathbf{k}$ also shows degenerate structure in the ground state for larger system size. The dispersion is vanishingly small in the NMI phase.

The origin of the gapless and dispersionless excitations is not completely clarified for the moment. It may arise either from coherent or from incoherent modes. The

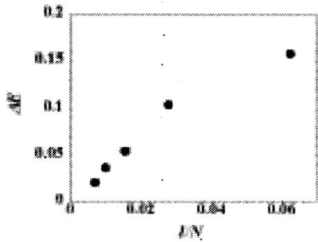

FIGURE 6. Size scalings of the $S=1$ excitation gaps in the nonmagnetic insulator phase. The parameter is at $U-5.7, t=1.0$, and $t'=0.5$.

first one would be the existence of essentially free one-body excitations as in the electron-hole excitations in the usual metal. In the present case, this mechanism corresponds to the possible coherent spinon excitations with a spinon Fermi surface. However, such a Fermi surface should generate a gapless excitations at momenta sensitively depending on the shape of the Fermi surface and with the finite-size gap scaled by $1/\sqrt{N}$. The whole and quick collapse of dispersion with increasing system size does not fit this expectation. Although the coherence of the spin excitations must be more carefully examined before definite conclusion, the present result supports that an unbound spin triplet does not propagate coherently due to strong scattering by other weakly bound RVB singlets. This remarkable result means that the quantum melting of the AF (or any other) order occurs through the divergence of the magnon (or Goldstone mode)mass.

3.2. Results from CPM

At the second-order level, CPM [14] enables the study of the bandwidth-controlled metal-insulator transition at $T > 0$. Now we review these results below. By applying CPM, the phase diagram for the 2D half-filled Hubbard model was obtained for the first time at nonzero temperature in the parameter space of the local Coulomb repulsion U, and the second-neighbor transfer t' scaled by the nearest-neighbor transfer t. This formalism has also enabled the calculation of the single-particle dynamics.

In the calculated single-particle spectra $A(\omega, \mathbf{k}) \equiv -\frac{1}{\pi}\mathrm{Im}\,G(\omega, \mathbf{k})$, a Mott gap Δ separating the two Hubbard bands where Δ seems to grow from 0 with increasing U is obtained. At low temperatures, generated Mott gap together with growth of AF or singlet correlations yield AF shadow structure and flat dispersions around the $(\pi, 0)$ and $(0, \pi)$ momenta at the Mott gap edge in general agreement with previous theoretical and experimental indications [15, 11, 43, 44, 45, 46]. The spectra in the second-order CPM are strongly modified from the conventional DMFA [16]. This is due to a proper consideration of spatial correlations mainly through superexchange interaction. Besides, present results reproduce a direct gap for $t'=0$ and $A(\omega, \mathbf{k})$

shows a remarkable similarity to the QMC results [11].

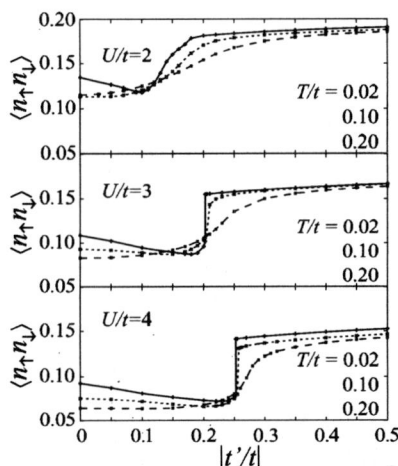

FIGURE 7. $\langle n_\uparrow n_\downarrow \rangle$ versus $|t'/t|$ at several U's and T's.

The metal-insulator transition was also studied in detail as a function of the frustration parameter t'/t at half filling [14]. $n_D \equiv \langle n_\uparrow n_\downarrow \rangle$ is plotted against $|t'/t|$ in Fig. 7. With decreasing T, the maximum slope of $\langle n_\uparrow n_\downarrow \rangle$ versus $|t'/t|$ increases. For $U/t = 3$ and 4, $\langle n_\uparrow n_\downarrow \rangle$ exhibits a jump at $|t'| = t'_{\text{MIT}}(U,T)$ below the critical temperature $T_{\text{cr}}(U)$. The jump indicates a first-order transition. The double occupation increases with decreasing t'/t in the insulating phase at low temperatures. One might think it strange. However, this is well accounted by the fact that the electrons can move more coherently when the antiferromagnetic correlation becomes more robust. The qualitative feature of the transition is consistent with that obtained in the PIRG results discussed in the previous section. This first-order metal-insulator transition below T_{cr} accompanies a discontinuous change in single-particle dispersions, while the dispersions continuously evolve at $T \geq T_{\text{cr}}$. The insulating dispersion with a choice of $t = 0.25$ eV agrees with that of La_2CuO_4 [46].

In this formalism, spatial correlations are restored as the **k** dependence of $\Sigma_1(\omega, \mathbf{k})$. Neglecting the **k** dependence of $\Sigma_n(\omega, \mathbf{k})$ with finite $n \geq 2$ does not automatically guarantee the Luttinger sum rule except the trivial case of $n = 1$. However, deviation from the Luttinger theorem becomes visible only near the Mott transition and stays small. This framework essentially satisfies the Fermi-liquid properties with a large Fermi surface and vanishing damping rates toward the Fermi energy and momenta for metals without a symmetry breaking. This is in contrast with previous equation-of-motion approaches [30, 31, 32], where the Fermi volume becomes small as a typical feature of the Hubbard approximation. In contrast to cluster methods [37, 38], the CPM is free from cluster-size effects.

The phase diagram for the metal-insulator transition obtained for the 2D half-filled Hubbard model is shown in Fig. 8 with the first-order metal-insulator transition surface (hatched surface), and the critical end line (bold solid curve). The metal-insulator transition phase boundary at $T \to 0$ agrees with the PIRG result

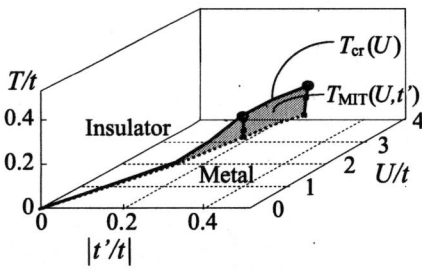

FIGURE 8. Phase diagram of the 2D half-filled Hubbard model obtained by the CPM. The critical curve exists as illustrated by the bold solid line. The hatched surface represents a first-order metal-insulator phase boundary.

discussed in the previous section [17].

4. FILLING-CONTROL TRANSITION

By using the GPIRG method, the ground-state phase diagram of the square-lattice Hubbard model is constructed in the plane of μ and U [25]. Figure 9 presents the phase diagram in the plane of U/t and the chemical potential μ for $t'/t = 0.2$. At this parameter, only the paramagnetic metal and antiferromagnetic insulator are visible in the phase diagram. As in the filling control, other long-range ordered phases with a symmetry breaking are not obtained. However, as mentioned in the introduction, this result needs a reservation, because of the instability through the enhanced compressibility as we discuss below. The phase boundary shows that the charge gap opens around $U/t = 3.25$ for $t = 1$ and $t' = -0.2$. It is remarkable that the gap increases very linearly with increasing U/t. Although the first-order metal-insulator transition occurs at half filling when U/t is increased, analyses of the carrier-density dependence on the chemical potential indicate that the phase separation does not occur for carrier doping at least at the doping concentration δ larger than 6% for $U/t = 4.0$ at $t'/t = 0.2$. Instead, a critical enhancement of the charge compressibility as $\kappa \propto \delta^{-1}$ is obtained in agreement with the previous quantum Monte Carlo study [1, 2].

Much progress has been made on understandings of filling-control metal-insulator transition also by the operator projection method (OPM) [15]. This theory reproduces numerically inferred key elements [1, 43, 44, 11]; the diverging compressibility towards the metal-insulator transition [1], the four-band-like structure [44] and flat dispersions around the momentum $(\pi,0)$ [11]. The results are consistent with the GPIRG results.

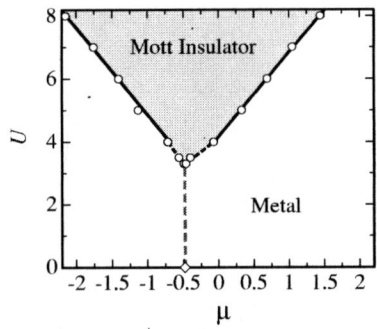

FIGURE 9. Ground-state phase diagram in the plane of μ and U for $t = 1.0$ and $t' = -0.2$ on the square lattice in the thermodynamic limit. The solid lines represent the least-square fit of metal-insulator transition for $U = 4.0, 5.0, 6.0, 7.0$ and 8.0. Shaded area represents the Mott-insulator phase in the thermodynamic limit. The grey dashed curve represents half filling in the metallic phase.

5. DISCUSSION

We have seen that the path-integral renormalization-group method and the correlator projection method give consistent results on the nature of the Mott transitions for the two-dimensional Hubbarad models. In the bandwidth-control transition, a clear first-order transition with jumps of the averaged double occupation n_D was identified. This first-order character also agrees with recent experimental results in κ-ET type organic compounds. The first-order transition appears to become more continuous type with smaller jump in n_D when the frustration parameter t'/t is increased. In the filling-control transition, physical properties show large fluctuations indicating the underlying continuous character of the transition. Especially the charge compressibility shows a strong enhancement with presumable critical divergence if some additional factor does not drive the system into a symmetry broken state induced by the diverging compressibility. This continuous character with diverging charge density of states at the transition is consistent with the recent experiment on the formation of the Mott insulator for ^{3}He on a substrate.

Another interesting feature of the Mott transition is the emergence of a nonmagnetic insulating phase near the metal-insulator phase boundary if the frustration becomes large. The character of the nonmagnetic insulating phase shows unexpected feature. The spin excitation appears to be gapless and the excitation spectra seem to show a vanishingly small dispersion indicating the lack of coherent propagation of the spin excitations.

Here, we discuss a possible mechanism of the first-order transition for the bandwidth-control coexisting with continuous character in the filling-control case. From the phenomenological point of view, the first-order transition is caused when the relevant excitations to drive the transition have attractive interaction. A typical example of the first-order transition is the gas-liquid transition, where the particle-particle attraction is crucial in realizing the liquid phase. When the particle-particle interaction is attractive, we have the negative coefficient for the n^2 term of the free energy expansion with respect to the density n. This results in the negative curvature in the free energy as a function of n and this necessarily causes the phase separation and the first-order transition.

In the case of the bandwidth-control metal-insulator transition, the transition is controlled by U or by the conjugate quantity n_D. The Mott insulating state is realized when the doubly occupied site (doublon) and the empty site (holon) form the bound state while the metal is stabilized when these become unbound. It is rather clear that the doublon and the holon strongly attract each other because of the electrostatic potential. This means that the term proportional to $n_D n_H$ with n_H being the holon density has a negative coefficient in the free energy. If the filling is fixed at half filling, we have a constraint $n_H = n_D$. In this case, the negative curvature of the free energy as a function of the double occupation n_D appears. This naturally leads to the first-order transition for the bandwidth control (phase separation does not occur because n_D is not a conserved quantity but U is controlled in real systems. On the other hand, the filling-control transition is realized by controlling the doping concentration, $\delta = n_H - n_D$. The interaction between two holons is a nontrivial issue and should be quite different from the doublon-holon interaction. The diverging compressibility as $\kappa \propto \delta^{-1}$ indicates that the δ^2 term in the phenomenological Landau free-energy expansion in terms of δ vanishes or very small. When we define $\nu = n_H + n_D$, the calculated results suggest, for example for $\delta > 0$, the Free-energy form $F = \mu\delta + A\nu + V\delta^2 - W\nu^2 + B\nu^3 + C\delta^3$ would satisfy the requirement with $V = 0, W > 0, B > 0$ and $C > 0$. This form oversimplifies the real structure because even the insulating phase has a nonzero ν. Nevertheless the basic point may be still valid. A more complicated form may also satisfy the requirement with additional mixed low-order terms. Actually, the term $-D\nu\delta$ with $D > 0$ may exist because the doping may cause the increase of ν. Now from these forms, the compressibility divergence is solely from $V = 0$.

The critical divergence of the compressibility may be truncated and induce an instability to various kinds of symmetry breakings, if additional tiny interaction beyond the Hubbard model is introduced or the lower energy scale beyond the accuracy of this calculation exists. It is rather clear that V may be controlled to a negative value if some additional interaction can be generated. This may easily lead to the phase separation. This may happen, for example, when the AF order is overestimated than the real one because it introduces an additional attraction of holes by hand. In other words, a mobile hole disturbs the AF correlation and raise the energy while the energy is lowered when such disturbance is prohibited by the phase separation of the holes. The similar situation may appear when one takes the t-J model with a rather large J. The interaction of the holons is determined in a self-consistent fashion with very subtle balance in the realistic electronic structure near half filling while the J term controlled by hand very easily destroy this requirement. This is also true for the stability of the superconductivity. The J term explicitly introduces an attraction of two singlet electrons, while an overestimate of the role of the antiferromagnetic correlations may also overestimate the stability of superconductivity. Another example is found in a small correlated hopping term added to the Hubbard model, which stabilizes the d-wave superconducting state [47, 48]. All these instabilities induced by a small perturbation or crudeness of the approximation show the significance of underlying instability originated from the criticality of the Mott transition with the diverging compressibility. The system becomes *almost* unstable to various directions.

We have to be very careful in concluding which type of the instability eventually wins since the calculation suggests a strong degeneracy of various different phases under quantum fluctuations. Near the critical points whichever at $T=0$ and $T\neq 0$, the electronic state may have strong instabilities to various symmetry breakings and inhomogeneities. In particular, if the critical point is suppressed to lower and lower temperatures, the instability is enhanced because of the interplay with the Fermi degeneracy. Even an emergence of a complicated hierarchy is expected with lowering energy after more severe competitions of various fluctuations. Without clarifying this subtle cascade-type instabilities and competitions under the compressibility divergence, the final result at zero temperature would not be reliable. The lower energy hierarchy, competitions and even emergence are an intriguing open issue. However, at least the diversity and complexity of the phenomena in experiments in this region is understood from the underlying uniqueness of statistical mechanical properties, namely combination of the Fermi degeneracy and the compressibility divergence caused by a unique quantum criticality of the Mott transition as reviewed in this article.

We have discussed only for the region near the Mott insulator at half filling of the Hubbard model. However, in principle, a similar structure appears near any simple commensurate fillings of particle density. It is known in many correlated electron systems that the Mott insulating state called the charge order is universally observed at a simple commensurate fillings such as $n = 1/3$ and $1/4$, while it quantum mechanically melts away from such simple fractional densities [10]. With the long-ranged Coulomb interaction, the melting is also driven by decreasing m^*/ϵ with m^* and ϵ being the effective mass and the dielectric constant. With increasing m^*/ϵ, charge orderings are stabilized at more and more complicated fractional fillings and a structure of the devil's staircase appears as clarified recently by the PIRG method [22]. In principle, a similar structure with divergent compressibility to the half-filled Mott insulator reviewed in this article may appear around each charge ordered states, which generates another complexity and hierarchy structure.

We have reviewed the substantial progress achieved for the understanding of correlated insulators and metals. The unique critical character of the Mott transitions requires us to introduce a new concept of cascading criticality. Although this hierarchy structure formation has been discussed as an emergence in *classical* complex systems many times, the present example provides us with a new aspect taking place near *quantum* phase transitions. The whole structure of instabilities is, however, a fundamental problem of quantum many-body systems to be clarified further and its understanding will also help controls of various instabilities.

REFERENCES

1. N.Furukawa and M. Imada, J. Phys. Soc. Jpn. **61**, 3331 (1992).

2. N. Furukawa and M. Imada, J. Phys. Soc. Jpn. **63**, 2557 (1993).
3. M. Imada, J. Phys. Soc. Jpn. **64**, 2954 (1995).
4. F.F. Assaad, and M. Imada, Phys. Rev. Lett.**76**, 3176 (1996).
5. H. Tsunetsugu and M.Imada, J. Phys. Soc. Jpn. **67**, 1864 (1998).
6. H. Nakano and M.Imada, J. Phys. Soc. Jpn. **68**, 1458 (1999).
7. F.F. Assaad, and M. Imada, Phys. Rev. Lett.**74**, 3868 (1995).
8. M. Imada and S. Onoda, J. Phys. Chem. Solids. **62**, 47 (2001).
9. M. Imada and S. Onoda, *"Open Problems in Strongly Correlated Electron Systems"* eds. J. Bonca et al. (Kluwer Academic Publishers 2001) p.69-80,
10. For a review, see M. Imada, A. Fujimori and Y. Tokura, Rev. Mod. Phys. **69**, 1039 (1998).
11. F. F. Assaad and M. Imada, Eur. Phys. J. B **10**, 595 (1999).
12. M. Imada and T. Kashima, J. Phys. Soc. Jpn. **69**, 2723 (2000).
13. T. Kashima and M. Imada, J. Phys. Soc. Jpn. **70**, 2287 (2001).
14. S. Onoda and M. Imada, Phys. Rev. B **67**, 161102 (2003).
15. S. Onoda and M. Imada, J. Phys. Soc. Jpn. **70**, 632 (2001); ibid. **70**, 3398 (2001);
16. W. Metzner and D. Vollhardt, Phys. Rev. Lett. **62**, 324 (1989); A. George et al., Rev. Mod. Phys. **68**, 13 (1996).
17. T. Kashima et al., J. Phys. Soc. Jpn. **70**, 3052 (2001).
18. H. Morita et al., J. Phys. Soc. Jpn. **71**, 2109 (2001).
19. A. Casey, H. Patel, J. Nyeki, B.P. Cowan, J. Saunders, Phys. Rev. Lett.**90**, 115301 (2003).
20. D. Fournier, M. Poirier, M. Castonguay, et al., Phys. Rev. Lett.**90**, 127002 (2003).
21. K. Oshima et al., Phys. Rev. B **38**, 938 (1988).
22. Y. Noda and M. Imada, Phys. Rev. Lett. **89**, 176803 (2002).
23. T. Mizusaki et al., Phys. Rev. C **65** 064319 (2002).
24. T. Mizusaki and M. Imada, unpublished.
25. S. Watanabe and M. Imada, unpublished.
26. S. Nakajima, Prog. Theor. Phys. **20**, 948 (1958); R. Zwanzig, *Lectures in Theoretical Physics*, Vol. 3, (Interscience, New York, 1961); H. Mori, Prog. Theor. Phys. **33**, 423 (1965); Prog. Theor. Phys. **34**, 399 (1965).
27. G. Baym, Phys. Rev. **127**, 1391 (1962); L. P. Kadanoff and G. Baym, *Quantum Statistical Mechanics*, (Benjamin, Menlo Park, 1962).
28. N. E. Bickers and D. J. Scalapino, Ann. Phys. (N.Y.) **193**, 206 (1989).
29. Y. M. Vilk and A.-M. S. Tremblay, J. Phys. (France) I **7**, 1309 (1997); B. Kyung, cond-mat/9802129.
30. J. Hubbard, Proc. Roy. Soc. A **276**, 238 (1963); ibid. **281**, 401 (1964).
31. L. M. Roth, Phys. Rev. **184**, 451 (1969).
32. C. Gors, Phys. Rev. B **50**, 7295 (1994).
33. N. F. Mott, *Metal-Insulator Transitions*, (Taylor and Francis, London/Philadelphia, 1990).
34. W. F. Brinkman and T. M. Rice, Phys. Rev. B **2**, 4302 (1970).
35. J. C. Slater, Phys. Rev. **82**, 538 (1951).
36. S. Onoda and M. Imada, unpublished.
37. Th. Maier et al., Eur. Phys. J. B **13**, 613 (2000); Jarrell et al., Phys. Rev. B **64**, 195130 (2001); Th. Maier et al., cond-mat/0111368.
38. G. Kotliar et al., Phys. Rev. Lett. **87**, 186401 (2001).
39. H. Matsumoto and F. Mancini, Phys. Rev. B **55**, 2095 (1997).
40. Y. Shimizu, M. Maeda, G. Saito, K. Miyagawa and K. Kanoda, private commun.
41. M. Tamura and R. Kato, J. Phys. Cond. Matt. **14**, L729 (2002).
42. M. Imada, T. Mizusaki and S. Watanabe, unpublished
43. N. Bulut et al., Phys. Rev. Lett. **72**, 705 (1994); ibid. **73**, 748 (1994); Phys. Rev. B **50**, 7215 (1994).
44. R. Preuss et al., Phys. Rev. Lett. **75**, 1344 (1995); C. Gröber et al., Phys. Rev. B **62**, 4336 (2000).
45. Z.-X. Shen and D. S. Dessau, Phys. Rep. **253**, 1 (1995).
46. A. Ino et al., Phys. Rev. B **62**, 4137 (2000).
47. F. F. Assaad, M. Imada and D. J. Scalapino, Phys. Rev. Lett. **77**, 4592 (1996).
48. H. Tsunetsugu and M. Imada, J. Phys. Soc. Jpn. **68**, 3162 (1999).

Antiferromagnetic exchange and spin-fluctuation pairing mechanisms in cuprates

Nikolay M. Plakida

Joint Institute for Nuclear Research, 141980 Dubna, Russia

Abstract. Superconducting pairing mediated by antiferromagnetic (AFM) exchange and spin-fluctuations is considered within the effective two-band Hubbard model for a copper-oxide plane. It is proved that retardation effects for the AFM exchange induced by interband hopping are unimportant that results in pairing of all the carries in the conduction band and high T_c proportional to the Fermi energy. The kinematic interaction induced by intraband hopping gives an additional spin-fluctuation contribution to the d-wave pairing. T_c dependence on pressure and oxygen isotope effect are explained.

I INTRODUCTION

A unique property of cuprates is that they belong to charge-transfer insulators with a small splitting energy between $3d$ copper and $2p$ oxygen levels and a large Coulomb correlations in $3d$ copper states. That results in a huge antiferromagnetic (AFM) superexchange interaction of the order of $J \simeq 1500$ K which brings to the long-range AFM order in the undoped regime and strong AFM dynamical spin fluctuations in the superconducting state. The AFM spin fluctuations can be also responsible for anomalous normal state properties of cuprates (see, e.g. [1]) and the superconducting pairing as has been first proposed by P.W. Anderson [2]. In a number of studies of the reduced one-band t-J model (see, e.g. [3]- [6] and the references therein) it was shown that the instantaneous AFM exchange interaction results in d-wave pairing with quite high superconducting temperature T_c. However, to prove the AFM pairing mechanism one has to consider the original two-band p-d model for CuO_2 layer [7] without reducing the interband hopping to the exchange interaction in one subband of the t-J model.

In this paper we present a microscopic theory of superconductivity within the effective p-d Hubbard model [8–10]. By applying a projection technique for the matrix Green function in terms of the Hubbard operators the Dyson equation is derived [11]. It is proved that in the mean field approximation (MFA) the d-wave superconducting pairing mediated by the AFM exchange interaction occurs as in the t-J model. The self-energy is calculated in the non-crossing or the self-consistent Born approximation which gives an additional contribution to the d-wave pairing. It is mediated by the spin-fluctuations caused by the kinematic interaction in the intraband hopping. We give the results of numerical solution of the gap equation for the superconducting T_c depending the on hole concentration and the superconducting gap as a function of the wave-vector [11]. A remarkable for cuprates increasing of T_c with pressure and the oxygen isotope effect are also explained [6].

II DYSON EQUATION

Starting from the original two-band p-d model for CuO_2 layer [7] we can reduce it to the effective two-band Hubbard model by applying the cell-cluster perturbation theory [8–10]:

$$H = E_1 \sum_{i,\sigma} X_i^{\sigma\sigma} + E_2 \sum_i X_i^{22} \qquad (1)$$
$$+ \sum_{i\neq j,\sigma} \{t_{ij}^{11} X_i^{\sigma 0} X_j^{0\sigma} + t_{ij}^{22} X_i^{2\sigma} X_j^{\sigma 2} + 2\sigma t_{ij}^{12}(X_i^{2\bar\sigma} X_j^{0\sigma} + \text{H.c.})\},$$

where $X_i^{nm} = |in\rangle\langle im|$ are the Hubbard operators for the four states $n,m = |0\rangle, |\sigma\rangle, |2\rangle = |\uparrow\downarrow\rangle$, $\sigma = \pm 1/2 = (\uparrow, \downarrow)$, $\bar\sigma = -\sigma$. Here $E_1 = \epsilon_d - \mu$ is the energy of the one-hole Cu d-like state in the lower Hubbard subband and $E_2 = 2E_1 + \Delta$ is the two-hole p-d singlet state in the upper Hubbard subband, μ is the chemical potential and $\Delta = \epsilon_p - \epsilon_d$ is the charge transfer energy (see [8]). The hopping integrals with the superscripts 2 and 1 refering to the singlet and one-hole subbands, respectively, are given by $t_{ij}^{\alpha\beta} = K_{\alpha\beta} 2t\nu_{ij}$ where t is the p-d hybridization parameter and ν_{ij} are estimated as: $\nu_1 = \nu_{j\ j\pm a_{x/y}} \simeq -0.14$, $\nu_2 = \nu_{j\ j\pm a_x \pm a_y} \simeq -0.02$. The coefficients $K_{\alpha\beta} < 1$, e.g., for the singlet subbands we have $t_{eff} \simeq K_{22} 2t\nu_1 \simeq 0.14t$ and the bandwidth $W = 8t_{eff}$. Since the ratio $\Delta/W \simeq 2$, the Hubbard model (2) corresponds to the strong correlation limit. The Hubbard operators entering (2) obey the completeness relation

$$X_i^{00} + X_i^{\sigma\sigma} + X_i^{\bar\sigma\bar\sigma} + X_i^{22} = 1, \qquad (2)$$

which rigorously preserves the constraint of no double occupancy of any quantum state $|in\rangle$ at each lattice site i. So we do not need to impose any constraint usually introduced in the slave-field technique (see, e.g., [12] and the references therein).

To discuss the superconducting pairing within the model Hamiltonian (2), we introduce the four-component Nambu operators $\hat{X}_{i\sigma}$ and $\hat{X}_{i\sigma}^\dagger$ and define the 4×4 matrix Green function (GF) in Zubarev notation [13]

$$\tilde{G}_{ij\sigma}(t-t') = \langle\langle \hat{X}_{i\sigma}(t) | \hat{X}_{j\sigma}^\dagger(t') \rangle\rangle, \quad \tilde{G}_{ij\sigma}(\omega) = \begin{pmatrix} \hat{G}_{ij\sigma}(\omega) & \hat{F}_{ij\sigma}(\omega) \\ \hat{F}_{ji\sigma}^\dagger(\omega) & -\hat{G}_{ji\bar\sigma}(-\omega) \end{pmatrix}, \qquad (3)$$

where $\hat{X}_{i\sigma}^{\dagger} = (X_i^{2\sigma}\ X_i^{\bar{\sigma}0}\ X_i^{\bar{\sigma}2}\ X_i^{0\sigma})$ and $\hat{G}_{ij\sigma}$ and $\hat{F}_{ij\sigma}$ are the normal and anomalous 2×2 matrix components, respectively. By applying the projection technique for the equation of motion method for GF (3) we derive the Dyson equation in $(\mathbf{q},\omega)$-representation [14]:

$$\left(\tilde{G}_\sigma(\mathbf{q},\omega)\right)^{-1} = \left(\tilde{G}^0_\sigma(\mathbf{q},\omega)\right)^{-1} - \tilde{\Sigma}_\sigma(\mathbf{q},\omega), \tag{4}$$

where the zero-order GF

$$\tilde{G}^0_\sigma(\mathbf{q},\omega) = \left(\omega\tilde{\tau}_0 - \tilde{E}_\sigma(\mathbf{q})\right)^{-1}\tilde{\chi}, \tag{5}$$

defines the spectrum of one-particle excitations in the mean field approximation (MFA) by the matrix

$$\tilde{E}_{ij\sigma} = \tilde{\mathcal{A}}_{ij\sigma}\tilde{\chi}^{-1}, \quad \tilde{\mathcal{A}}_{ij\sigma} = \langle\{[\hat{X}_{i\sigma},H],\hat{X}_{j\sigma}^{\dagger}\}\rangle, \quad \tilde{\chi} = \langle\{\hat{X}_{i\sigma},\hat{X}_{i\sigma}^{\dagger}\}\rangle. \tag{6}$$

The self-energy operator in the Dyson equation (4) in the projection technique method is defined by a *proper* part (having no single zero-order GF) of the many-particle GF in the form

$$\tilde{\Sigma}_\sigma(\mathbf{q},\omega) = \tilde{\chi}^{-1}\langle\!\langle \hat{Z}_\sigma^{(ir)} | \hat{Z}_\sigma^{(ir)\dagger}\rangle\!\rangle^{(prop)}_{\mathbf{q},\omega}\tilde{\chi}^{-1}. \tag{7}$$

Here the *irreducible* $\hat{Z}$-operator is given by the equation: $\hat{Z}_\sigma^{(ir)} = [\hat{X}_{i\sigma},H] - \sum_l \tilde{E}_{il\sigma}\hat{X}_{l\sigma}$ which follows from the orthogonality condition: $\langle\{\hat{Z}_\sigma^{(ir)},\hat{X}_{j\sigma}^{\dagger}\}\rangle = 0$. The equations (4)-(7) provide an exact representation for the GF (3). However, to calculate the self-energy matrix (7) which describes finite lifetime effects (inelastic scattering of electrons on spin and charge fluctuations) one has to apply some approximations.

III MEAN-FIELD APPROXIMATION

In the MFA the electronic spectrum and superconducting pairing are described by the zero-order GF in Eq. (5). By applying the commutation relations for the Hubbard operators we get for the frequency matrix (6):

$$\tilde{\mathcal{A}}_{ij\sigma} = \begin{pmatrix} \hat{\omega}_{ij\sigma} & \hat{\Delta}_{ij\sigma} \\ \hat{\Delta}^*_{ji\sigma} & -\hat{\omega}_{ji\bar{\sigma}} \end{pmatrix}, \tag{8}$$

where $\hat{\omega}_{ij\sigma}$ and $\hat{\Delta}_{ij\sigma}$ are 2×2 matrices for the normal and anomalous components, respectively. The normal component defines quasiparticle spectra of the model in the normal state which have been studied in details in [8]. The anomalous component defines the gap functions for the singlet and one-hole subbands. By disregarding the single-site terms, $i=j$, which give no contribution to the d-wave pairing we obtain the following representation for the gap functions for lattice sites $(i \neq j)$ in terms of the following anomalous correlation functions:

$$\Delta_{ij\sigma}^{22} = -2\sigma t_{ij}^{12} \langle X_i^{02} N_j \rangle, \quad \Delta_{ij\sigma}^{11} = -2\sigma t_{ij}^{12} \langle (2-N_j) X_i^{02} \rangle, \tag{9}$$

where the number operator is $N_i = \sum_\sigma X_i^{\sigma\sigma} + 2X_i^{22}$. If we define formally the Fermi annihilation operator: $c_{i\sigma} = X_i^{0\sigma} + 2\sigma X_i^{\bar{\sigma}2}$ then we can write the anomalous correlation function as follow: $\langle X_i^{02} N_j \rangle = \langle X_i^{0\downarrow} X_i^{\downarrow 2} N_j \rangle = \langle c_{i\downarrow} c_{i\uparrow} N_j \rangle$ since other products of the Hubbard operators vanish according to the multiplication rules: $X_i^{\alpha\beta} = X_i^{\alpha\gamma} X_i^{\gamma\beta}$. Therefore the anomalous correlation functions describe the pairing at one lattice site but in different subbands.

The same anomalous correlation functions were obtained in MFA for the original Hubbard model in Refs. [15–17]. To calculate the anomalous correlation function $\langle c_{i\downarrow} c_{i\uparrow} N_j \rangle$ in [15,17] the Roth procedure was applied based on decoupling of the operators on the same lattice site in the time-dependent correlation function: $\langle c_{i\downarrow}(t) | c_{i\uparrow}(t') N_j(t') \rangle$. However, the decoupling of the Hubbard operators on the same lattice site is not unique (as has been really observed in Refs. [15,17]) and unreliable. To escape uncontrollable decoupling, in Ref. [16] kinematical restrictions to the Hubbard operators were imposed which, however, also have not resulted in a unique solution for superconducting equations.

In our approach we perform a direct calculation of the correlation function $\langle X_i^{02} N_j \rangle$ without *any decoupling* by writing equation of motion for the corresponding commutator GF $L_{ij}(t-t') = \langle \langle X_i^{02}(t) | N_j(t') \rangle \rangle$:

$$(\omega - E_2) L_{ij}(\omega) \simeq 2\delta_{ij} \langle X_i^{02} \rangle$$
$$+ \sum_{m \neq i, \sigma} 2\sigma t_{im}^{12} \left\{ \langle \langle X_i^{0\bar{\sigma}} X_m^{0\sigma} | N_j \rangle \rangle_\omega - \langle \langle X_i^{\sigma 2} X_m^{\bar{\sigma} 2} | N_j \rangle \rangle_\omega \right\}, \tag{10}$$

where we have neglected intraband hopping ($\propto t_{ij}^{\alpha\alpha}$) which gives only small renormalization for the large excitation energy $E_2 \simeq \Delta$. After applying the spectral theorem and neglecting exponentially small term of the order of $\exp(-\Delta/T)$ which comes from the pole $\omega = E_2$ we obtain the following representation for the correlation function at sites $i \neq j$ for the singlet subband in the case of hole doping [11]:

$$\langle X_i^{02} N_j \rangle = -\frac{1}{\Delta} \sum_{m \neq i, \sigma} 2\sigma t_{im}^{12} \langle X_i^{\sigma 2} X_m^{\bar{\sigma} 2} N_j \rangle \simeq -\frac{4 t_{ij}^{12}}{\Delta} 2\sigma \langle X_i^{\sigma 2} X_j^{\bar{\sigma} 2} \rangle. \tag{11}$$

The approximate value is obtained in the two-site approximation, $m = j$, usually applied for the t-J model. Here the identity for the Hubbard operators $X_j^{\bar{\sigma}2} N_j = 2X_j^{\bar{\sigma}2}$ was used. This finally allows us to write the expression of the gap function in Eq. (9) in the case of hole doping as following

$$\Delta_{ij\sigma}^{22} = -2\sigma t_{ij}^{12} \langle X_i^{02} N_j \rangle = J_{ij} \langle X_i^{\sigma 2} X_j^{\bar{\sigma} 2} \rangle. \tag{12}$$

This result recovers the AFM exchange interaction contribution to the pairing in the t-J model with the exchange energy $J_{ij} = 4 (t_{ij}^{12})^2 / \Delta$. In the case of electron doping analogous calculations for the anomalous correlation function of the one-hole subband $\langle(2-N_j)X_i^{02}\rangle$ gives for the gap function $\Delta_{ij\sigma}^{11} = J_{ij}\langle X_i^{0\bar\sigma}X_j^{0\sigma}\rangle$.

We may therefore conclude that the anomalous contributions in the zero-order GF (5) are given by the conventional anomalous quasiparticle pairs in one subband and there are no new "composite operator excitations – cexons" proposed in [17]. The quasiparticle pairing in MFA is mediated by the AFM exchange interaction with negligible retardation effects due to the large interband hopping energy, $|E_2| \gg |t_{ij}^{\alpha\alpha}|$ that proves the results obtained in the t-J model [3,5].

IV SELF-ENERGY

The self-energy matrix (7) can be written in the form

$$\tilde\Sigma_{ij\sigma}(\omega) = \tilde\chi^{-1}\begin{pmatrix}\hat M_{ij\sigma}(\omega) & \hat\Phi_{ij\sigma}(\omega) \\ \hat\Phi_{ji\sigma}^\dagger(\omega) & -\hat M_{ji\bar\sigma}(-\omega)\end{pmatrix}\tilde\chi^{-1}, \tag{13}$$

where the 2×2 matrices $\hat M$ and $\hat\Phi$ denote the normal and anomalous contributions to the self-energy, respectively. They are essentially the many-particle GF of the type $\langle\langle B_{i\sigma\sigma'}(t)\hat X_{i'\sigma'}(t)|B_{j\sigma\sigma'}^\dagger(t')\hat X_{j'\sigma'}^\dagger(t')\rangle\rangle$ where $B_{l\sigma\sigma'}$ are the Bose-like operators which describe spin and charge fluctuations.

We calculate the self-energy (13) in the non-crossing or the self-consistent Born approximation (SCBA) which assumes that the propagation of the Fermi-like $\hat X_{i'\sigma'}(t)$ and the Bose-like $B_{i\sigma\sigma'}(t)$ excitations in the many-particle GF in (13) occurs independently of each other. This approximation is described by the decoupling of the corresponding operators in the time-dependent correlation functions for lattice sites $(i\neq i', j\neq j')$ as follows :

$$\langle B_{i\sigma\sigma'}(t)\hat X_{i'\sigma'}(t)|B_{j\sigma\sigma'}^\dagger(t')\hat X_{j'\sigma'}^\dagger(t')\rangle \simeq \langle\hat X_{i'\sigma'}(t)\hat X_{j'\sigma'}^\dagger(t')\rangle\langle B_{i\sigma\sigma'}(t)B_{j\sigma\sigma'}^\dagger(t')\rangle. \tag{14}$$

Using the spectral representation for the correlation functions we get a closed system of equations for the GF (3) and the self-energy components (13) [11]. Below we write down only the anomalous part of the self-energy for the singlet subband which gives the spin-fluctuation contribution to the superconducting pairing:

$$\Phi_\sigma^{22}(\mathbf{q},\omega) = \frac{1}{N}\sum_{\mathbf k}|t(\mathbf k)|^2 \int_{-\infty}^{+\infty}\int_{-\infty}^{+\infty}\frac{d\omega_1 d\omega_2}{\omega-\omega_1-\omega_2}\frac{1}{2}\left(\tanh\frac{\omega_1}{2T}+\coth\frac{\omega_2}{2T}\right)$$
$$\times \chi_s''(\mathbf q-\mathbf k,\omega_2)\{-(1/\pi)\mathrm{Im}[K_{22}^2 F_\sigma^{22}(\mathbf k,\omega_1) - K_{21}^2 F_\sigma^{11}(\mathbf k,\omega_1)]\}. \tag{15}$$

The kinematic interaction is given by the hopping integral for the nearest and the second neighbors $t(\mathbf k) = 8t[\nu_1\gamma(\mathbf k) + \nu_2\gamma'(\mathbf k)]$, where $\gamma(\mathbf k) = (1/2)(\cos k_x + \cos k_y)$ and $\gamma'(\mathbf k) = \cos k_x \cos k_y$. The spin-fluctuation pairing is defined by the spin susceptibility $\chi_s''(\mathbf q,\omega) = -(1/\pi)\mathrm{Im}\langle\langle \mathbf S_q | \mathbf S_{-q}\rangle\rangle_{\omega+i\delta}$ which comes from the correlation functions $\langle B_{i\sigma\sigma'}(t)B_{j\sigma\sigma'}^\dagger(t')\rangle$.

For the hole doped case at frequencies $|\omega, \omega_1| \ll \omega_s \ll W$ close to the Fermi surface (FS) ($\omega_s \lesssim J$ is a characteristic spin-fluctuation energy) we can use the weak coupling approximation (WCA) for calculation of the first term in the self-energy (15). The contribution from the second term $F_\sigma^{11}(\mathbf{k}, \omega_1)$ is rather small since the one-hole band lies below the FS at an energy of the order $\Delta \gg W$. Neglecting it and taking into account the contribution from the exchange interaction in MFA, Eq.(12), we arrive at the following equation for the singlet gap in the WCA:

$$\Phi_\sigma^{22}(\mathbf{q}) = \frac{1}{N} \sum_{\mathbf{k}} [J(\mathbf{k} - \mathbf{q}) - K_{22}^2 \lambda(\mathbf{k}, \mathbf{q} - \mathbf{k})] \frac{\Phi_\sigma^{22}(\mathbf{k})}{2 E_2(\mathbf{k})} \tanh \frac{E_2(\mathbf{k})}{2T}, \qquad (16)$$

where $\lambda(\mathbf{k}, \mathbf{q} - \mathbf{k}) = |t(\mathbf{k})|^2 \chi_s(\mathbf{q} - \mathbf{k}, \omega = 0) > 0$. The quasiparticle energy in the singlet band is given by $E_2(\mathbf{k}) = [\Omega_2(\mathbf{k})^2 + \Phi_\sigma^{22}(\mathbf{k})^2]^{1/2}$ where $\Omega_2(\mathbf{k})$ is the quasiparticle energy in the normal state [8]. Similar considerations hold true for an electron doped system, $n \leq 1$, when the chemical potential is in the one-hole subband, $\mu \simeq 0$. In that case, the WCA equation for the gap $\Phi^{11}(\mathbf{q})$ is quite similar to the Eq. (16).

V NUMERICAL RESULTS AND DISCUSSION

To solve the gap equation (16) we used the following model for the static spin-fluctuation susceptibility:

$$\chi_s(\mathbf{q}, \omega = 0) = \frac{\chi_0}{1 + \xi^2[1 + \gamma(\mathbf{q})]}, \qquad (17)$$

where ξ is the AFM correlation length. The constant $\chi_0 = 3(2-n)/(2\pi\omega_s C_1)$ with $C_1 = (1/N) \sum_{\mathbf{q}} \{1 + \xi^2[1 + \gamma(\mathbf{q})]\}^{-1}$ is defined from the normalization condition: $(1/N) \sum_i \langle \mathbf{S_i S_i} \rangle = (3/4)(1 - |1 - n|)$.

Let us at first estimate the superconducting transition temperature T_c by solving the gap equation (16) for a model d-wave gap function $\Phi^{22}(\mathbf{q}) = \phi^{22}(\cos q_x - \cos q_y) = \phi^{22} \eta(\mathbf{q})$. Integrating over $\mathbf{q}$ both sides of Eq. (16) multiplied by $\eta(\mathbf{q})$ we get the following equation for T_c:

$$1 = \frac{1}{N} \sum_{\mathbf{k}} \frac{1}{2\Omega_2(\mathbf{k})} \tanh \frac{\Omega_2(\mathbf{k})}{2T_c} \left[J \eta(\mathbf{k})^2 + \lambda_s (4\gamma(\mathbf{k}))^2 \eta(\mathbf{k})^2 \right], \qquad (18)$$

where the effective coupling constant for $\xi \gg 1$ is given by $\lambda_s \simeq t_{eff}^2/\omega_s$. For the AFM exchange interaction mediated by the interband hopping with large energy transfer $\Delta \gg W$ the retardation effects are negligible that results in coupling of all electrons in the conduction subband of the bandwidth W. In the standard logarithmic approximation in the case of weak coupling we get for T_c [6]:

$$T_c \simeq \sqrt{\mu(W - \mu)} \exp(-1/\lambda_{ex}), \qquad (19)$$

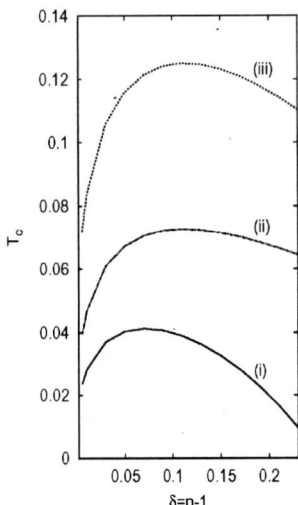

FIGURE 1. $T_c(\delta)$ (in units of $t_{eff} \simeq 0.2$ eV) for (i) spin-fluctuation interaction (solid line), (ii) exchange interaction (dashed line), (iii) for the both contributions (dotted line) [11].

The effective coupling constant $\lambda_{ex} \simeq J N_{ex}(\delta)$ is defined by the exchange interaction J and the averaged over the Fermi surface (FS), with the function $\eta(\mathbf{k})^2$ electronic density of states $N_{ex}(\delta)$ for the doping δ. By taking into account the both contributions we can write the following estimation for T_c:

$$T_c \simeq \omega_s \exp(-1/\tilde{\lambda}_{sf}), \quad \tilde{\lambda}_{sf} = \lambda_{sf} + \frac{\lambda_{ex}}{1 - \lambda_{ex}\ln(\mu/\omega_s)}, \quad (20)$$

where $\lambda_{sf} \simeq \lambda_s N_{sf}(E_F)$ is the coupling constant for the spin-fluctuation pairing. It should be pointed out that the effective electronic density of states for the spin-fluctuation pairing $N_{sf}(E_F)$ should be smaller then $N_{ex}(\delta)$ since it is averaged over the FS with the function $\gamma(\mathbf{k})^2$ which gives no contribution at the van Hove singularity at the $(\pm\pi, 0), 0, \pm\pi)$ points of the Brillouin zone (BZ). By taking for estimation $\mu = W/2 \simeq 0.35$ eV, $\omega_s \simeq J \simeq 0.13$ eV and $\lambda_{sf} \simeq \lambda_{ex} = 0.2$ we get $\tilde{\lambda}_{sf} \simeq 0.2 + 0.25 = 0.45$ and $T_c \simeq 160$ K, while only the spin-fluctuation pairing gives $T_c^0 \simeq \omega_s \exp(-1/\lambda_{sf}) \simeq 10$ K. Therefore, the exchange interaction is essential for achieving high-T_c.

Direct numerical solution of the gap equation (16) in **k**-space proves the analytical estimations given above. The following parameters were used: $\xi = 3$, $J = 0.4 t_{eff}$, $\omega_s = 0.15$ eV and $t_{eff} = K_{22} 2t\nu_1 \simeq 0.2$ eV. On Fig. 1 the superconducting transition temperature $T_c(\delta)$ is shown as a function of hole doping δ [11]. The maximum T_c is achieved for the chemical potential $\mu = E_F \simeq W/2$ at the optimal doping $\delta_{opt} \simeq 0.12$. The spin-fluctuation interaction produces much lower T_c since

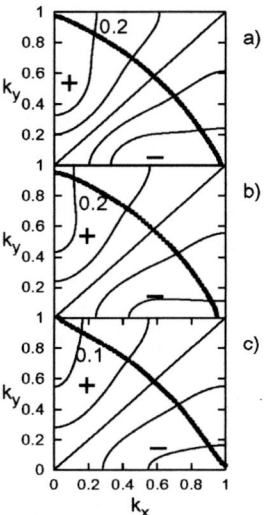

FIGURE 2. Gap function $\Phi^{22}(\mathbf{k})$ at $\delta_{opt} \simeq 0.12$ for $(0 < k_x, k_y < \pi)$ and temperatures: $T = 0$ (a), $0.5\,T_c$ (b), $0.9\,T_c$ (c). The circles plot the Fermi surface [11].

it couples the holes in a narrow energy shell, $\omega_s \ll E_F$, near the Fermi surface (FS) and it is rather weak at the FS close to the lines $|k_x| + |k_y| = \pi$ since along these lines the main contribution due to the nearest neighbor hopping in $t(\mathbf{k})$ vanishes: $\gamma(\mathbf{k}) = 0$ as pointed out before. The $\mathbf{k}$-dependence of the gap $\Phi^{22}(\mathbf{k})$ shown on Fig. 2 in the quarter of BZ, $(0 < k_x, k_y < \pi)$, proves the $d_{x^2-y^2}$ symmetry with zeroes along the lines $|k_x| = |k_y|$ and positive $(+)$ and negative $(-)$ maxima close to the BZ points $(\pi, 0)$ and $(0, \pi)$.

The AFM pairing mechanism is further proved by considering T_c dependence on the ptressure or the lattice constant in the plane a. While in electron-phonon superconductors T_c usually decreases with pressure in cuprates T_c increases. In particular, in mercury superconductors it was observed a strong increase of T_c with decreasing the distance Cu-O: $dT_c/da \simeq -1.35 \times 10^3$ K/Å [18]. For Hg-1201 compound we get for the dimensionless coefficient $d\ln T_c/d\ln a \simeq -54$. From Eq. (19) we obtain the estimation:

$$\frac{d\ln T_c}{d\ln a} = \frac{d\ln T_c}{d\ln J}\frac{d\ln J}{d\ln a} \simeq -\frac{14}{\lambda} \simeq -47 \qquad (21)$$

for $\lambda = JN_{ex}(\delta) \simeq 0.3$, which is quite close to the experimentally observed one. Here we take into account that $J(a) \propto t_{pd}^4$ and $t_{pd}(a) \propto 1/a^{7/2}$.

Concerning the oxygen isotope effect in cuprates on substitution ^{18}O oxygen for ^{16}O which is small, $\alpha_c = -d\ln T_c/d\ln M \leq 0.1$, we can estimate it also from Eq. (19). By using the experimentally observed isotope shift for the Néel temperature in La$_2$CuO$_4$ [19]: $\alpha_N = -d\ln T_N/d\ln M \simeq 0.05$ with $T_N \propto J$ we get

$$\alpha_c = -\frac{d\ln T_c}{d\ln M} = -\frac{d\ln T_c}{d\ln J}\frac{d\ln T_N}{d\ln M} \simeq \frac{\alpha_N}{\lambda} \simeq 0.16, \qquad (22)$$

for $\lambda \simeq 0.3$ which is close to experiments.

To conclude, the present investigation proves the existence of a singlet $d_{x^2-y^2}$-wave superconducting pairing for holes or electrons in the two-band Hubbard model mediated by the antiferromagnetic exchange interaction and spin-fluctuation scattering induced by the kinematic interaction, characteristic to the Hubbard model. These mechanisms of superconducting pairing are absent in the fermionic models (for a discussion, see [20]). Therefore we can suggest that the generic for cuprates huge AFM exchange interaction is responsible for high-$T_c \geq 100$ K which has not been observed so far in any other materials.

REFERENCES

1. Plakida, N.M., *High-Temperature Superconductivity*, Springer, Berlin-Heidelberg, 1995.
2. Anderson, P.W., *Science*, **235**, 1196 (1987);Anderson, P.W., *The theory of superconductivity in the high-T_c cuprates*, Princeton University Press, Princeton, 1997.
3. Plakida, N.M., Yushankhai, V.Yu., and Stasyuk, I.V. *Physica C*, **160** 80 (1989); Yushankhai, V.Yu., Plakida, N.M. and Kalinay P., *Physica C*, **174**, 401 (1991).
4. Izyumov, Yu.A. and Letfulov, B.M., *Intern. J. Modern Phys. B*, **6**, 321 (1992).
5. Plakida, N.M. and Oudovenko, V.S., *Phys. Rev. B*, **59**, 11949 (1999).
6. Plakida, N.M., *JETP Letters*, **74**, 36 (2001).
7. Emery, V.J., *Phys. Rev. Lett.* **58**, 2794 (1987); Varma, C.M., Schmitt-Rink, S., and Abrahams, E., *Solid State Commun.*, **62**, 681 (1987).
8. Plakida, N.M., Hayn, R., and Richard, J.-L., *Phys. Rev. B*, **51**, 16599 (1995).
9. Feiner, L.F., Jefferson, J.H., and Raimondi, R., *Phys. Rev. B*, **53**, 8751 (1996).
10. Yushankhai, V.Yu., Oudovenko, V.S., and Hayn, R., *Phys. Rev. B*, **55**, 15562 (1997).
11. Plakida, N.M., Anton, L., Adam, S., and Adam Gh., *JETP*, **124**, No.1 (2003); arXiv:cond-mat/0104234.
12. Plakida, N.M., *Condensed Matter Phys. (Ukraine)*, **5**, 1 (2002); arXiv:cond-mat/0210385.
13. Zubarev, D.N., *Sov. Phys. Usp.*, **3**, 320 (1960).
14. Plakida, N.M., *Physica C*, **282–287**, 1737 (1997).
15. Beenen, J. and Edwards, D.M., *Phys. Rev. B*, **52**, 13636 (1995).
16. Avella, A., Mancini, F., Villani, D., and Matsumoto, H, *Physica C*, **282-287**, 1757 (1997); Matteo, T. Di., Mancini, F., Matsumoto, H., and Oudovenko, V.S., *Physica B*, **230 - 232**, 915 (1997).
17. Stanescu, T.D., Martin, I., and Phillips, Ph., *Phys. Rev. B*, **62**, 4300 (2000).
18. Lokshin, K.A., Pavlov, D.A., Putilin, S.N., Antipov, E.N., Sheptyakov, D.V., and Balagurov, A.M., *Phys. Rev. B* **63**, 064511 (2001) .
19. Zhao, G.-M., Singh, K.K., and Morris, D.E., *Phys. Rev. B*, **50**, 4112 (1994).
20. Anderson,. P.W., *Adv. in Physics*, **46**, 3 (1997).

Magnetic fluctuations and resonant peak in doped antiferromagnets

P. Prelovšek, I. Sega, and J. Bonča

J. Stefan Institute and University of Ljubljana, SI-1000 Ljubljana, Slovenia

Abstract. A theory of the dynamical spin susceptibility is presented, relevant for the normal-state magnetic response and the resonant magnetic peak in superconducting (SC) cuprates. The analysis is based on the equations of motion for spins within the t-J model and on the memory-function representation of magnetic response. An evidence from numerical studies of the model confirms that the damping of the spin collective model is large in the normal state being due to the decay into fermionic degrees of freedom. Assuming the saturation of equal-time correlations at low T this leads to the anomalous ω/T scaling, explaining neutron scattering experiments on cuprates at low doping. In the SC phase a d-wave SC gap leads to a sharp resonant peak with reduced intensity and downward dispersion.

INTRODUCTION

Magnetic properties of cuprates have been and remain the subject of intensive experimental and theoretical investigations. Particularly challenging theoretical task is to find an explanation and a corresponding theoretical technique to describe the evolution of properties as they develop by doping the reference antiferromagnetic (AFM) insulator into a high-T_c superconductor.

One of quite long standing puzzles waiting for a proper theoretical explanation is the anomalous ω/T scaling behavior of the magnetic response [1], observed by the neutron scattering experiments on cuprates, mostly in the regime of low doping. Such a scaling applies to $La_{2-x}Sr_xCuO_4$ (LSCO) with $x = 0.02, x = 0.04$ and $YBa_2Cu_3O_{6+x}$ (YBCO) with $x = 0.5, x = 0.6$ [2], and is even more pronounced in Zn-substituted YBCO, where T_c is supressed [3]. In particular, mentioned experiments indicate that one can represent results for the local (**q**-integrated) susceptibility in a broad range of ω and T as $\chi_L''(\omega,T) = I(\omega)f(\omega/T)$ where $I(\omega) = \chi_L''(\omega, T=0)$ and $f(\omega/T)$ represents the scaling function. The largest response in this regime is at the AFM wavevector $\mathbf{Q} = (\pi,\pi)$. At the same time, the inverse AFM correlation length $\kappa = 1/\xi$, as extracted from the **q**-dependent $\chi''(\mathbf{q},\omega)$, is found to saturate at low ω and T. It is quite evident that the ω/T scaling is inconsistent with the concept of the usual Fermi liquid where $\chi''(\mathbf{q},\omega)$ should be essentially T-independent. The concept of the 'marginal' Fermi liquid has been introduced [4] to explain scaling of the magnetic response as well as of other anomalous electronic properties. Another or possibly related explanation intensively considered recently is the vicinity of the quantum critical point [5]. It should be noted that also numerical investigations of the two-dimensional t-J model [6, 7] confirm the ω/T scaling behavior of the local $\chi_L''(\omega)$, although restricted to rather high

T as compared to experiments.

On the other hand, the central feature of the magnetic response in the superconducting (SC) phase is the resonant peak, first found in optimally doped YBCO $x = 1$ [8], has been systematically followed in YBCO into the underdoped regime [9] where the resonant frequency ω_r decreases while the peak intensity increases. Although the scenario of a resonant mode as a collective magnetic mode [10, 11, 12] seems to correspond well to experimental facts, we are still missing a theoretical description starting from a strongly correlated model and explaining in particular the development of the resonant response with doping.

The general aim is to develop a theory of the dynamical spin susceptibility $\chi_\mathbf{q}(\omega)$ applicable to models of strongly correlated electrons. The prototype model, which seems to describe quite well at least several normal state properties of cuprates [7], is the well known t-J model,

$$H = -\sum_{i,j,s} t_{ij} \tilde{c}_{js}^\dagger \tilde{c}_{is} + J \sum_{\langle ij \rangle} (\mathbf{S}_i \cdot \mathbf{S}_j - \frac{1}{4} n_i n_j), \qquad (1)$$

with the nearest neighbor $t_{ij} = t$ and (for more complete comparison with experiments) the next-nearest-neighbor hopping $t_{ij} = t'$. Strong correlations among electrons are incorporated via the projected operators, e.g. $\tilde{c}_{is}^\dagger = (1 - n_{i,-s}) c_{is}^\dagger$, which do not allow for the double occupancy of sites. The latter poses an essential problem in most analytical considerations and plays also the crucial role in our further anaytical analysis.

MEMORY-FUNCTION REPRESENTATION

Within the memory function approach the dynamical spin susceptibility can be expressed in the form [13, 14]

$$\chi_\mathbf{q}(\omega) = -\langle\langle S_\mathbf{q}^z; S_\mathbf{q}^z \rangle\rangle_\omega = \frac{-\eta_\mathbf{q}}{\omega^2 + \omega M_\mathbf{q}(\omega) - \omega_\mathbf{q}^2}, \qquad (2)$$

suitable for the analysis of the collective magnetic response. Here, $\omega_\mathbf{q}$ is related to the dispersion of the collective mode provided that the mode damping is small enough, i.e. $\gamma_\mathbf{q} \sim M_\mathbf{q}''(\omega_\mathbf{q}) < \omega_\mathbf{q}$. In the opposite case, we are dealing with an overdamped mode, as seems to be generally the case for the magnetic response near the AFM wavevector $\mathbf{q} \sim \mathbf{Q}$ in the normal state of cuprates. Constants entering Eq. (2) can be expressed as

$$\eta_\mathbf{q} = -i\langle [S_{-\mathbf{q}}^z, \dot{S}_\mathbf{q}^z] \rangle, \quad \omega_\mathbf{q}^2 = \eta_\mathbf{q}/\chi_\mathbf{q}^0, \qquad (3)$$

where $\chi_\mathbf{q}^0 = \chi_\mathbf{q}(\omega = 0)$ is the static susceptibility.

For the chosen microscopic model and corresponding equations of motion (EQM) for the operator $S_\mathbf{q}^z$ the 'spin stiffenss' $\eta_\mathbf{q}$ can be expressed in terms of the static correlation functions. For the particular case of the t-J model, Eq.(1), it is directly related to the internal energy density $\eta_{\mathbf{q} \sim \mathbf{Q}} \sim -\langle H \rangle/N$. Let us suppose for a moment that we know the damping function $M_\mathbf{q}(\omega)$. Still, static $\chi_\mathbf{q}^0$ (or $\omega_\mathbf{q}$) remains to be determined. Both quantities are quite sensitive and delicate to calculate directly. Therefore, we rather fix them via the sum rule

$$\frac{1}{\pi}\int_0^\infty d\omega\, \text{cth}\frac{\omega}{2T}\chi_q''(\omega) = \langle S_{-q}^z S_q^z\rangle = C_q, \tag{4}$$

Namely, equal time correlations C_q are expected to be much more robust, i.e., less T-dependent and smoothly varying with doping. Moreover, within the t-J model they are bound by the local constraint $(1/N)\sum_q C_q = (1-c_h)/4$, where c_h is an effective hole doping.

Most evident illustration of the usefullness of the relation (4) is the underdamped situation with $\gamma, T \ll \omega_q$ where one can express the frequency of the underdamped mode as $\omega_q \sim \eta_q/(2C_q)$. Note that such an analogous relation has been used to determine the roton dispersion within the famous Feynman-Bijl approach to the theory of He II. In cuprates, such an underdamped mode and the related analysis can partly apply to the resonant peak in the SC state as elaborated later. In the normal-state, however, the mode is generally overdamped, so the consequences of the sum rule (4) are qualitatively different.

Within the memory-function formalism $M_q(\omega)$ can be expressed in terms of dynamic correlations of $\ddot{S}_q^z = -\mathcal{L}^2 S_q^z$,

$$M_q(\omega) = (Q\mathcal{L}^2 S_q^z | \frac{1}{\mathcal{L}_Q - \omega}|Q\mathcal{L}^2 S_q^z)/\eta_q, \tag{5}$$

where $\mathcal{L}A \equiv [H,A]$ and Q is a projector which removes from an operator any component proportional to either S_q^z or $\dot{S}_q^z$ [13].

For an undoped AFM at $T \sim 0$ the damping of spin excitations - AFM magnons is small. The argument can be extended to doped systems, so that the contribution to damping from the spin exchange part $\mathcal{L}_J^2$ should remain subleading. So we argue that the main contribution to the damping $M_q''(\omega)$ in the regime of interest, i.e. at $q \sim Q$ and low T, stems from the kinetic term $\mathcal{L}_t^2 S_q^z$. The evaluation of the latter requires explicit consideration of the projection of fermionic operators in the hopping part H_t, Eq.(1), and is quite tedious [14]. Its complicated form reflects the well-known involved nature of the correlated hopping in a strongly correlated system, representing a reshuffling of spins along the hole path. Since the main goal is to get the coupling to nonlocal fermionic degrees, one can replace the spin-exchange operators by their thermodynamical averages, to get

$$Q\mathcal{L}_t^2 S_q^z \sim \frac{1}{2\sqrt{N}}\sum_{ks} w_{kq s}\tilde{c}_{ks}^\dagger \tilde{c}_{k+q,s} \qquad w_{kq} = (\varepsilon_k^0 - \varepsilon_{k+q}^0)(\tilde{\varepsilon}_k^0 - \tilde{\varepsilon}_{k+q}^0) - \zeta_q, \tag{6}$$

where ε_k^0 is the 'free' band dispersion with the hopping t_{ij}, whereas $\tilde{\varepsilon}_k^0$ is defined with renormalized hopping parameters $\tilde{t}_{ij} \approx t_{ij}T_{ij}$, whereby in the region of interest $0.1 < c_h < 0.25$ we get approximately $T_1 \sim c_h$ [14]. ζ_q is determined by the condition $\sum_k w_{kq} = 0$.

Eq.(6) represents a decay of local spin variables into fermions in a doped (metallic) system. Within the lowest approximation $\Gamma_q(\omega) = M_q''(\omega)$ can be then expressed as a convolution of electron spectral functions $A_{\mathbf{k}}(\omega)$,

$$\Gamma_{\mathbf{q}}(\omega) = \frac{\pi}{2N\eta_{\mathbf{q}}\omega} \sum_{\mathbf{k}} w_{\mathbf{k}\mathbf{q}}^2 \int d\omega'[f(\omega') - f(\omega+\omega')]A_{\mathbf{k}}(\omega')A_{\mathbf{k}+\mathbf{q}}(\omega+\omega'). \qquad (7)$$

The consequence of the fermionic decay is that the damping in the normal state is large and nearly constant, both as a function of ω and T, i.e., $M_{\mathbf{q}}''(\omega) \sim \gamma$, provided that the spectral functions roughly behave as in the Fermi liquid and the Fermi surface crosses the AFM zone boundary (e.g. in the tight-binding band involving n.n.n. hopping t' near half filling). This is consistent with experiments on cuprates showing always an overdamped collective mode for $\mathbf{q} \sim \mathbf{Q}$ [9].

ω/T SCALING

In order to support the simplification of constant damping γ, we performed a numerical evaluation of $\chi(\mathbf{q},\omega)$ within the t-J model, using the finite-T Lanczos method (FTLM) [6] for a system of $N = 20$ sites on a square lattice. We use $J/t = 0.3$ as relevant for cuprates. For comparison with experiments note that $t \sim 400$meV. Data can be used to extract via Eq.(2) the damping function $M_{\mathbf{q}}''(\omega)$ [15]. For $T > T_{fs}$, where the latter represents the 'finite-size' temperature, one can conclude that in the presented case we are clearly dealing with an overdamped spin dynamics for all $\mathbf{q} \sim \mathbf{Q}$. In spite of widely different $\chi_{\mathbf{q}}''(\omega)$ the damping function $M_{\mathbf{q}}''(\omega)$ is nearly constant in a broad range of $\omega < t$ and almost independent of $\mathbf{q}$. Hence, it is meaningful to extract $\gamma_{\mathbf{q}} = M_{\mathbf{q}}''(\omega = 0)$, which we present in Fig. 1 for $\mathbf{q} = \mathbf{Q}$ and various doping $c_h = N_h/N \leq 0.15$ as a function of T.

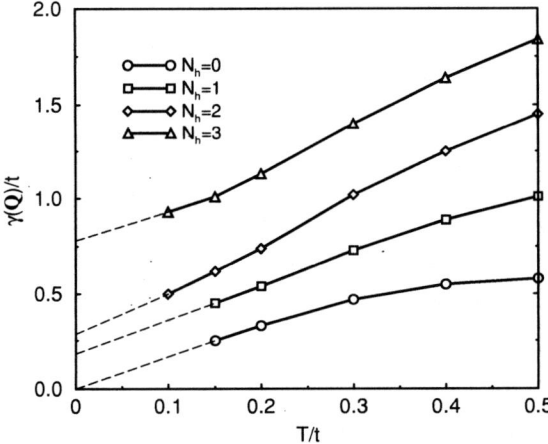

FIGURE 1. Low-frequency damping $\gamma_{\mathbf{Q}}$ within the t-J model as a function of T for various doping $c_h = N_h/N$.

Our results in an undoped system are consistent with the vanishing $\gamma_{\mathbf{Q}}(T \to 0)$. On the other hand, for doped systems with $c_h > 0$ data indicate clearly a finite extrapolated value $\gamma_{\mathbf{Q}}(T \to 0)$. For the T window of interest our results confirm that in a doped system the simplification of constant γ is sensible, leading to a resonance form

$$\chi_\mathbf{q}''(\omega) = \frac{\eta\gamma\omega}{(\omega^2 - \omega_\mathbf{q}^2)^2 + \gamma^2\omega^2}. \tag{8}$$

Assuming that static correlations take the standard Lorentzian form, $C_\mathbf{q} = C/(\kappa^2 + |\mathbf{q} - \mathbf{Q}|^2)$ where κ saturates at low T [1, 3], we arrive to an unusual behavior governed by the parameter $\zeta = C\pi\gamma/(\eta\kappa^2)$. It appears that realistic situations at low- to intermediate-doped cuprates correspond to $\zeta \gg 1$. In this case we can show that the sum rule (4) requires strong T-variation of $\omega_\mathbf{Q}$ and consequently leads to a nearly universal scaling behavior of the local spin response $\chi_L''(\omega, T) = (1/N)\sum_\mathbf{q}\chi_\mathbf{q}''(\omega) \sim \chi_L''(\omega, 0) f(\omega/T)$. The calculated scaling function, presented in Fig. 2, indeed remains essentially the same in the broad range of $T_p < T < \gamma$ and corresponds well to the experimentally observed behavior in YBCO [3]. A nontrivial quantity which is also the consequence of the presented analysis is the local $\chi_L''(\omega, 0) = I(\omega)$. The latter is related to the measured 'normalization' function [1, 3], which was so far not explained in any way while our results [15] qualitatively reproduce the 'anomalous' increase at $\omega \to 0$.

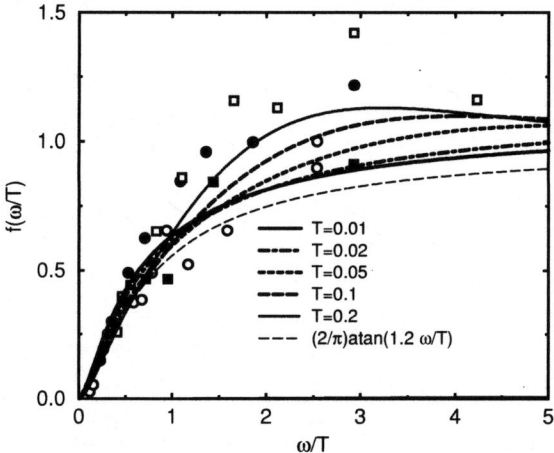

FIGURE 2. Scaling function $f(\omega/T)$ for $\zeta = 8$ for different T (in units of t). For comparison results for Zn-substitued YBCO are also plotted, as measured for different energies ω, taken from Ref.[3].

Let us further discuss the behavior at low $T \to 0$. In this case the l.h.s. of Eq.(4) can be explicitly integrated, and for $\omega_\mathbf{q} < \gamma$ we get $C_\mathbf{q} \sim (2\eta/\pi\gamma)\ln(\gamma/\omega_\mathbf{q})$. The relevant quantity is the peak frequency $\omega_p = \Gamma_\mathbf{Q}(T \to 0)$. We see that the crucial parameter is ζ which exponentially renormalizes ω_p,

$$\omega_p \sim \gamma e^{-2\zeta}, \tag{9}$$

Since C is fixed by the total sum rule, i.e., $C \sim (1 - c_h)\pi/(2\ln(\pi/\kappa)) \sim O(1)$ and $\eta \sim 0.2\,t$ at low doping, ζ is effectively governed by the ratio γ/κ^2. Our results for the t-J model, as presented above as well as the analysis of experiments on cuprates, indicate that generally $\zeta \gg 1$ in the regime of low to intermediate doping. This means that the crossover to the 'normal' Fermi liquid behavior happens at very low $T_p \sim \omega_p$, which is either not reachable or masked by a phase transition, e.g., to SC or charge (or stripe) ordering. On the other hand, in the regime beyond the optimum doping one can plausibly expect a steady decrease of ζ hence also a crossover to a more 'normal' Fermi liquid where finally $\chi''(\mathbf{q}, \omega)$ is rather T-independent for low T, hence T_p is large.

RESONANT PEAK

Let us turn to the SC phase. To get un underdamped resonant mode at ω_r one needs a depleted damping $\Gamma_\mathbf{Q}(\omega_r)$. The latter can evidently arise from the SC gap [11, 12] which we introduce via an effective d-wave gap $\Delta_\mathbf{k} = \Delta_0(\cos k_x - \cos k_y)/2$ into the spectral functions $A_\mathbf{k}(\omega)$ entering Eq.(7). Such an analysis is well suited close to *optimum doping* where spectral functions appear more coherent already in the normal state. The SC gap eventually leads to the vanishing of damping for $\omega < \omega_\mathbf{Q}^* < 2\Delta_0$. At $\mathbf{q} = \mathbf{Q}$ the resonant mode is undamped since $2\Delta_0 > \omega_r$. However, due to quite large $\Gamma_\mathbf{Q}(\omega > \omega_\mathbf{Q}^*)$ the resonant frequency is significantly renormalized,

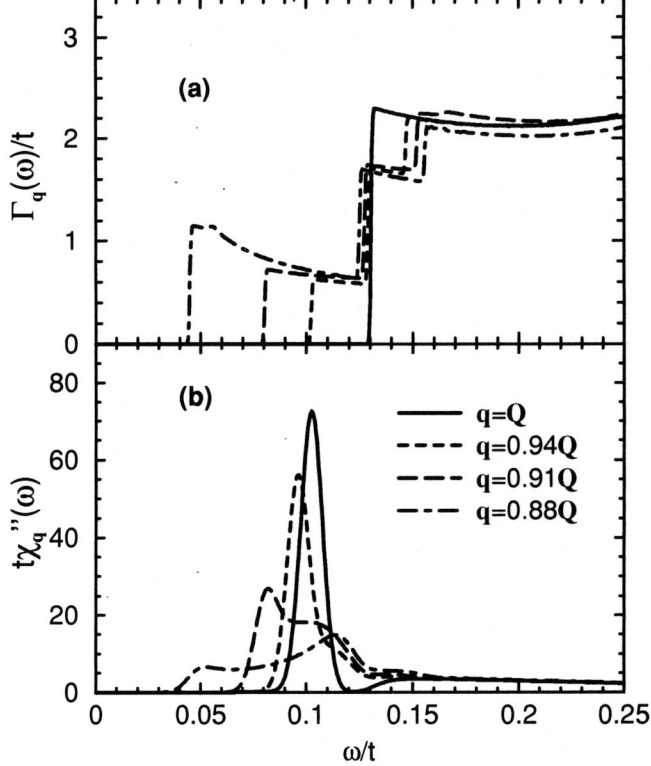

FIGURE 3. (a) Damping function $\Gamma_\mathbf{q}(\omega)$ at optimal doping $c_h \sim 0.2$ for various $\mathbf{q} \parallel \mathbf{Q}$ for a d-wave SC, (b) corresponding spin response $\chi_\mathbf{q}''(\omega)$.

$$\omega_r = \omega_Q \left[1 + M'_Q(\omega)/\omega|_{\omega=\omega_r}\right]^{-1/2}, \tag{10}$$

i.e. $\omega_r < \omega_Q$, while its intensity $I_Q \sim C_Q \omega_r/\omega_Q$ is simultaneously reduced. Characteristic results for $\Gamma_Q(\omega)$ and $\chi''_q(\omega)$ in the optimum doping regime are presented in Fig. 3. Moving away from **Q** the mode gets overdamped and we establish as well a downward dispersion of the resonant peak [14]. Presented spectra are well consistent with experiments on YBCO [9].

Analysing the resonant peak at *low doping* it is crucial that the normal-state γ also decreases with doping. Still it seems clear that here the experimental data [9] cannot be explained with a single gap only. In this regime we can argue that instead of coherent approximation (7) it makes more sense to assume the **k**- incoherence of spectral functions, i.e., $A(\mathbf{k}, \omega) \sim \mathcal{N}(\omega)$. Consequently the damping behaves very similarly to the c-axis conductivity $\sigma_c(\omega)$ [16, 14]. In the latter quantity the pseudogap ω_{pg} is clearly observed at low $T < T_{pg}$ in the underdoped regime. We can therefore state some particular features in the magnetic response in underdoped systems [14], which differ from optimum doping: a) The resonant peak is damped even for $T < T_c$, but still underdamped. b) The spin response and the sum rule for $\chi''_Q(\omega)$ are essentially exhausted within $\omega < \omega_{pg}$.

REFERENCES

1. B. Keimer *et al.*, Phys. Rev. Lett. **67**, 1930 (1991); Phys. Rev. B **46**, 14034 (1992).
2. for a review see M. A. Kastner, R. J. Birgeneau, G. Shirane, and Y. Endoh, Rev. Mod. Phys. **70**, 897 (1998).
3. K. Kakurai *et al.*, Phys. Rev. B **48**, 3485 (1993).
4. C. M. Varma, P. B. Littlewood, S. Schmitt-Rink, E. Abrahams, and A. E. Ruckenstein, Phys. Rev.
5. S. Chakravarty, B. I. Halperin, and D. R. Nelson, Phys. Rev. Lett. **60**, 1057 (1988); Phys. Rev. B **39**, 2344 (1989).
6. J. Jaklič and P. Prelovšek, Phys. Rev. Lett. **75**, 1340 (1995).
7. J. Jaklič and P. Prelovšek, Adv. Phys. **49**, 1 (2000).
8. J. Rossat-Mignod *et al.*, Physica C **185 - 189**, 86 (1991).
9. P. Bourges, in *The Gap Symmetry and Fluctuations in High Temperature Superconductors*, Ed. J. Bok *et al.* (Plenum Press, New York, 1998).
10. M. Lavagna and G. Stemmann, Phys. Rev. B **49**, 4235 (1994); D. Z. Liu, Y. Zha, and K. Levin, Phys. Rev. Lett. **75**, 4130 (1995); I. I. Mazin and V. M. Yakovenko, Phys. Rev. Lett. **75**, 4134 (1995).
11. D. K. Morr and D. Pines, Phys. Rev. Lett. **81**, 1086 (1998).
12. A. Abanov and A. V. Chubukov, Phys. Rev. Lett. **83**, 1652 (1999).
13. H. Mori, Prog. Theor. Phys. **33**, 423 (1965).
14. I. Sega, P. Prelovšek, and J. Bonča, cond-mat/0211090.
15. P. Prelovšek, I. Sega, and J. Bonča, cond-mat/0306366.
16. P. Prelovšek, A. Ramšak, and I. Sega, Phys. Rev. Lett. **81**, 3745 (1998).

A criterion for the transition of a three dimensional Bravais lattice from bulk to molecular behaviour

M Ghanashyam Krishna and V Srinivasan

School of Physics, University of Hyderabad, Hyderabad-500 046, India

Abstract This paper addresses the question of "how large is large enough for a crystallite to be a three dimensional Bravais lattice?" Based on the premise that the ratio of the bulk volume to that of the volume of the unit cell for a given material should determine the transition to a large molecule, it is proposed that for values of the ratio $< 10^6$ crystallites can be considered as large molecules. The universality of this criterion has been established by examples from magnetism, optics, ferroelectrics and superconductivity. It is expected, due to the wide variety of materials considered, that other classes of inorganic materials would also obey this rule leading to a quantitative definition for a nanocrystal or nanostructure.

INTRODUCTION

In solid state physics/crystal structure, when group theory is applied, an oft-repeated standard theorem based on the fact that a translation and a rotation together should lead to another allowed translation, only allows rotations C_1, C_2, C_3, C_4 and C_6[1]. This result became questionable in the 80's when quasi-crystals were discovered. Rotational symmetries of the order $2\pi/n$ where n is none of the above were also observed. Various explanations were given, including that of two intertwining Bravais lattices or Penrose tiling[2]. In this paper we examine this phenomenon purely on an operational/experimental point of view. We notice in the usual proofs that the lattices of infinite size are used whereas the crystals are of a finite size. It is therefore, worthwhile examining as to when the system is large enough for this theorem to be applicable To that end we suggest that the ratio of the volume of unit cell to that of the sample used in the experiment is a good criterion. Molecular behaviour is exhibited below a critical value of this ratio. This can be regarded as a universal parameter, if, as will be demonstrated in this paper, several observed phenomena based on published experimental data obey this rule. The present criterion will not place the quasicrystals as a peculiar feature but demonstrate that they can be regarded as large molecules. Indeed any substance should exhibit such behaviour if this criterion is adopted and samples are made of the appropriate size.

Examples for phenomena such as superparamagnetism, ferroelectric-superparaelectric transition, quantum confinement, occurrence of single domain particles in magnetic materials will be given in the following section to establish the ratio. The values used are based on reported experimental data and can therefore be considered realistic estimates.

EXPERIMENTAL DATA

Superparamagnetism is a well-documented phenomenon wherein a previously ferromagnetic material with a finite coercivity undergoes a transition to a state of zero coercivity below a certain critical diameter of particles[3]. This behaviour is observed when single domain particles are of a size that even small thermal fluctuations can cause reversal of magnetization. As a consequence, in the presence of an applied field, though the moments tend to align themselves with the direction of the field, thermal fluctuations will tend to misalign them, a behaviour reminiscent of a normal paramagnetic substance. The behaviour of such particles is therefore described by the term, superparamagnetism. It has been extensively investigated for many years and has seen a resurgence of interest due to its importance in nanotechnology. Superparamagnetism has been observed in compounds as well as alloys. A recent example is of a study on $CoFe_2O_4$ that is a partially inverse spinel type ferrite and a soft magnetic substance that has found application in a number of devices[4]. It has a lattice parameter[5] of 8.39 Å and the volume of the unit cell is 590.59 Å^3. In this study Rajendran et al have found that below a crystallite size of the order of 30 nm the material shows a transition to superparamagnetic behaviour. Figure 1 shows the variation in coercivity as a function crystallite size replotted using data reported in this work. Quite clearly, the coercivity goes to zero below 30 nm.

From these values the ratio of the bulk volume to that of the unit cell can easily be calculated.

Bulk $CoFe_2O_4$/ Unit cell of $CoFe_2O_4$ = $27 \times 10^{-18} / 590.59 \times 10^{-24}$
$$= 0.046 \times 10^6$$

Therefore, for values of the ratio less than 10^6 the transition to molecular behaviour is observed.

The next example is the observation of quantum confinement in CdTe thin films[6]. Quantum confinement is a purely size dependent phenomenon that has been observed in a variety of semiconducting materials including Si, CdS and CdTe[7]. The confinement effect is manifested in a so-called "blue shift" of the optical band gap of the semiconductors.

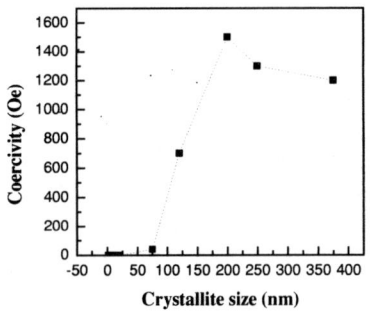

FIGURE 1. Variation in coercivity as a function of crystallite size for $CoFe_2O_4$. (Replotted from data presented in reference 4). Dotted line is only a guide to the eye.

This behaviour has been modeled, most frequently, using the effective mass approximation. (EMA) given in equation (1)[6].

$$\Delta E_g = (\hbar^2\pi^2/2\mu R^2) - 1786(e^2/\varepsilon R) - 0.248 R_x \qquad (1)$$

EMA is used to estimate the radius and binding energies of shallow trapped electrons or holes and excitons and uses the effective mass instead of the free electron mass to account for the interactions with the lattice. The effective mass is related to the curvature of the conduction and valence bands, that below a critical size begin to overlap causing an increase in the optical band gap. It has been shown in this study and in figure 2 using data from the study, that below a crystallite radius of 10 nm onset of the blue shift is observed. CdTe has a cubic structure[8] with a unit cell parameter of 6.41 Å and a unit cell volume of 263.37 Å^3 From this and the point of onset, as in the earlier case it can be shown that the ratio of the bulk volume to unit cell at onset is

$$(200 \times 10^{-8})^3 / 163 \times 10^{-24} = 0.03 \times 10^6$$

The third example is from the size dependence of ferroelectric behaviour in $PbTiO_3$[9]. It has been observed that the ferroelectric state undergoes a transition to a superparoic state characterised by a zero spontaneous polarization and disappearance of hysteresis. The critical size for the onset of this behaviour was found to be 80 Å. $PbTiO_3$ possesses a tetragonal crystal structure[10] with a= b =3.904 Å and c= 4.152 Å. The unit cell volume is 63.28 Å^3. Again using ratio of bulk to unit cell volume at the point of onset of molecular behaviour it can be shown that value is

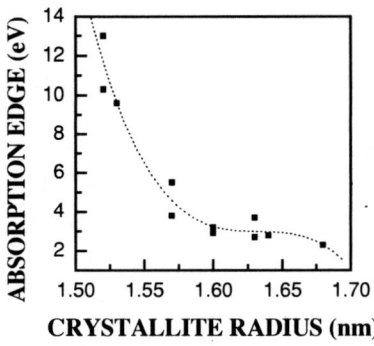

FIGURE 2. Variation in absorption edge as a function of crystallite radius for CdTe films. (Replotted from data presented in reference 6). Dotted line is only a guide to the eye

$$512 \times 10^{-21} / 63.28 \times 10^{-24} = 0.008 \times 10^6$$

A further example of transition to single domain particles is from a study on γ-Fe_2O_3 where the onset was found to be at 30 nm[11]. Haematite[12] is rhombohedral in structure and its lattice parameters are a=b= 5.0355 Å and c=13.7471 Å. Unit cell volume is 301.87 Å^3. The ratio of bulk to unit cell volume at onset is
$$27 \times 10^{-18} / 301.27 \times 10^{-24} = 0.09 \times 10^6$$

Similarly in the case of YBCO high T_c thin films suppression of superconductivity has been demonstrated for 4-6 unit cell thick layers on PBCO films[13]. Whilst it is difficult to isolate proximity effects from these observations, it is clear that one can again expect deviation to molecular behaviour at a similar ratio. Finite size effects have, however, been reported in YBCO powders at around 75 nm[14]. The bulk to unit cell volume ratio in this case would be approximately 10^6

DISCUSSION

We again address the question as to how large is a three dimensional molecule. The strongest bond known in molecular/quantum physics is the covalent bond. The maximum radius over which it is effective is of the order of 60 Å, so the volume is of the order $\leq 10^6$ Å^3. Significantly, the unit cell in the 3D Bravais lattice is of the same order of magnitude in volume. In the formation of a lattice, electrostatic potential plays a major role; however, this interaction is much smaller in strength than that of a covalent bond. Since there is a limit to the radius of a covalent bond, the size of the molecule is also restricted. Therefore, the ratio of the size of the sample to that of the unit cell, i.e. 10^6, from experiments indeed agrees with this semi-quantitative argument. For sizes $\leq 10^6$ covalent bond can be formed and the molecular nature takes over; this also reminds us of nuclear forces that take over at distances $\leq 10^{-13}$ cms. The atomic size is of the order of 10^{-8} cms. The ratio in this case turns out to be 10^5, a fact used in the scale and formation of structures by Calogero[15]. In the present work we see that even in the formation of nanostructures such a ratio of 10^6 emerges. We regard this ratio as an independent universal parameter. The criterion given here can be used for preparing any nanostructured material. This is one of the predictions that arise from our observations.

CONCLUSIONS

From the examples given in the previous section it is clear that for values of the ratio of the bulk volume to unit cell volume less than 10^6 a three dimensional Bravais lattice will exhibit molecular behaviour independent of the type of material and physical phenomenon studied. This, we believe, should lead to a universal definition that when the solid has less than 100 molecules or crystallites the transition from a 3D Bravais lattice to molecular behaviour occurs.

REFERENCES

1. See for example Introduction to Solid State Physics by Kittel C (john Wiley, New York, USA, 1996)
2. See for example Quasicrystals: The State of the Art (Series on Directions in Condensed Matter Physics, Vol. 16), Edited by DiVincenzo D P and Steinhardt P J (World Scientific Publishing, Singapore, 1999).
3. See for example "Introduction to Magnetic Materials" by Cullity B D (Addison Wesley, USA, 1972)
4. Rajendran M, Pullar R C, Bhattacharya A K, Das D, Chintalapudi S N and Majumdar C K, J. Mag.Mag.Mater., **232**, 71 (2001).
5. Powder diffraction file JCPS No.22-1086
6. Arizpe-Chavez H.,.Ramirez-Bon R, Espinoza-Beltran F.J., Zelaya-Angel O., Marin J.L.and Riera R., J.Phys.Chem.Solids, **61**, 511 (2000).
7. Freedhoff M.I and Marchetti A P, in Handbook of Optical Properties Vol.II Optical Properties of Small particles Interfaces and Surfaces edited by.Hummel R.E, Wisbmann P. (CRC Press, USA, 1997).
8. Powder Diffraction files JCPDS No. **75-2086**
9. Ishikawa K Nagareda, K, J. Kor.Phys. Soc., **32**, 56 (1998).
10. Powder Diffraction Files JCPDS No.**72-1135**
11. Berkowitz A E, Schule W J, Flanders P J, J.Appl.Phys., **39**, 1261 (1968).
12. Powder Diffraction files JCPDS No. **86-0550**
13. Triscone J M, Fischer O, Antognazza L., Brunner O., Kent A D, Mieville L,

Karkut M G in, Science and technology of Thin Film Superconductors 2, Edited by McConnell and Noufi R (Plenum Press, New York, 1990).
14. Multani M S, Guptasarma P, Physics Letters, **142**, 293 (1989).
15. Calogero F., Phys. Lett. A, **228**, 335 (1997).

Bose-Einstein Condensation in Thermo Field Dynamics

Hideki Matsumoto

Institute of Physics, University of Tsukuba, Ibaraki 305-8571, Japan

Abstract. In the development of Thermo Field Dynamics (the quantum field theory at finite temperature and density), Prof. Mancini acted as one of the originators and played the important role for its birth through the work of superconductivity. Recent development of the Bose-Einstein condensation in the dilute alkaline atomic gas enables us to investigate deeply the quantum coherence in bose and fermi particle systems, and gives us a plenty of experimental data even for the problem in nonequilibrium formation of the condensate. There the relaxation time is reasonably slow to make the laboratory observation possible. In this talk, we will show that Thermo Field Dynamics gives us a reasonable approximation scheme to describe time-dependent ordered states with spatial inhomogeniety by use of a single-time quasi-particle field equation, which contains the information of the spatial and temporal dependent order parameter, particle energy and particle distribution.

```
http://www.px.tsukuba.ac.jp/tcm/matumoto/matumoto-E.html.
```

INTRODUCTION

It is my pleasure to give a talk in the conference honoring the Prof. F. Mancini's 60th birthday. I met him in 1973 in the first time at the University of Wisconsin-Milwaukee. Since then Prof. Mancini and I have been good friends and collaborators of research; I can count the co-authored papers with him as 35.

Before starting my talk, I would like to present a short summary of Prof. Mancini's work. His PhD thesis was on the Boson method in superconductivity under the supervision of Prof. H. Umezawa, and its work was summarized in ref.[1]. There they summarized their field theoretical deviation of the Maxwell equation in superconductivity. They obtained the Pippard-type kernel called the C-function, and it has turned out in later works that this C-function describes the spatial distribution of the magnetic flux around the single superconducting vortex, and is useful to study propertied of vortex lattices[2]. In the same time, this issue of Phys. Rep. is important in the sense that it contains the original idea of finite temperature quantum field theory in real time, now so-called Thermo Field Dynamics (TFD) [3, 4, 5]. In 1973-1985, Prof. Mancini's works were naturally on the superconductivity. The boson method of superconductivity worked as a powerful tool to analyze the newly discovered magnetic superconductors at 1978. The interplay between magnetism and superconductivity was extensively studied by use of the boson method[6]. In 1986-1988, I and Prof. Mancini extended TFD to nonequilibrium cases [7, 8, 9, 10, 11, 12] and my talk is based on the formulation developed in this period. In 1986, the high Tc superconductors were discovered. From 1987, when the LT conference was held at Kyoto (Japan), we started to work on the high T_c superconductor and highly electron systems, but real works on this area have started in Mancini's group from 1994. They had worked on this problem by the method, called the composite operator method[13, 14]. He proposed the idea of using constraints on operators [15, 16]. About 55 papers were produced after that in the Salerno group on highly correlated electron systems, and the content of the development will be talked by Prof. Mancini, himself, in this conference. Therefore I will present in the present talk, another aspect of his work, that is, the finite temperature field theory, TFD. I will show that TFD works as a powerful tool to analyze nonequilibrium phenomena in systems of Bose-Einstein condensation (BEC), recently realized in dilute alkaline atomic gases.

NONEQUILIBRIUM PHENOMENA IN BEC

After the experimental realization of the Bose-Einstein condensation (BEC) in dilute alkaline atomic gasses [17, 18, 19, 20], a large number of theoretical and experimental studies of the quantum nature of bose systems have been performed[21, 22]. In those systems, the interaction among atoms are very small, and the thermalization process of condensate formation following the fast evaporation of laser-cooled alkaline atomic gases has been observed [23] in a reasonably long time scale of 10-100ms, and it has been shown that one can obtain detailed experimental information on time-dependence of the spatial particle distribution and formation of the condensate. This gives us a good laboratory system for checking nonequilibrium theories in the Bose-Einstein condensed state. A magnetically trapped alkaline bose gas can be described by a simple model in which a single component complex quantum field interacts through a very weak repulsive force. But the presence of the trapping potential inevitably introduces a strong spatial dependence of the particle distribution. Therefore, it is necessary to develop a nonequilibrium theory that can treat the strong space-time dependence of particle distributions. We will show that TFD provides us a convenient theoretical framework to handle such a situation and that one can describe in a reasonable approximation the quasi-particle field equation that contains the information of time-dependent energy, decay width and particle distribution.

The scenario of condensate formation has been discussed by several authors [24, 25, 26, 27, 28, 29, 30, 31, 32]. The Schwinger-Keldysh path-ordered formalism is developed in Ref. [25], quantum kinetic theory is used in Refs. [30, 31, 32] and some successes have been reported. A field theoretical method is used in Refs. [33, 34]. Here in this talk, we will follow totally a field theoretical method[34].

For definiteness, we consider the single self-interacting complex field with a time-dependent disturbance,

$$H = \int d^3x \left[\Psi^\dagger(\vec{x})\varepsilon_0(-i\vec{\nabla},\vec{x};t)\Psi(\vec{x}) + \frac{1}{2}V\Psi^\dagger(\vec{x})^2\Psi(\vec{x})^2 \right], \quad (1)$$

where $\varepsilon_0(-i\vec{\nabla},\vec{x};t)$ contains the kinetic energy, time-dependent trapped potential, and the chemical potential. The interaction coupling constant V is originated from the atom-atom s-wave scattering, and is given by $V = \frac{4\pi\hbar^2 a}{m}$, where a is the s-wave scattering length, m is the mass of atoms. For a harmonic trapped potential $U(\vec{x}) = \frac{m}{2}\omega_h^2 \vec{x}^2$, the length scale is $a_h = \sqrt{\frac{\hbar}{m\omega_h}}$. The typical ratio of a and a_h is $\frac{a}{a_{h0}} \approx 0.5 \times 10^{-3}$, and the effective boson-boson interaction is very small[22].

The Heisenberg field operator $\Psi(\vec{x},t)$ is rewritten as

$$\Psi(\vec{x},t) = v(\vec{x},t) + \psi(\vec{x},t) \tag{2}$$

with

$$v(\vec{x},t) = <\Psi(\vec{x},t)> . \tag{3}$$

Then we can ask the question how the quantum excitation $\psi(\vec{x},t)$ is modified in the background $v(\vec{x},t)$ and vice versa. This kind of question appears also in other field of physics, such as time-development of chiral symmetry spontaneous breakdown in elementary particle physics (relating multi-particle production in heavy-ion collision), time-evolution after big-bang, explosion of supernova in cosmology, and time-dependent control of quantum devices, and so on. In the present BEC system, one has to handle strong spatial dependence due to the presence of a trapped potential. However, since the self-interaction is small, and the time-variation is slow, one can expect that a quasi-particle picture may still work. We seek the approach to describe the phenomena in terms of coupled equations of the order parameter $v(\vec{x},t)$ and a suitably defined quasi-particle field $\varphi(\vec{x},t)$.

The formalism used in this talk was developed in refs. [8, 9, 11, 12, 34]. The assumed situation is the following. We set up a state at $t = 0$, and let the system to develop by the Hamiltonian H_t, which is given by field operators at $t = 0$. The initial state is specified by a certain hermitian operator Ω, which, if we consider a correspondence with the density matrix ρ, is given by $\rho = e^{-\Omega}/Tr[e^{-\Omega}]$[10].

The Heisenberg operator $\Psi(\vec{x},t)$ is defined by

$$\Psi(\vec{x},t) = U(t,0)^{-1}\Psi(\vec{x},0)U(t,0) \tag{4}$$

with

$$U(t,0) = T\exp\{-\frac{i}{\hbar}\int_0^t dt' H_{t'}\} . \tag{5}$$

The field operator Ψ satisfies the Heisenberg equation

$$i\hbar\frac{\partial}{\partial t}\Psi(\vec{x},t) = [\Psi(\vec{x},t), H(t)] \tag{6}$$

with

$$H(t) = U(t,0)^{-1} H_t U(t,0) . \tag{7}$$

Since the order parameter contains an explicit time-dependence, it is dangerous to set up the unperturbed part only for the quantum fluctuation part $\psi(\vec{x},t)$. In fact, the time-dependence of $\psi(\vec{x},t)$ is not fully generated by the Hamiltonian,

$$i\frac{\partial}{\partial t}\psi(\vec{x},t) \neq [\psi(\vec{x},t).H(t)]. \tag{8}$$

This consideration leads us the following conditions to set up the unperturbed Hamiltonian H_{t0}[34]; 1) H_{t0} should be written in terms of the canonical variables $\Psi(\vec{x},0)$ and $\Psi^\dagger(\vec{x},0)$, and 2) $\Psi(\vec{x},t)$ generated by H_{t0} should allow the decomposition Eq. (2), giving the full $v(\vec{x},t)$ and $\psi(\vec{x},t)$ in the interaction representation. As the most general expression of H_{t0} constructed from the bilinear form of $\Psi(\vec{x},0)(=\Psi(\vec{x}))$, we have

$$\begin{aligned}H_{t0} &= \int d^3x[\Psi^\dagger(\vec{x})C(\vec{x};t)+C^*(\vec{x};t)\Psi(\vec{x})] + \int d^3x d^3y[\Psi^\dagger(\vec{x})\varepsilon(\vec{x},\vec{y};t)\Psi(\vec{y})\\ &+ \frac{1}{2}(\Psi(\vec{x})\Delta^*(\vec{x},\vec{y};t)\Psi(\vec{y})+\Psi^\dagger(\vec{x})\Delta(\vec{x},\vec{y};t)\Psi^\dagger(\vec{y}))].\end{aligned} \tag{9}$$

Then for the field $\Psi(\vec{x},t)$ in the interaction representation,

$$\Psi(\vec{x},t) = U_0(t)^{-1}\Psi(\vec{x},0)U_0(t) \quad , \quad U_0(t) = T\exp\{-i\int_0^t dt' H_{t'0}\}, \tag{10}$$

we have

$$i\hbar\frac{\partial}{\partial t}\Psi(\vec{x},t) = \int d^3y[\varepsilon(\vec{x},\vec{y};t)\Psi(\vec{y},t)+\Delta(\vec{x},\vec{y};t)\Psi^\dagger(\vec{y},t)]+C(\vec{x};t). \tag{11}$$

The above equation allows us to define the decomposition $\Psi(\vec{x},t) = v(\vec{x},t)+\psi(\vec{x},t)$ in the interaction representation, leading to the equations

$$i\hbar\frac{\partial}{\partial t}v(\vec{x},t) = \int d^3y[\varepsilon(\vec{x},\vec{y}:t)v(\vec{y},t)+\Delta(\vec{x},\vec{y};t)v^*(\vec{y},t)]+C(\vec{x};t), \tag{12}$$

$$i\hbar\frac{\partial}{\partial t}\psi(\vec{x},t) = \int d^3y[\varepsilon(\vec{x},\vec{y}:t)\psi(\vec{y},t)+\Delta(\vec{x},\vec{y};t)\psi^\dagger(\vec{y},t)]. \tag{13}$$

The interaction Hamiltonian, $H_I(t) = U_0(t)^{-1}(H_t - H_{0t})U_0(t)$, is given by

$$\begin{aligned}H_I(t) &= \int d^3x V\left(v(\vec{x},t)^*\psi^\dagger(\vec{x},t)\psi(\vec{x},t)^2 + v(\vec{x},t)\psi^\dagger(\vec{x},t)^2\psi(\vec{x},t)\right)\\ &+ \int d^3x \frac{1}{2}V\psi^\dagger(\vec{x},t)^2\psi(\vec{x},t)^2 + \int d^3x d^3y\left[\psi^\dagger(\vec{x},t)\delta\varepsilon(\vec{x},\vec{y};t)\psi(\vec{y},t)\right]\\ &+ \int d^3x d^3y\left[\frac{1}{2}\left(\delta\Delta(\vec{x},\vec{y};t)^*\psi(\vec{x},t)\psi(\vec{y},t)+\delta\Delta(\vec{x},\vec{y};t)\psi^\dagger(\vec{x},t)\psi^\dagger(\vec{y},t)\right)\right]\\ &+ \int d^3x\left[\delta C(\vec{x};t)^*\psi(\vec{x},t)+\psi^\dagger(\vec{x},t)\delta C(\vec{x};t)\right],\end{aligned} \tag{14}$$

where

$$\delta\varepsilon(\vec{x},\vec{y};t) = \left[-\frac{\hbar^2}{2m}\vec{\nabla}^2 + U(\vec{x}) - \mu(t) + 2Vv(\vec{x},t)^*v(\vec{x},t)\right]\delta(\vec{x}-\vec{y})$$
$$-\varepsilon(\vec{x},\vec{y};t), \qquad (15)$$
$$\delta\Delta(\vec{x},\vec{y};t) = Vv(\vec{x},t)^2\delta(\vec{x}-\vec{y}) - \Delta(\vec{x},\vec{y};t), \qquad (16)$$
$$\delta C(\vec{x},t) = \left[-\frac{\hbar^2}{2m}\vec{\nabla}^2 + U(\vec{x}) - \mu(t)\right]v(t,\vec{x}) - \varepsilon(t)v(\vec{x},t) - \Delta(t)v^*(\vec{x},t)$$
$$+ Vv(\vec{x},t)^*v(\vec{x},t)^2 - C(\vec{x};t). \qquad (17)$$

INCLUSION OF PARTICLE DISTRIBUTION AND TFD

The origin of the temporal and spatial variation of the order parameter $v(\vec{x},t)$ is the particle distribution in space-time. To obtain the quasi-particle field in nonequilibrium quantum field theory, TFD provides a powerful theoretical framework. Thermo field dynamics was developed as a finite temperature quantum field theory in refs. [1, 3, 4]. Its nonequilibrium extension has been studied in refs. [35, 7, 8, 9, 10, 11, 36]. We follow the formulation developed in refs. [8, 9, 11, 12].

In order to go from the usual quantum field theory to nonequilibrium quantum field theory with particle distribution, that is TFD, the following two frameworks are set as axioms. The first is the rules of the doubling of the field freedom: Any operator set$\{A\}$ is doubled as $\{A\}$ and $\{\tilde{A}\}$ and tilde and nontilde operators are independent,

$$[A,\tilde{B}]_\pm = 0. \qquad (18)$$

The doubling follows the tilde conjugation rule;

1. $[AB]^\sim = \tilde{A}\tilde{B}$
2. $[c_1 A_1 + c_2 A_2]^\sim = c_1^* \tilde{A}_1 + c_2^* \tilde{A}_2$
3. $[\tilde{A}]^\sim = A$, $[\tilde{A}^\dagger]^\sim = A^\dagger$

It is convenient to use the thermal doublet notation for this doubled operators,

$$A^a = A \ (a=1), \ \tilde{A}^\dagger \ (a=2). \qquad (19)$$

The other is the condition on the vacuum. The bra- and ket vacuums $<1|$ and $|\rho>$ are introduced. They are invariant under the tilde conjugation,

$$[<1|]^\sim =<1| \quad , \quad [|\rho>]^\sim =|\rho>. \qquad (20)$$

The state of the vacuum is set as the thermal state condition, which relates tilde and nontilde operators operating on the vacuum,

$$A|\rho> = \sigma e^{\hat{\Omega}}\tilde{A}^\dagger|\rho> \quad , \quad <1|A^\dagger =<1|\sigma^*\tilde{A}, \qquad (21)$$

where $\sigma = 1$(for boson) and i(for fermion). The operator $\hat{\Omega}$ is given by

$$\hat{\Omega} = \Omega - \tilde{\Omega}, \tag{22}$$

and it is related the initial density matrix as[11]

$$<A> = <1|A|\rho> = Tr[e^{-\Omega}A]/Tr[e^{-\Omega}]. \tag{23}$$

When the initial state is set by the thermal state condition, Eq. (21), we can develop a perturbation theory by using the unperturbed vacuum specified by Ω_0 and the time-dependent Hamiltonian[8, 9],

$$\hat{\mathcal{H}}_t = \hat{H}_t \theta(t) + \theta(-t)\beta^{-1}\hat{\Omega}, \tag{24}$$

where

$$\hat{H}_t = H_t - \tilde{H}_t \tag{25}$$

and β is a suitably chosen temperature parameter. We have the following formula of the perturbation theory,

$$<T\Psi^{a_1}(t_1)\cdots\Psi^{\dagger a_n}(t_n)>_H$$
$$= \frac{<T exp\{-i\int_{-\infty}^{\infty} dt \hat{H}_I(t)\}\Psi^{a_1}(t_1)\cdots\Psi^{\dagger a_n}(t_n)T_c exp\{-i\int_{-\infty}^{-\infty+i\beta} dzH_I(z)\}>}{<T_c exp\{-i\int_{-\infty}^{-\infty+i\beta} dzH_I(z)\}>} \tag{26}$$

Let us introduce a bosonic doubled Φ defined by

$$\Phi = \begin{pmatrix} \Psi \\ \tilde{\Psi}^\dagger \end{pmatrix} \tag{27}$$

in order to handle Eq. (13). Its thermal extension is

$$\Phi^a = \Phi \ (a=1), \ \tilde{\Phi}^\dagger \ (a=2). \tag{28}$$

The canonical relation is

$$[\Phi^a(\vec{x},t), \Phi^{\dagger b}(\vec{y},t)] = \tau^{ab}\rho_3\delta(\vec{x}-\vec{y}), \tag{29}$$

where

$$\tau = \begin{pmatrix} 1 & 0 \\ 0 & -1 \end{pmatrix}, \quad \rho_3 = \begin{pmatrix} 1 & 0 \\ 0 & -1 \end{pmatrix}. \tag{30}$$

We write

$$\Psi^a(\vec{x},t) = v(\vec{x},t) + \psi^a(\vec{x},t), \quad \Phi^a(\vec{x},t) = v_\phi(\vec{x},t) + \phi^a(\vec{x},t) \tag{31}$$

with

$$v_\phi(\vec{x},t) = \begin{pmatrix} v(\vec{x},t) \\ v^*(\vec{x},t) \end{pmatrix}. \tag{32}$$

In terms of ϕ^a, Eq. (13) is rewritten as

$$i\hbar\frac{\partial}{\partial t}\phi^a(\vec{x},t) = \int d^3y \rho_3 \begin{pmatrix} \varepsilon(\vec{x},\vec{y};t) & \Delta(\vec{x},\vec{y};t) \\ \Delta^\dagger(\vec{x},\vec{y};t) & \varepsilon(\vec{x},\vec{y};t) \end{pmatrix} \phi^a(\vec{y},t). \tag{33}$$

Its solution is obtained in the form

$$\phi^a(x) = \sum_\ell \left[u_\ell(x)\alpha_\ell^a + v_\ell(x)\alpha_\ell^{a\dagger} \right], \tag{34}$$

where

$$u_\ell(x) = \begin{pmatrix} f_\ell(x) \\ g_\ell(x) \end{pmatrix}, \quad v_\ell(x) = \begin{pmatrix} g_\ell^*(x) \\ f_\ell^*(x) \end{pmatrix}. \tag{35}$$

Since $\{u_\ell, v_\ell\}$ form a complete set at each time t, We can express

$$\phi^a(\vec{x},t_x) = \int d^3y D(\vec{x},t_x;\vec{y},t_y) \rho_3 \phi^a(\vec{y},t_y), \tag{36}$$

where

$$D(\vec{x},t_x;\vec{y},t_y) = \sum_\ell \left[u_\ell(\vec{x},t_x) u_\ell^\dagger(\vec{y},t_y) - v_\ell(\vec{x},t_x) v_\ell^\dagger(\vec{y},t_y) \right]. \tag{37}$$

QUASI-PARTICLE FIELD EQUATION

Let us now obtain the formula to determine the time-dependent energy and the quasi-particle equation including the information of change of particle distribution. By use of the perturbation formula Eq. (26), we can calculate the self-energy part, $\Sigma^{ab}(x,y)$, and the propagator is expressed as

$$<T\phi^a(x)\phi^{b\dagger}(y)>_H$$
$$= <T\phi^a(x)\phi^{b\dagger}(y)\exp\{-\frac{i}{\hbar}\frac{1}{2}\int d^4z_1 d^4z_2 \phi^{c\dagger}(z_1)(\tau\Sigma(z_1,z_2)\tau)^{cd}\phi^d(z_2)\}>. \tag{38}$$

For $t_x > t_y$, we have

$$<T\phi^a(x)\phi^{b\dagger}(y)>_H = <\phi^a(x)\Phi^{b\dagger}(t_x;y)> \tag{39}$$

with

$$\Phi^{b\dagger}(t_x;y) = T\left[\exp\left\{-\frac{i}{\hbar}\frac{1}{2}\int_{t_0}^{t_x} d^4z_1 d^4z_2 \phi^{c\dagger}(z_1)(\tau\Sigma(z_1,z_2)\tau)^{cd}\phi^d(z_2)\right\}\phi^{b\dagger}(y)\right]. \tag{40}$$

We have

$$i\hbar\frac{\partial}{\partial t_x}\Phi^{b\dagger}(t_x;y) = T[\hat{K}_I(t_x)\Phi^{b\dagger}(t_x;y)], \tag{41}$$

where
$$\hat{K}_I(t) = \frac{1}{2}\int d^3x \int_{t_0}^{t} d^4z [\phi^{a\dagger}(\vec{x},t)(\tau\Sigma(\vec{x},t;\vec{z},t_z)\tau)^{ab}\phi^b(\vec{z},t_z) \\
+ \phi^b(\vec{y},t)(\tau\Sigma(\vec{z},t_z;\vec{y},t)\tau)^{ab}\phi^{a\dagger}(\vec{z},t_z)] . \quad (42)$$

When we use the approximation
$$i\hbar\frac{\partial}{\partial t_x}\Phi^{b\dagger}(t_x;y) \approx \hat{K}_I(t_x)\Phi^{b\dagger}(t_x;y) , \quad (43)$$

we have
$$\hat{K}_I(t_x) = \frac{1}{2}\int d^3x d^3y \phi^{a\dagger}(\vec{x},t)(\tau K_I(\vec{x},\vec{y};t)\tau)^{ab}\phi^b(\vec{y},t) \quad (44)$$

with
$$K_I^{ab}(\vec{x},\vec{y};t) = \int_{t_0}^{t} d^4z \Sigma^{ab}(\vec{x},t;\vec{z},t_z)D(\vec{z},t_z;\vec{y},t)\rho_3 + \int_{t_0}^{t} d^4z \rho_3 D(\vec{x},t;\vec{z},t_z)\Sigma^{ab}(\vec{z},t_z;\vec{y},t) . \quad (45)$$

Define a quasiparticle field $\varphi^a(\vec{x},t)$ by
$$\varphi^a(\vec{x},t) = \hat{W}(t)^{-1}\phi^a(\vec{x},t)\hat{W}(t) \quad (46)$$

with
$$\hat{W}(t) = T\exp\{-\frac{i}{\hbar}\int_{t_0}^{t} dt' \hat{K}_I(t')\} . \quad (47)$$

Then the propagator is approximately given by
$$<T\phi^a(\vec{x},t_x)\phi^{b\dagger}(\vec{y},t_y)>_H \\
\approx \theta(t_x - t_y)<\varphi^a(\vec{x},t_x)\varphi^{b\dagger}(\vec{y},t_y)> + \theta(t_y - t_x)<\varphi^{b\dagger}(\vec{y},t_y)\varphi^a(\vec{x},t_x)> . \quad (48)$$

The quasi-particle field $\varphi^a(\vec{x},t)$ satisfies the following equation,
$$i\hbar\frac{\partial}{\partial t}\varphi^a(\vec{x},t) = \rho_3(E(t)\delta^{ab} + (K_I(t)\tau)^{ab})\varphi^b(\vec{x},t) \quad (49)$$

with
$$E(t) = \begin{pmatrix} \varepsilon(t) & \Delta(t) \\ \Delta^*(t) & \varepsilon(t) \end{pmatrix} . \quad (50)$$

Here and hereafter the matrix nature in the coordinates is understood.

The matrix $K_I(t)$ is obtained from the self-energy as
$$K_I^{ab}(\vec{x},\vec{y};t) = \int_{t_0}^{t} d^4z [\Sigma^{ab}(\vec{x},t;\vec{z},t_z)D(\vec{z},t_z;\vec{y},t)\rho_3 + \rho_3 D(\vec{x},t;\vec{z},t_z)\Sigma^{ab}(\vec{z},t_z;\vec{y},t)] . \quad (51)$$

From the thermal state condition, we can write Σ^{ab} in general as
$$\Sigma^{ab}(x,y) = \delta(t_x - t_y)\tau^{ab}\delta E(t_x;\vec{x},\vec{y}) \\
+ \theta(t_x - t_y)\begin{pmatrix} \Sigma_1(x,y) & \Sigma_2(x,y) \\ \Sigma_1(x,y) & \Sigma_2(x,y) \end{pmatrix} + \theta(t_y - t_x)\begin{pmatrix} \Sigma_2(x,y) & \Sigma_2(x,y) \\ \Sigma_1(x,y) & \Sigma_1(x,y) \end{pmatrix} \quad (52)$$

Explicitly we have for the local part and for the nonlocal part as

$$\delta E = \begin{pmatrix} \delta\varepsilon + 2V<\psi^\dagger\psi> & \delta\Delta + V<\psi^2> \\ \delta\Delta^* + V<\psi^{\dagger 2}> & \delta\varepsilon + 2V<\psi^\dagger\psi> \end{pmatrix}, \tag{53}$$

$$\Sigma_1(x,y) = -i<j_\phi(x)j_\phi^\dagger(y)>_I, \quad \Sigma_2(x,y) = -i<j_\phi^\dagger(y)j_\phi(x)>_I, \tag{54}$$

where

$$j_\phi = \begin{pmatrix} j_\psi \\ j_\psi^\dagger \end{pmatrix}, \quad j_\psi = V(v^*\psi^\dagger\psi^2 + v\psi^{\dagger 2}\psi + \frac{1}{2}\psi^{\dagger 2}\psi^2) \tag{55}$$

and "I" indicates the one-particle irreducible part. By use of the expression of Σ^{ab}, Eq. (51) is rewritten as

$$K_I^{ab}(\vec{x},\vec{y};t) = \delta E(\vec{x},\vec{y};t)\tau^{ab} + \begin{pmatrix} k_1(\vec{x},\vec{y};t) - k_2^\dagger(\vec{x},\vec{y};t) & k_2(\vec{x},\vec{y};t) - k_2^\dagger(\vec{x},\vec{y};t) \\ k_1(\vec{x},\vec{y};t) - k_1^\dagger(\vec{x},\vec{y};t) & k_2(\vec{x},\vec{y};t) - k_1^\dagger(\vec{x},\vec{y};t) \end{pmatrix} \tag{56}$$

with

$$k_1(\vec{x},\vec{y};t) = \int_{t_0}^t d^4z \Sigma_1(\vec{x},t;\vec{z},t_z) D(\vec{z},t_z;\vec{y},t)\rho_3, \tag{57}$$

$$k_2(\vec{x},\vec{y};t) = \int_{t_0}^t d^4z \Sigma_2(\vec{x},t;\vec{z},t_z) D(\vec{z},t_z;\vec{y},t)\rho_3. \tag{58}$$

Defining

$$k(\vec{x},\vec{y};t) = k_1(\vec{x},\vec{y};t) - k_2(\vec{x},\vec{y};t) \tag{59}$$

and

$$k_r = \frac{1}{2}(k+k^\dagger), \quad \gamma = \frac{i}{2}(k-k^\dagger), \quad \gamma_n = \frac{i}{2}(k_2 - k_2^\dagger), \tag{60}$$

we have

$$K_I^{ab}(t) = (\delta E(t) + k_r(t))\tau^{ab} - i\begin{pmatrix} \gamma(t) + 2\gamma_n(t) & 2\gamma_n(t) \\ 2(\gamma(t) + \gamma_n(t)) & \gamma(t) + 2\gamma_n(t) \end{pmatrix}^{ab}. \tag{61}$$

The counter term δC is obtained by evaluating diagrams corresponding

$$<\phi(\vec{x},t)>_H = 0. \tag{62}$$

We can summarize the equation of the order parameter as

$$i\hbar\frac{\partial}{\partial t}v(\vec{x},t)$$
$$= \left(-\frac{\hbar^2}{2m}\vec{\nabla}^2 + U(\vec{x}) - \mu(t) + V(|v(\vec{x},t)|^2 + 2<\psi^\dagger(\vec{x},t)\psi(\vec{x},t)>)\right)v(\vec{x},t)$$
$$+ V<\psi(\vec{x},t)^2> v(\vec{x},t)^* - \delta C_2(\vec{x},t), \tag{63}$$

where δC_2 is the correction arising from the higher than two-loop approximation. Since the quantum correction k_r in Eq. (61) is a function of the single time t, our choice of the starting unperturbed Hamiltonian H_{t0} with coefficients depending only on the single time t is justified. We set the renormalization condition

$$\delta E(\vec{x},\vec{y};t)+k_r(\vec{x},\vec{y};t)=0. \tag{64}$$

Then the equation of the quasi-particle field $\varphi^a(\vec{x},t)$ is obtained as

$$i\hbar\frac{\partial}{\partial t}\varphi^a(\vec{x},t) = \rho_3\left(\begin{pmatrix}\varepsilon(t) & \Delta(t) \\ \Delta^*(t) & \varepsilon(t)\end{pmatrix}\delta^{ab}\right.$$
$$\left. - i\left[\begin{pmatrix}\gamma(t)+2\gamma_n(t) & 2\gamma_n(t) \\ 2(\gamma(t)+\gamma_n(t)) & \gamma(t)+2\gamma_n(t)\end{pmatrix}\tau\right]^{ab}\right)\varphi^b(\vec{x},t), \tag{65}$$

where

$$\varepsilon(\vec{x},\vec{y};t) = \left(-\frac{\hbar^2}{2m}\vec{\nabla}^2+U(\vec{x})-\mu(t)+2V(v(\vec{x},t)^*v(\vec{x},t)\right.$$
$$\left. + <\psi^\dagger(\vec{x},t)\psi(\vec{x},t)>)\right)\delta(\vec{x}-\vec{y})+[k_{r1}(\vec{x},\vec{y};t)+k_{r2}(\vec{x},\vec{y};t)]_{11} \tag{66}$$
$$\Delta(\vec{x},\vec{y};t) = V(v(\vec{x},t)^2+<\psi(\vec{x},t)^2>)\delta(\vec{x}-\vec{y})+[k_{r1}(\vec{x},\vec{y};t)]_{12}. \tag{67}$$

Here k_{r1} and k_{r2} are contributions from one-loop and two-loop approximation. We have shown that we can define the quasi-particle equation with time-dependent energy $\varepsilon(t)$ and decay width $\gamma(t)$. From the derivation, we can see that the renormalization of the energy is a generalization of the on-shell renormalization.

BOLTZMANN-LIKE EQUATION

The remaining task is to show the quasi-particle equation contains the information of change of particle distribution. Consider

$$S_>^{ab}(x,y) = <\varphi^a(x)\varphi^{b\dagger}(y)>\theta(t_x-t_y) \tag{68}$$
$$S_<^{ab}(x,y) = <\varphi^{b\dagger}(y)\varphi^a(x)>\theta(t_y-t_x) \tag{69}$$

From the quasi-particle equation, Eq.(49) and its conjugate

$$-i\hbar\frac{\partial}{\partial t}\varphi^{\dagger a} = \varphi^{\dagger b}\left(E-i\tau\begin{pmatrix}\gamma+2\gamma_n & 2\gamma_n \\ 2(\gamma+\gamma_n) & \gamma+2\gamma_n\end{pmatrix}\right)^{ba}\rho_3, \tag{70}$$

we have a damping behavior

$$(i\hbar\frac{\partial}{\partial t_x}-E(t_x)+i\gamma(t_x))S_>^{ab}(x,y) = i\delta(t_x-t_y)<\varphi^a(\vec{x},t_y)\varphi^{\dagger b}(\vec{y},t_y)> \tag{71}$$
$$S_<^{ab}(x,y)(-i\hbar\frac{\partial}{\partial t_y}-E(t_y)-i\gamma(t_y)) = i\delta(t_x-t_y)<\varphi^{\dagger b}(\vec{y},t_x)\varphi^a(\vec{x},t_x)>. \tag{72}$$

For the equal time quantities, we define for example

$$< \varphi^{\dagger b}(\vec{y},t)\varphi^a(\vec{x},t) > = \begin{pmatrix} N(\vec{x},\vec{y};t) & N(\vec{x},\vec{y};t) \\ \rho_3\delta(\vec{x}-\vec{y})+N(\vec{x},\vec{y};t) & \rho_3\delta(\vec{x}-\vec{y})+N(\vec{x},\vec{y};t) \end{pmatrix}, \quad (73)$$

where

$$N_{ss'}(\vec{x},\vec{y},t) = < \phi_{s'}^\dagger(\vec{y},t)\phi_s(\vec{x},t) > . \quad (74)$$

By use of the quasi-particle equation, we have a Boltzmann-like equation

$$i\hbar\frac{\partial}{\partial t}N(\vec{x},\vec{y},t) = \rho_3(E_x(t)-i\gamma_x)N(\vec{x},\vec{y},t) - N(\vec{x},\vec{y},t)(E_y(t)+i\gamma_y)\rho_3$$
$$+ 2i\rho_3\gamma_n(\vec{x},\vec{y};t)\rho_3, \quad (75)$$

which shows how the particle distribution changes in space-time. Although we have used here the double-time function in order to show the damping nature and time variational equation of the particle distribution, such information is included in the quasi-particle equation of φ, that is Eq. (65). We can analyze nonequilibrium phenomena in the condensate, by use of the coupled equations with the single time, Eq. (63) and Eq. (65). Within the present scheme, the phase invariant approximation is performed, and such a matter is discussed in ref.[37].

SUMMARY

Prof. Mancini participated as one of the originators of Thermo Field Dynamics, the quantum field theory at finite temperature. In this talk, I have shown that the framework of TFD enables us to derive quasi-particle field equations which describes space-time dependent energy, decay width and particle distribution in nonequilibrium phenomena. It provides us a useful tool to analyze nonequilibrium phenomena in Bose-Einstein condensation of dilute alkaline atomic gas.

ACKNOWLEDGMENTS

The author would like to express his sincere congratulation for Prof. Mancini's 60th birthday, and his thanks to his long friendship and hospitality in research. He also expresses his thanks for members of the department of Physics, University of Salerno, of their hospitality, especially during the period of 1986-1988, when the author stayed in Salerno, and worked on the formulation used in this paper. This work was supported by the Special Fund of University Project, University of Tsukuba, Japan.

REFERENCES

1. Umezawa, H., Leplae, L., and Mancini, F., *Physics Reports*, **10C**, 151 (1974).
2. Mancini, F., Tachiki, M., and Umezawa, H., *Physica*, **94B**, 1 (1978).

3. Takahashi, T., and Umezawa, H., *Collective Phenomena*, **2**, 55 (1975).
4. Matsumoto, H., *Fortschritte der Physik*, **25**, 1 (1977).
5. Umezawa, H., Matsumoto, H., and Tachiki, M., *Thermo Field Dynamics and Condensed States*, North Holland Pub., Amsterdam, 1982.
6. See as a review article; Matsumoto, H., and Umezawa, H., *Cryogenics*, **January**, 37 (1983).
7. Matsumoto, H., *Z. Physik C-Particle and Field*, **33**, 201 (1986).
8. Matsumoto, H., Mancini, F., and Marinaro, M., *Europhys.Lett.*, **4**, 153 (1987).
9. Matsumoto, H., Mancini, F., and Marinaro, M., *J. Phys. A*, **30 Gen& Math**, 6543 (1987).
10. Matsumoto, H., *Prog. Theor. Phys.*, **79**, 373 (1988).
11. Matsumoto, H., *Prog. Theor. Phys.*, **80**, 57 (1988).
12. Matsumoto, H., *Physica A*, **158**, 291 (1989).
13. Matsumoto, H., Sasaki, M., Ishihara, S., and Tachiki, M., *Phys.. Rev. B*, **46**, 3009 (1992).
14. Ishihara, S., Matsumoto, H., Odashima, S., Tachiki, M., and Mancini, F., *Phys. Rev. B*, **49**, 1350 (1994).
15. Mancini, F., Marra, S., and Matsumoto, H., *Physica C*, **244**, 49 (1995).
16. Mancini, F., Marra, S., and Matsumoto, H., *Physica C*, **250**, 184 (1995).
17. Anderson, M. H., Ensher, J. R., Mathews, M. R., Wieman, C. E., and Cornnell, E. A., *Science*, **269**, 198 (1995).
18. Davis, K. B., Mewes, M. O., Andrews, M. R., van Druten, N. J., Durfee, D. S., Kurn, D. M., and Ketterle, W., *Phys. Rev. Lett.*, **75**, 3969 (1995).
19. Bradley, C. C., Sackette, C. A., Tollet, J. J., and Hulet, R. G., *Phys. Rev. Lett.*, **75**, 1687 (1995).
20. Bradley, C. C., Sackette, C. A., and Hulet, R. G., *Phys. Rev. Lett.*, **78**, 985 (1997)).
21. See the following review paper; Parkins, A. S., and Walls, D. F., *Phys. Rep.*, **303**, 1 (1998), and references cited there.
22. See the following review paper; Dalfovo, F., Giorgini, S., Pitaevskii, L. P., and Stringari, S., *Rev. Mod. Phys.*, **71**, 463 (1999), and references cited there.
23. Miesner, H. J., Stamper-Kurn, D. M., Andrews, M. R., Durfee, D. S., Inoue, S., and Ketterle, K., *Science*, **279**, 1005 (1998).
24. Stoof, H. T., *Phys. Rev. Lett.*, **66**, 3148 (991).
25. Stoof, H. T., *Phys. Rev.*, **A45**, 8398 (1992).
26. Stoof, H. T., *Phys. Rev. Lett.*, **78**, 768 (1997).
27. Kagan, Y., Svistunov, B. V., and Shlyapnikov, G. V., *Sov. Phys. -JETP*, **75**, 387 (1992), [Zh. Eksp. Teor. Fiz. **101** (1992), 528].
28. Kagan, Y., and Svistunov, B. V., *Sov. Phys. -JETP*, **78**, 187 (1994), [Zh. Eksp. Teor. Fiz. **105** (1994), 353].
29. Kagan, Y., and Svistunov, B. V., *Phys. Rev. Lett.*, **79**, 3331 (1997).
30. Gardiner, C. W., Zoller, P., Ballagh, R. J., and Davis, M. J., *Phys. Rev. Lett.*, **79**, 1793 (1997).
31. Gardiner, C. W., Zoller, P., Ballagh, R. J., and Davis, M. J., *Phys. Rev. Lett.*, **81**, 5266 (1998).
32. Lee, M. D., and Gardiner, C. W., *Phys. Rev. A*, **62**, 33606 (2000).
33. Barci, D. G., Fraga, E. S., and Ramos, R. O., *Phys. Rev. Lett.*, **85**, 479 (2000).
34. Matsumoto, H., and Sakamoto, S., *Prog. Thoer. Phys.*, **105**, 573 (2001).
35. Arimitsu, T., and Umezawa, H., *Prog. Theor. Phys.*, **74**, 429 (1985).
36. Arimitsu, T., Umezawa, H., and Yamanaka, Y., *J. Math. Phys.*, **28**, 2741 (1987).
37. Matsumoto, H., and Sakamoto, S., *Prog. Thoer. Phys.*, **107**, 689 (2002).

Coherent structures of Bose-Einstein condensates in optical lattices

B. B. Baizakov*, V. V. Konotop† and M. Salerno*

*Dipartimento di Fisica "E.R. Caianiello" and Istituto Nazionale di Fisica della Materia (INFM), Universitá di Salerno, I-84081 Baronissi (SA), Italy
†Departamento de Física and Centro de Física da Materia Condensada, Universidade de Lisboa, Complexo Interdisciplinar, Av. Prof. Gama Pinto 2, Lisboa 1649-003, Portugal

Abstract. We propose to employ the phenomenon of modulational instability in order to create regularly arranged localized excitations in arrays of Bose-Einstein condensates. These excitations are narrow tubes in 2D and small hollows in 3D arrays filled in with the condensed atoms of much greater density compared to surrounding array sites. As the regions with high atomic concentration develop due to the modulational instability, they can be preserved by increasing the strength of the optical lattice. Theoretical model, based on the multiple scale expansion, describes the main features of the phenomenon. Analytical predictions are confirmed by numerical simulations of the Gross-Pitaevskii equation.

INTRODUCTION

The experimental realization of Bose-Einstein condensation (BEC) in dilute atomic gases [1, 2] founded a rapidly progressing new field of research [3]. In recent years considerable attention has been focused on the properties of BEC confined to periodic potentials, particularly using laser standing waves. Progress in optical manipulation with BEC arrays resulted in observation of diverse new phenomena, such as coherent emission of Bose-condensed atoms [4], Bloch oscillations and Landau-Zener tunneling [5], atomic Josephson effect [6], reversible Mott insulator - superfluid transition [7]. Realization of BEC arrays in 2D and 3D optical lattices [7, 8] opened new perspectives for investigation of fundamental properties of quantum gases in lower dimensions. Theoretical studies on the properties of BEC in optical lattices are often considered in the weak interaction limit or small amplitude approximation for which concepts of the Schrödinger linear theory of solids such as Bloch states, band structure, effective masses etc., can be safely used [9, 10, 11]. Particularly, the existence of stable 2D and 3D solitons with the energies located in the band gaps of the underlying linear system was shown [12, 13, 14]. These solitons are similar to the gap solitons found in nonlinear optics [15]

An important subject relevant to BEC dynamics in optical lattices is the creation and control of quasistable localized excitations. Continuous BEC with repulsive interatomic forces does not support spatially localized humps of atomic concentration. However, superimposing a periodic potential on the effectively 1D continuous repulsive BEC creates special conditions, which make the existence of bright solitons possible [11, 16, 17]. Out of existing studies on the properties of soliton-like BEC structures in optical lattices, little attention has been devoted to methods of creation of such structures, so far.

It has recently been suggested that the modulational instability, which constitutes one of the fundamental effects in the physics of nonlinear waves, can be employed to create bright solitons in a repulsive BEC confined to a 1D optical lattice [11]. Dynamics of these solitons was shown to obey the nonlinear Schrödinger equation (NLSE). Natural development of this route would be the investigation of modulational instability in 2D and 3D BEC arrays. However, the major difference here is that, as opposed to the 1D case NLSE in 2D and 3D does not support stable solitonic solutions, and therefore one cannot expect the analogous outcome as in 1D. Nevertheless, as the modulational instability results in formation of spatially localized excitations in 2D and 3D cases, they could be stabilized by external means, e.g. changing the parameters of the periodic potential. Particularly in this paper we show the possibility to create regularly spaced localized excitations in BEC arrays, which present narrow tubes in 2D and small hollows in 3D cases, filled in with BEC atoms of much greater density, compared to surrounding array sites.

THE MODEL AND MULTIPLE SCALE ANALYSIS

Our model is based on the following dimensionless 3D Gross-Pitaevskii (GP) equation

$$i\frac{\partial \psi}{\partial t} = -\Delta\psi + V(\mathbf{r})\psi + \chi|\psi|^2\psi, \qquad (1)$$

describing the properties of the ground state wavefunction in the mean field approximation of a BEC in a trapping potential $V(\mathbf{r})$ (for a derivation of this equation see Ref. [18, 19]. Here $\mathbf{r} \equiv (r_x, r_y, r_z)$ and the potential $V(\mathbf{r})$ is assumed to be separable, i.e. of the form $V(\mathbf{r}) = \sum_j V_j(r_j)$, $j = x, y, z$ (which corresponds to the majority of experimental settings), and periodic in each of the spatial directions: $V_j(r_j) = V_j(r_j + a_j)$, with a_j the period in the direction r_j. For convenience Eq. (1) is considered subject to periodic boundary conditions $\psi(\mathbf{r}) = \psi(r_x + L_x, r_y, r_z)$, etc., where $L_j = N_j a_j$ with N_j and L_j respectively, the number of primitive cells and the length of the system in the direction r_j. The theory is developed for the small amplitude limit, when the multiple scale analysis [20] is applicable. Therefore, we look for a solution of Eq. (1) in the form

$$\psi = \varepsilon\psi_1 + \varepsilon^2\psi_2 + \varepsilon^3\psi_3 + \cdots, \qquad (2)$$

where the ψ_j are functions of the scaled independent variables $\tau_p = \varepsilon^p t$, $\xi_p = \varepsilon^p \mathbf{r}$, $p = 0, 1, 2, ...$, with ε a small parameter to be specified later. Denoting with $\omega_{\alpha_j}(q_j)$, and $\Phi_{\alpha_j}(r_j) \equiv |\alpha_j, q_j\rangle$, the eigenvalues and eigenfunctions of the periodic operators $\Lambda_{r_j} = -\partial_{r_j}^2 + V_j(r_j)$, we have that the solution of the linear part of Eq. (1), $\Lambda\psi = 0$, with $\Lambda = i\partial_t - \sum_j \Lambda_{r_j}$, can be written in the form $|m_x m_y m_z\rangle = \prod_j \Phi_{m_j}(r_j)e^{i\omega_{\alpha_j}(q_j)t}$, with $\Phi_{m_j}(r_j)$ Bloch states of the corresponding 1D linear problems. Here m_j denotes the couple of quantum numbers $\{\alpha_j, q_j\}$, with α_j the band index and q_j the component of the wave vector in the j direction (note that the imposed boundary conditions obviously imply that $q_j \equiv q_{j,n} = \frac{2\pi}{L_j}n$ so that the extension of the Brillouin zone in the j direction is $[-\pi/a_j, \pi/a_j])$.

Substituting Eq. (2) into (1), and collecting terms of equal powers in ε, one arrives at the set of equations $\Lambda \psi_n = M_n$, where $M_1 = 0$, $M_2 = -i\partial_{\tau_1}\psi_1 - 2\nabla_0\nabla_1\psi_1$,

$$M_3 = -i\partial_{\tau_2}\psi_1 - i\partial_{\tau_1}\psi_2 - \Delta_1\psi_1 - 2\nabla_0\cdot\nabla_1\psi_2 - 2\nabla_0\cdot\nabla_2\psi_1 + \chi|\psi_1|^2\psi_1,$$

where ∇_p denotes the gradient with respect to ξ_p.

Since we are interested in instabilities of the condensate wavefunction, we explore the influence of the nonlinear term in Eq. (1) on the Bloch states of the underlying linear problem. To this end we start with

$$\psi_1 = A(\xi_1, ...; \tau_1, ...)e^{-i\omega_0\tau_0}|m_{0x}m_{0y}m_{0z}\rangle, \tag{3}$$

with $\omega_0 \equiv \sum_j \omega_{\alpha_0,j}(q_j)$ (the subscript zero refers to the chosen band, below we consider the two lowest ones). Then the first order equation is automatically satisfied by ψ_1, while the equation of the second order can be solved in the form

$$\psi_2 = \sum_\alpha{}' B_\alpha e^{-i\omega_0\tau_0}|\alpha_x, q_{0,x}; \alpha_y, q_{0,y}; \alpha_z, q_{0,z}\rangle, \tag{4}$$

where the prime denotes $\alpha \neq \alpha_0$ in the sum and has taken into account that the terms with $\mathbf{q} \neq \mathbf{q}_0$ give zero contribution. An analysis similar to the one in Ref. [11] shows that $A = A(\mathbf{R}; \xi_2...; \tau_2, ...)$ with $\mathbf{R} = \xi_1 - \mathbf{v}\tau_1$ and $\mathbf{v} = -\langle\alpha_{0x}\alpha_{0y}\alpha_{0z}|2i\nabla|\alpha_{0x}\alpha_{0y}\alpha_{0z}\rangle$ is the group velocity of the carrier wave. The coefficients B_α are found as

$$B_\alpha = \frac{\Gamma^{(y,z)}_{\alpha_x,\alpha_{0,x}}\partial_{x_1}A + \Gamma^{(x,z)}_{\alpha_y,\alpha_{0,y}}\partial_{y_1}A + \Gamma^{(x,y)}_{\alpha_z,\alpha_{0,z}}\partial_{z_1}A}{\omega_0 - \omega_\alpha(\mathbf{q}_0)}, \tag{5}$$

where $\omega_\alpha(\mathbf{q}_0) = \sum_j \omega_{\alpha_j}(q_{0j})$, $\Gamma^{(y,z)}_{\alpha_x,\alpha_{0,x}} = -\langle\alpha_x, q_{0,x}|2\partial_{x_0}|\alpha_{0,x}, q_{0,x}\rangle\delta_{\alpha_y,\alpha_{0,y}}\delta_{\alpha_z,\alpha_{0,z}}$ (other coefficients Γ are obtained by cyclic permutations of x, y, z). Finally, considering the orthogonality of M_3 to $|m_{0x}m_{0y}m_{0z}\rangle$ we obtain the following 3D NLS equation for the slowly varying envelope

$$i\frac{\partial A}{\partial \tau_2} + \frac{1}{2}\sum_{j=x,y,z}M^{-1}_{\alpha_j,jj}\frac{\partial^2 A}{\partial R_j^2} - \tilde\chi|A|^2A = 0, \tag{6}$$

where we assumed A not depending on ξ_2, and introduced the inverse of the effective mass tensor

$$\frac{1}{2}M^{-1}_{\alpha,jj} = 1 + \sum_{\alpha_x}\frac{|\Gamma^{(y,z)}_{\alpha_x\alpha_{0,x}}|^2}{\omega_{\alpha_x}(q_{0,x}) - \omega_{\alpha_{0,x}}(q_{0,x})} = \frac{1}{2}\partial^2_{q_j}\omega_{\alpha_j}(\mathbf{q}),$$

and the effective nonlinearity

$$\tilde\chi = \chi \prod_{j=x,y,z}\int_0^{L_j}|\Phi_{m_{0j}}|^4 dr_j.$$

To discuss the small parameter ε we consider a cubic box with $L_j = L$ and $a_j = l$, l being the characteristic scale of the potential (e.g. smallest of its periods). On the one hand, the physical order parameter is normalized to the total number of atoms N_0, while the formal wave function ψ must be normalized to one: $\int |\psi|^2 d\mathbf{r} = 1$. On the other hand all the parameters in Eq. (6) must be of order one. Taking into account that $\chi = 8\pi N_0 a_s/l$ (see e.g. [11]), and the oscillatory character of the Bloch functions, leads to the fact that the integrals in the last expression for $\tilde{\chi}$ give a numerical smallness (see e.g. the examples below), and one can then define $\varepsilon = N_0 a_s l/L^2$. Consider now a condensate with $N_0 = 10^5$ of ^{87}Rb atoms homogeneously distributed over a cubic box with $L = 100\,\mu$m. Then we compute $\varepsilon \approx 0.086$. In the experiment [21] with a cigar-shaped BEC ε was 0.018 well supporting the weakly nonlinear limit. A physical situation when ε is not small enough (say in experiments [8] it can be identified as $\varepsilon \approx 0.257$) the multiple scale expansion, strictly speaking, is not valid. For this reason below we employ numerical simulations, which, however, clearly illustrates that the small amplitude limit gives remarkably good estimates for the characteristic scales of the problem and allows to understand the symmetry of developed patterns.

Let us analyze the stability problem within the framework of Eq. (6), i.e. look for a solution of the form Eq. (3) with $A = \left(\rho + ae^{i(\Omega\tau_2 - \mathbf{KR})} + be^{-i(\Omega\tau_2 - \mathbf{KR})}\right) e^{-i\rho^2 \tau_2}$, where $|a|, |b| \ll \rho$. This solution is unstable if

$$Z(Z + 4\tilde{\chi}\rho^2) < 0, \quad Z = \sum_{j=x,y,z} M_{jj}^{-1} K_j^2. \qquad (7)$$

NUMERICAL SIMULATIONS

For the numerical study we have used a 2D problem with the potential $V_j(r_j) = -V_0 \cos(kr_j)$ for $j = x, y$ and $V_z(r_z) = 0$, which is motivated by the recent experiments [8]. The minus sign in front of V_0 is selected to have the minimum of the potential at $r = 0$. Numerical solution of Eq. (1) was performed using the two-dimensional fast Fourier transform [22]. The spatial square domain $x, y \in [-\frac{L}{2}..\frac{L}{2}]$ (i.e. $L_x = L_y = L$) was represented by an array of 256 x 256 points.

To be specific, we concentrate on the case of the positive scattering length, $\chi = 1.0$, choose $k = 2.0$ (i.e. $a_x = a_y = \pi$), $\rho = 0.5$, $L = 14\pi$ and consider the points $\mathbf{q}_0 = (\pm 1, \pm 1)$ at the boundary of the BZ. Then, restricting consideration to the two lowest bands ($\alpha = 1, 2$), one can distinguish three different cases:

Case 1. Both eigenfunctions $|m_{0,x}\rangle$ and $|m_{0,y}\rangle$ belong to the first lowest zone: $m_{0,x} = m_{0,y} = (1, \pm 1)$. Then $\mathbf{M}_{1,xx}^{-1} = \mathbf{M}_{1,yy}^{-1} = \mathbf{M}_1^{-1} < 0$ and the wave is unstable. BEC dynamics in this case is presented in Fig. 1. The most interesting feature of the modulational instability development is that it evolves in a *regular* structure which represents symmetrically spaced, localized in space (we call them soliton-like), distributions (see Fig. 1a). Each of the humps shown in the figure represents a tightly confined tube along the z-direction. The number of the tubes is proportional to the size of the box. In order to understand this behavior we notice that from Eq. (7) we get that the excitations with characteristic scales $\lambda > \lambda_{min} = \frac{2\pi}{K_{max}} = \frac{\pi}{\rho}\sqrt{\frac{|M_1^{-1}|}{\tilde{\chi}}}$ are unstable. The largest increment

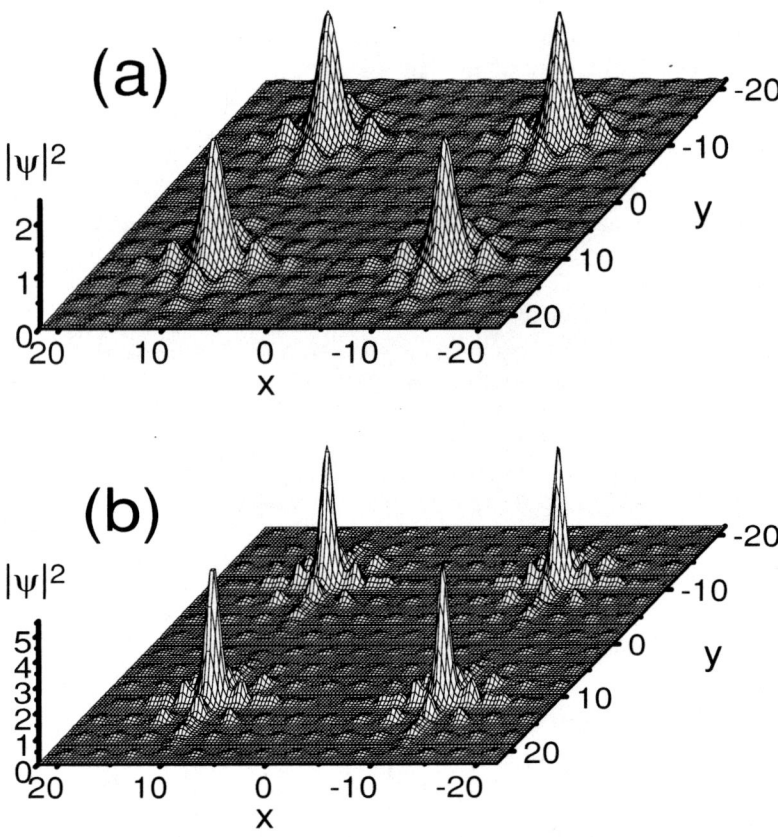

FIGURE 1. Snapshots of BEC distribution in a 2D optical lattice with $V_0 = 1.0$, $L = 14\pi$. The initial waveform $\psi(x,y,0) = 0.5\cos(x)\cos(y)$ is modulated by $\delta\psi(x,y) = 0.05\cos(0.286x)\cos(0.286y)$. (a) Soliton-like excitations are formed due to modulational instability at t=35. (b) The distribution (a) remains stable for long times when the strength of the trap potential is increased up to $V_0 = 4.0$ during $35 < t < 40$. The snapshot is shown at $t = 100$ (stability is verified up to t=1000).

(i.e. the large $\text{Im}|\Omega|$) is achieved for $\lambda_0 = \sqrt{2}\lambda_{min}$. This has two consequences. First the symmetry group of the developed structure must be of C_n type with the symmetry axis coinciding with that of the condensate, and second, an effective scale $\lambda_{eff} \sim \lambda_0$ must be a characteristic scale of the largest excitation at the beginning of the evolution. To estimate the value of λ_0 we take into account that for a chosen point of BZ the effective group velocity dispersion tensor is $|M_1^{-1}| \approx 6$ and for the solutions studied numerically the effective nonlinearity is $\tilde{\chi} \approx 0.1935$ (the respective normalized eigenfunctions are approximated by $\frac{2}{\pi}\sin(x)$ [11], which gives $\lambda_0 \approx 20.053$. This result corroborates with the distances between the humps along the radial direction measured from the direct numerical simulations: $\lambda_{eff} \approx 23.0$. Next we take into account that the carrier wave mode is chosen at the point $\mathbf{q} = (\pm 1, \pm 1)$ placed at the corner of the BZ which corresponds to waves whose phases propagate in the directions $x = \pm y$. This immediately specifies

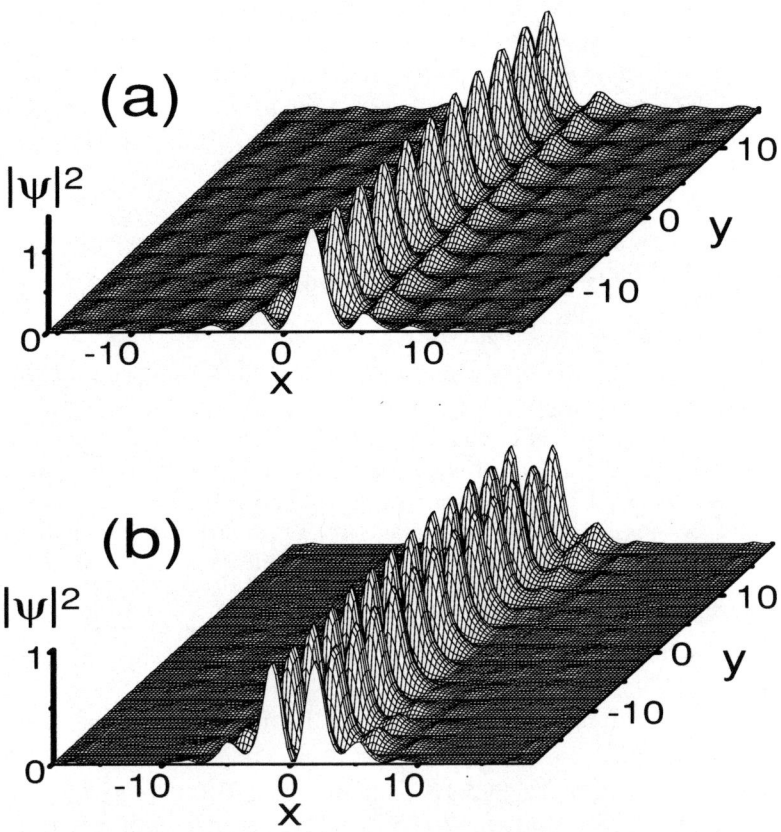

FIGURE 2. Coherent structures in a 2D optical lattice $V(x,y) = \cos(2x) + \cos(2y)$, $V_0 = 1.0$, $\chi = 1.0$. The initial waveform $\psi(x,y,0) = 0.5\cos(x)\sin(y)$ is modulated by $\psi(x,y) = 0.01\sin(0.4x)\sin(0.4y)$ which leads to generation of a regular pattern of soliton-like excitations. (a) The mode localized in a minimum of the potential well. (b) Bound state of two localized modes occupying the neighboring cells.

the symmetry C_4. In other words, one can specify the points where the humps (confined tubes) should appear: in the plane (x,y) these are intersections of lines $x = \pm y$ with the circles of radii $\left(\frac{1}{2} + p\right)\lambda_{eff}$ where $p = 0, 1, \ldots$. This in particular gives that in a square box of a size L one will observe $\frac{L^2}{\lambda_{eff}^2}$ humps. This estimate being rather rough (it does not take into account boundary effects) was confirmed by numerical simulations. Also one can predict that the characteristic diameters of the humps should be less than λ_{min} (≈ 7.1 in our case). This gives an estimate for the BEC density in a tube n_t vs. the initial density n_0: $n_t = \frac{L_x L_y}{\lambda_{min}^2} n_0$, which in our case gives $n_t \approx 29 n_0$ (the result very well confirmed by the numerical simulations). In a 3D case, relevant estimate for the density of BEC matter in hollows n_h is modified as $n_h = \frac{L_x L_y L_z}{\lambda_{min}^3} n_0$.

Finally we notice that the proposed approach can be verified by applying to the results of the recent experiment [21]. Indeed, in the absence of the lattice the most unstable plane wave corresponds to $k_{max} = \sqrt{\chi}/a_0$ (where a_0 is a transverse radius of the BEC) with the effective nonlinearity $\chi = \sqrt{2}N_0 a_s/a_0 \approx 52$ (we have taken into account the factor $2^{-3/2}\pi^{-1}$ due to the 3D to 1D reduction). Then for $N_0 = 3 \cdot 10^5$ this gives $n_s = 16$ for the number of solitons, which corroborates with the experimental observations.

Case 2. The eigenfunctions $\Phi_{m_{0,x}}$ and $\Phi_{m_{0,y}}$ belong to different zones, say $\Phi_{m_{0,x}}$ belongs to the first lowest zone: $m_{0,x} = (1, \pm 1)$ and $\Phi_{m_{0,y}}$ belongs to the second lowest zone: $m_{0,y} = (2, \pm 1)$. Then $\mathbf{M}_{1,xx}^{-1} < 0$ and $\mathbf{M}_{1,yy}^{-1} > 0$, and the condensate is unstable. In this case the instability condition takes the form $0 < \mathbf{M}_{2,yy}^{-1} K_y^2 - |\mathbf{M}_{1,xx}^{-1}| K_x^2 < 4\tilde{\chi}\rho^2$, and the most unstable excitations have $K_x^2 < \frac{\mathbf{M}_{2,yy}^{-1}}{|\mathbf{M}_{1,xx}^{-1}|} K_y^2$ (which is related to the fact that an eigenfunction $\Phi_{m_{0,x}}$ belongs to the "unstable" branch). That is why the main instability results in a pattern having different symmetry: it develops in the x-direction. The instability develops also along the y-direction, but at much larger time scales. Fig. 2 represents two different modes of localized excitations induced by the modulational instability at strength of the optical lattice $V_0 = 1.0$ and the domain size $L = 10\pi$.

To estimate the number of humps, we take into account that the periodic boundary conditions impose a characteristic scale $K_x = 2\pi/L$ which leads to the following estimate for the most unstable scale λ_0, and thus to λ_{eff}: $\lambda_0 \approx 2\pi/\sqrt{\frac{2\pi}{L} - \frac{2\tilde{\chi}\rho^2}{\mathbf{M}_1^{-1}}}$. For the case illustrated in Fig. 2, we obtain $\lambda_0 \approx 24$, which yields for the number of humps in the x direction $LK_{max}/2\pi \sim L/\lambda_0 \sim 1$.

Case 3. Both eigenfunctions $\Phi_{m_{0,x}}$ and $\Phi_{m_{0,y}}$ belong to the second lowest zone: $m_{0,x} = m_{0,y} = (2, \pm 1)$. Then $\mathbf{M}_{2,xx}^{-1} = \mathbf{M}_{2,yy}^{-1} > 0$ and the wave is stable, which was confirmed numerically using the initial conditions $\psi(x, y, 0) = 0.5 \sin(x)\sin(y)$.

Hence, the modulational instability results in the formation of regular pattern of soliton-like excitations in arrays of BEC. However, they eventually decay in accordance with Eq. (6), which does not support stable solitonic solutions in 2D and 3D. A simple way to retain these excitations would be the increasing the strength of the periodic trap potential, when the excitations are formed. Particularly, in the numerical simulation of Fig.1a the localized excitations are formed at $t \sim 35$, $V_0 = 1.0$, after that the strength of the optical lattice is increased to $V_0 = 4.0$. A high potential barrier between lattice sites suppresses the atomic tunneling, providing strong confinement (Fig.1b).

To understand the stabilization phenomenon we notice that λ_{min} is of the order of $2a_x$ ($2a_y$), which means that most of the BEC atoms are concentrated in a unique cell (resembling the intrinsic localized modes of the discrete model [23, 24]). By increasing the potential amplitude one increases the interwell barriers, which in turn leads to a decreasing of both the probability of tunneling of atoms from the most populated cell to neighboring cells, and a "number" of atoms in the classically forbidden zone. As a consequence the BEC density in the most populated cell is growing, which is illustrated by Fig. 1b. Changing the strength of the optical lattice is also accompanied by modification of the band structure and shifting the energy position of localized excitations with respect to the energy bands. The energy of quasi-stationary 2D solitons

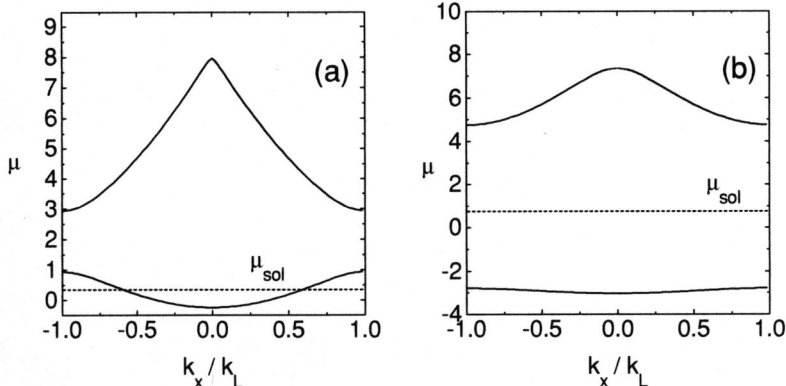

FIGURE 3. The energy level μ_{sol} (dashed line) of a 2D soliton of repulsive BEC shown in Fig. 1 with respect to the band structure of the underlying linear problem Eq.(1) shown at $k_y = 0$ section. The strength of the optical lattice: (a) initial $V_0 = 1.0$ (Fig.1a), (b) final $V_0 = 4.0$ (Fig.1a).

in Eq.(1) $\psi(x,y,t) = u(x,y)\exp(-i\mu t)$ can be associated with the chemical potential

$$\mu = \frac{1}{N}\int_{-\infty}^{\infty}\left(|\nabla u|^2 - V(x,y)\cdot|u|^2 + \chi\cdot|u|^4\right)dxdy, \qquad (8)$$

where $N = \int_{-\infty}^{\infty}|u(x,y)|^2 dxdy$ is the norm. When the strength of the optical lattice $V(x,y) = V_0\cdot[\cos(kx)+\cos(ky)]$ is changed, the energy of the soliton is changed accordingly.

The energy of the configuration shown in Fig. 1a calculated from Eq.(8) at parameters $V_0 = 1.0$, $\chi = 1.0$ is $\mu = 1.44$, which is situated in the first band gap (Fig. 3a). The energy of an individual soliton (if all 4 solitons are identical) appears to be within the band, since $\mu_{sol} = 1.44/4 = 0.36$. Therefore the quasi-stationary soliton emerged from the modulational instability can recurrently transform into an extended state without energy constraint. However, when the strength of the optical lattice is increased to $V_0 = 4.0$, the band structure and energy of solitons are modified. Now the energy of an individual soliton $\mu_{sol} = 0.64$ is located in the band gap (see Fig. 3b), and therefore transformation to an extended state (in the 1st or 2nd band) requires the energy loss or gain. For this set of parameters the soliton is stable.

Different initial Bloch states can be prepared experimentally, by loading the BEC into moving optical lattices [25]. As we observed from numerical simulations, the features of the developed pattern of soliton-like excitations depend on the initial waveform. In this paper we considered the simplest situation (perhaps also from the experimental viewpoint), when the initial Bloch states are represented by trigonometric functions.

Numerical simulations in 3D case show in a similar behaviour of the modulational instability with regard to generation of localized excitations. We expect the same phenomena to appear in other physical systems described by similar (periodic NLS-type) equations, such as optically induced photonic crystals.

CONCLUSION

In conclusion, we have illustrated the possibility to create and preserve regularly spaced soliton-like excitations in 2D and 3D BEC arrays employing the phenomenon of modulational instability. The main features of their spatial arrangement are described by the theory based on the multiple scale expansion.

ACKNOWLEDGMENTS

B. B. B. thanks the Physics Department of the University of Salerno, Italy, for a two-year research grant during which this work was done. V.V.K. acknowledges support from the Programme "Human Potential - Research Training Networks", contract No. HPRN-CT-2000-00158. M. S. acknowledges partial support from a MURST-PRIN-2000 Initiative, and from the European grant LOCNET no. HPRN-CT-1999-00163.

REFERENCES

1. M. H. Anderson, J. R. Eshner, M. R. Matthiews, C. E. Wieman, and E. A. Cornell, Science, **269**, 198 (1995).
2. K. B. Davis, M. -O. Mewes, M. R. Andrews, N. J. van Druten, D. S. Durfee, D. M. Kurn, and W. Ketterle, Phys. Rev. Lett. **75**, 3969 (1995).
3. F. Dalfovo, S. Giorgini, L. P. Pitaevskii, and S. Stringary, Rev. Mod. Phys. **71**, 463 (1999).
4. B. P. Anderson and M. Kasevich, Science **282**, 1686 (1998).
5. O. Morsch, J. H. Müller, M. Cristiani, D. Ciampini, and E. Arimondo, Phys. Rev. Lett. **87**, 140402 (2001).
6. F. S. Cataliotti, S. Burger, C. Fort, P. Maddaloni, F. Minardi, A. Trombettoni, A. Smerzi, M. Inguscio, Science, **293**, 843 (2001).
7. M. Greiner, O. Mandel, T. Esslinger, T.W. Hänsch, and I. Bloch, Nature, **415**, 6867 (2002).
8. M. Greiner, I. Bloch, O. Mandel, T.W. Hänsch, and T. Esslinger, Phys. Rev. Lett. **87**, 160405 (2001).
9. B. Wu, R. B. Diener, and Q. Niu, Phys. Rev. A **65**, 025601 (2002).
10. M. J. Steel and W. Zhang, cond-mat/9810284.
11. V. V. Konotop and M. Salerno, Phys. Rev. A **65**, 021602 (2002).
12. B. B. Baizakov, V. V. Konotop, and M. Salerno, J. Phys. B **35**, 5105 (2002).
13. B. B. Baizakov, B. A. Malomed, and M. Salerno, Europhys. Lett. **63**, 642 (2003).
14. E. A. Ostrovskaya and Y. S. Kivshar, Phys. Rev. Lett. **90**, 160407 (2003).
15. W. Chen and D. Mills, Phys. Rev. Lett. **58**, 160 (1987).
16. S. Pötting, P. Meystre, and E. M. Wright, cond-mat/0009289.
17. G. L. Alfimov, V. V. Konotop, and M. Salerno, Europhys. Lett. **58**, 7 (2002).
18. E. P. Gross, Nuovo Cimento 20, 454 (1961).
19. L. P. Pitaevskii, Zh. Eksp. Teor. Fiz. 40, 646 (1961) [Sov. Phys. JETP 13, 451 (1961)].
20. A.H. Nayfeh, *Introduction to Perturbation Techniques*, (Wiley, N.Y., 1981).
21. K. E. Strecker, G. B. Partridge, A. G. Truscott, and R. G. Hulet Nature, **417**, 150 (2002).
22. W. H. Press, S. A. Teukolsky, W. T. Vetterling, and B. P. Flannery, *Numerical Recipes. The Art of Scientific Computing.* (Cambridge University Press, 1996).
23. A. Trombettoni and A. Smerzi, Phys. Rev. Lett. **86**, 2353 (2001).
24. F. K. Abdullaev, B. B. Baizakov, S. A. Darmanyan, V. V. Konotop, and M. Salerno, Phys. Rev. A **64**, 043606 (2001).
25. J. H. Denschlag, J. E. Simsarian, H. Häffner, C.McKenzie, A. Browaeys, D. Cho, K. Helmerson, S. L. Rolson, and W. D. Phillips, J. Phys. B **35**, 3095, (2002).

Multifractal Analysis of Various PDF in Turbulence based on Generalized Statistics: A Way to Tangle in Superfluid He

Toshihico Arimitsu*[†] and Naoko Arimitsu**

*Institute of Physics, University of Tsukuba, Ibaraki 305-8571, Japan
[†]E-mail: arimitsu@cm.ph.tsukuba.ac.jp
**Graduate School of EIS, Yokohama Nat'l. University, Yokohama 240-8501, Japan

Abstract. By means of the multifractal analysis (MFA), the expressions of the probability density functions (PDFs) are unified in a compact analytical formula which is valid for various quantities in turbulence. It is shown that the formula can explain precisely the experimentally observed PDFs both on log and linear scales. The PDF consists of two parts, i.e., the *tail* part and the *center* part. The structure of the tail part of the PDFs, determined mostly by the intermittency exponent, represents the intermittent large deviations that is a manifestation of the multifractal distribution of singularities in physical space due to the scale invariance of the Navier-Stokes equation for large Reynolds number. On the other hand, the structure of the center part represents small deviations violating the scale invariance due to thermal fluctuations and/or observation error.

INTRODUCTION

In this paper, we derive the unified formula for various probability density functions (PDFs) in fully developed turbulence by means of the *multifractal analysis* (MFA) [1, 2, 3, 4, 5, 6, 7, 8, 9, 10, 11, 12], and analyze the PDFs observed in two experiments, i.e., the PDFs of velocity fluctuations, of velocity derivatives and of fluid particle accelerations at $R_\lambda = 380$ that was extracted by Gotoh et al. from the DNS of the size 1024^3 [13], and the PDF of fluid particle accelerations at $R_\lambda = 690$ obtained in the Lagrangian measurement of particle accelerations that was realized by Bodenschatz and co-workers [14, 15, 16] by raising dramatically the spatial and temporal measurement resolutions with the help of the silicon strip detectors. The MFA of turbulence is a unified self-consistent approach for the systems with large deviations, which has been constructed based on the Tsallis-type distribution function [17, 18] that provides an extremum of the *extensive* Rény [19] or the *non-extensive* Tsallis entropy [17, 18, 20] under appropriate constraints. The analysis rests on the scale invariance of the Navier-Stokes equation for high Reynolds number, and on the assumptions that the singularities due to the invariance distribute themselves multifractally in physical space. The MFA is a generalization of the log-normal model [21, 22, 23]. It has been shown [5] that the MFA derives the log-normal model when one starts with the Boltzmann-Gibbs entropy.

For high Reynolds number $\text{Re} \gg 1$, or for the situation where effects of the kinematic viscosity ν can be neglected compared with those of the turbulent viscosity, the Navier-Stokes equation, $\partial \mathbf{u}/\partial t + (\mathbf{u} \cdot \nabla)\mathbf{u} = -\nabla(p/\rho) + \nu \nabla^2 \mathbf{u}$, of an incompressible fluid is invariant under the scale transformation [24, 25, 26] $\mathbf{r} \to \lambda \mathbf{r}$, $\mathbf{u} \to \lambda^{\alpha/3} \mathbf{u}$, $t \to \lambda^{1-\alpha/3} t$ and $(p/\rho) \to \lambda^{2\alpha/3}(p/\rho)$ where the exponent α is an arbitrary real quantity. The quantities ρ and p represent, respectively, mass density and pressure. The Reynolds number Re of the system is given by $\text{Re} = \delta u_{\text{in}} \ell_{\text{in}} / \nu = (\ell_{\text{in}}/\eta)^{4/3}$ with the Kolmogorov scale $\eta = (\nu^3/\epsilon)^{1/4}$ [27] where ϵ is the energy input rate at the input scale ℓ_{in}. Here, we introduced $\delta u_{\text{in}} = |u(\bullet + \ell_{\text{in}}) - u(\bullet)|$ with the definition of the velocity fluctuation (difference) $\delta u_n = |u(\bullet + \ell_n) - u(\bullet)|$ where u is a component of velocity field $\mathbf{u}$, and ℓ_n is a distance between two points. The *pressure* (divided by the mass density) difference $\delta p_n = |p/\rho(\bullet + \ell_n) - p/\rho(\bullet)|$ between two points separated by the distance ℓ_n is another important observable quantity. We are measuring distance by the discrete units $\ell_n = \delta_n \ell_0$ with $\delta_n = 2^{-n}$ ($n = 0, 1, 2, \cdots$). The non-negative integer n can be interpreted as the *multifractal depth*. However, we will treat it as positive real number in the analysis of experiments. The multifractal depth n is related to the number of steps within the energy cascade model.[1] At each step of the cascade, say at the nth step, eddies break up into two pieces producing an energy cascade with the energy-transfer rate ϵ_n that represents the rate of transfer of energy per unit mass from eddies of size ℓ_n to those of size ℓ_{n+1}.

SINGULARITIES AND SCALING EXPONENTS

Let us consider the quantity $\delta x_n = |x(\bullet + \ell_n) - x(\bullet)|$ having the scaling property $|x_n| \equiv |\delta x_n / \delta x_0| = \delta_n^{\phi \alpha / 3}$. Its spatial derivative is defined by $|x'| = \lim_{\ell_n \to 0} \delta x_n / \ell_n \propto \lim_{n \to \infty} \ell_n^{\phi \alpha / 3 - 1}$ which becomes singular for $\alpha < 3/\phi$. The values of exponent α specify the degree of singularity. We see that the scale invariance provides us with $\delta u_n / \delta u_0 = \delta_n^{\alpha/3}$ and $\delta p_n / \delta p_0 = (\ell_n / \ell_0)^{2\alpha/3}$ giving, respectively, $\phi = 1$ for the velocity fluctuation and $\phi = 2$ for the pressure fluctuation. The velocity derivative and the fluid particle acceleration may be estimated, respectively, by $|u'| = \lim_{n \to \infty} u'_n$ and by $|\mathbf{a}| = \lim_{n \to \infty} a_n$ where we introduced the nth velocity derivative $u'_n = \delta u_n / \ell_n$ and the nth fluid particle acceleration $a_n = \delta p_n / \ell_n$ corresponding to the characteristic length ℓ_n. Note that the acceleration $\mathbf{a}$ of a fluid particle is given by the substantive time derivative of the velocity: $\mathbf{a} = \partial \mathbf{u} / \partial t + (\mathbf{u} \cdot \nabla) \mathbf{u}$. We see that the velocity derivative and the fluid particle acceleration become singular for $\alpha < 3$ and $\alpha < 1.5$, respectively, i.e., $|u'| \propto \lim_{\ell_n \to 0} \ell_n^{(\alpha/3) - 1} \to \infty$ and $|\mathbf{a}| \propto \lim_{\ell_n \to 0} \ell_n^{(2\alpha/3) - 1} \to \infty$. We also see that the energy dissipation rate becomes singular in the limit $n \to \infty$ for $\alpha < 1$, i.e., $\lim_{n \to \infty} \epsilon_n / \epsilon_0 = \lim_{n \to \infty} (\ell_n / \ell_0)^{\alpha - 1} \to \infty$ giving $\phi = 3$.

The MFA rests on the multifractal distribution of singularities that is a manifestation of the scale invariance of the Navier-Stokes equation for large Re as mentioned above. The probability $P^{(n)}(\alpha) d\alpha$ to find, at a point in physical space, a singularity labeled by an exponent in the range $\alpha \sim \alpha + d\alpha$ is given by [2, 3, 4, 5]

---

[1] The definition of the number of steps $\bar{n}$ within the energy cascade model is given by $\bar{n} = -\log_2(r/\ell_{\text{in}})$ for the eddies whose diameter is equal to r. By putting $r = \ell_n$, this gives us the relation between $\bar{n}$ and n in the form

$$\bar{n} = n - \log_2(\ell_0 / \ell_{\text{in}}). \tag{1}$$

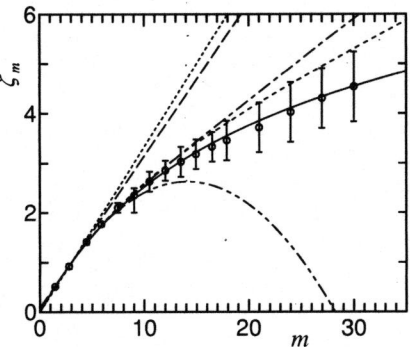

FIGURE 1. Comparison of the present scaling exponents ζ_m for $\mu = 0.238$ (solid curve) with the experimental results plotted by circles at $R_\lambda = 110$ (Re = 32000) [30], and with other theories with the same value of μ. K41 is given by the dotted line, β-model by the dashed line, p-model by the dotted dashed line, log-Poisson model by the short dashed curve, and log-normal by the two dotted dashed curve.

$P^{(n)}(\alpha) = [1 - (\alpha - \alpha_0)^2/(\Delta\alpha)^2]^{n/(1-q)}/Z_\alpha^{(n)}$ with an appropriate partition function $Z_\alpha^{(n)}$ and $(\Delta\alpha)^2 = 2X/[(1-q)\ln 2]$. This is consistent with the relation [26, 5] $P^{(n)}(\alpha) \propto \delta_n^{1-f(\alpha)}$ that reveals how densely each singularity, labeled by α, fills physical space. In the present model, the multifractal spectrum $f(\alpha)$ is given by [2, 3, 4, 5] $f(\alpha) = 1 + (1-q)^{-1}\log_2[1 - (\alpha - \alpha_0)^2/(\Delta\alpha)^2]$. The range of α is $\alpha_{\min} \le \alpha \le \alpha_{\max}$ with $\alpha_{\min} = \alpha_0 - \Delta\alpha$, $\alpha_{\max} = \alpha_0 + \Delta\alpha$. The distribution function $P^{(n)}(\alpha)$ is determined by taking an extremum of generalized entropies, i.e., the *extensive* Rényi entropy or the *non-extensive* Tsallis entropy, under the condition that the information one has for the system is only the value of the intermittency exponent. In spite of the different characteristics of the entropies, the distribution functions $P^{(n)}(\alpha)$ giving their extremum have the common structure.[2]

The dependence of the parameters α_0, X and q on the intermittency exponent μ is determined, self-consistently, with the help of the three independent equations, i.e., the energy conservation: $\langle \epsilon_n \rangle = \epsilon$, the definition of the intermittency exponent μ: $\langle \epsilon_n^2 \rangle = \epsilon^2 \delta_n^{-\mu}$, and the scaling relation[3]: $1/(1-q) = 1/\alpha_- - 1/\alpha_+$ with $\alpha_\pm$ satisfying $f(\alpha_\pm) = 0$. The average $\langle \cdots \rangle$ is taken with $P^{(n)}(\alpha)$.

The scaling exponents ζ_m of the mth order velocity fluctuations, defined by $\langle |u_n|^m \rangle = \langle \delta_n^{m\alpha/3} \rangle \propto \delta_n^{\zeta_m}$, are given in the analytical form [2, 3, 4, 5]

$$\zeta_m = \alpha_0 m/3 - 2Xm^2/[9(1 + C_{m/3}^{1/2})] - [1 - \log_2(1 + C_{m/3}^{1/2})]/(1-q) \quad (2)$$

with $C_{\bar{q}} = 1 + 2\bar{q}^2(1-q)X\ln 2$. The formula (2) is independent of n, that is a manifestation of the scale invariance.

---

[2] Within the present formulation, the decision cannot be pronounced which of the entropies is underlying the system of turbulence.

[3] The scaling relation is a generalization of the one derived first in [28, 29] to the case where the multifractal spectrum has negative values.

The derived scaling exponents (2) are shown in Fig. 1 by the solid curve for the case $\mu = 0.238$, and are compared with experimental data [30] and with the curves given by other theories, i.e., K41 [27], log-normal [21, 22, 23], β-model [31], p-model [32, 26] and log-Poisson [33, 34].

VARIOUS PROBABILITY DENSITY FUNCTIONS

It has been shown that the probability $\Pi_{\phi,S}^{(n)}(x_n)dx_n$ to find a physical quantity x_n in the range $x_n \sim x_n + dx_n$ is given in the form

$$\Pi_\phi^{(n)}(x_n)dx_n = \Pi_{\phi,S}^{(n)}(x_n)dx_n + \Delta\Pi_\phi^{(n)}(x_n)dx_n \quad (3)$$

with the normalization $\int_{-\infty}^{\infty} dx_n \Pi_\phi^{(n)}(x_n) = 1$. The first term represents the contribution by the singular part of the quantity x_n stemmed from the multifractal distribution of its singularities in physical space. This is given by $\Pi_{\phi,S}^{(n)}(|x_n|)dx_n \propto P^{(n)}(\alpha)d\alpha$ with the transformation of the variables, $|x_n| = \delta_n^{\phi\alpha/3}$. Whereas the second term $\Delta\Pi_\phi^{(n)}(x_n)dx_n$ represents the contribution from the dissipative term in the Navier-Stokes equation, and/or the one from the errors in measurements. The dissipative term has been discarded in the above investigation for the distribution of singularities since it violates the invariance under the scale transformation. The contribution of the second term provides a correction to the first one. Note that each term in (3) is a multiple of two probability functions, i.e., the one to determine the portion of the contribution among the above mentioned two independent origins, and the other to find x_n in the range $x_n \sim x_n + dx_n$. Note also that the values of x_n originated in the singularity are rather large representing intermittent large deviations, and that those contributing to the correction terms are of the order of or smaller than its standard deviation.

The mth moment of the variable $|x_n|$ is given by $\langle\langle |x_n|^m \rangle\rangle_\phi \equiv \int_{-\infty}^{\infty} dx_n |x_n|^m \Pi_\phi^{(n)}(x_n) = 2\gamma_{\phi,m}^{(n)} + (1 - 2\gamma_{\phi,0}^{(n)})a_{\phi m} \delta_n^{\zeta_{\phi m}}$ where $2\gamma_{\phi,m}^{(n)} = \int_{-\infty}^{\infty} dx_n |x_n|^m \Delta\Pi_\phi^{(n)}(x_n)$, and $a_{3\bar{q}} = \{2/[\sqrt{C_{\bar{q}}}(1+\sqrt{C_{\bar{q}}})]\}^{1/2}$.

We now derive the PDF, $\hat{\Pi}_\phi^{(n)}(\xi_n)$, defined by the relation $\hat{\Pi}_\phi^{(n)}(\xi_n)d\xi_n = \Pi_\phi^{(n)}(x_n)dx_n$ with the variable $\xi_n = x_n/\langle\langle |x_n|^2 \rangle\rangle_\phi^{1/2}$ normalized by the standard deviation $\langle\langle x_n^2 \rangle\rangle_\phi^{1/2}$. This PDF is to be compared with the observed PDFs. The variable is related with α by $|\xi_n| = \bar{\xi}_n \delta_n^{\phi\alpha/3 - \zeta_{2\phi}/2}$ with $\bar{\xi}_n = [2\gamma_{\phi,2}^{(n)}\delta_n^{-\zeta_{2\phi}} + (1 - 2\gamma_{\phi,0}^{(n)})a_{2\phi}]^{-1/2}$. It is reasonable to imagine that the origin of intermittent rare events is attributed to the first singular term in (3), and that the contribution from the second term is negligible. We then have for the tail part, i.e., $\xi_n^* \leq |\xi_n| \leq \xi_n^{\max}$,

$$\begin{aligned}\hat{\Pi}_\phi^{(n)}(\xi_n)d\xi_n &= \Pi_{\phi,S}^{(n)}(x_n)dx_n \\ &= \bar{\Pi}_\phi^{(n)} \frac{\bar{\xi}_n}{|\xi_n|}\left[1 - \frac{1-q}{n}\frac{(3\ln|\xi_n/\xi_{n,0}|)^2}{2\phi^2 X |\ln\delta_n|}\right]^{n/(1-q)} d\xi_n \quad (4)\end{aligned}$$

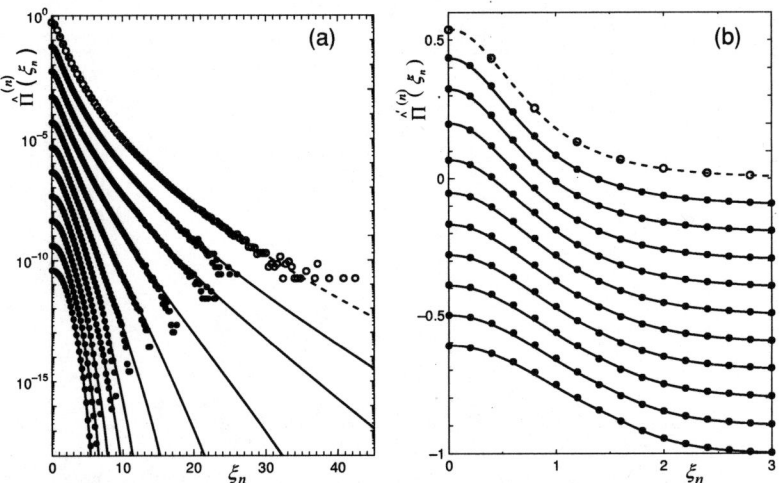

FIGURE 2. Analyses of the PDFs of *velocity fluctuations* (closed circles) and of *velocity derivatives* (open circles) measured in the DNS by Gotoh et al. at $R_\lambda = 380$ by the present theoretical PDFs $\hat{\Pi}^{(n)}(\xi_n)$ for *velocity fluctuations* (solid lines) and for *velocity derivatives* (dashed line) are plotted on (a) log and (b) linear scales. The DNS data points are symmetrized by taking averages of the left and the right hand sides data. The measuring distances, $r/\eta = \ell_n/\eta$, for the PDF of velocity fluctuations are, from the second top to bottom: 2.38, 4.76, 9.52, 19.0, 38.1, 76.2, 152, 305, 609, 1220. For the theoretical PDFs of velocity fluctuations, $\mu = 0.240$ ($q = 0.391$), from the second top to bottom: $(n, \bar{n}, q') = (20.7, 14.6, 1.60), (19.2, 13.1, 1.60), (16.2, 10.1, 1.58), (13.6, 7.54, 1.50), (11.5, 5.44, 1.45), (9.80, 3.74, 1.40), (9.00, 2.94, 1.35), (7.90, 1.84, 1.30), (7.00, 0.94, 1.25), (6.10, 0.04, 1.20)$, $\xi_n^* = 1.10 \sim 1.43$ ($\alpha^* = 1.07$), and $\xi_n^{\max} = 204 \sim 6.63$. For the theoretical PDF of velocity derivatives, $(n, \bar{n}, q') = (22.4, 16.3, 1.55)$, $\xi_n^* = 1.06$ ($\alpha^* = 1.07$), and $\xi_n^{\max} = 302$. For better visibility, each PDF is shifted by -1 unit along the vertical axis.

with $\bar{\Pi}_\phi^{(n)} = 3(1 - 2\gamma_0^{(n)})/(2\phi\bar{\xi}_n\sqrt{2\pi X|\ln\delta_n|})$, $\xi_{n,0} = \bar{\xi}_n \delta_n^{\phi\alpha_0/3 - \zeta_{2\phi}/2}$, $\xi_n^{\max} = \bar{\xi}_n \delta_n^{\phi\alpha_{\min}/3 - \zeta_{2\phi}/2}$. On the other hand, for the center part, the contribution to the PDF comes, mainly, from thermal fluctuations or measurement error. It may be described by the Tsallis distribution function with respect to the variable ξ_n itself, i.e., $|\xi_n| \leq \xi_n^*$,

$$\hat{\Pi}_\phi^{(n)}(\xi_n)d\xi_n = \left[\hat{\Pi}_{\phi,S}^{(n)}(x_n) + \Delta\hat{\Pi}_\phi^{(n)}(x_n)\right]dx_n$$
$$= \bar{\Pi}_\phi^{(n)}\left\{1 - \frac{1-q'}{2}\left(1 + \frac{3f'(\alpha^*)}{\phi}\right)\left[\left(\frac{\xi_n}{\xi_n^*}\right)^2 - 1\right]\right\}^{1/(1-q')}d\xi_n. \quad (5)$$

This specific form of the Tsallis distribution function is determined by the condition that the two PDFs (4) and (5) should have the same value and the same slope at ξ_n^* which is defined by $\xi_n^* = \bar{\xi}_n \delta_n^{\phi\alpha^*/3 - \zeta_{2\phi}/2}$ with α^* being the smaller solution of $\zeta_{2\phi}/2 - \phi\alpha/3 + 1 - f(\alpha) = 0$. It is the point at which $\hat{\Pi}_\phi^{(n)}(\xi_n^*)$ has the least n-dependence for large n.

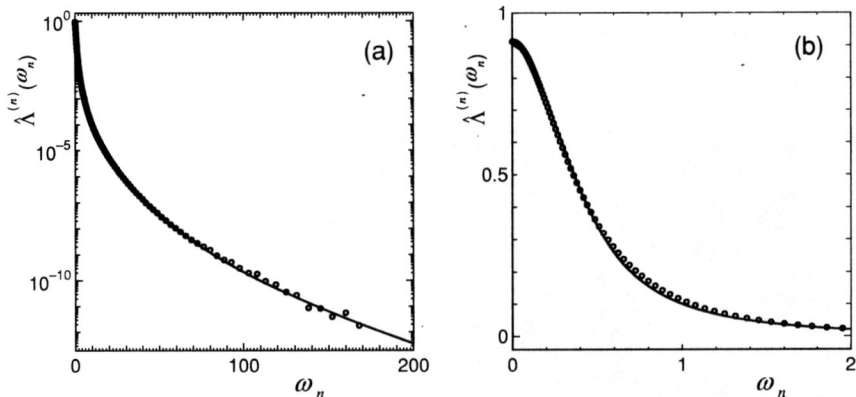

FIGURE 3. Comparison between the PDF of *fluid particle accelerations* measured in the DNS by Gotoh et al. at $R_\lambda = 380$ and the present theoretical PDF $\hat{\Lambda}^{(n)}(\omega_n)$ are plotted on (a) log and (b) linear scales. Closed circles are the DNS data points both on the left and right hand sides of the PDF. Solid lines represent the curves given by the present theory with $\mu = 0.240$ ($q = 0.391$), $(n, \bar{n}, q') = (17.5, 11.4, 1.70)$, $\omega_n^* = 0.622$ ($\alpha^* = 1.01$), and $\omega_n^{\max} = 2530$.

With the help of the second equality in (5), we obtain $\Delta\Pi_\phi^{(n)}(x_n)$, and have the formula to evaluate $\gamma_{\phi,m}^{(n)}$ in the form $2\gamma_{\phi,m}^{(n)} = \left(K_{\phi,m}^{(n)} - L_{\phi,m}^{(n)}\right) / \left(1 + K_{\phi,0}^{(n)} - L_{\phi,0}^{(n)}\right)$ where

$$K_{\phi,m}^{(n)} = \frac{3\,\delta_n^{\phi(m+1)\alpha^*/3 - \zeta_{2\phi}/2}}{\phi\sqrt{2\pi X |\ln\delta_n|}} \int_0^1 dz\, z^m \left[1 - \frac{1-q'}{2}\left(1 + \frac{3f'(\alpha^*)}{\phi}\right)(z^2 - 1)\right]^{1/(1-q')} \quad (6)$$

$$L_{\phi,m}^{(n)} = \frac{3\,\delta_n^{\phi m \alpha^*/3}}{\phi\sqrt{2\pi X |\ln\delta_n|}} \int_{z_{\min}^*}^1 dz\, z^{m-1} \left[1 - \frac{1-q}{n}\frac{(3\ln|z/z_0^*|)^2}{2\phi^2 X |\ln\delta_n|}\right]^{n/(1-q)} \quad (7)$$

with $z_{\min}^* = \xi_{\min}/\xi_n^* = \delta_n^{\phi(\alpha_{\max} - \alpha^*)/3}$, $z_0^* = \xi_{n,0}/\xi_n^* = \delta_n^{\phi(\alpha_0 - \alpha^*)/3}$. We see that the tail part of the PDF, given by (4), is mostly determined by the intermittency exponent μ and the multifractal depth n which gives a length scale ℓ_n. On the other hand, the center part of the PDF, (5), is mainly controlled by q'.

The PDFs both for velocity fluctuations and for velocity derivatives are given by the common formula $\hat{\Pi}^{(n)}(\xi_n) \equiv \hat{\Pi}_{\phi=1}^{(n)}(\xi_n)$ in their normalized variables $\xi_n = \delta u_n / \langle\!\langle (\delta u_n)^2 \rangle\!\rangle^{1/2}$. On the other hand, the PDFs both for pressure differences and for fluid particle accelerations are given by the common formula $\hat{\Lambda}^{(n)}(\omega_n) \equiv \hat{\Pi}_{\phi=2}^{(n)}(\omega_n)$ in their normalized variables $\omega_n = \delta p_n / \langle\!\langle (\delta p_n)^2 \rangle\!\rangle^{1/2}$. The PDF for energy dissipation rates is given with $\phi = 3$.

The PDFs extracted by Gotoh et al. from their DNS data [13] at $R_\lambda = 380$ are shown, on log and linear scales, in Fig. 2 both for *velocity fluctuations* and for *velocity derivatives*, and in Fig. 3 for *fluid particle accelerations*. We found the value $\mu = 0.240$ by analyzing the measured scaling exponents ζ_m of velocity structure function with the

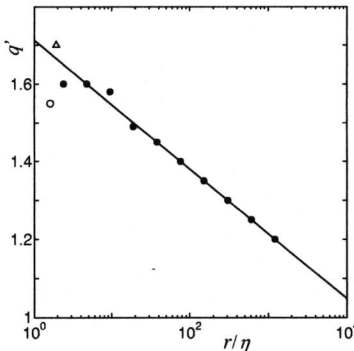

FIGURE 4. Dependence of q' on the distance r/η extracted from the analyses of the PDFs for velocity fluctuations (closed circles), for velocity derivatives (open circle) and for fluid particle accelerations (open triangle). The line represents $q' = -0.05\log_2(r/\eta) + 1.71$.

formula (2), which gives the values $q = 0.391$, $\alpha_0 = 1.14$ and $X = 0.285$. Through the analyses of the PDFs for velocity fluctuations in Fig. 2, we extracted the formula for the dependence of n on r/η: [7, 8]

$$n = -0.989 \times \log_2 r/\eta + 16.1 \quad (\text{for } \ell_c \leq r), \tag{8}$$
$$n = -2.40 \times \log_2 r/\eta + 24.0 \quad (\text{for } r < \ell_c). \tag{9}$$

This shows that the inertial range is divided into two scaling regions separated by the characteristic length $\ell_c/\eta = 48.7$ which is close to the Taylor microscale $\lambda/\eta = 38.3$ of the system. The equation (8) is consistent with the picture of the energy cascade model in which each eddy breaks up into 2 pieces at every cascade steps, whereas (9) indicates that, for $r < \ell_c$, each eddy breaks up, effectively, into $1.33 \approx 4/3$ [8] pieces at every cascade steps. This fact may be attributed to a manifestation of structural difference of eddies, which can be checked by visualizing DNS eddies. Actually, one observes that DNS eddies with larger diameters than Taylor microscale λ have rather round shapes, whereas eddies with smaller diameters compared with λ have rather stretched shapes [35]. The energy input scale for this DNS is estimated as the longest scale available in the lattice with cyclic boundary condition, i.e., $\ell_{in}/\eta = \pi/\eta \approx 1220$ with $\eta \approx 0.258 \times 10^{-2}$ [13] which gives the number of steps $\bar{n}$ within the energy cascade model through the formula (1) with $\ell_0/\eta \approx 81300$ determined by (8).

For the analysis of the PDF for velocity derivatives in Fig. 2, we chose the value $(n, \bar{n}, q') = (22.4, 16.3, 1.55)$. The length corresponding to n is calculated by (9) to give $r/\eta = 1.61$, which may provide us with an estimate for the effective shortest length in processing the DNS data to extract velocity derivatives. Note that it is about the same order of the mesh size $\Delta r/\eta = 2\pi/(1024 \times \eta) \approx 2.38$ [13] of the DNS lattice.

For the PDFs for *fluid particle accelerations* in Fig. 3, we have $(n, \bar{n}, q') = (17.5, 11.4, 1.70)$. Substitution of this value into (9) gives the corresponding characteristic length $r/\eta = 7.91$ [11]. This may be the effective minimum resolution in cooking the DNS data to distill accelerations.

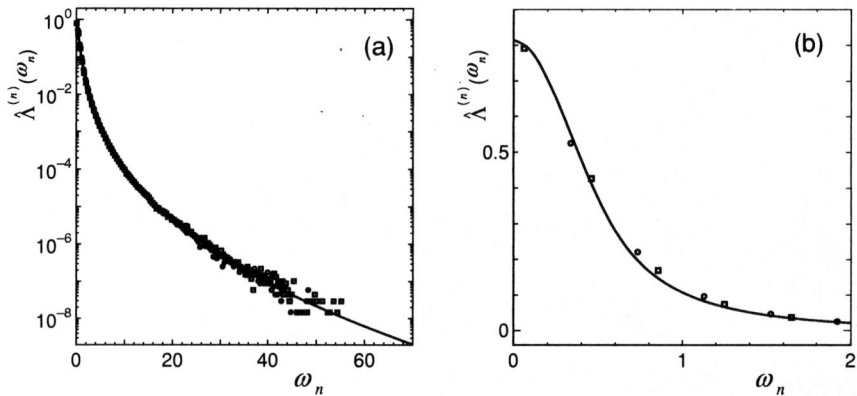

FIGURE 5. Comparison between the experimentally measured PDF of *fluid particle accelerations* by Bodenschatz et al. at $R_\lambda = 690$ (Re = 31 400) and the present theoretical PDF $\hat{\Lambda}^{(n)}(\omega_n)$ are plotted on (a) log and (b) linear scales. Open squares are the experimental data points on the left hand side of the PDF, whereas open circles are those on the right hand side. Solid lines represent the curves given by the present theory with $\mu = 0.240$ ($q = 0.391$), $(n, q') = (17.1, 1.45)$, $\omega_n^* = 0.605$ ($\alpha^* = 1.01$), and $\omega_n^{\max} = 2040$.

In Fig. 4, we plotted the dependence of q' on the distance r/η extracted from the analyses of the DNS data, i.e., the closed circles are extracted from the PDF of velocity fluctuations, the open circle is from the PDF of velocity derivatives and the open triangle from the PDF of fluid particle accelerations. The line represents $q' = -0.05 \log_2(r/\eta) + 1.71$. The points for $r/\eta > 20$ (closed circles) and for the accelerations (open triangle) are quite sensitive and easy to be determined. Other points are insensitive and have a rather wide range in deciding the values q'.

The PDFs for fluid particle accelerations measured by Bodenschatz et al. at $R_\lambda = 690$ [16] are given in Fig. 5. We determined the value $n = 17.1$ by substituting the reported value $\ell_0 = \ell_{\text{in}} = 7.1$ cm and the spatial resolution 0.5 μm of the measurement for ℓ_n into its definition, $n = \log_2(\ell_0/\ell_n)$. The values $\mu = 0.240$ and $q' = 1.45$ are extracted by the analysis of the experimental PDF with the derived theoretical formula [11]. Then, we have the values of parameters: $q = 0.391$, $\alpha_0 = 1.14$ and $X = 0.285$. The flatness of the PDF turns out to be [11] $F_a^{(n)} \equiv \langle\langle a_n^4 \rangle\rangle / \langle\langle a_n^2 \rangle\rangle^2 = \langle\langle \xi_n^4 \rangle\rangle = 56.9$ which is compatible with the value of the flatness $\sim 55 \pm 4$ reported in [16].

DISCUSSIONS AND PROSPECTS

In this paper, the various experimental PDFs in turbulence are analyzed precisely with the formulae (4) and (5) of the corresponding PDFs derived by the MFA. It is revealed that there are two distinct mechanisms underlying the dynamics of turbulence. One contributes to the *tail* part of PDFs and the other to the *center* part. The structure of the tail part of the PDFs is determined by the global structure representing the intermittent large deviations that is manifestations of the multifractal distribution of singularities in physical space due to the scale invariance of the Navier-Stokes equation for large Reynolds number. The specific form of the tail part comes from the assumption that the probability to find a singularity exponent α within the range $\alpha \sim \alpha + d\alpha$ at a point in physical space is given by the Tsallis-type distribution function with the Tsallis parameter q. The relation between α and an observing variable is given by the scale transformation. On the other hand, the structure of the center part represents small deviations violating the scale invariance due to thermal fluctuations and/or observation error. The center part is assumed to be given by the Tsallis-type distribution function with the Tsallis parameter q' for the observing variable itself. The value of q' may be determined by a local structure of the system, e.g., the dynamics of a vortex, the mutual interaction between vortices and so on, and depends on the distance of two measuring points in contrast to q. The latter parameter q does not depend on the distance, and is determined once the value of the intermittency exponent is given. It is one of the attractive future problems to derive two different dynamics taking care of the tail part controlled by q and the center part by q', and will be reported in the near future.

The success of the MFA in analyzing turbulence in high accuracy may provide us with a good tool to see what is the origin of the singularities and why their distribution is multifractal. In order to investigate them, the vortex tangle is one of the attractive candidates as Feynman proposed [36], since the vorticity in superfluid ^{4}He and ^{3}He is quantized and the normal component within the sense of the two fluid model can be negligible at very low temperature. If the singularity originates from the core of vortex, the multifractality of turbulence in normal fluid can be related to various values of vorticities in the fluid. In this case, the vortex tangle may be uni-fractal, and does not exhibit intermittency. If the singularity originates from the reconnection of vortices, the multifractality of turbulence in normal fluid is related to the distribution of reconnection points in the fluid. Then, the vortex tangle may be also multifractal, and does exhibit intermittency. A temperature-independent vortex decay mechanism below $T \sim 70$ mK has been observed in superfluid ^{4}He [37], and the Kolmogorov spectrum (K41) is extracted from the simulation within the vortex filament model for non-frictional suferfluid ^{4}He [38]. In tangle, the quantized vortex crosses the stream lines of superfluid velocity field, which may result in the decay mechanism at low temperature. We expect that, through the analysis of the local dynamics controlled by the Tsallis parameter q', we can extract some information about the mutual friction between the superfluid and normal components.

A circular vortex lattice is observed in the simulation of fast rotating Bose-Einstein condensate confined in a 2-dimensional quadratic-plus-quartic potential [39]. A possibility of generation of tangle phase in this system is one of the attractive problems. The situation may be quite similar to the one in the generation of the Taylor-Couette turbulence in normal fluid.

Let us close this paper by mentioning that the vortex tangle can be an important stationary phase of *quLSI* (quantum LSI) consisting of, for example, a huge number of superconducting loops of flux qubit [40]. The direction of current in a loop may change so quickly under an operation of quantum computer. The congeries of the flux qubits can produce a tangle phase of flux quantums. It may be important to see if the tangle phase benefits quantum entangled states or not.

ACKNOWLEDGMENTS

The authors would like to thank Prof. R.H. Kraichnan and Prof. C. Tsallis for their fruitful and enlightening comments with encouragement, and are grateful to Prof. E. Bodenschatz and Prof. T. Gotoh for the kindness to show their data prior to publication.

REFERENCES

1. T. Arimitsu and N. Arimitsu, *Phys. Rev. E* **61**, 3237-3240 (2000).
2. T. Arimitsu and N. Arimitsu, *J. Phys. A: Math. Gen.* **33**, L235-L241 (2000) [CORRIGENDUM: **34**, 673-674 (2001)].
3. T. Arimitsu and N. Arimitsu, *Chaos, Solitons and Fractals* **13**, 479-489 (2002).
4. T. Arimitsu and N. Arimitsu, *Prog. Theor. Phys.* **105**, 355-360 (2001).
5. T. Arimitsu and N. Arimitsu, *Physica A* **295**, 177-194 (2001).
6. N. Arimitsu and T. Arimitsu, *J. Korean Phys. Soc.* **40**, 1032-1036 (2002).
7. T. Arimitsu and N. Arimitsu, *Physica A* **305**, 218-226 (2002).
8. T. Arimitsu and N. Arimitsu, *J. Phys.: Condens. Matter* **14**, 2237-2246 (2002).
9. N. Arimitsu and T. Arimitsu, *Europhys. Lett.* **60**, 60-65 (2002).
10. T. Arimitsu and N. Arimitsu, *Condenced Matter Physics* **6** 85-92 (2003).
11. T. Arimitsu and N. Arimitsu, cond-mat/0210274 (2002).
12. T. Arimitsu and N. Arimitsu, cond-mat/0301516 (2003).
13. T. Gotoh, D. Fukayama and T. Nakano, *Phys. Fluids* **14**, 1065-1081 (2002).
14. A. La Porta, et al., *Nature* **409**, 1017-1019 (2001).
15. G. A. Voth, et al., *J. Fluid Mech.* **469**, 121-160 (2002).
16. A. M. Crawford, N. Mordant and E. Bodenschatz, (2002) physics/0212080.
17. C. Tsallis, *J. Stat. Phys.* **52**, 479-487 (1988).
18. C. Tsallis, *Braz. J. Phys.* **29**, 1-35 (1999); See recent progresses at http://tsallis.cat.cbpf.br/biblio.htm.
19. A. Rényi, "On measures of entropy and information" in *Proc. 4th Berkeley Symp. Maths. Stat. Prob.* **1**, 547 (1961).
20. J.H. Havrda and F. Charvat, *Kybernatica* **3**, 30-35 (1967).
21. A.M. Oboukhov, *J. Fluid Mech.* **13**, 77-81 (1962).
22. A.N. Kolmogorov, *J. Fluid Mech.* **13**, 82-85 (1962).
23. A.M. Yaglom, *Sov. Phys. Dokl.* **11**, 26-29 (1966).
24. S. S. Moiseev, A. V. Tur and V. V. Yanovskii, *Sov. Phys. JETP* **44**, 556-561 (1976).
25. U. Frisch and G. Parisi, "On the singularity structure of fully developed turbulence" in *Turbulence and Predictability in Geophysical Fluid Dynamics and Climate Dynamics*, ed. by M. Ghil, R. Benzi and G. Parisi, North-Holland, New York, 1985, p84.
26. C. Meneveau and K. R. Sreenivasan, *Nucl. Phys. B (Proc. Suppl.)* **2**, 49-76 (1987).
27. A.N. Kolmogorov, *Dokl. Akad. Nauk SSSR* **30**, 301-305 (1941); ibid. **31**, 538-540 (1941).
28. U.M.S. Costa, M.L. Lyra, A.R. Plastino and C. Tsallis, *Phys. Rev. E* **56**, 245-250 (1997).
29. M.L. Lyra and C. Tsallis, *Phys. Rev. Lett.* **80**, 53-56 (1998).
30. C. Meneveau and K. R. Sreenivasan, *J. Fulid Mech.* **224**, 429-484 (1991).
31. U. Frisch, P-L. Sulem and M. Nelkin, *J. Fluid Mech.* **87**, 719-736 (1978).
32. C. Meneveau and K. R. Sreenivasan, *Phys. Rev. Lett.* **59**, 1424-1427 (1987).
33. Z-S. She and E. Leveque, *Phys. Rev. Lett.* **72**, 336-339 (1994).
34. Z-S. She and E. Waymire, *Phys. Rev. Lett.* **74**, 262-265 (1995).
35. Private communication with Prof. M. Tanahashi at Tokyo Institute of Technology.
36. R. P. Feynman, "Application of quantum mechanics to liquid helium" in *Progress in Low Temperature Physics*, ed. by C. J. Gorter, North Holland, Amsterdam, 1955, p17.
37. S.I. Davis, P.C. Hendry and P.V.E. McClintock, *Physica B* **280**, 43-44 (2000).
38. T. Araki, M. Tsubota and S.K. Nemirovskii, *Phys. Rev. Lett.* **89**, 145301 (2002).
39. K. Kasamatsu, M. Tsubota and M. Ueda, *Phys. Rev. A* **66**, 053606 (2002).
40. I. Chiorescu et al., *Science* **299**, 1869-1871 (2003).

Combinatorial aspects of exclusion and parastatistics

S. Chaturvedi[1]

School of Physics, University of Hyderabad, Hyderabad 500 046, India

Abstract.
Combinatorial aspects of all statistics based on the permutation group are analyzed by imposing the requirements of indistinguishability in the permutation group sense on the Hilbert space describing N identical particles. Compact expressions for the grand canonical partition functions are given wherever possible. The theory of symmetric functions is found to play a significant role in this development. An analysis of the semion statistics of Haldane is also presented from this perspective together with some recent developments in the field of exclusion statistics.

INTRODUCTION

Since the advent of Bose and Fermi statistics, numerous statistics other than Bose and Fermi have been proposed [1]. These statistics can be divided into two broad categories:

- Generalizations of Bose and Fermi statistics based on exploring all the possibilities permitted by the mathematical principles underlying Bose and Fermi statistics.
- Interpolations between Bose and Fermi statistics motivated largely by the observed behavior of the quasi-particles of interacting quantum systems.

The purpose of the present work is to consider, within the framework of non relativistic quantum mechanics, some combinatorial aspects of (a) statistics based on the permutation group [2]-[8] from the first category and (b) Haldane's exclusion statistics [9] from the second.

GENERALITIES

To set the notation, we begin with some generalities regarding the structure of the canonical partition function for an assembly of a fixed number of non interacting identical particles obeying any statistics. Consider a system of N identical particles each of which can occupy M states corresponding to energies $\varepsilon_1, \cdots, \varepsilon_M$. Let n_i denote the occupation number of the state corresponding to the energy ε_i. The canonical partition function for

---

[1] e-mail: scsp@uohyd.ernet.in

such a system has the following structure:

$$Z_N(x) = \sum_{\substack{n_i \\ \Sigma n_i = N}} f(n_1, \cdots, n_M) \, x_1^{n_1} x_2^{n_2} \cdots x_M^{n_M} \, , \qquad (1)$$

where $x \equiv x_1, \cdots, x_M$; $x_i \equiv \exp(-\beta \varepsilon_i)$ and $\beta \equiv 1/kT$. The function $f(n_1, \cdots, n_M)$ which weights various sequences of occupation numbers is required to be positive

$$f(n_1, \cdots, n_M) \geq 0 \, . \qquad (2)$$

If $f(n_1, \cdots, n_M)$ is integer valued we may think of it as giving the number of independent N particle states corresponding to the occupation numbers $n_1, \cdots, n_M$.
Setting $x_i = 1$ we obtain the counting formula

$$W_N^M = \sum_{\substack{n_i \\ \Sigma n_i = N}} f(n_1, \cdots, n_M), \qquad (3)$$

which gives the number of independent states of a N particle system under consideration.

The function $f(n_1, \cdots, n_M)$ may either be

- a symmetric function of $n_1, \cdots, n_M$ or
- a non-symmetric function of $n_1, \cdots, n_M$.

If former is the case, then $Z_N(x)$ becomes a symmetric function of $x_1, \cdots, x_M$ and (1) may compactly be written as

$$Z_N(x) = \sum_\lambda f(\lambda_1, \cdots, \lambda_M) \, m_\lambda(x) \, . \qquad (4)$$

Here $\lambda \equiv (\lambda_1, \lambda_2, \cdots, \lambda_M)$, $\lambda_1 \geq \lambda_2 \geq \lambda_3 \cdots \geq \lambda_M$ is a partition of N and $m_\lambda(x)$ denotes the monomial symmetric function corresponding to the partition λ [10]

$$m_\lambda(x) = \sum x_1^{\lambda_1} x_2^{\lambda_2} \cdots x_M^{\lambda_M} \, . \qquad (5)$$

The sum on the rhs of (5) is over all distinct permutations of $(\lambda_1, \cdots, \lambda_M)$. The functions $m_\lambda(x)$ provide a basis for all symmetric polynomials in the variables $x_1, \cdots, x_M$ of degree N.

If $f(n_1, \cdots, n_M)$ is not only symmetric but also factorizable as

$$f(n_1, n_2, \cdots, n_M) = p(n_1) p(n_2) \cdots p(n_M) \, , \qquad (6)$$

then the grand canonical partition function

$$\mathscr{Z}(X) = \sum_N e^{\mu \beta N} Z_N(x) \, , \qquad (7)$$

also factorizes:

$$\mathscr{Z}(X) = \prod_{i=1}^M z(X_i) \, , \qquad (8)$$

where $X_i \equiv \exp(-\beta(\varepsilon_i - \mu))$ and

$$z(X) = \sum_{n=0}^{\infty} p(n) X^n \quad . \tag{9}$$

Bose and Fermi (and also Maxwell-Boltzmann) statistics belong to this special category:

$$p(n) = 1 \quad \text{(Bose)} \tag{10}$$
$$p(n) = 1 \; for \; n_i = 0, \; 1 \; \text{zero otherwise} \quad \text{(Fermi)}. \tag{11}$$

IDENTICAL PARTICLES: INFINITE STATISTICS

Let H denote the single particle Hilbert space, and let $|1> \cdots |M>$ the states corresponding to the single particle energies $\varepsilon_1, \cdots, \varepsilon_M$. The N-particle Hilbert state $\mathcal{H}$ is built by an N-fold tensor product of H. The M^N basis vectors of $\mathcal{H}$ correspond to each term in the product

$$\frac{(|1> + \cdots + |M>)(|1> + \cdots + |M>) \cdots (|1> + \cdots + |M>)}{N \text{ factors}}. \tag{12}$$

If one regards all of them as independent then one has what is called the infinite statistics of Greenberg [11]. Decomposing this set of M^N states according to occupation numbers by grouping together states which have the same number of 1's, 2's $\cdots$ etc. regardless of their location in the product, one finds that $f(n_1, \cdots, n_M) = N!/n_1! \cdots n_M!$ and hence, on using the multinomial theorem

$$Z_N^{(I)}(x) = (x_1 + \cdots + x_M)^N \quad . \tag{13}$$

Note that $f(n_1, \cdot, n_M)$ here though symmetric is not factorizable. It becomes so upon division by $N!$ and when this ('correction') is done one obtains the Maxwell-Boltzmann statistics.

The M^N states could also be viewed as the carrier space for an M^N dimensional representation of the permutation group S_N[12]. This reducible representation can be decomposed into the irreducible representations of S_N which are in one to one correspondence with the partitions λ of N. All features of this decomposition are encapsulated in

$$Z_N^{(I)}(x) = \sum_{\lambda} n(\lambda) s_\lambda(x), \tag{14}$$

where $n(\lambda)$ denotes the dimension of the irreducible representation λ of S_N and $s_\lambda(x)$ denote Schur functions [10].

INDISTINGUISHABILITY IN THE PERMUTATION GROUP SENSE

The Hilbert space $\mathcal{H}_{phy}$ describing identical and indistinguishable particles is a generalized ray space constructed out of $\mathcal{H}$ by [8] (a) admitting only those operators on $\mathcal{H}$

which are permutation symmetric i.e. those operators which treat all the factors in the tensor product democratically. (b) identifying those states in $\mathcal{H}$ which have the same expectation values for all permutation symmetric operators.

These assumptions, by Schur's Lemma, imply that all states in $\mathcal{H}$ belonging to an irreducible representation λ of S_N count as one state of $\mathcal{H}_{phy}$.

IDENTICAL AND INDISTINGUISHABLE PARTICLES

From the considerations given above, it is clear that partition function for statistics describing identical and indistinguishable particles is obtained by setting $n(\lambda) = 1$ in Eq. (14) [13]

$$Z_N^{(II)}(x) = {\sum_\lambda}' s_\lambda(x), \qquad (15)$$

where the prime over the summation indicates possible restrictions on λ. Different statistics describing identical and indistinguishable particles correspond to different restrictions on λ's. Some well known cases, for which a Fock-space formulation is also available, are listed below.

[1] Bose Statistics: $\lambda = (N)$

$$Z_N^{(B)}(x) = s_{(N)}(x) \quad , \qquad (16)$$

$$\mathscr{L}^B(X) = \prod_{i=1}^{M} \frac{1}{(1-X_i)} \quad . \qquad (17)$$

The Fock space for Bosons is characterized by the following commutation relations

$$[a_k, a_l] = 0 \,;\, [a_k, a_l^\dagger] = \delta_{kl} \,;\, k,l = 1 \cdots M, \qquad (18)$$

together with

$$a_k |0> = 0 \,;\quad k = 1 \cdots M. \qquad (19)$$

[2] Fermi Statistics: $\lambda = (1^N)$

$$Z_N^{(F)}(x) = s_{(1^N)}(x) \quad , \qquad (20)$$

$$\mathscr{L}^F(X) = \prod_{i=1}^{M}(1+X_i) \quad . \qquad (21)$$

The Fock space for Fermions is characterized by the following anti commutation relations

$$\{a_k, a_l\} = 0 \,;\, \{a_k, a_l^\dagger\} = \delta_{kl} \,;\, k,l = 1 \cdots M, \qquad (22)$$

together with

$$a_k |0> = 0 \,;\quad k,l = 1 \cdots M. \qquad (23)$$

[3] para Bose Statistics of order $p : l(\lambda) \leq p$

$$Z_N^{(PB)}(x;p) = \sum_{l(\lambda) \leq p} s_\lambda(x). \tag{24}$$

Here $l(\lambda)$ denotes the length i.e. the number of non zero parts of the partition λ. A closed form expression for $\mathscr{Z}^B(X)$ is not available.

The Fock space for the para Bose systems is characterized by the following trilinear relations [3]

$$[a_k, \{a_l, a_m\}] = 0 \; ; \; [a_k, \{a_l^\dagger, a_m^\dagger\}] = 2\delta_{kl} a_m^\dagger + 2\delta_{km} a_l^\dagger \; ; \; [a_k, \{a_l^\dagger, a_m\}] = 2\delta_{kl} a_m, \tag{25}$$

and the supplementary conditions

$$a_k a_l^\dagger |0> = p\delta_{kl} |0>. \tag{26}$$

[4] para Fermi Statistics of order $p : \lambda_1 \leq p$

$$Z_N^{(PB)}(x;p) = \sum_{\lambda_1 \leq p} s_\lambda(x). \tag{27}$$

The grand canonical partition function in this case can be given in a closed form as [14]

$$\mathscr{Z}^{(PF)}(X;p) = = \frac{\det(X_j^{2M+p+1-i} - X_j^i)}{\det(X_j^{2M+1-i} - X_j^i)} \; ; \; 1 \leq i,j \leq M. \tag{28}$$

The relations characterizing a para Fermi system are as follows [3]

$$[a_k, [a_l, a_m]] = 0 \; ; \; [a_k, [a_l^\dagger, a_m^\dagger]] = 2\delta_{kl} a_m^\dagger - 2\delta_{km} a_l^\dagger \; ; \; [a_k, [a_l^\dagger, a_m]] = 2\delta_{kl} a_m, \tag{29}$$

together with the supplementary conditions

$$a_k a_l^\dagger |0> = p\delta_{kl} |0>. \tag{30}$$

It may be remarked that finding the Fock space description of statistics for identical and indistinguishable particles other than those listed above is still an open problem.

In the case of the statistics for identical and indistinguishable particles with no restrictions on λ (which could be thought of as the $p \to \infty$ limit of para Bose or para Fermi statistics), the grand canonical partition function has a nice closed form expression [13]:

$$\mathscr{Z}(X) = \prod_i \frac{1}{(1-X_i)} \prod_{i<j} \frac{1}{(1-X_i X_j)}. \tag{31}$$

We thus have a complete picture of all the statistics based on the permutation group in the sense defined above. If in addition one requires them to satisfy cluster decomposition property then the choice gets restricted only to parastatistics [8]. If one goes a step further and requires that the f's be factorizable then one has only two possibilities- Bose and Fermi.

HALDANE'S FRACTIONAL EXCLUSION STATISTICS

Drawing inspiration from his work on certain one dimensional spin chains, Haldane proposed the following interpolation between Bose and Fermi statistics [9] at the level of the counting formula:

$$W_N^{(g)}(M) = \binom{d_N^g + N - 1}{N} , \quad (32)$$

where $d_N^g = M - g(N-1)$ is to be thought of as the effective dimension of the single particle Hilbert space available to the Nth particle after $N-1$ particles have been put into the available M states. For $g=0$ ($g=1$) one recovers the counting formula for Bose (Fermi) statistics. Starting from this counting formula, standard procedures of statistical mechanics yield the following result for the mean number of particles $\bar{n}_i$ in the state i corresponding to the single particle energy ε_i:

$$\bar{n}_i = \frac{1}{w(X_i, g) + g} , \quad (33)$$

where $X_i \equiv e^{-\beta(\varepsilon_i - \mu)}$ and $w(X)$ satisfies the Ouvry equation[15]

$$w^g(1+w)^{1-g} = \frac{1}{X} . \quad (34)$$

These equations imply that at $T = 0$ the maximum number of particles that can be put in a state is $1/g$.

The expression for $\bar{n}_i$ given above can also be derived from a factorized form for the grand canonical partition function:

$$\mathscr{Z}^{(g)}(X) = \prod_{i=1}^{M}(1 + 1/w(X_i)) , \quad (35)$$

with the help of the relation

$$\bar{n}_i = X_i \frac{\partial}{\partial X_i} \log \mathscr{Z}^{(g)}(X) . \quad (36)$$

The assumption that $\mathscr{Z}^{(g)}(X)$ for this statistics can be written as a product in a similar fashion as for Bose and Fermi statistics leads to difficulties except for the limiting cases $g = 0, 1$. It gives rise to negative probabilities- the coefficients $p(n)$ in (9) become negative. This question was resolved in [16]-[17] by explicitly determining the function $f(n_1, \cdots, n_M)$ consistent with Haldane's counting formula (with appropriate restrictions on g and N so that $W_N^{(g)}(M)$ is a whole number) and showing the f's thus obtained are not factorizable and thereby making the assumed factorizable form for $\mathscr{Z}^{(g)}(X)$ untenable. In particular, assuming that $f(n_1, \cdot, n_M)$ is a symmetric function of its arguments, it was shown that for $g = 1/2$ and $N = 2p+1$ one finds that

$$Z_{2p+1}^{(1/2)}(x) = \sum_{q=0}^{p} \binom{p}{q} \binom{2p-q+1}{q}^{-1} m_{\lambda(q)}(x) , \quad (37)$$

where the sum is over partitions $\lambda(q)$ of the type $(2^q, 1^{2(p-q)+1})$; $q = 1, \cdots, p$. (Here we use 2^q as a shorthand for an array of q 2's). As one can see, no negative f's appear anywhere. However, unlike Bose and Fermi, the f's in this case are not factorizable.

The line of thought pursued in [16]-[17] was further developed by Bergére who, in a remarkably beautiful work [18], analyzed the general case $g = \ell/m, N = mp + r, r \leq m$ and showed, without assuming symmetric f's, that the canonical partition function in this can be written as

$$Z_N^{(g)}(x) = \sum_{\{n_i\} \in \Lambda} x_1^{n_1} x_2^{n_2} \cdots x_M^{n_M} , \qquad (38)$$

where the set Λ of permitted n_i's is constrained by the conditions:

- $\sum_i^M n_i = N$
- $\{m, 2m, \cdots, pm\} \subseteq \{\pi_i\}$ where $\pi_i = \sum_{i \leq j}$.
- For the indices $(i_1, \cdots, i_p)$ such that $\pi_{i_q} = qm$ then $p_{i_q+1} = p_{i_q+2} = \cdots = p_{i_q+\ell-1} = 0$; $q = 1, \cdots, p$.

These results reproduce the cases considered in [17] and upon symmetrization, those found in [16] for the semion case. The occupancy restrictions given above also lend themselves to providing a composite particle picture for the exclusion statistics.

It is a pleasure to dedicate this work to Prof F. Mancini on the occasion of his sixtieth birthday.

REFERENCES

1. For an excellent overview on this subject and related matters, see, for instance, A. P. Polychronakos in *Topological aspects of low dimensional systems* Session XIX, Les Houches, A. Comtet, T. Jolicoeur, S. Ouvry and F. David, (Springer, 1999).
2. P. A. M. Dirac *The Principles of Quantum Mechanics*, (Oxford University Press, Oxford, 1967), Chapter IX.
3. H.S. Green, Phys. Rev. **90**, 270 (1953).
4. S.Doplicher, R. Haag and J.E. Roberts, Comm. Math. Phys. **23**, 199 (1971).
5. Y. Ohnuki and S. Kamefuchi, *Quantum field theory and parastatistics* (Springer Verlag, Berlin, 1982); S.N. Biswas in *Statistical Physics*, eds. N. Mukunda, A.K. Rajagopal and K.P. Sinha, Proceedings of the symposium on fifty years of Bose statistics, I.I.Sc. Bangalore (India).
6. O.W. Greenberg Phys.Rev.Lett. **13**, 598 (1964).
7. .M.L. Messiah and O.W. Greenberg, Phys. Rev. **B136**, 248 (1964); O.W. Greenberg and A.M.L. Messiah, B**138**, 1155 (1965).
8. J.B. Hartle and J.R. Taylor, Phys. Rev. **178**, 2043 (1969); R.H. Stolt and J.R. Taylor, Phys. Rev. D**1**, 2226 (1970); J.B. Hartle, R.H. Stolt and J.R. Taylor, Phys. Rev. D**2**, 1759 (1970).
9. F.D.M. Haldane Phys.Rev.Lett. **67**, 937 (1991).
10. I.G. Macdonald, *Symmetric functions and Hall polynomials*(Clarendon, Oxford, 1979).
11. O. W. Greenberg, Phys. Rev. Lett. **64**, 705 (1990).
12. See, for instance, B.E. Sagan *The symmetric group* (Brooks / Cole , Pacific Grove, California, 1991).
13. S, Chaturvedi, Phys. Rev. E**54**, 1378 (1996).
14. S. Chaturvedi and V, Srinivasan, Phys. Lett. A**224**, 249 (1997).
15. A. Dasniére de Veigy and S.Ouvry Phys.Rev.Lett. **72**, 600(1994).
16. S. Chaturvedi and V. Srinivasan, Phys. Rev. Lett. 78, 4316 (1997).
17. M. V. N. Murthy and R. Shankar, Phys. Rev. B**60**, 6517 (1999).
18. M. C. Bergère, J. Math. Phys. **41**, 7252 (2000).

Charge and Phase Dynamics in a Stack of Intrinsic Josephson Junctions

Tomio Koyama

Institute for Materials Research, Tohoku University, Katahira 2-1-1, Sendai 980-8577, Japan

Abstract. Our recent studies for the intrinsic Josephson effect are reviewed. We propose a simple phenomenological model based on the time-dependent Ginzburg-Landau model at $T = 0K$ for describing the phase dynamics of a stack of intrinsic Josephson junctions. In this model the capacitive and the inductive couplings between junctions dominate the dynamics of the gauge-invariant phase differences. It is shown that our model well explains the intrinsic Josephson effects observed in strongly anisotropic layered high-T_c superconductors.

INTRODUCTION

Strongly-anisotropic layered high-T_c superconductors such as $Bi_2Sr_2CaCu_2O_8$ show the Josephson effect in the c-axis transport [1,2] and optical properties [3,4] even in single crystals, which indicates that the superconducting CuO_2 layers are coupled by the Josephson effect. The Josephson effect observed in high-T_c superconductors is called the *intrinsic Josephson effect*. In this paper we review our theoretical works on the intrinsic Josepson effect, focusing on the phase and charge dynamics of the intrinsic Josephson-junction systems. The high-T_c superconductors may be considered as a stack of plainer Josephson junctions formed by thin superconducting layers with atomic-scale layer thickness. The superconducting layers are so thin in these superconductors that the electromagnetic field generated at a single junction site cannot be screened out inside the superconducting layers forming the junction, so that the dynamics of the phase differences at different junction sites is strongly affected by this electromagnetic field. This effect produces inductive and capacitive couplings peculiar to the intrinsic Josephson-junction systems among the phase differences [5-8].

In this paper we investigate the phase and charge dynamics of a stack of intrinsic Josephson junctions on the basis of a simple phenomenological model derived from the time-dependent Ginzburg-Landau model at $T = 0K$ [9-11]. The dynamical equation for the gauge-invariant phase differences in a stack of intrinsic Josephson junctions is presented. We solve this equation in several cases. In the weak capacitive coupling case our equation predicts the multiple-branch structure in the $I - V$ characteristics. The multi-branch structure similar to our numerical results has been observed in the c-axis $I - V$ characteristics in $Bi_2Sr_2CaCu_2O_8$ [1]. We also calculate the small phase oscillation modes in the Josephson vortex lattice.

The obtained field dependence of the resonant frequencies is in good agreement with that observed in microwave absorption experiments for $Bi_2Sr_2CaCu_2O_8$ [12]. Finally, we formulate a theory for the Coulomb blockade effect in a stack of intrinsic Josephson junctions with small in-plane area on the basis of our model. Our theory predicts the Coulomb blockade effect in a whicker of $Bi_2Sr_2CaCu_2O_8$ with submicron size in-plane area.

MODEL

In this paper, as a simple phenomenological model describing the time-evolution of the superconducting order parameter $\Psi(\mathbf{r},t)$, we utilize the time-dependent Ginzburg-Landau Model at $T = 0K$ [9-11]. This model is described by the Lagrangian,

$$L = \int dV \left\{ \frac{\gamma}{2m^*} \left| \left(i\frac{\hbar}{c}\partial_t - \frac{e^*}{c}A^0 \right)\Psi \right|^2 - \frac{1}{2m^*} \left| \left(\frac{\hbar}{i}\nabla - \frac{e^*}{c}\mathbf{A} \right)\Psi \right|^2 - \alpha|\Psi|^2 - \frac{\beta}{2}|\Psi|^4 \right\} + L_{EM} \tag{1}$$

where L_{EM} is the Lagrangian for the electromagnetic field. In strongly-anisotropic layered superconductors such as high-T_c cuprates the order parameter can be defined only on superconducting layers and its spatial variation along the z-direction ($\parallel$ c-axis) can be neglected within each superconducting layer, i.e., $\Psi(\mathbf{r},z,t) \to \Psi_\ell(\mathbf{r}_\parallel,t)$, ℓ being the layer index. Then, the integral in the z-direction in Eq.(1) is replaced with the discrete sum. Thus, on the basis of the Lagrangian (1) we can assume the Lagrangian for the strongly-layered superconductors as follows,

$$L = \sum_{\ell=-\infty}^{\infty} \int d\mathbf{r}_\parallel \left\{ \frac{s\gamma}{2m^*} \left| \left(i\frac{\hbar}{c}\partial_t - \frac{e^*}{c}A_\ell^0 \right)\Psi_\ell(\mathbf{r}) \right|^2 - \frac{s}{2m^*} \left| \left(\frac{\hbar}{i}\nabla_\parallel - \frac{e^*}{c}\mathbf{A}_\ell^\parallel \right)\Psi_\ell(\mathbf{r}) \right|^2 \right.$$

$$\left. -\frac{\hbar^2}{2Md^2} \left| \Psi_{\ell+1}(\mathbf{r})\exp(-i\frac{e^*}{\hbar c}\int_{\ell d}^{(\ell+1)d} dz A_z) - \Psi_\ell(\mathbf{r}) \right|^2 - \alpha|\Psi_\ell|^2 - \frac{\beta}{2}|\Psi_\ell|^4 \right\} + L_{EM}, \tag{2}$$

where s (d) is the thickness of the superconducting (insulating) layers. In this paper, for simplicity, the amplitude of the order parameter is assumed to be constant (phase-only model), i.e., $\Psi_\ell(\mathbf{r},t) \simeq \Delta e^{i\varphi_\ell(\mathbf{r},t)}$, with $\varphi_\ell(\mathbf{r},t)$ being the phase of the order parameter. Then, the Lagrangian in our model has the following form,

$$L = \sum_{\ell=-\infty}^{\infty} \int d\mathbf{r}_\parallel \left\{ \frac{s}{8\pi\mu^2} \left(A_\ell^0 + \frac{e^*}{\hbar}\partial_t\varphi_\ell \right)^2 - \frac{s}{8\pi\lambda_{ab}^2} \left(\mathbf{A}_\ell^\parallel - \frac{\hbar c}{e^*}\nabla_\parallel\varphi_\ell \right)^2 \right.$$

$$\left. -\frac{\hbar}{e^*}j_c\left(1 - \cos\theta_{\ell+1,\ell}\right) \right\} + L_{EM}, \tag{3}$$

where $\theta_{\ell+1,\ell}(x,t)$ is the gauge-invariant phase difference defined as

$$\theta_{\ell+1,\ell}(x,t) = \varphi_{\ell+1}(x,t) - \varphi_\ell(x,t) - \frac{e^*}{\hbar c}\int_{\ell d}^{(\ell+1)d} dz A^z(x,z,t). \tag{4}$$

In Eq.(3) μ, λ_{ab} and j_c are identified with the Debye length, the in-plane London penetration depth and the Josephson critical current density, respectively. Furthermore, we assume $\mathbf{E} \parallel z$-axis and $\mathbf{B} \parallel y$-axis for the electromagnetic field. Then, L_{EM} is expressed as

$$L_{EM} = \sum_{\ell=-\infty}^{\infty} \int d\mathbf{r}_\parallel \frac{d}{8\pi}\left(\epsilon E^{z2}_{\ell+1,\ell} - B^{y2}_{\ell+1,\ell}\right), \tag{5}$$

where ϵ is the dielectric constant of the insulating layers.

From the Lagrangian (3) it follows the equations of motion,

$$E^z_{\ell+1,\ell} - E^z_{\ell,\ell-1} = -\frac{s}{\epsilon\mu^2}(A^0_\ell + \frac{e^*}{\hbar}\partial_t\varphi_\ell), \tag{6}$$

$$\partial_x B^y_{\ell+1,\ell} = \frac{4\pi}{c}j_c \sin\theta_{\ell+1,\ell} + \frac{\epsilon}{c}\partial_t E^z_{\ell+1,\ell}, \tag{7}$$

$$B^y_{\ell+1} - B^y_{\ell,\ell-1} = -\frac{s}{\lambda^2_{ab}}(A^x_\ell - \frac{\hbar c}{e^*}\partial\varphi_\ell). \tag{8}$$

Note that Eqs.(6), (7) and (8) are the Maxwell equations in the discrete layered system. Then, from Eq.(6) we have the equation for the charge density on the ℓth superconducting layer, ρ_ℓ,

$$\rho_\ell = -\frac{1}{4\pi\mu^2}(A^0_\ell + \frac{e^*}{\hbar}\partial_t\varphi_\ell). \tag{9}$$

On the other hand, from Eqs.(7) and (8) one finds that the current flowing on the superconducting layers is described by the London equation,

$$j^x_\ell = -\frac{sc}{4\pi\lambda^2_{ab}}(A^x_\ell - \frac{\hbar c}{e^*}\partial\varphi_\ell), \tag{10}$$

and the inter-plane current is given by the Josephson current,

$$j^z_{\ell+1,\ell} = j_c \sin\theta_{\ell+1,\ell}, \tag{11}$$

in our model. It is noted that the in-plane current-current interaction (inductive coupling) and the electrostatic interaction between superconducting charges (capacitive coupling or charging effect) are incorporated in our model in addition to the Josephson coupling between superconducting layers.

It is possible to derive the dynamical equation for the gauge-invariant phase difference from Eqs.(6), (7) and (8) by using the Josephson relations. In the present intrinsic Josephson-junction systems, in which the inductive and capacitive couplings exist between junctions, the Josephson relations are generalized as

$$\frac{\hbar c}{e^*}\partial_x\theta_{\ell+1,\ell} = \Phi_{\ell+1,\ell} - \eta\Delta^{(2)}\Phi_{\ell+1,\ell}, \tag{12}$$

$$\frac{\hbar}{e^*}\partial_t\theta_{\ell+1,\ell} = V_{\ell+1,\ell} - \alpha\Delta^{(2)}V_{\ell+1,\ell}, \tag{13}$$

where $\Delta^{(2)}$ is the 2nd-rank difference operator ($\Delta^{(2)} F_\ell \equiv F_{\ell+1} - 2F_\ell + F_{\ell-1}$) and $\Phi_{\ell+1,\ell} \equiv dB^y_{\ell+1,\ell}$ and $V_{\ell+1,\ell} \equiv dE^z_{\ell+1,\ell}$ are, respectively, the flux line-density and the voltage difference between ℓth and $(\ell+1)$th superconducting layers [5-7]. Note that the second terms on the right hand sides of Eqs.(12) and (13) gives correction to the conventional Josephson relations in single Josephson-junction systems. These correction terms arise from the inductive and capacitive couplings between junctions. The parameters, η and α, are the inductive and capacitive coupling constants defined as

$$\eta = \frac{\lambda_{ab}^2}{sd}, \quad \alpha = \frac{\epsilon \mu^2}{sd}. \tag{14}$$

In high-T_c superconductors η takes a very large value of $O(10^5)$. On the other hand α is small, $O(1)$ [13-15]. Substituting Eqs.(12) and (13) into Eq.(7), we obtain the equation for $\theta_{\ell+1,\ell}$,

$$\frac{1}{1-\alpha\Delta^{(2)}}\frac{\epsilon}{c^2}\partial_t^2 \theta_{\ell+1,\ell} - \frac{1}{1-\eta\Delta^{(2)}}\partial_x^2 \theta_{\ell+1,\ell} + \frac{1}{\lambda_c^2}\sin\theta_{\ell+1,\ell} = 0, \tag{15}$$

where λ_c is the c-axis penetration depth, $\lambda_c^2 = \frac{\hbar c^2}{4\pi e^* d j_c}$.

Let us now present explicit expressions for Eq.(15) in several specific cases. First, we consider the case of $\partial_x \theta_{\ell+1,\ell} = 0$, that is, $\theta_{\ell+1,\ell}$ is uniform in the in-plane direction, which is realized in the absence of an external magnetic field. In this case we have the equation,

$$\frac{1}{\omega_{pl}^2}\partial_t^2 \theta_{\ell+1,\ell} + \sin\theta_{\ell+1,\ell} - \alpha(\sin\theta_{\ell+2,\ell+1} - 2\sin\theta_{\ell+1,\ell} + \sin\theta_{\ell,\ell-1}) = 0, \tag{16}$$

where $\omega_{pl} = \frac{c}{\sqrt{\epsilon}\lambda_c}$. Equation (16) has a plane wave solution for small phase oscillations and the dispersion relation is given as

$$\omega_L(q) = \omega_{pl}\sqrt{1 + 2\alpha[1 - \cos qd]}. \tag{17}$$

This plane-wave mode is the charge density mode propagating in the stacking direction of the Josephson junctions and is called the *longitudinal Josephson plasma* [7]. This plasma mode has been observed in the microwave absorption spectrum of Bi-2212 in the so-called longitudinal configuration [4]. Equation (15) can also describe the in-phase motion propagating in the in-plane direction. Since $\cdots = \theta_{\ell+2,\ell+1} = \theta_{\ell+1,\ell} = \theta_{\ell,\ell-1} = \cdots$ in this motion, Eq.(15) is reduced to

$$\frac{\epsilon}{c^2}\partial_t^2 \theta_{\ell+1,\ell} - \partial_x^2 \theta_{\ell+1,\ell} + \frac{1}{\lambda_c^2}\sin\theta_{\ell+1,\ell} = 0, \tag{18}$$

which is the usual sine-Gordon equation, the same as in the single Josephson junction systems. This equation has a solution corresponding to the plasma mode with the dispersion relation,

$$\omega_T(k) = \omega_{pl}\sqrt{1 + (\lambda_c k)^2}. \tag{19}$$

This plane wave mode is called the *transverse Josephson plasma*, since the transverse electromagnetic field accompanies this plasma mode.

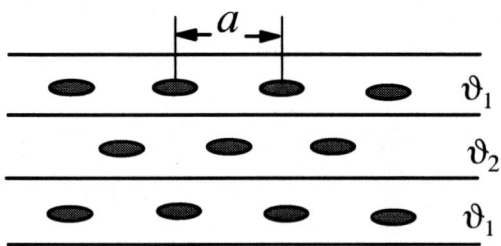

FIGURE 1: Josephson vortex lattice

In the presence of an in-plane magnetic field quantized vortices enter the junctions and form a periodic lattice. Suppose that the Josephson vortices form a triangular lattice as shown in Fig.1. The small phase oscillation mode in this triangular Josephson vortex lattice state can be calculated, using Eq.(15). The phase differences can be split into the static and small oscillatory parts as

$$\theta_{\ell+1,\ell}(x,t) = \Theta_0^{(1)}(x) + \chi_\ell^{(1)}(x,t), \quad \text{for } \ell = 2m, \tag{20}$$

$$\theta_{\ell+1,\ell}(x,t) = \Theta_0^{(2)}(x) + \chi_\ell^{(2)}(x,t), \quad \text{for } \ell = 2m+1, \tag{21}$$

where $\Theta_0^{(1)}(x)$ ($\Theta_0^{(2)}(x)$) is the static phase difference at the junction sites with odd (even) layer indices and $\chi_\ell^{(1)}(x,t)$ and $\chi_\ell^{(2)}(x,t)$ denotes the small fluctuating parts. Substituting Eqs.(20) and (21) into Eq.(15), we find the coupled equations for the static parts,

$$\partial_x^2 [\Theta_0^{(1)}(x) - \Theta_0^{(2)}(x)] = \frac{4}{\lambda_J^2}[\sin \Theta_0^{(1)}(x) - \sin \Theta_0^{(2)}(x)], \tag{22}$$

$$\partial_x^2 [\Theta_0^{(1)}(x) + \Theta_0^{(2)}(x)] = \frac{1}{\lambda_c^2}[\sin \Theta_0^{(1)}(x) + \sin \Theta_0^{(2)}(x)], \tag{23}$$

where $\lambda_J \equiv \lambda_c/\sqrt{\eta}$ is the Josephson length. The small phase oscillation parts are Fourier-transformable in the z-direction as $\chi_\ell^{(a)} = \frac{1}{\sqrt{N}} \sum_q \chi_q^{(a)}(x,t) e^{iq\ell 2d}$. The Fourier components, $\chi_q^{(1,2)}(x,t)$, with $q = 0$ obey the linearized equation,

$$\left[-\frac{\epsilon}{c^2}\partial_t^2 + \partial_x^2 - \frac{\cos \Theta_0^{(1)}(x) + \cos \Theta_0^{(2)}(x)}{2\lambda_c^2} \right] \chi^{(+)}(x,t)$$

$$= \frac{\cos \Theta_0^{(1)}(x) - \cos \Theta_0^{(2)}(x)}{2\lambda_c^2} \chi^{(-)}(x,t), \tag{24}$$

$$\left[-\frac{1}{1+4\alpha}\frac{\epsilon}{c^2}\partial_t^2 + \frac{1}{1+4\eta}\partial_x^2 - \frac{\cos \Theta_0^{(1)}(x) - \cos \Theta_0^{(2)}(x)}{2\lambda_c^2} \right] \chi^{(-)}(x,t)$$

$$= \frac{\cos \Theta_0^{(1)}(x) + \cos \Theta_0^{(2)}(x)}{2\lambda_c^2} \chi^{(+)}(x,t), \quad (25)$$

where

$$\chi^{(\pm)}(x,t) = \chi_0^{(1)}(x,t) \pm \chi_0^{(2)}(x,t). \quad (26)$$

The phase oscillation modes in the triangular Josephson-vortex lattice state can be determined by solving these coupled equations as an eigen-value problem. The numerical results will be given in the next section.

COMPARISON WITH EXPERIMENTS

$I - V$ Characteristics in Mesa Samples

The $I - V$ characteristics of a stack of intrinsic Josephson junctions can be calculated on the basis of our present model. Consider a mesa crystal of intrinsic Josephson junctions in the absence of an external magnetic field. The dynamics of the phase differences in this system is described by Eq.(16). In the presence of a bias current and the Ohmic quasi-particle tunneling current Eq.(16) is extended as

$$\frac{1}{\omega_{pl}^2}\partial_t^2 \theta_\ell + \frac{\beta}{\omega_{pl}}\partial_t \theta_\ell + \sin\theta_\ell = \frac{I}{j_c} - \alpha[\sin\theta_{\ell+1} - 2\sin\theta_\ell + \sin\theta_{\ell-1}], \quad (27)$$

as shown in [7,14], where $\beta \equiv 1/\sqrt{\beta_c}$ is the damping constant, β_c being the Mc-Cumber constant. This equation can be solved as an initial value problem and the $I - V$ characteristics is calculated from the relation,

$$V = (\frac{\hbar}{e^*})\sum_{\ell=1}^{N} \lim_{T\to\infty} \frac{\theta_{\ell+1,\ell}(T) - \theta_{\ell+1,\ell}(0)}{T}. \quad (28)$$

It is interesting to note that Eq.(27) is eqauivalent to Newton's equation for a 1D array of coupled-pendulums with friction, in which the pendulums are nonlinearly coupled with each other, having the coupling constant α. By the analogy with the motion of a pendulum one understands that two kinds of motion are possible in this system, i.e., oscillation and rotation. Note that the dc-voltage appears only on the rotating sites.

Let us now present numerical results for the $I - V$ characteristics. First we investigate the weak capacitive coupling case ($\alpha \ll 1$). Figure 2 shows the calculated hysteresis-loop of the $I - V$ curves for the system with $\alpha = 0.1$ composed of 10 junctions under the periodic boundary condition. As seen in this figure, the overall I-V characteristics is similar to the conventional one. However, we observe anomalous jumps on the resistive branch in the small voltage region. When the bias current is increased again in this anomalous region, the system shows hysteresis and new branches of the $I - V$ curves appears as shown in Fig.3 (a). The full $I - V$ characteristics obtained in our calculations is presented in Fig.3 (b). Thus, our theory predicts the multiple branch structure in the $I - V$ characteristics in

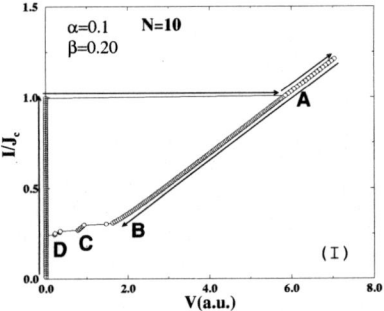

FIGURE 2: The I-V characteristics. The bias current I is increased from 0 to a value above j_c and then decreased back to 0 as indicated by arrows. Anomalous jumps appear at B, C, and D in the current decreasing process.

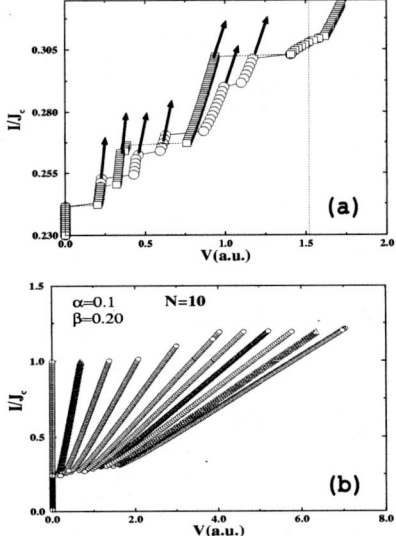

FIGURE 3: (a) The enlarged view of the anomalous region in the resistive branch. New branches appear when the current is increased in this region. (b) The full I-V characteristics in the weak capacitive coupling region. The number of resistive branches is nearly equal to the number of junctions forming a stack of intrinsic Josephson junctions.

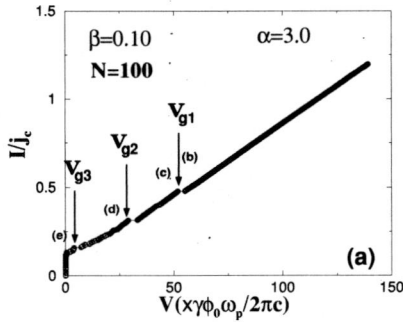

FIGURE 4: The I-V characteristics in the moderate coupling case.

the small capacitive coupling case. The similar multiple branch structure has been observed in the $I-V$ characteristics of Bi-2212 mesa crystals [1].

Let us next investigate the moderate coupling case ($\alpha \geq 1$). Figure 4 shows the numerical results in the case of $\alpha = 3.0$. In this coupling region only a single resistive branch appears in the $I-V$ characteristics. We do not observe any hysteretic behavior on this resistive branch, though small jumps are found at voltage values indicated by arrows in this figure. The analysis of the voltage pattern in this system indicates that the voltage distribution changes discontinuously at these voltages, that is, the dynamical transitions take place at these voltages. Thus, one understands that the $I-V$ characteristics does not show the multi-branch structure in the moderate capacitive coupling region. We mention that the $I-V$ characteristics shown in Fig.4 is very close to that observed in LSCO [16]. From the above results one may conclude that Eq.(27) well describes the c-axis $I-V$ characteristics in high-T_c superconductors.

JPR in the Triangular Josephosn Vortex Lattice

Let us next investigate the Josephson plasma resonance in a Josephson vortex lattice. In the microwave absorption experiments under an in-plane magnetic field Kadowaki et al. observed two resonant absorption modes below and above the zero-field Josephson plasma frequency, ω_{pl} [12]. The frequency of the lower (higher) resonant mode decreases (increases) with increasing the external magnetic field as shown in Fig.5. These resonant modes may be identified with the small phase oscillation modes in the Josephson vortex lattice. We assume that Josephson vortices enter all the junctions and form a triangular lattice as shown in Fig.1. The normal modes of the small phase oscillations in a triangular Josephson vortex lattice can be obtained by solving the eigen-value problem for Eqs.(24) and (25). We calculated the eigen-frequencies as a function of the lattice constant a in the in-plane direction (see Fig.1). In Fig.6 we plot the lowest two eigen-frequencies, ω_1 and ω_2 as a function of $2\pi/a \propto H$. As seen in this figure, ω_1 is located below ω_{pl}

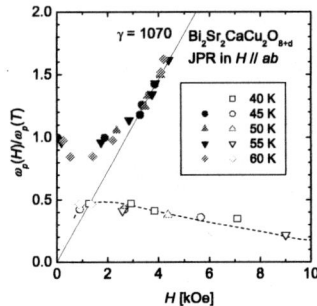

FIGURE 5: Field dependence of the two resonant frequencies observed in microwave absorption spectrum.

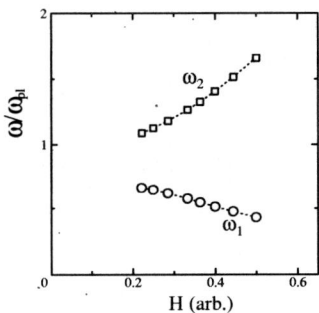

FIGURE 6: Field dependence of the eigen-frequencies ω_1 and ω_2.

and is a decreasing function of H. On the other hand ω_2 increases with increasing the external magnetic field. Hence, one may identify these two normal modes with the two resonant modes observed in the microwave absorption spectrum.

Coulomb blockade Effect in a Whisker Crystal of High-T_c Cuprates

In a whisker crystal of Bi-2212 with submicron size in-plane area the Coulomb blockade effect has been observed in the c-axis transport properties [18]. In this section we construct a theory for the Coulomb blockade effect in a stack of intrinsic Josephson junctions. Suppose that the phase differences, $\theta_{\ell+1,\ell}(\mathbf{r}_\|,t)$'s, are uniform in the in-plane directions. In this case the integral with respect to $\mathbf{r}_\|$ in Eq.(3) simply gives the in-plane area W. Then, our model is described by the Lagrangian in the absence of an external magnetic field,

$$L = W \sum_{\ell=-\infty}^{\infty} \left\{ \frac{s}{8\pi\mu^2}\left(A_\ell^0 + \frac{e^*}{\hbar}\partial_t\varphi_\ell\right)^2 - \frac{\hbar}{e^*}j_c\left(1 - \cos\theta_{\ell+1,\ell}\right) + \frac{d\epsilon}{8\pi}E_{\ell+1,\ell}^{z2}\right\}. \quad (29)$$

Let us now move to the canonical form in the Coulomb gauge. Note that the gauge-invariant phase difference $\theta_{\ell+1,\ell}$ is reduced to $\theta_{\ell+1,\ell} = \varphi_{\ell+1} - \varphi_\ell$ in the Coulomb gauge in the absence of an external magnetic field. Then, from the Lagrangian (20) it follows the Hamiltonian [17],

$$H = \sum_\ell \{ \frac{e^{*2}}{\hbar^2} \frac{2\pi d}{\epsilon W}[(1+2\alpha)u_\ell^2 - 2\alpha u_\ell u_{\ell+1}] + \frac{\hbar j_c}{e^*} W(1 - \cos\theta_{\ell+1,\ell}) \}, \quad (30)$$

where u_ℓ is the canonical momentum for $\theta_{\ell+1,\ell}$, i.e., $u_\ell = \partial L/\partial \dot\theta_{\ell+1,\ell}$. The first and the second terms in this Hamiltonian correspond to the charging and the Josephson coupling energy, respectively. It is noted that the charging energy is inversely proportional to W, while the Josephson coupling energy is proportianl to W. As a result, the charging energy exceeds the Josephson coupling energy at a certain small in-plane area, W_c. Note that Hamilton's equations of motion derived from the Hamiltonian (30) ($\dot\theta_{\ell+1,\ell} = \partial H/\partial u_\ell$, $\dot u_\ell = -\partial H/\partial \theta_{\ell+1,\ell}$) yield Eq.(16), which does not contain the scale parameter W, that is, the size effect emerges as a macroscopic quantum effect in the present system. This fact manifests the quantum nature of the Coulomb blockade effect. Then, we assume the canonical commutation relation, $[\theta_{\ell+1,\ell}, u_m] = i\hbar \delta_{lm}$, in the following calculations.

The critical in-plane area, W_c, at which the Coulomb blockade effect appears, can be calculated on the basis of a renormalization group analysis. The Hamiltonian (30) yields the effective action in the long wavelength limit,

$$S = \int_0^\beta d\tau \int dz \{ \frac{1}{2}\theta(\tau,z)[-(1+u)\partial_\tau^2 - u\partial_z^2]\theta(\tau,z) - g\cos\frac{\theta(\tau,z)}{\sqrt{g_0}} \}, \quad (31)$$

where $g_0 = \frac{\epsilon W}{4\pi e^{*2} d} \cdot \hbar\omega_{pl}$. This action contains two renormalized coupling constant, u and g. The renormalization group analysis gives the following flow equations,

$$\frac{dg}{dx} = \left(2 - \frac{1}{4\pi g_0 \sqrt{u(1+u)}}\right)g, \quad \frac{du}{dx} = \frac{g^2}{48\pi^5 g_0^2 \sqrt{u(1+u)}}, \quad (32)$$

with the initial condition: $u = 0, g = g_0$. From the solution of these flow equations, one can derives the explicit expression for the critical in-plane area as follows,

$$W_c = (\frac{\pi^2}{8}) \frac{4\pi d e^{*2} \lambda_c}{\hbar \sqrt{\epsilon c}}. \quad (33)$$

In $Bi_2Sr_2CaCu_2O_8$ Eq.(33) gives a value, $W_c \sim 0.1\mu m^2$, that is, our theory predicts the Coulomb blockade effect in a whisker with submicron size in-plane area for $Bi_2Sr_2CaCu_2O_8$. This prediction is consistent with the experimental observation by Latyshev et al. [18].

SUMMARY

In this paper we have presented a simple phenomenological model describing the dynamical properties of a stack of intrinsic Josephson junctions on the basis of the time-dependent Ginzburg-Landau model at $T = 0K$. In this model the capacitive and inductive couplings between junctions dominate the phase dynamics in this system. The multiple branch structure observed in the c-axis $I - V$ characteristics of $Bi_2Sr_2Ca\ Cu_2O_8$ is well explained by our model in the weak capacitive coupling case. Two resonant modes observed in the microwave absorption experiments under an in-plane magnetic field are identified with the small phase oscillation modes in the Josephson vortex lattice state. We have also formulated a theory for the Coulomb blockade effect in a whisker crystal of high-T_c superconductors.

ACKNOWLEDGMENTS

This paper is dedicated to Prof. F. Mancini on the occasion of his sixtieth birthday. This paper is based on the studies done in collaboration with M. Machida (Japan Atomic Enrgy Research Institute), M. Tachiki (National Research Institute for Metals) and H. Matsumoto (Tsukuba University). The author would like to thank I. Kakeya and K. Kadowaki (Tsukuba University) for permitting him the use of the experimental data (Fig.5) prior to publication.

REFERENCES

1. Kleiner, R., Steinmeyer, F., Kunkel, G., and Muller, P., Phys. Rev. Lett. **68**, 2394-2397 (1992).

2. Oya, G., Aoyama, N., Irie, A., Kishida, S., and Tokutaka, H., Jpn. J. Appl. Phys. **31**, L829-834 (1992).

3. Tamasaku,. K., Nakamura, Y., Uchida S., Phys. Rev. Lett. **69**, 1455-1458 (1992).

4. Matsuda, Y., Gaifullin, M.B., Kumagai, K., Kadowaki, K., and Mochiku, T., Phys. Rev. Lett. **75**, 4512-4515 (1995).

5. Bulaevskii, L.N., Zamora, M., Baeriswyl, D., Beck, H., and Clem, J.R., Phys. Rev. **B 50**, 12831-12834 (1994).

6. Sakai, S., Bodin, P., and Pedersen, J. Appl. Phys. **73**, 2411- (1993).

7. Koyama, T. and Tachiki, M., Phys. Rev. **B 54**, 16183-16191 (1996).

8. Machida, M., Koyama, T., Tanaka, A., and Tachiki, M., Physica C, **331**, 85-96 (2000).

9. Stephen, M.J. and Suhl, H., Phys. Rev. Lett. **13**, 797-800 (1964).

10. Abrahams, E. and Tsuneto, T., Phys. Rev. **152**, 416-432 (1966).

11. Castro, C. Di and Young, W, IL Nuovo Cimento **62 B**, 273-300 (1969).

12. Kadowaki, K., Wada, T., and Kakeya, I., Physica C **362**, 71-77 (2001).

13. Preis, C., Helm, C., Keller, J., Sergeev, A., and Kleiner, R., Proceeding of SPIE, **3480**, 236-244 (1998).

14. Machida, M., Koyama, T., and Tachiki, M., Phys. Rev. Lett., **83**, 4618-4621 (1999).

15. Matsumoto, H., Sakamoto, S., Wajima, F., Koyama, T., and Machida, M., Phys. Rev. B **60**, 3666-3672 (1999).
16. Uematsu, Y., Sasaki, N., Mizugaki, Y., Nakajima, K., Yamashita, T., Watauchi, S., and Tanaka, I., Physica C **362**, 290-295 (2001).
17. Koyama, T., J. Phys. Soc. Jpn., **70**, 2114-2123 (2001).
18. Latyshev, Y.I., Kim, S.J., and Yamashita, T., JETP Lett. **69**, 84-90 (1999).

Self-Consistent Mean-Field Theory for Frustrated Josephson Junction Arrays

F.P. Mancini*, P. Sodano* and A. Trombettoni[†]

*Dipartimento di Fisica and Sezione I.N.F.N., Università di Perugia,
Via A. Pascoli, I-06123 Perugia, Italy
[†]I.N.F.M. and Dipartimento di Fisica, Università di Parma,
parco Area delle Scienze 7A, I-43100, Parma, Italy

Abstract. We review the self-consistent mean-field theory for charge-frustrated Josephson junction arrays. Using $\langle\cos\varphi\rangle$ (φ is the phase of the superconducting wavefunction) as order parameter and imposing the self-consistency condition, we compute the phase boundary line between the superconducting region ($\langle\cos\varphi\rangle \neq 0$) and the insulating one ($\langle\cos\varphi\rangle = 0$). For a uniform offset charge $q = e$ the superconducting phase increases with respect to the situation in which $q = 0$. We generalize the self-consistent mean-field theory to include the effects induced by a random distribution of offset charges and/or of diagonal self-capacitances. We find results in agreement with the ones obtained in studies using the path-integral approach.

I. INTRODUCTION

The first artificially fabricated Josephson junction arrays (JJA's) were realized twenty years ago at IBM [1] as an effort to develop an electronics based on superconducting devices. Immediately after, it became clear that JJA's provided an ideal model to investigate classical phase transitions, frustration effects and relevant aspects of non-linear dynamics [2, 3]. JJA's are built by placing on the sites of a lattice islands of superconducting material coupled by a Josephson junction. The huge variety of behaviors of the system is rather simply described by the competition between the Josephson energy E_J and the charging energy E_C: the former being responsible for the Josephson tunneling of Cooper pairs between the sites of the lattice, while the latter measures the effects of the electrostatic repulsion between Cooper pairs. The superconductor-insulator transition typical of JJA's, for instance, depends crucially on the ratio between these two energy scales.

In many situations, it is relevant to analyze the effect of a background of external charges on the superconductor-insulator transition of a quantum JJA. Offset charges arise in real physical systems as a result of charged impurities or by the application of a gate voltage between the array and the ground. In the former situation, offset charges are naturally randomly distributed on the lattice while in the latter situation they play the role of a sort of chemical potential and, then, their distribution may also be uniform. Offset charges may be regarded as effective charges q_i, located at the sites of the lattice and, when $q_i \neq 2e$, they cannot be eliminated by Cooper pair tunneling; in general, offset charges frustrate the attempts of the system to minimize the energy of the charge distribution of the ground state (for this reason they are also called *frustration charges*).

A large number of studies has by now been devoted to the analysis of the effects induced by offset charges both on the zero-temperature phase transition [4-7] and on the phase transition at finite temperature [8-10]. In this paper we shall use the self-consistent mean-field theory (SCMFT) to investigate the finite temperature phase diagram for the self-charging (SC) model of JJA's [2]; in this model it is assumed that the potential at site i depends only on the charge at the same site and, thus, the capacitance matrix describing the charge effects of the array is diagonal. In particular, we shall investigate situations in which offset charges (both uniform and random) are present.

Although quantum corrections are expected to be very relevant for $d \leq 2$, the SCMFT has the merit of providing a rather intuitive and physically transparent approach to the analysis of some general features of the superconductor-insulator transition in these systems. The results we obtained are, in fact, in good qualitative agreement with the outcome of numerical simulations (see [3] and references therein) and consistent with other analytical approaches not relying on mean-field theory [11].

The plan of the paper is as follows: in Section II we review SCMFT for the SC model of quantum JJA's with a uniform distribution of offset charges [2, 10]. We study the eigenvalue equation of the mean-field Hamiltonian and for a uniform offset charge $q_i = e$ we show that there is superconductivity for all values of E_J/E_C. In Section III we discuss how to extend the self-consistent mean-field approach to situations in which there is capacitive disorder: one has to impose the self-consistency condition with a double average, the quantum one and the average over the disorder. The results are in agreement with the ones obtained with the path-integral approach [12, 13] and, at very low temperatures, are consistent with the phase diagram obtained in Ref. [14]. Section IV is devoted to our concluding remarks.

II. MEAN-FIELD THEORY FOR JJA'S WITH OFFSET CHARGES

The Hamiltonian commonly used to describe Cooper pairs tunneling in superconducting quantum networks defines the so-called quantum phase model (QPM):

$$H = \frac{1}{2}\sum_{ij}(Q_i+q_i)C_{ij}^{-1}(Q_j+q_j) - E_J\sum_{\langle ij\rangle}\cos(\varphi_i - \varphi_j), \tag{1}$$

where φ_i is the phase of the superconducting order parameter at the grain i. Its conjugate variable $n_i = -i\partial/\partial\varphi_i$ (with $[\varphi_i, n_i] = i\delta_{ij}$) describes the number of excess Cooper pairs on the i-th superconducting grain and C_{ij} is the capacitance matrix. The symbol $\langle ij\rangle$ indicates a sum over nearest-neighbor grains only.

The first term in the Hamiltonian (1) determines the electrostatic coupling between the Cooper pairs while the second term describes the hopping of Cooper pairs between neighboring sites (E_J is the Josephson energy). The charging energy E_C is defined as $E_C = e^2 C_{ii}^{-1}/2$. An external gate voltage V_i provides a contribution to the energy via the offset charge $q_i = \sum_j C_{ij}V_j$; this external voltage can be either applied to the ground plane or it may be induced by charges trapped in the substrate. The former situation leads to the appearance of a uniform frustration charge, while the latter naturally induces a random offset charge. In this paper we shall limit our investigation only to the SC model described by the Hamiltonian (1) with $C_{ij} = \delta_{ij} C_{ii}$.

With a uniform distribution of offset charges $q_i = q$, the Hamiltonian of the array is given by:

$$H = 4E_C \sum_i \left(i\frac{\partial}{\partial \varphi_i} - \frac{q}{2e} \right)^2 - E_J \sum_{\langle i,j \rangle} \cos(\varphi_i - \varphi_j). \qquad (2)$$

Mean-field theory for the SC model for quantum JJA's was first used by Symanek [15, 2]. The approximation consists in replacing the Josephson coupling on the link $i-j$ by an average coupling so that $E_J \sum_{\langle ij \rangle} \cos(\varphi_i - \varphi_j) \approx zE_J \langle \cos \varphi \rangle \sum_i \cos \varphi_i$, where z is the coordination number. Requiring the order parameter to be real, leads to the assumption that $\langle \sin \varphi_i \rangle = 0$; it is also assumed that $\langle \cos \varphi \rangle$ does not depend on the island index i. In the mean-field approximation the Hamiltonian is given by a sum of single site Hamiltonians H_i describing a quantum particle in the potential $\cos \varphi_i$:

$$H_{MFA} = \sum_i H_i = \sum_i \left[-4E_C \frac{\partial^2}{\partial \varphi_i^2} - 8i\frac{q}{2e} E_C \frac{\partial}{\partial \varphi_i} + 4\left(\frac{q}{2e}\right)^2 E_C - zE_J \langle \cos \varphi \rangle \cos \varphi_i \right]. \qquad (3)$$

The pertinent Schrödinger equation to be solved is then

$$\left[-4E_C \frac{d^2}{d\varphi^2} - 8i\frac{q}{2e} E_C \frac{d}{d\varphi} + 4\left(\frac{q}{2e}\right)^2 E_C - zJ \langle \cos \varphi \rangle \cos \varphi \right] \psi_m(\varphi) = E_m \psi_m(\varphi). \qquad (4)$$

Due to the periodicity of the phase φ, the eigenfunctions should be 2π-periodic functions of φ, i.e.,

$$\psi_n(\varphi) = \psi_n(\varphi + 2\pi). \qquad (5)$$

Furthermore, since the Hamiltonian (1) is invariant under the shift $q \to q + 2le$, where l is an integer, it is only relevant to analyze the situations corresponding to $q = 0$ and $q = e$.

The order parameter $\langle \cos \varphi \rangle$ is evaluated in terms of the eigenfunctions of Eq. (4) through the self-consistency equation

$$\langle \cos \varphi \rangle = \frac{\sum_m e^{-E_m/K_B T} \langle \psi_m | \cos \varphi | \psi_m \rangle}{\sum_m e^{-E_m/K_B T}}. \qquad (6)$$

From Eq. (6) one immediately sees that, for high temperatures or low E_J, only the solution $\langle \cos \varphi \rangle = 0$ exists and, thus, there is no superconductivity; for low temperatures or high E_J instead, $\langle \cos \varphi \rangle \neq 0$ and the system as a whole behaves as a superconductor. Solving then the eigenvalue equation (4) provides us with all the tools needed to investigate the finite temperature phase diagram of the SC model of frustrated JJA's.

Defining $v = -zE_J \langle \cos \varphi \rangle / 2E_C$, $\lambda'_m = [E_m - 4E_C(q/2e)^2]/E_C$, and $K = 2i(q/2e)$, one finds

$$\frac{d^2\psi_m}{d\varphi^2} + K\frac{d\psi_m}{d\varphi} + \left(\frac{\lambda'_m}{4} - \frac{v}{2}\cos\varphi\right)\psi_m = 0. \qquad (7)$$

Eq. (7) is a Mathieu equation with a term proportional to a first derivative: setting

$$\psi_m(\varphi) = e^{-\frac{1}{2}K\varphi}\rho_m(\varphi), \tag{8}$$

one gets an equation for ρ_m, namely

$$\frac{d^2\rho_m}{d\varphi^2} + \left(\frac{\lambda_m'}{4} - \frac{K^2}{4} - \frac{v}{2}\cos\varphi\right)\rho_m = 0. \tag{9}$$

If one puts $\lambda_m = E_m/E_C$ and $\varphi = 2x$, the eigenvalue equation (7) is usefully recast in the standard form of the Mathieu equation [16]:

$$\frac{d^2\rho_m}{dx^2} + (\lambda_m - 2v\cos 2x)\rho_m = 0. \tag{10}$$

It is well known [16] that the Mathieu equation admits the following periodic solutions:

1. $ce_{2n}(x,v)$, even solutions with period π corresponding to the eigenvalues $a_{2n}(v)$;
2. $se_{2n+2}(x,v)$, odd solutions with period π corresponding to the eigenvalues $b_{2n+2}(v)$;
3. $ce_{2n+1}(x,v)$, even solutions with period 2π corresponding to the eigenvalues $a_{2n+1}(v)$;
4. $se_{2n+1}(x,v)$, odd solutions with period 2π corresponding to the eigenvalues $b_{2n+1}(v)$.

Since the eigenfunctions of the Schrödinger equation (4) should satisfy the periodic boundary condition (5), from Eq. (8) one immediately sees that one should treat differently the situations where $q/2e$ is integer or half-integer. In fact, for integer $q/2e$ one has to consider only π-periodic solution of Eq. (10): in this way $\rho_m(\varphi)$ are 2π-periodic, which in turns leads to 2π-periodic $\psi_m(\varphi) = e^{-il\varphi}\rho_m(\varphi)$ (where l is an integer). The solutions of Eq. (10) with period π are the Mathieu eigenfunctions $ce_{2n+2}(x)$, $se_{2n}(x)$ (with $n = 0, 1, \ldots$) [16]. If, instead, $q/2e$ is half-integer, Eqs. (5) and (8) require the use of solutions of Eq. (10) which are π-anti-periodic. Then $\rho_m(\varphi)$ are 2π-anti-periodic and the eigenfunctions $\psi(\varphi) = e^{-il\varphi/2}\rho_m(\varphi)$ (with integer l) are 2π-periodic. Thus, if $q/2e$ is half-integer, it is pertinent to use the π-anti-periodic Mathieu eigenfunctions $ce_{2n+1}(x)$, $se_{2n+1}(x)$ (with $n = 0, 1, \ldots$).

Since the phase transition is expected to be second order [18], near the transition temperature the order parameter $\langle\cos\varphi\rangle$ and the parameter v are small: this allows one to use the expansion for small v of the Mathieu functions [16]. As a result, one finds that, at first order in v and apart from the phase factor $e^{-i\varphi/2}$, the normalized eigenfunctions of the Schrödinger equation (4) satisfying the condition (5) are given by

$$\begin{aligned}\psi_1^e(\varphi) &= \tfrac{1}{\sqrt{\pi}}\left(\cos\tfrac{\varphi}{2} - \tfrac{v}{8}\cos\tfrac{3\varphi}{2}\right)\\ \psi_1^o(\varphi) &= \tfrac{1}{\sqrt{\pi}}\left(\sin\tfrac{\varphi}{2} - \tfrac{v}{8}\sin\tfrac{3\varphi}{2}\right),\end{aligned} \tag{11}$$

and, for $n = 1, 2, \ldots$,

$$\psi^e_{2n+1}(\varphi) = \frac{1}{\sqrt{\pi}}\left\{\cos\frac{(2n+1)\varphi}{2} - v\left[\frac{\cos\frac{(2n+3)\varphi}{2}}{4(2n+2)} - \frac{\cos\frac{(2n-1)\varphi}{2}}{8n}\right]\right\}$$
$$\psi^o_{2n+1}(\varphi) = \frac{1}{\sqrt{\pi}}\left\{\sin\frac{(2n+1)\varphi}{2} - v\left[\frac{\sin\frac{(2n+3)\varphi}{2}}{4(2n+2)} - \frac{\sin\frac{(2n-1)\varphi}{2}}{8n}\right]\right\},$$
(12)

where e (o) stands for *even* (*odd*). The corresponding energy eigenvalues are

$$\begin{aligned}E^e_1 &= E_C(1+v)/2,\\ E^o_1 &= E_C(1-v)/2,\\ E^e_{2n+1} &= E^o_{2n+1} = E_C(2n+1)^2/2\\ n &= 1, 2, \ldots\end{aligned}$$
(13)

The expectation values $\langle \psi_m | \cos\varphi | \psi_m \rangle$ at the order v are given by

$$\begin{aligned}\langle \psi^e_1 | \cos\varphi | \psi^e_1 \rangle &= 1/2 - v/8\\ \langle \psi^o_1 | \cos\varphi | \psi^o_1 \rangle &= -1/2 - v/8\\ \langle \psi^e_{2n+1} | \cos\varphi | \psi^e_{2n+1} \rangle &= \langle \psi^o_{2n+1} | \cos\varphi | \psi^o_{2n+1} \rangle = v/[8n(n+1)]\\ n &= 1, 2, \ldots\end{aligned}$$
(14)

Upon inserting the above eigenfunctions and eigenvalues in Eq. (6) and keeping only the terms proportional to $v \sim \langle \cos\varphi \rangle$, one obtains the following equation for the critical temperature T_c:

$$\frac{1}{\alpha} = g(q = e, y)$$
(15)

where $y = \frac{K_B T_c}{E_C}$, $\alpha = \frac{zJ}{4E_C}$ and

$$g(q=e,y) = \frac{\frac{4+y}{4y}e^{-1/y} - \sum_{n=1}^{\infty} \frac{1}{1-4(n+1/2)^2}e^{-(4/y)(n+1/2)^2}}{e^{-1/y} + \sum_{n=1}^{\infty} e^{-(4/y)(n+1/2)^2}}.$$
(16)

From Eq. (16) one can easily show that in the presence of charge frustration $\pm(2n+1)e$ on the lattice sites, for each value of α there is a insulator-superconductor transition. Indeed, $g(e,y) \to \infty$ for $y \to 0$ and $g(e,y) \to 0$ for $y \to \infty$; also $dg/dy < 0$ for all $y > 0$. It follows that Eq. (16) has a unique solution for each value of α. Moreover, since $g \approx 1/y = 1/\alpha$, the critical temperature at which the transition occurs is given by $T_c \approx zE_J/4$.

For $q = 0$, the solutions of the Mathieu equation (10) with period π are the Mathieu eigenfunctions $ce_{2n+2}(x)$, $se_{2n}(x)$ (with $n = 0, 1, \ldots$): in this case one finds

$$\frac{1}{\alpha} = g(q = 0, y)$$
(17)

where

$$g(q=0,y) = \frac{1 - 2\sum_{n=1}^{\infty} \frac{e^{-4n^2/y}}{4n^2 - 1}}{1 + 2\sum_{n=1}^{\infty} e^{-4n^2/y}}.$$
(18)

From Eq. (18) one sees that for $\alpha > 1$ the system has an insulator-superconductor phase transition, while for $\alpha < 1$ there is no evidence for a transition. Indeed $g(e,y) \to \infty$ for $y \to 0$ and $g(e,y) \to 1$ for $y \to \infty$; also $dg/dy < 0$ for all $y > 0$. Therefore the self-consistency equation (6) does not have solutions for $\alpha < 1$ and it has a unique solution for $\alpha > 1$. For small values of y, Eq. (18) gives $1/\alpha \approx (1 - 2e^{-4/y}/3)/(1 + 2e^{-4/y})$, from which

$$\frac{K_B T_c}{E_C} \approx \frac{4}{\log \frac{2(\alpha+3)}{3(\alpha-1)}}. \tag{19}$$

It is worth noting that the eigenfunctions $\psi_m^{(0)}$ of the Schrödinger equation (4) without the periodic potential are

$$\psi_m^{(0)} = \frac{1}{\sqrt{2\pi}} e^{\pm im\varphi/2}. \tag{20}$$

These wavefunctions are also eigenfunctions of the number operator with eigenvalues $\mathcal{N} = \langle \psi_m^{(0)} | -i\partial/\partial\varphi | \psi_m^{(0)} \rangle = \pm m/2$. When $q = 0$, the wave functions ψ_{2n}^e and ψ_{2n}^o are, respectively, the even and odd combinations resulting from the splitting of the eigenfunctions $\psi_{2n}^{(0)}$ and are related to the expectation value of the half-integer number of Cooper pairs ($\mathcal{N} = 0, \pm 1, \pm 2, \ldots$). On the other hand, when $q = e$, the expectation value of the number operator on the eigenfunctions ψ_{2n+1}^e and ψ_{2n+1}^o is half-integer and is equal to $1/2$ in the ground state. Therefore an offset charge $q = e$ favors the the Cooper pairs tunneling, making possible the insulator-superconductor transition also when $E_C \gg E_J$.

If one should use both periodic and anti-periodic solutions, the general solution of Eq. (4) would not have a definite periodicity and, consequently, the charges n_i can take any value; this situation is expected to be relevant in the description of continuous flows of currents due, for instance, to ohmic shunt resistances [19, 20]. Unless there is dissipation, the use of both periodic and anti-periodic solution is unwarranted. However, if in the self-consistency equation (6) one includes also the 2π-anti-periodic eigenfunctions, one would find - for small critical temperatures - the following equation

$$1 = \alpha \frac{1 + (\frac{2}{y} + \frac{1}{2})e^{-1/y} - \frac{2}{3}e^{-4/y}}{1 + 2e^{-1/y} + 2e^{-4/y}}. \tag{21}$$

Eq. (21) for α less than a critical value α_c ($\alpha_c \approx 0.79$) does not have solutions, for $\alpha_c < \alpha < 1$ has two solutions and for $\alpha > 1$ has just one solution. This behavior is called *reentrant*.

In Fig. 1 we plot T_c as a function of α for $q = 0$ and $q = e$ according to Eqs. (16) and (18). For $q = 0$ one sees that there is no superconductivity for $\alpha < 1$. From $q = e$ superconductivity is attained for all the values of α: a uniform offset charge $q = e$ always *favors* superconductivity.

We conclude this Section observing that the phase boundary line obtained within the mean-field approximation in the path-integral approach is $1/\alpha = g(q,y)$ with [9, 10]

$$g(q,y) = \frac{\sum_n e^{-\frac{4}{y}(n+\frac{q}{2e})^2} \frac{1}{1-4(n+\frac{q}{2e})^2}}{\sum_m e^{-\frac{4}{y}(m+\frac{q}{2e})^2}}. \tag{22}$$

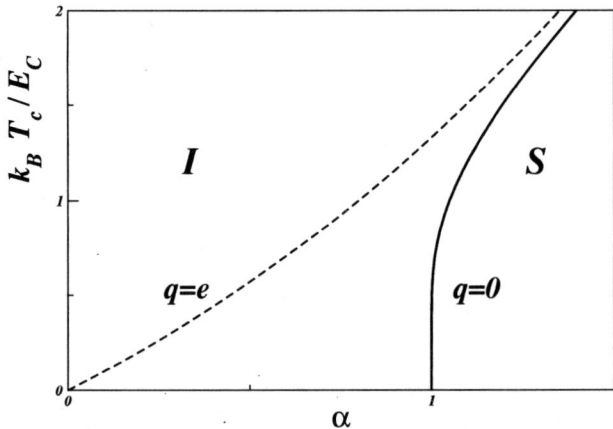

FIGURE 1. Phase diagram for the SC model without charge frustration (solid line) and with charge frustration $q = e$ (dashed line). α stands for the ratio $zE_J/4E_C$. The **I** and **S** indicate, respectively, insulating and superconducting phase.

Eq. (22) coincides with Eqs. (16) and (18) for $q = e$ and $q = 0$, respectively.

III. CAPACITIVE DISORDER

In practical realizations of Josephson devices [3], one has to deal with disorder caused by offset charge defects in the junctions or in the substrate [22]. Random offset charges cannot be made to vanish by using a gate for each superconducting island since in large arrays too many electrodes would be necessary, making impossible the cooling of the system at the desired temperatures. In Ref. [23] it was observed a sensible variation ($\sim 40\%$) of the resistance between the unfrustrated and the fully frustrated array. Moreover, it may also happen that the network's parameters are not uniform across the whole array: despite recent advances in fabrication techniques, variation of junction parameters associated to the shape of the islands can be also of 20% [3]. Thus, it is relevant in many practical situations to study JJA's with randomly distributed self-capacitances: this corresponds to have a random diagonal charging energy [12, 24].

In this Section, we shall determine the finite temperature phase diagram of JJA's with capacitive disorder (i.e., with random offset charges and/or random self-capacitances). To derive the phase boundary between the insulating and the superconducting phase, we shall use the mean-field approach for quantum JJA's with offset charges and diagonal capacitance matrices reviewed in the previous Section. One has to impose now the self-consistency condition with a double average: the quantum average and the one over the disorder.

As we shall see, charge disorder supports superconductivity; furthermore, the relative changes of the insulating and superconducting regions of the phase diagram depend crucially on the weights of the δ-like charge probability distribution. In the physical relevant situation of two charge distributions peaked at the values $q = 0$ and $q = e$, increasing the frustrated weight favors the superconducting phase. Also the randomness of the self-capacitances leads to remarkable effects, namely, the superconducting phase increases with respect to the case where disorder is not present. In the following, we shall provide a quantitative analysis of these phenomena.

A pertinent extension of SCMFT in the presence of on-site disorder may be obtained introducing an order parameter averaged *also* over the disorder. In the following $\langle \cdots \rangle$ denotes only the quantum average while $[\cdots]_{av}$ an average over the random variables. The single-site Hamiltonians of Eq. (3) then become

$$H_i = -4E_C^{(i)} \frac{\partial^2}{\partial \varphi_i^2} - 8iE_C^{(i)} \frac{q_i}{2e} \frac{\partial}{\partial \varphi_i} - zE_J [\langle \cos \varphi \rangle]_{av} \cos \varphi_i. \tag{23}$$

The Hamiltonian (23) depends on a random variable X, which can be either q_i or $E_C^{(i)}$. Thus, its eigenfunctions and eigenvalues depend either on q_i or $E_C^{(1)}$:

$$H_i \psi_n(\varphi_i; X) = E_n(X) \psi_n(\varphi_i; X). \tag{24}$$

The self-consistency condition is given by

$$[\langle \cos \varphi \rangle]_{av} = \int dX P(X) \frac{\sum_n e^{-\beta E_n(X)} \langle \psi_n(X) | \cos \varphi | \psi_n(X) \rangle}{\sum_n e^{-\beta E_n(X)}} \tag{25}$$

where $P(X)$ is the probability distribution of X.

The phase boundary line is obtained from Eq. (25) by requiring $[\langle \cos \varphi \rangle]_{av}$ to be small and by keeping only terms proportional to it. The self-consistency condition yields a mean-field phase boundary line in agreement with the results obtained by the path-integral approach [12]. The low temperature behavior, obtained by a pertinent extrapolation of our finite T results, is consistent with the phase diagram obtained in Ref. [14].

Random Offset Charges

In this Section we shall study JJA's at finite temperature with random charge frustration q and with probability distribution given by a sum of δ-like distributions

$$P(q_i) = \sum_n p_n \delta(q_i - ne) \tag{26}$$

with $\sum_n p_n = 1$. This corresponds to a random distribution of charges which are integer multiples of e and, actually, this is the most realistic situation for a random distribution. Inserting the probability distribution (26) in Eq. (25), one has

$$\frac{1}{\alpha} = \int dq \sum_n p_n \delta(q - ne) g(q, y) = \sum_{odd} p_n g(ne, y) + \sum_{even} p_n g(ne, y) \tag{27}$$

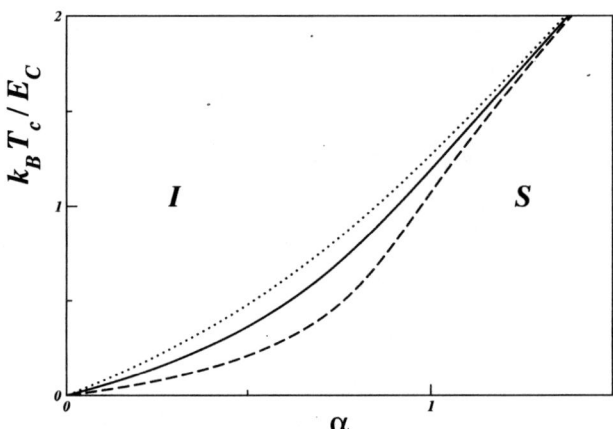

FIGURE 2. Phase diagram for random offset charges with the probability distribution given by Eq. (26). In the plot we take $p_e = 1/4$ (dashed line), $1/2$ (solid line), and $3/4$ (dotted line).

where Σ_{odd} (Σ_{even}) is a sum restricted to odd (even) integer.

Since the thermodynamical properties of the system are invariant under the shift $q \to q + 2ne$, one should note that $g(q+2ne,y) = g(q,y)$, where n is an integer. As a consequence, Eq. (27) leads to

$$\frac{1}{\alpha} = p_0 g(0,y) + p_e g(e,y) \qquad (28)$$

where $p_0 = \Sigma_{even} p_n$ ($p_e = \Sigma_{odd} p_n$) is the probability that the offset charge q is an even (odd) integer multiple of e. The results obtained from Eq. (27) with the probability distribution (26) are displayed in Fig. 2, where we plot the phase boundary line for $p_e = 1/4, 1/2, 3/4$. One observes that increasing p_e leads to an enlargement of the superconducting phase. It is worth noting that applying the SCMFT approximation with the probability distribution (26) it is possible to find exactly the same result obtained in a path-integral approach for a JJA model with diagonal capacitance matrices [12].

Random Self-Capacitances

In this Section we shall study JJA's at finite temperature with uniform charge frustration q and random self-capacitance C_{ii}. The diagonal charging energy terms U_{ii} are related to the self-capacitances C_{ii} via $U_{ii} = 4e^2 C_{ii}^{-1}$ and are assumed to be independently distributed according to a probability distribution with mean U_0. The average charging energy is defined as $E_C^0 = U_0/8$. By averaging the self-consistency equation (25) over the random variables $E_C^{(i)}$, the equation for the phase boundary becomes

$$\frac{1}{\alpha} = \int_0^\infty dU \, \frac{P(U)}{U} g(q,U,y) \qquad (29)$$

where now $\alpha = zE_J/4E_C^0$, $y = k_B T_c/E_C^0$, and $U = U_{ii}/U_0$. Eq. (29) can be also obtained by using the path-integral approach [12]. The function $g(q = e, U, y)$ is given by

$$g(q=e,U,y) = \frac{\frac{4U+y}{4y}e^{-U/y} + \sum_{n=1}^{\infty} \frac{1}{1-4(n+1/2)^2} e^{-(4U/y)(n+1/2)^2}}{e^{-U/y} + \sum_{m=1}^{\infty} e^{-(4U/y)(m+1/2)^2}}, \qquad (30)$$

whereas the function $g(q = 0, U, y)$ is given by

$$g(q=0,U,y) = \frac{\sum_{n=-\infty}^{\infty} e^{-4U/y} \frac{1}{1-4n^2}}{\sum_{m=-\infty}^{\infty} e^{-(4U/y)m^2}}. \qquad (31)$$

If one, for instance, considers a bimodal distribution of the U's, then

$$P(U) = p\,\delta(U_1 - U) + (1-p)\,\delta(U_2 - U), \qquad (32)$$

where U_1 and U_2 are positive numbers. Inserting the probability distribution (32) in Eq. (29), one gets:

$$\frac{1}{\alpha} = p\frac{1}{U_1}g(q,U_1,y) + (1-p)\frac{1}{U_2}g(q,U_2,y). \qquad (33)$$

The phase boundary line given by Eq. (33) is plotted in Fig. 3. One observes that when $q = 0$, the superconducting phase increases in comparison to the nonrandom case: this is due to the factors $1/U_1$ and $1/U_2$ in Eq. (33), which makes larger the contribution of junctions with charging energies less than U_0. The increase of the superconducting phase is thus due to a decrease of the effective value of E_c. This phenomenon is largely independent from the specific choice of the distribution [12].

It is pertinent to observe that, when $q = e$ (maximum frustration induced by the external offset charges), the randomness does not modify considerably the phase diagram. This should be compared with the unfrustrated case ($q = 0$), where randomness sensibly affects the phase diagram.

IV. CONCLUDING REMARKS

In this paper, we reviewed the use of the self-consistent mean-field theory to analyze the effects induced by offset charges on the finite temperature phase diagram of Josephson junction arrays.

We reviewed, for a diagonal Coulomb interaction matrix, the explicit derivation of the equation for the phase boundary line between the insulating and superconducting phase. The resulting phase diagram is drawn for a uniform offset charge distribution q: with $q = e$, the superconducting phase increases with respect to $q = 0$, and the model exhibits superconductivity for all the values of $\alpha = zE_J/4E_C$. An offset charge $q = e$

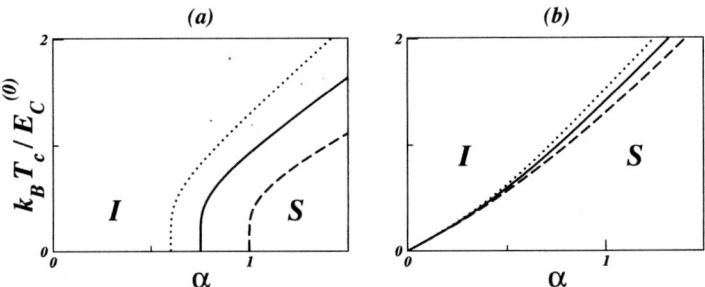

FIGURE 3. Phase diagram in the $T_c - \alpha$ plane for random diagonal charging energies distributed according to Eq. (32) and uniform offset charge $q = 0$ (a), $q = e$ (b). $E_C^{(0)}$ is the average charging energy. We take $U_1 = 1/2$ and $U_2 = 3/2$, while p is $1/4$ (dashed line), $1/2$ (solid line) and $3/4$ (dotted line).

tends therefore to decrease the charging energy and thus favors the superconducting behavior even for small Josephson energy E_J.

Using a pertinent extension of the self-consistent mean-field approach, we obtained also the phase diagram at finite temperature of JJA's with capacitive disorder. For a random distribution of offset charges which are integer multiples of e one has that the superconducting phase increases in comparison with the unfrustrated case. For arrays with random charging energies, the superconducting phase increases with respect to the situation in which all self-capacitances are equal.

The mean-field analysis is, of course, not the whole story: for example, within the approach presented here, it may not be possible to capture the existence of other phases. The full physical picture can be much richer than the one extracted from mean-field. A consistent improvement of the results provided by mean-field theory is still a topic of ongoing investigations [25].

ACKNOWLEDGMENTS

It is our pleasure to contribute with this paper to the volume to honor the 60th birthday of Prof. Ferdinando Mancini. We are very glad to have had the chance to benefit from many stimulating discussions during the years of our friendship. We are grateful to F. Cooper, G. Grignani, A. Mattoni, S. R. Shenoy and A. Tagliacozzo for enlightening discussions. We acknowledge financial support by M.I.U.R. through grant No. 2001028294.

REFERENCES

1. Voss, R.F., and Webb, R.A., Phys. Rev. B **25**, 3446-3449 (1982); Webb, R.A., Voss, R.F., Grinstein, G., and Horn, P.M., Phys. Rev. Lett. **51**, 690-693 (1983).
2. Simànek, E., *Inhomogeneous Superconductors*, Oxford University Press, New York, 1994.
3. Fazio, R., and van der Zant, H., Phys. Rep. **355**, 235-334 (2001).
4. Roddick, E., and Stroud, D., Phys. Rev. B **48**, 16600-16606 (1993).

5. Luciano, G., Eckern, U., and Kissner, J.G., Europhys. Lett. **32**, 669-674 (1995); Luciano, G., Eckern, U., Kissner, J.G., and Tagliacozzo, A., J. Phys.: Condens. Matter **8**, 1241-1255 (1996).
6. Larkin, A.I., and Glazman, L.I., Phys. Rev. Lett. **79**, 3736-3739 (1997).
7. Choi, M.Y., Rhee, S.W., Lee, M., and Choi, J., Phys. Rev. B **63**, 094516 (2001).
8. Bruder, C., Fazio, R., Kampf, A., van Otterlo, A., and Schön, J., Phys. Scri. **42**, 159-170 (1992).
9. van Otterlo, A., Wagenblast, K.H., Fazio, R., and Schön, G., Phys. Rev. B **48**, 3316-3326 (1993).
10. Grignani, G., Mattoni, A., Sodano, P., and Trombettoni, A., Phys. Rev. B **61**, 11676-11688 (2000).
11. Diamantini, M.C., Sodano, P., and Trugenberger, C.A., Nucl. Phys. B **474** 641-677 (1996).
12. Mancini, F.P., Sodano, P., and Trombettoni, A., Phys. Rev. B, **67** 014518 (2003).
13. Mancini, F.P., Sodano, P., and Trombettoni, A., in *Horizons in Superconductivity Research*, Nova Science Publishers, New York, 2003, in press.
14. Fisher, M.P.A., Weichman, P.B., Grinstein, G., and Fisher, D.S., Phys. Rev. B **40**, 546-570 (1989).
15. Simànek, E., Solid State Commun. **31**, 419 (1979).
16. Abramowitz, M., and Stegun, I.A., *Handbook of Mathematical Functions*, Dover, New York, 1964.
17. Wang, Z.X., and Guo, G.R., *Special Functions*, World Scientific, Singapore, 1989.
18. Simkin, M.V., Physica C **267**, 161-163 (1996).
19. Likharev, K.K., and Zorin, A.B., J. Low. Temp. Phys. **59**, 347 (1985).
20. Schön, G., and Zaikin, A.D., Physica B **152**, 203-206 (1988).
21. Simànek, E., Phys. Rev. B **22**, 459-462 (1980); Phys. Rev. B **23**, 5762-5768 (1981); Phys. Rev. B **32** 500-502 (1985).
22. Krupenin, V.A., Presnov, D.E., Zorin, A.B., and Niemeyer, J., J. Low Temp. Phys. **118**, 287-296 (2000).
23. Lafarge, P., Meindersma, J.J., and Mooji, J.E., in *Macroscopic Quantum Phenomena and Coherence in Superconducting Networks*, edited by C. Giovanella, and M. Tinkham, World Scientific, Singapore, 1995, pg. 94.
24. Al-Saidi, W.A., and Stroud, D., Phys. Rev. B **67**, 024511 (2003).
25. Cooper, F., Sodano, P., Trombettoni, A., and Chodos, A., hep-th/0304112.

Magnetic Interactions in Transition Metal Oxides with Orbital Degrees of Freedom

Andrzej M. Oleś

Department of Condensed Matter Theory, M. Smoluchowski Institute of Physics, Jagellonian University, Reymonta 4, PL-30059 Kraków, Poland

Abstract. We review the frustrated magnetic interactions in spin-orbital models which describe superexchange in transition metal oxides with orbital degeneracy, and analyze the reasons for the symmetry breaking in cubic perovskites. The superexchange in e_g systems is dominated by orbital interactions responsible for the orbital ordering, and the A-type antiferromagnetic ordering follows at lower temperatures. Instead, a generic tendency towards dimerization, found already in the degenerate Hubbard model, occurs in t_{2g} systems. In this case the quantum orbital fluctuations may stabilize orbital liquid states along one directions even in some undoped t_{2g} systems, leading to the C-type antiferromagnetic order. The orbital liquid in manganites is triggered by doping. The present understanding of the spectroscopic parameters provides reliable information on the magnetic interactions, as shown on the example of magnons in ferromagnetic cubic and bilayer manganites.

The physical properties of transition metal oxides are dominated by large on-site Coulomb interactions $\propto U$ which suppress charge fluctuations. Therefore, such systems are either Mott or charge-transfer insulators, and the metallic behavior might occur only as a consequence of doping. Here we will discuss first the undoped systems with localized d electrons which interact by effective superexchange (SE) interactions. An interesting situation occurs when d electrons occupy partly degenerate orbital states, and one has to consider *orbital degrees of freedom* in the SE at equal footing with electron spins [1]. Competition between different states is then possible, holes may couple to orbital excitations [2], and the quantum effects are enhanced already in undoped systems [3]. The first models of SE in such situations were proposed almost three decades ago [4], either by considering the degenerate Hubbard model [5, 6], or more realistic situations encountered in cuprates ($KCuF_3$ and K_2CuF_4) and in V_2O_3 [7]. Then it was realized that the SE which is usually antiferromagnetic (AF) might become ferromagnetic (FM) when Hund's exchange interaction J_H is finite, but only in recent years the phenomena which originate from *the orbital physics* were investigated in a more systematic way.

It is intriguing why the FM SE interactions occur either in all three directions in a cubic structure, or two, or only along one direction [1]. We will demonstrate that the answer to this question depends on the type of orbitals involved, and in particular on the tendency towards orbital ordering or fluctuations. A possibility of orbital fluctuations was recently pointed out for t_{2g} systems — it provides a beautiful example how the magnetic behavior may be driven by quantum phenomena in the orbital sector. Such fluctuations are generic in the models when two orbitals of different symmetry participate in the SE processes and may thus interchange on a given bond $\langle ij \rangle$.

The simplest spin-orbital model which illustrates already the physical consequences

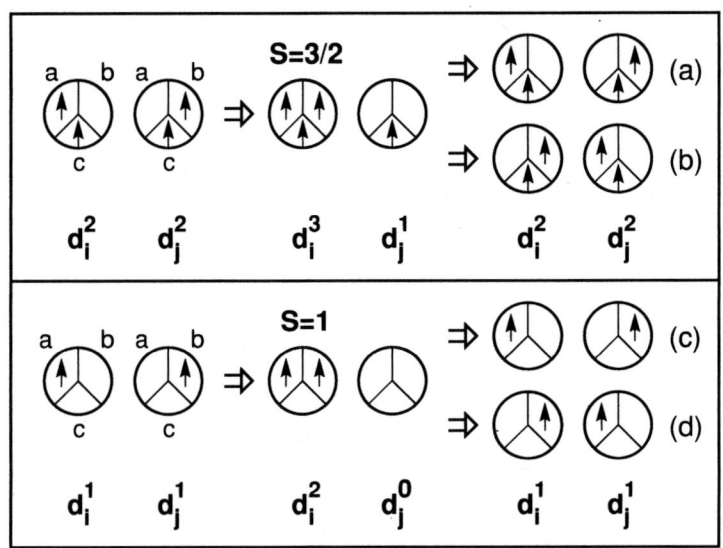

FIGURE 1. Schematic representation of virtual charge excitations $d_i^n d_j^n \to d_i^{n+1} d_j^{n-1} \to d_i^n d_j^n$ contributing to the superexchange on the bond $\langle ij \rangle$ along c axis in t_{2g} systems with: V^{3+} ions at $n_{ic} = 1$ (top) and Ti^{3+} ions at $n_{ic} = 0$ (bottom), for the high-spin $S = (n+1)/2$ state. Next to the (usual) retraceable transitions (a) and (c), the orbital fluctuations (b) and (d) happen. Such terms trigger the FM interactions even in absence of Hund's exchange ($J_H = 0$) and enhance them at finite J_H [20].

of orbital degrees of freedom in strongly correlated systems may be derived for the density of one electron per site ($n = 1$) for a strongly correlated doubly degenerate Hubbard model. We assume that the hopping is isotropic and diagonal between two orbitals α and β at sites i and j, $t_{\mu\nu} = t\delta_{\mu\nu}$, and include only the on-site interaction elements: Coulomb U and Hund's exchange J_H,

$$H = -\sum_{ij\mu\nu\sigma} t_{\mu\nu} a_{i\mu\sigma}^\dagger a_{j\nu\sigma} + U\sum_{i\mu} n_{i\mu\uparrow} n_{i\mu\downarrow} + (U - \frac{5}{2}J_H)\sum_i n_{i\alpha} n_{i\beta}$$
$$- 2J_H \sum_i \vec{S}_{i\alpha} \cdot \vec{S}_{i\beta} + J_H \sum_i (a_{i\alpha\uparrow}^\dagger a_{i\alpha\downarrow}^\dagger a_{i\beta\downarrow} a_{i\beta\uparrow} + a_{i\beta\uparrow}^\dagger a_{i\beta\downarrow}^\dagger a_{i\alpha\downarrow} a_{i\alpha\uparrow}), \quad (1)$$

with the spin, $\{S_{i\mu}^+, S_{i\mu}^-, S_{i\mu}^z\} = \{a_{i\mu\uparrow}^\dagger a_{i\mu\downarrow}, a_{i\mu\downarrow}^\dagger a_{i\mu\uparrow}, (n_{i\mu\uparrow} - n_{i\mu\downarrow})/2\}$, and density operators, $n_{i\mu} = n_{i\mu\uparrow} + n_{i\mu\downarrow}$, at orbital $\mu = \alpha, \beta$ defined in the usual way. The interactions in Eq. (1) are rotationally invariant in the orbital space [11]. Here we neglect a small anisotropy of the interorbital Coulomb and exchange elements, which leads to some quantitative changes in the excitation spectra of real transition metal ions [12]. If $U \gg t$, the electrons localize and the low-energy physics is described by the SE interactions which follow from the virtual $d_i^1 d_j^1 \rightleftharpoons d_i^0 d_j^2$ processes on individual bonds $\langle ij \rangle$. They lead either a high-spin $S = 1$ state, or to a double occupancy in either orbital which has to be subsequently projected onto two low-spin $S = 0$ eigenstates. The case of cuprates with d^9 (Cu^{2+}) ions occupied by one e_g hole each is analogous, and the respective states

TABLE 1. Excitation energies due to $d_i^n d_j^n \to d_i^{n+1} d_j^{n-1}$ processes leading either to high-spin $S = (m+1)/2$ states [ε_{HS}], or to low-spin $S = (m-1)/2$ states [$\varepsilon_{LS}^{(i)}$], where $m = \min(n, 10-n)$. The spin excitation energy $\Delta E = \varepsilon_{LS}^{(1)} - \varepsilon_{HS}$ is largest for the half-filled configuration d^5 obtained as an excited state in manganites ($n = 4$).

n	d^n	excitation type	ε_{HS}	$\varepsilon_{LS}^{(1)}$	$\varepsilon_{LS}^{(2)}$	$\varepsilon_{LS}^{(3)}$	ΔE
1	d^1	Hubbard model Eq. (1)	$U - 3J_H$	$U - J_H$	$U - J_H$	$U + J_H$	$2J_H$
1	d^1	t_{2g} electron	$U - 3J_H$	$U - J_H$	$U - J_H$	$U + 2J_H$	$2J_H$
2	d^2	t_{2g} electron	$U - 3J_H$	U	U	$U + 2J_H$	$3J_H$
4	d^4	e_g electron	$U - 3J_H$	$U + 2J_H$	$U + 2J_H$	$U + 4J_H$	$5J_H$
9	d^9	e_g hole	$U - 3J_H$	$U - J_H$	$U - J_H$	$U + J_H$	$2J_H$

of a d^8 (Cu^{3+}) ion are: $^3A_2$, ^1E and $^1A_1$. Both in the present example of the Hubbard model with one electron per site and for d^9 ions, the excitation spectrum is equidistant (Table 1), with the excitation energies: $\varepsilon(^3A_2) = U - 3J_H$, $\varepsilon(^1E) = U - J_H$, and $\varepsilon(^1A_1) = U + J_H$ [9]. The SE Hamiltonian derived from Eq. (1) takes the form,

$$H_I = Jr_1 \sum_{\langle ij \rangle} \left(\vec{S}_i \cdot \vec{S}_j + \frac{3}{4} \right) \left(\vec{T}_i \cdot \vec{T}_j - \frac{1}{4} \right) + \frac{1}{2} J \sum_{\langle ij \rangle} \left(\vec{S}_i \cdot \vec{S}_j - \frac{1}{4} \right)$$
$$\times \left[r_2(T_i^+ T_j^- + T_i^- T_j^+ + 1) + r_3 \left(2T_i^z T_j^z + \frac{1}{2} \right) + (r_2 - r_3)(T_i^+ T_j^+ + T_i^- T_j^-) \right], \quad (2)$$

where $J = 4t^2/U$ is the energy unit for the SE interaction, and the coefficients $r_1 = 1/(1 - 3\eta)$, $r_2 = 1/(1 - \eta)$, $r_3 = 1/(1 + \eta)$ follow from the above charge excitations, where $\eta = J_H/U$ measures the Hund's exchange. Similar to spin, the pseudospin operators are defined by: $\{T_i^+, T_i^-, T_i^z\} = \{\sum_\sigma a_{i\alpha\sigma}^\dagger a_{i\beta\sigma}, \sum_\sigma a_{i\beta\sigma}^\dagger a_{i\alpha\sigma}, (n_{i\alpha} - n_{i\beta})/2\}$. It is important to use the accurate form of the electron-electron interactions given by Eq. (1) [5, 9], and for this reason some early work led to inaccurate expressions for the effective spin-orbital SE models [4, 6].

Spin SE interactions have SU(2) symmetry, while this symmetry is always removed in the orbital channel by the multiplet structure which splits spin-singlet excited states. We will see below that this feature is general and concerns all spin-orbital models discussed so far. The first term in Eq. (2) is simple and follows from the charge excitations of spin-triplet and interorbital singlet state. The low-spin terms $\propto r_{2(3)}$ are more involved and include not only orbital-flip processes, but also pair hopping terms at both sites $\propto (T_i^+ T_j^+ + T_i^- T_j^-)$. This demonstrates that the anisotropy in the orbital sector is a feature which follows from the multiplet spectra of transition metal ions [12], where the orbital triplet state never occurs at $J_H > 0$.

The generic feature of the SE in transition metal oxides with orbital degeneracy described by spin-orbital models is the frustration of magnetic and orbital interactions [3]: the FM terms occur next to the AF ones, and it is impossible to satisfy simultaneously all of them. The balance between these interaction terms depends on physical parameters of the model which decide what kind of orbital and magnetic order finally wins and is stabilized at low temperature. This frustration is visible already in the limit of $J_H \to 0$ of

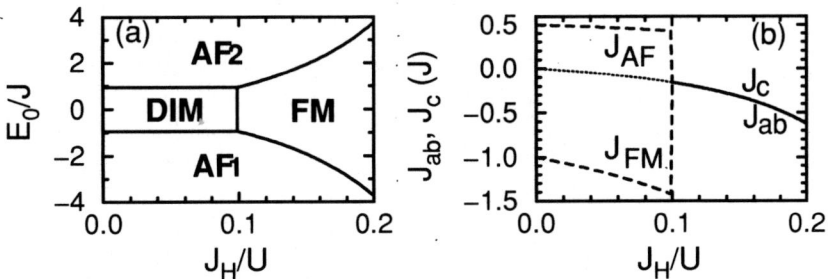

FIGURE 2. (a) Mean-field phase diagram in (J_H, E_0) plane of the isotropic spin-orbital model (2) with four different magnetic phases: OVB, FM, AF1 and AF2 (with either α or β orbitals occupied); (b) exchange constants J_{ab} and J_c for increasing J_H at $E_0 = 0$ as obtained in the OVB phase (J_{FM} and J_{AF}) and in the FM phase ($J_{ab} = J_c$). This figure is reprinted from Ref. [15].

the present spin-orbital model given by Eq. (2), which in this limit represents a superposition of excitations involving either spin-triplet and orbital-singlet, or spin-singlet and orbital-triplet, and the SU(2) symmetry of the orbital interactions is restored:

$$H_I = J \sum_{\langle ij \rangle} \left[\left(\vec{S}_i \cdot \vec{S}_j + \frac{3}{4} \right) \left(\vec{T}_i \cdot \vec{T}_j - \frac{1}{4} \right) + \left(\vec{S}_i \cdot \vec{S}_j - \frac{1}{4} \right) \left(\vec{T}_i \cdot \vec{T}_j + \frac{3}{4} \right) \right]. \quad (3)$$

The result Eq. (3) is just a different way of writing the SU(4) symmetric spin-orbital model [13]. In this case the spin and orbital correlations obey full SU(4) symmetry, and the correlations functions: $\langle \vec{S}_i \cdot \vec{S}_j \rangle$, $\langle \vec{T}_i \cdot \vec{T}_j \rangle$, $\frac{4}{3} \langle (\vec{S}_i \cdot \vec{S}_j)(\vec{T}_i \cdot \vec{T}_j) \rangle$, are all identical [14]. This condition is violated when the mean-field approximation (MFA) is used and the spin and orbital operators are decoupled in the composite correlation function $\frac{4}{3} \langle (\vec{S}_i \cdot \vec{S}_j)(\vec{T}_i \cdot \vec{T}_j) \rangle$, so the results of the MFA might be unreliable in some cases.

Although some qualitative arguments were given, the classical phase diagram of the spin-orbital model (2) was not investigated until recently [15]. We included the orbital splitting at every site, $\sim E_0(n_{i\alpha} - n_{i\beta})/2$, and compared the energies of four different three-dimensional (3D) phases: (i) AF long-range order (LRO) with either α or β orbital occupied at every site, (ii) FM phase with alternating α/β orbitals on two sublattices, and (iii) a dimer phase (DIM) characterized by orbital valence bond (OVB) states, with orbital singlets at every second bond along c axis (or any other, as the present problem is isotropic). When the orbital singlet is formed on a single bond, the energy gain due to the first term in Eq. (2) is maximized and the FM interaction follows. This leads at small J_H to a DIM state in the models for titanates or vanadates under certain conditions, as we will discuss below. On the contrary, the orbitals are uncorrelated at all other bonds, and the AF terms win as long as the Hund's interaction is weak (Fig. 2). The AF states with either α or β orbital occupied are stabilized by the orbital splitting E_0 which has to counterbalance the energy gained by the orbital fluctuations on the FM bonds. Of course, it is hard to imagine that the DIM state with ordered orbital singlets might be realized as such, but the alternation of FM/AF bonds is plausible, so the present phase diagram should rather be viewed as demonstrating a generic competition between different signs of the SE interactions. It shows that one may indeed expect enhanced quantum fluctuations close to the orbital degeneracy when J_H is small [3].

The simplest *realistic* spin-orbital model describes d^9 ions interacting on a cubic lattice, as in KCuF$_3$. The interactions are the same as in Eq. (1), but the hopping term $t_{\mu\nu}$ is now nondiagonal and allows for orbital excitations [2]. In the limit of $U \gg t$ the charge excitations $d_i^9 d_j^9 \rightleftharpoons d_i^8 d_j^{10}$ lead again to the same energies of excited states as those considered above for the degenerate Hubbard model (Table 1), reproducing the exact spectrum of d^8 ions [12]. We define the SE $J = 4t^2/U$ by the largest hopping element t between two $|z\rangle = |3z^2 - r^2\rangle$ orbitals along the c axis, leading to:

$$\mathcal{H}(d^9) = \frac{1}{4} J \sum_{\gamma} \sum_{\langle ij \rangle \| \gamma} \left[\left(\vec{S}_i \cdot \vec{S}_j + \frac{1}{4} \right) \hat{J}_{ij}^{(\gamma)}(d^9) + \hat{K}_{ij}^{(\gamma)}(d^9) \right], \quad (4)$$

where $\gamma = a, b, c$, and $\vec{S}_i$ are spin $S = 1/2$ operators. The operator expressions:

$$\hat{J}_{ij}^{(\gamma)}(d^9) = (2 + \eta r_2 - \eta r_3) \mathscr{P}_{\langle ij \rangle}^{\zeta\zeta} - \eta (3r_1 - r_2) \mathscr{P}_{\langle ij \rangle}^{\zeta\xi}, \quad (5)$$

$$\hat{K}_{ij}^{(\gamma)}(d^9) = -[1 + \eta(3r_1 + r_2)/2] \mathscr{P}_{\langle ij \rangle}^{\zeta\zeta} - [1 + \eta(r_2 - r_3)/2] \mathscr{P}_{\langle ij \rangle}^{\zeta\zeta}, \quad (6)$$

describe spin and orbital SE, and the coefficients r_i are defined again in the same way as in Eq. (2). The present model is however quite different from that obtained from the degenerate Hubbard model as e_g orbitals are directional. The orbital states on the bonds $\langle ij \rangle$ are described by the projection operators:

$$\mathscr{P}_{\langle ij \rangle}^{\zeta\xi} = (1/2 + \tau_i^{\gamma})(1/2 - \tau_j^{\gamma}) + (1/2 - \tau_i^{\gamma})(1/2 + \tau_j^{\gamma}), \quad (7)$$

$$\mathscr{P}_{\langle ij \rangle}^{\zeta\zeta} = 2(1/2 - \tau_i^{\gamma})(1/2 - \tau_j^{\gamma}), \quad (8)$$

which project either on two orthogonal orbital states, being parallel to the bond $\langle ij \rangle$ direction at one site ($P_{i\zeta} = 1/2 - \tau_i^{\gamma}$) and perpendicular at the other ($P_{j\xi} = 1/2 + \tau_j^{\gamma}$), or two parallel parallel orbital states at both sites. They are represented by the orbital operators τ_i^{γ} for the three cubic axes:

$$\tau_i^{a(b)} = (-\sigma_i^z \pm \sqrt{3}\sigma_i^x)/4, \qquad \tau_i^c = \sigma_i^z/2, \quad (9)$$

where the σ's are Pauli matrices acting on: $|x\rangle = \begin{pmatrix} 1 \\ 0 \end{pmatrix}$, $|z\rangle = \begin{pmatrix} 0 \\ 1 \end{pmatrix}$, which transform as $|x\rangle \propto x^2 - y^2$ and $|z\rangle \propto (3z^2 - r^2)/\sqrt{3}$.

In LaMnO$_3$ the SE is more involved and couples *total spins* $S = 2$ at the Mn^{3+} ions. It originates from the charge excitations, $d_i^4 d_j^4 \rightleftharpoons d_i^3 d_j^5$ [10]. The e_g part, following from $d_i^4 d_j^4 \rightleftharpoons d_i^3(t_{2g}^3) d_j^5(t_{2g}^3 e_g^2)$ processes, involves again FM terms due to the high-spin $^6A_1$ state, and three AF terms due to the low-spin states: $^4A_1$, ^4E, and $^4A_2$ (Table 1), and has analogous orbital dependence as in the cuprate case. In contrast, the t_{2g} part follows only from low-spin excitations $d_i^4 d_j^4 \rightleftharpoons d_i^3(t_{2g}^3) d_j^5(t_{2g}^4 e_g)$ and is therefore AF and almost orbital independent. Both terms are given explicitly in Ref. [10].

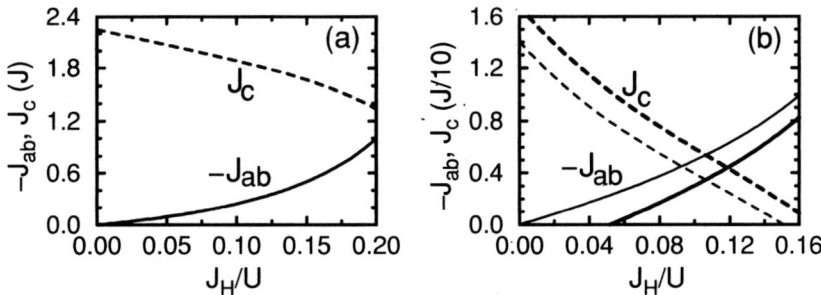

FIGURE 3. Exchange constants: FM $-J_{ab}$ (solid lines) and AF J_c (dashed lines), as obtained in the A-AF phase for e_g systems at orbital degeneracy ($E_z = 0$) as functions of J_H/U for: (a) cuprates (KCuF$_3$); (b) manganites (LaMnO$_3$). In case (b) thin lines show the SE due to e_g electrons only, while heavy lines show the total SE which includes also the SE due to t_{2g} electrons. This figure is reprinted from Ref. [8].

Both the cuprate model (4) and the e_g term in the manganite model describe strongly frustrated SE interactions, which take a universal form in the limit of $J_H \to 0$,

$$\mathcal{H}_e^{(0)} = \frac{1}{4}J \sum_\gamma \sum_{\langle ij \rangle \| \gamma} \left[\left(\frac{1}{S^2} \vec{S}_i \cdot \vec{S}_j + 1 \right) \left(\frac{1}{2} - \tau_i^\gamma \right) \left(\frac{1}{2} - \tau_j^\gamma \right) - 1 \right]. \quad (10)$$

Several classical phases have the same energy of $-3J/4$ per site [3]: the G-AF phases with arbitrary occupation of orbitals, and A-AF phases with $\langle (1/2 - \tau_i^\gamma)(1/2 - \tau_j^\gamma) \rangle = 0$, as obtained for staggered planar orbitals, e.g. for $x^2 - y^2/y^2 - z^2$ orbitals staggered in (a,b) planes. We emphasize that the model (10) is *qualitatively different* from the idealized SU(4) symmetric case (3) due to the directionality of e_g orbitals. In fact, the e_g orbitals order easier, may couple to the lattice and thus appear to be more classical than the isotropic case described by Eq. (2). Their ordering supports magnetic phases with coexisting FM [in (a,b) planes] and AF (along c axis) interactions.

The exchange constants favor the A-AF phase for degenerate e_g orbitals ($E_z = 0$), supported by the orbital ordering. The FM interactions in (a,b) planes increase gradually with J_H, and are even stronger that the AF ones along c axis in LaMnO$_3$ ($J_H/U \simeq 0.117$ [18]) due to the large splitting between the high- and low-spin states ΔE (Fig. 3).

The transition metal oxides with partly filled t_{2g} orbitals exhibit different and even more interesting phenomena. In this case the JT coupling is much weaker, and (unlike for the e_g orbitals) the orbital quantum number is conserved in the hopping processes. This leads to qualitatively different physics realized in t_{2g} systems, somewhat more similar to the isotropic case, Eq. (2). Each t_{2g} orbital is orthogonal to one of the cubic axes, so we label them as a, b, and c (for instance, xy orbitals are labelled as c). The models for titanates and vanadates follow from the $d_i^n d_j^n \rightleftharpoons d_i^{n-1} d_j^{n+1}$ processes [19, 20]:

$$\mathcal{H}(d^n) = J \sum_\gamma \sum_{\langle ij \rangle \| \gamma} \left[(\vec{S}_i \cdot \vec{S}_j + S^2) \hat{J}_{ij}^{(\gamma)}(d^n) + \hat{K}_{ij}^{(\gamma)}(d^n) \right], \quad (11)$$

with the exchange constants $J_{ij}^{(\gamma)}(d^n)$ between $S = 1/2$ spins for titanates ($n = 1$) and $S = 1$ spins for vanadates ($n = 2$), and purely orbital interactions $\hat{K}_{ij}^{(\gamma)}(d^n)$. In titanates these interactions depend on the Hund's rule splittings of d^2 ions [12] (Table 1) via the coefficients: $r_1 = 1/(1 - 3\eta)$, $r_2 = 1/(1 - \eta)$, $r_3 = 1/(1 + 2\eta)$, and are given by [21]:

$$J_{ij}^{(\gamma)}(d^1) = (r_1 + r_2)X_{ij}^{(\gamma)} - \frac{2}{3}(r_2 - r_3)Y_{ij}^{(\gamma)} - \frac{1}{4}(r_1 - r_2)(n_i + n_j)^{(\gamma)}, \quad (12)$$

$$K_{ij}^{(\gamma)}(d^1) = (r_1 - r_2)X_{ij}^{(\gamma)} + \frac{2}{3}(r_2 - r_3)Y_{ij}^{(\gamma)} - \frac{1}{4}(r_1 + r_2)(n_i + n_j)^{(\gamma)}, \quad (13)$$

$$X_{ij}^{(\gamma)} = \left(\vec{\tau}_i \cdot \vec{\tau}_j + \frac{1}{4}n_i n_j\right)^{(\gamma)}, \quad Y_{ij}^{(\gamma)} = \left(\vec{\tau}_i \otimes \vec{\tau}_j + \frac{1}{4}n_i n_j\right)^{(\gamma)}. \quad (14)$$

Apart from the scalar product for pseudospins, $\vec{\tau}_i \cdot \vec{\tau}_j$, occuring due to charge excitations to the configurations with different orbitals occupied, we introduce also a complementary expression, $\vec{\tau}_i \otimes \vec{\tau}_j = \tau_i^x \tau_j^x - \tau_i^y \tau_j^y + \tau_i^z \tau_j^z$, which stands for the orbital interactions resulting from the configurations with double occupancies of individual orbitals. Similar expressions are obtained for the vanadate model with $R = 1/(1 - 3\eta)$ and $r = 1/(1 + 2\eta)$ which describe the multiplet structure in this case (see Table 1) [20]:

$$J_{ij}^{(\gamma)}(d^2) = \frac{1}{2}(1 + 2\eta R)X_{ij}^{(\gamma)} - \eta r Y_{ij}^{(\gamma)} - \frac{1}{2}\eta R(n_i + n_j)^{(\gamma)}, \quad (15)$$

$$K_{ij}^{(\gamma)}(d^2) = \eta R X_{ij}^{(\gamma)} + \eta r Y_{ij}^{(\gamma)} - \frac{1}{4}(1 + \eta R)(n_i + n_j)^{(\gamma)}. \quad (16)$$

In both cases the structure of excited states given in Table 1 is faithfully reproduced with a model Hamiltonian containing U and J_H elements of Coulomb interaction for t_{2g} orbitals. The pseudospin operators $\vec{\tau}_i = \{\tau_i^x, \tau_i^y, \tau_i^z\}$ have here a different meaning in each cubic direction $\gamma = a, b, c$, and refer to the pair of t_{2g} orbital flavors which are active along a given direction γ, and contribute there to the SE via charge excitations [19, 20]; we give an example below. A priori, the magnetic interactions are anisotropic, and may be either AF or FM, depending on the intersite orbital correlations.

In the limit of $J_H/U = 0$ the Hamiltonian Eq. (11) takes the form,

$$\mathcal{H}^{(0)} = \frac{1}{2}J\sum_\gamma \sum_{\langle ij\rangle \| \gamma} \left[\left(\frac{1}{S^2}\vec{S}_i \cdot \vec{S}_j + 1\right)\left(\vec{\tau}_i \cdot \vec{\tau}_j + \frac{1}{4}n_i n_j\right)^{(\gamma)} - \frac{4}{3}S\right], \quad (17)$$

and shows again a strong frustration of SE interactions [19]. Although it resembles formally the spin-orbital model Eq. (3) with SU(4) symmetry [13] much more than Eq. (10), the pseudospin operators $\vec{\tau}_i = \{\tau_i^x, \tau_i^y, \tau_i^z\}$ correspond here not to global $\tau = 1/2$ pseudospins, but *have a different meaning for each cubic direction γ*. Thus, the model is again different from the idealized spin-orbital model with SU(4) symmetry Eq. (3).

Increasing η favors FM SE interactions, similar to e_g systems. In order to get some qualitative insight into the competition between the AF and FM terms at finite η, one may include an anisotropy term $\sim E_c[n_{ic} - (n_{ia} + n_{ib})/2]$, and evaluate the energy of an OVB phase with fluctuating a (yz) and b (xz) orbitals along c axis ($n_{ia} + n_{ib} = 1$), a FM phase with equally and randomly occupied orbitals ($n_{i\gamma} = 1/3$), and a two-dimensional

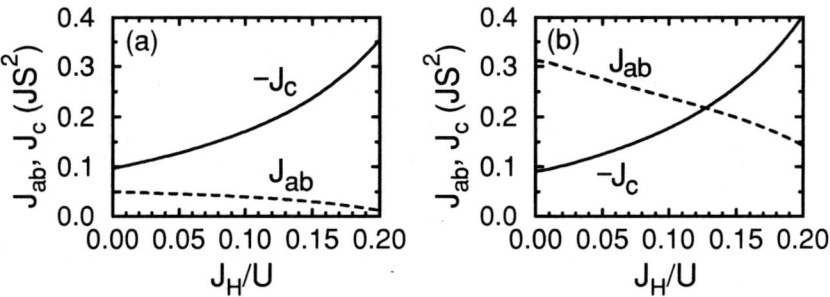

FIGURE 4. Exchange constants J_{ab} (AF, dashed lines) and J_c (FM, solid lines) as obtained for t_{2g} systems with anisotropic orbital occupancies, as functions of J_H/U: (a) titanate model Eqs. (18) and (19) with $n_{ic} = 0$; (b) vanadate model Eqs. (20) and (21) with $n_{ic} = 1$.

(2D) AF phase with only c (xy) orbitals occupied ($n_{ic} = 1$). Unlike for d^9 case, the FM phase is now stable in a broad regime of parameters, and the OVB phase is stabilized by $E_c \simeq 0.2J$ [15]. Of course, the MFA of Ref. [15] is oversimplified and large corrections due to quantum effects are expected. Indeed, a FM isotropic phase is realized in YTiO$_3$, but a closer inspection shows that the orbitals do order, but this ordering does not break the cubic symmetry [22]. A completely different state is realized in LaTiO$_3$, however, with isotropic AF interactions [23]; such interactions are explained by quantum resonance realized simultaneously in spin and orbital sector [19]. This shows provides a beautiful example of AF ordering triggered by quantum fluctuations in the orbital liquid.

However, if only two out of three orbitals are half occupied, and the third orbital is empty, one arrives at a physical realization of the 1D SU(4) model along c axis, with weak intersite interactions in (a,b) planes. This is a specific example of the titanate model Eqs. (12) and (13), if the orbital degeneracy is removed and the c orbital is empty, i.e., $n_{ic} = 0$ and $n_{ia} + n_{ib} = 1$. It is likely that so large distortions which would stabilize this phase cannot be realized in titanates, but we consider it here to get more insight into the physical consequences of orbital fluctuations. In this case one finds that the orbital dynamics drives the FM interactions along the c direction. The expressions for exchange constants within the (a,b) planes (J_{ab}) and along the c axis (J_c) are given by:

$$J_{ij}^{(c)}(d^1) = \left[(r_1+r_2) - \frac{2}{9}(r_2-r_3)\right]\left\langle \vec{\tau}_i \cdot \vec{\tau}_j + \frac{1}{4}\right\rangle - \frac{1}{2}r_1 + \frac{1}{3}r_2 + \frac{1}{6}r_3, \quad (18)$$

$$J_{ij}^{(a)}(d^1) = -\frac{1}{8}r_1 + \frac{3}{8}r_2 - \frac{1}{12}(r_2 - r_3). \quad (19)$$

The pseudospin operators in Eq. (18) may be represented by Schwinger bosons corresponding to two active orbitals: $\tau_i^x = (a_i^\dagger b_i + b_i^\dagger a_i)/2$, $\tau_i^y = i(a_i^\dagger b_i - b_i^\dagger a_i)/2$, $\tau_i^z = (n_{ia} - n_{ib})/2$, where $\{a_i^\dagger, b_i^\dagger\}$ are Schwinger bosons for a and b orbitals at site i. Here yz and zx orbitals are active along c axis, and we label them as a and b, as they lie in the planes orthogonal to these axes. While the SE constant along c axis depends on the a/b orbital fluctuations. The result obtained using a Bethe ansatz for the 1D pseudospin chain ($\langle \vec{\tau}_i \cdot \vec{\tau}_j + \frac{1}{4}\rangle = 1/2 - \ln 2$) is shown in Fig. 4(a) as a function of J_H/U. One finds that the FM interactions along the c axis are strong and increase with J_H/U. It is important to realize that the fluctuations of t_{2g} orbitals provide a new mechanism of FM SE which operates even in the absence of the Hund's exchange [20]. Indeed, at $J_H = 0$ one finds the FM exchange constant $J_c \simeq 0.38J$. On the contrary, the interactions in the (a,b) planes are AF and follow from the double occupancies of active orbitals along a or b axis. In fact, they decrease with increasing J_H which makes the FM chains almost decoupled from each other at realistic ratio $J_H/U \simeq 0.12$.

As next example of the SE interactions in transition metal oxides in t_{2g} systems let us consider cubic vanadates. Here the experiment suggests that the c orbitals are occupied [24], and one can indeed assume that $n_{ic} = 1$ and $n_{ia} + n_{ib} = 1$. Under these circumstances one finds the following expressions for the exchange constants J_c and J_{ab}:

$$J_{ij}^{(c)}(d^2) = (1+2\eta R)\left(\vec{\tau}_i \cdot \vec{\tau}_j + \frac{1}{4}\right) - \eta r\left(\vec{\tau}_i \otimes \vec{\tau}_j + \frac{1}{4}\right) - \eta R, \tag{20}$$

$$J_{ij}^{(a)}(d^2) = \frac{1}{2}\left[(1-\eta r)(1+n_{ib}n_{jb}) - \eta R(n_{ib} - n_{jb})^2\right]. \tag{21}$$

The orbital dynamics contributes to J_c in a similar way as in the d^1 model, while the SE within the (a,b) planes is driven by static correlations. First we consider the uniform pseudospin correlations along the c axis. The SE along the c axis [Fig. 4(b)] is similar to that found before in the titanate model [Fig. 4(a)]. It is remarkable that the values of $J_c S^2$ are almost the same, and thus the magnetic energy $\propto J_c \langle S_i \cdot S_{i+1} \rangle$ along c axis is almost independent of S. This demonstrates that the FM interactions which occur due to orbital fluctuations are quite common and may occur both in d^1 and d^2 configurations. In contrast, the AF interactions J_{ab} are now much enhanced by the charge excitations to doubly occupied c orbitals at each site. Therefore, the vanadate model shows a generic tendency towards the C-type AF order at finite J_H.

The magnetic ordering realized in cubic vanadates is indeed different from that in cubic titanates: C-type AF order is here robust and was observed both in LaVO$_3$ [25] and in YVO$_3$ at intermediate temperatures $77 < T < 116$ K, and G-type AF order is stable in YVO$_3$ for $T < 77$ [26]. As in V$_2$O$_3$ [27], the SE interactions between $S = 1$ spins follow from the $d_i^2 d_j^2 \rightleftharpoons d_i^1 d_j^3$ processes, leading to the effective spin-orbital model given by Eq. (11) with $n = 2$. When the electrons are condensed in c orbitals ($n_{ic} = 1$) due to the orbital splitting caused by the JT effect [24], the second electron occupies either a or b orbital at every site ($n_{ia} + n_{ib} = 1$), allowing for a resonance on the bonds $\langle ij \rangle$ along c axis, supporting the C-AF order.

The vanadate t_{2g}^2 model has an interesting classical phase diagram which unifies certain features we have already seen both for degenerate isotropic orbitals (Fig. 2) and in the t_{2g}^1 model [Fig. 5(a)]. At $\eta = 0$ and $E_c = 0$ one finds again the frustrated SE (17) between $S = 1$ spins. While the orbital liquid cannot stabilize in this case, orbital singlets may form along the c axis when c orbitals have condensed ($n_{ic} = 1$) and the a and b orbitals fluctuate. This favors the OVB state, with strong FM interactions alternating with weak AF ones along the one-dimensional (1D) chains [28]. At large J_H this state is unstable, however, and the orbital fluctuations support FM interactions along c axis and stabilize the C-AF phase [20]. By considering the energy in the MFA one finds a phase transition from the DIM to the C-AF phase at $\eta_c \simeq 0.09$ [29], as long as $n_{ic} = 1$. At large

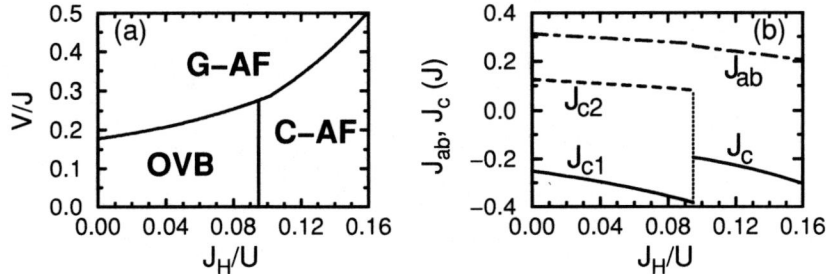

FIGURE 5. The mean-field phase diagram (a) of the vanadate spin-orbital model in $(J_H/U, V/J)$ plane with three magnitic phases stable at $T = 0$: OVB, C-AF, and G-AF. The superexchange interactions (b) in the OVB phase (J_{c1}, J_{c2} and J_{ab}) and in the C-AF phase (J_c and J_{ab}), as functions of J_H/U for $V = 0$. This figure is reprinted from Ref. [29].

uniform orbital splitting $E_c > 0$, the charge gets redistributed to $n_{ia} = n_{ib} = 1$, and an anisotropic G-AF state with strong AF bonds along c axis, and weaker ones within (a,b) planes, follows. To our knowledge, such a state has not been observed so far. The G-AF phase found in YVO$_3$ at $T < T_{N1}$ is characterized instead by large JT distortions, and thus can be explained by a field staggered in (a,b) planes which favors C-type orbital ordering [20], in agreement with recent experiments [26].

One may investigate also the SE interactions in the OVB states, realized both in the titanate [15] and in the vanadate [29] model. As an example we consider here the vanadate case, and present the results which follow from Eqs. (20) and (21). The orbital singlets form at every second bond, and $\langle \vec{\tau}_{2i} \cdot \vec{\tau}_{2i+1} \rangle = -3/4$, while the orbital correlations vanish between the singlets, $\langle \vec{\tau}_{2i} \cdot \vec{\tau}_{2i-1} \rangle = 0$. Therefore, the singlet bonds are strong and FM, while the ones between them are weak and AF. One finds that the OVB state is stable for $J_H/U < 0.095$ within the MFA [29]. This tendency towards bond alternation is quite robust, and its signatures are also found at finite temperatures [30]. The changes of the exchange constants with increasing J_H/U and the crossover between the OVB and the C-AF phase are shown in Fig. 5(b). It is interesting to note that the values of J_{ab} and $|J_c|$ are similar for a realistic value of $\eta \simeq 0.116$ [18], as the orbital fluctuations enhance sufficiently the FM interactions $\propto J_c$.

Let us come back to the question why the orbital liquid state cannot stabilize in LaMnO$_3$. In this case the orbitals have a *generic tendency* towards orbital order, and the orbital interactions are so strong that the orbital ordering would occur well above T_N even in the absence of the JT interaction [10]. However, the splitting between the high-spin $^6A_1$ state and low-spin states ΔE is $5J_H$ (Table 1), which explains the proximity of LaMnO$_3$ to the FM ordering. The manganites at $x < 0.15$ are insulating [17], and are orbital ordered, with either A-AF or FM insulating phase due to polaronic effects. In contrast, large doping $x > 0.15$ stabilizes the FM *metallic* state due to the double exchange (DE) for strongly correlated e_g orbitals [31]. This FM metallic state is nothing else than the realization of the orbital liquid in an e_g system. By considering the DE and SE together one arrives at a quantitative explanation of the spin-wave stiffness D increasing with x, both in cubic systems [32] and in the (a,b) planes of bilayer

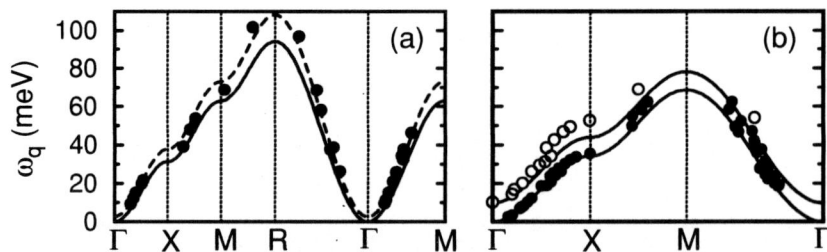

FIGURE 6. Magnon dispersion $\omega_{\vec{q}}$, as obtained in the manganite t-J model with $t = 0.48$ eV for: (a) cubic manganites with $x = 0.30$ doping (solid line) compared with the data points for $La_{0.7}Pb_{0.3}MnO_3$ (circles and dashed line) of Ref. [34]; (b) bilayer system with $x = 0.35$ (solid line), compared with the experimental points for $La_{2-2x}Sr_{1+2x}Mn_2O_7$ (full and empty circles) of Ref. [35]. The high-symmetry points are: $\Gamma = (0,0,0), X = (\pi,0,0), M = (\pi,\pi,0), R = (\pi,\pi,\pi)$.

manganites, which just reflects the gradual release of the kinetic energy by hole doping; in fact the magnons measure here the Gutzwiller renormalization factor for e_g electrons.

In order to determine the magnetic interactions from the kinetic energy of e_g electrons by the DE mechanism, it is crucial to include strong electron correlations. For this reason the band structure calculations fail in this case in a spectacular way and predict the largest DE contribution at no doping ($x = 0$) [33], precisely at the point where the kinetic energy vanishes and the magnetic interactions are completely controlled by the SE processes. The spin-wave stiffness $D_{\text{eff}} = 7.45$ meV, obtained by us at $x = 0.3$ without any fitting parameters [31], agrees well with $D_{\text{exp}} = 8.79$ meV measured in $La_{0.7}Pb_{0.3}MnO_3$ [34] and explains the observed magnon dispersion $\omega_{\vec{q}}$ [Fig. 6(a)].

A similar analysis of the DE and SE contributions allows to describe quantitatively the magnon dispersion $\omega_{\vec{q}}$ observed in bilayer manganites $La_{2-2x}Sr_{1+2x}Mn_2O_7$ [35]; an example for $x = 0.35$ is shown in Fig. 6(b). The total dispersion is lower in this case as the DE along the c axis is much weaker than that within the (a,b) planes, where the frustrated AF SE interactions reduce the FM coupling. The doping dependence of the DE and SE contributions allows to understand also the transition from the FM to the A-AF phase observed in the bilayer manganites at $x = 0.45$ [35]. As we have shown elsewhere [36], hole doping increases the anisotropy between electron densities in $|x\rangle$ and $|z\rangle$ orbitals which gradually reduces the DE contribution to the interlayer coupling. The transition from the FM to A-AF phase occurs when the DE term is overbalanced by the AF SE. This happens only along the c direction, as the large electron density within $|x\rangle$ orbitals induces strong FM interactions within the (a,b) planes.

In summary, the transition metal oxides with orbital degrees of freedom show a very fascinating behavior, with various types of *magnetic and orbital order*. While e_g orbitals usually order and explain A-AF phases, further stabilized by the JT effect, the t_{2g} orbitals have a generic tendency towards disorder, which leads to the *isotropic* orbital liquid in the G-AF phase in $LaTiO_3$, and to a 1D *anisotropic* orbital liquid in the C-AF phase in $LaVO_3$ and YVO_3. So strong orbital fluctuations in e_g systems and the orbital liquid state are triggered only by large doping in the manganites. Very interesting quantum effects might also be discovered in the orbital liquid states in doped t_{2g} systems quite soon.

Acknowledgments. This paper is dedicated to Professor Ferdinando Mancini on the occasion of his 60th birthday. It is a pleasure to express my warm thanks to Lou-Fe' Feiner, Peter Horsch, Giniyat Khaliullin, and Jan Zaanen for a very friendly collaboration on this subject and for numerous stimulating discussions. I thank also J. Bała, A. Fujimori, B. Keimer, G. A. Sawatzky, and Y. Tokura for valuable discussions. The author acknowledges the finacial support by the Polish State Committee of Scientific Research (KBN), Project No. 2 P03B 055 20.

REFERENCES

1. Y. Tokura and N. Nagaosa, Science **288**, 462 (2000).
2. J. Zaanen and A. M. Oleś, Phys. Rev. B **48**, 7197 (1993).
3. L. F. Feiner, A. M. Oleś, and J. Zaanen, Phys. Rev. Lett. **78**, 2799 (1997).
4. K. I. Kugel and D. I. Khomskii, Sov. Phys. JETP **37**, 725 (1973); Sov. Phys. Usp. **25**, 231 (1982).
5. M. Cyrot and C. Lyon-Caen, J. Phys. (Paris) **36**, 253 (1975).
6. S. Inagaki, J. Phys. Soc. Jpn. **39**, 596 (1975).
7. C. Castellani, C. R. Natoli, and J. Ranninger, Phys. Rev. B **18**, 4945, 4967 and 5001 (1978).
8. A. M. Oleś, Acta Phys. Polon. B **32**, 3303 (2001).
9. A. M. Oleś, L. F. Feiner, and J. Zaanen, Phys. Rev. B **61**, 6257 (2000).
10. L. F. Feiner and A. M. Oleś, Phys. Rev. B **59**, 3295 (1999).
11. A. M. Oleś, Phys. Rev. B **28**, 327 (1983).
12. J. S. Griffith, *The Theory of Transition Metal Ions* (Cambridge University Press, Cambridge, 1971).
13. Y. Q. Li, M. Ma, D. N. Shi, and F. C. Zhang, Phys. Rev. Lett. **81**, 3527 (1998).
14. B. Frischmuth, F. Mila, and M. Troyer, Phys. Rev. Lett. **82**, 835 (1999).
15. A. M. Oleś, Phys. Stat. Sol. (b) **236**, 281 (2003).
16. B. Lake, D. A. Tennant, and S. E. Nagler, Phys. Rev. Lett. **85**, 832 (2000).
17. F. Moussa, M. Hennion, C. Biotteau, J. Rodríguez-Carvajal, L. Pinsard, and A. Revcolevschi, Phys. Rev. B **60**, 12299 (1999).
18. T. Mizokawa and A. Fujimori, Phys. Rev. B **54**, 5368 (1996).
19. G. Khaliullin and S. Maekawa, Phys. Rev. Lett. **85**, 3950 (2000).
20. G. Khaliullin, P. Horsch, and A. M. Oleś, Phys. Rev. Lett. **86**, 3879 (2001).
21. G. Khaliullin, Phys. Rev. B **64**, 212405 (2001).
22. G. Khaliullin and S. Okamoto, Phys. Rev. Lett. **89**, 167201 (2002).
23. B. Keimer, D. Casa, A. Ivanov, J. W. Lynn, M.v. Zimmermann, J. P. Hill, D. Gibbs, Y. Taguchi, and Y. Tokura, Phys. Rev. Lett. **85**, 3946 (2000).
24. G. R. Blake, T. T. M. Palstra, Y. Ren, A. A. Nugroho and A. A. Menovsky Phys. Rev. Lett. **87**, 245501 (2001).
25. S. Miyasaka, T. Okuda, and Y. Tokura, Phys. Rev. Lett. **85**, 5388 (2000).
26. Y. Ren, T. T. M. Palstra, D. I. Khomskii, A. A. Nugroho, A. A. Menovsky, and G. A. Sawatzky, Phys. Rev. B **62**, 6577 (2000).
27. R. Shiina, F. Mila, F.-C. Zhang, and T. M. Rice, Phys. Rev. B **63**, 144422 (2001); S. Di Matteo, N. B. Perkins, and C. R. Natoli, Phys. Rev. B **65**, 054413 (2002).
28. S. Q. Shen, X. C. Xie, and F. C. Zhang, Phys. Rev. Lett. **88**, 027201 (2002).
29. A. M. Oleś, P. Horsch, and G. Khaliullin, Acta Phys. Polon. B **34**, 857 (2003).
30. J. Sirker and G. Khaliullin, Phys. Rev. B **67**, 100408(R) (2003).
31. A. M. Oleś and L. F. Feiner, Phys. Rev. B **65**, 052414 (2002).
32. Y. Endoh and K. Hirota, J. Phys. Soc. Jpn. **66**, 2264 (1997).
33. I. V. Solovyev and K. Terakura, Phys. Rev. Lett. **82**, 2959 (1999).
34. T. G. Perring, G. Aeppli, S. M. Hayden, S. A. Carter, J. P. Remeika, and S.-W. Cheong, Phys. Rev. Lett. **77**, 711 (1996).
35. T. G. Perring, D. T. Adroja, G. Chamboussant, G. Aeppli, T. Kimura, and Y. Tokura, Phys. Rev. Lett. **87**, 217201 (2001).
36. A. M. Oleś and L. F. Feiner, Phys. Rev. B **67**, 092407 (2003).

Orbital Physics versus Spin Physics

Louis Felix Feiner* and Andrzej M. Oleś[†]

*Institute for Theoretical Physics, Utrecht University,
Leuvenlaan 4, NL-3584 CC Utrecht, The Netherlands, and
Philips Research Laboratories, Prof. Holstlaan 4,
NL-5656 AA Eindhoven, The Netherlands
[†]Marian Smoluchowski Institute of Physics, Jagellonian
University, Reymonta 4, PL-30059 Kraków, Poland, and
Max-Planck-Institut für Festkörperforschung,
Heisenbergstrasse 1, D-70569 Stuttgart, Germany

Abstract. To elucidate the similarities and differences between the physics displayed by orbital and spin degrees of freedom, we analyze an orbital-Hubbard model with two orbital flavors, corresponding to pseudospin 1/2, and contrast its behavior with that of the familiar (spin-1/2) Hubbard model. The orbital-Hubbard model describes a partly filled spin-polarized e_g band on a cubic lattice, as occurs in ferromagnetic manganites.

We demonstrate that the absence of SU(2) invariance in orbital space has important implications — superexchange contributes in all orbital ordered states, the Nagaoka theorem does not apply, and the kinetic energy is enhanced as compared with the spin case. As a result orbital-ordered states are destabilized by doping, and instead a strongly correlated *orbital liquid* with disordered orbitals is realized.

1. INTRODUCTION

Recently there has been renewed interest in orbital degrees of freedom in Mott insulators [1]. Typically such systems are stoichiometric oxides in which the strong on-site Coulomb repulsion U on the (transition) metal ions eliminates charge fluctuations and replaces them by effective low-energy interactions of superexchange (SE) type. When the electrons occupy partly-filled degenerate e_g or t_{2g} orbitals, as in the perovskites such as $KCuF_3$, $LaMnO_3$, $LaTiO_3$, and $LaVO_3$, the orbital and spin degrees of freedom are equally important and interfere with one another [2, 3]. The SE interactions are then usually strongly frustrated. As a consequence the quantum fluctuations are enhanced, which might even lead to a *spin liquid* state, as possibly realized in $LiNiO_2$ [4]. Another possibility, that an *orbital liquid* (OL) is stabilized and coexists with long-range spin order, was pointed out recently for t_{2g} systems [5]. In contrast, in undoped e_g systems the SE usually favors alternating orbital (AO) order which coexists with antiferromagnetic spin (AS) order, as in $KCuF_3$ [2] and $LaMnO_3$ [6].

An interesting question is how such systems with orbital degrees of freedom behave when they are doped, especially in comparison with the more familiar doped spin systems. In this paper we therefore study a generic model of correlated electrons in a fully spin-polarized e_g band, with the two orbital flavors described by a pseudospin $1/2$ in the orbital Hilbert space, and consider its relation to the common (spin) Hubbard model for electrons with spin $s = 1/2$. We investigate in particular: (*i*) in what respect long-range order in *orbital systems* is different from that in *spin systems*, and (*ii*) whether the orbitals are *ordered* or rather form a *disordered* OL. These questions are clearly of fundamental nature, but they are also of immediate interest for understanding the metallic ferromagnetic phase of the manganites [3, 7]. Our purpose here is to elucidate the physical mechanisms which operate in the e_g band and are typical for orbital degeneracy, and we do so by contrasting them with the mechanisms in spin systems [8].

2. THE ORBITAL-HUBBARD MODEL

So we consider spinless e_g electrons [as in a fully saturated ferromagnetic spin (FS) state] on a cubic lattice with kinetic energy

$$H_t = -t \sum_\alpha \sum_{\langle ij \rangle \| \alpha} c^\dagger_{i\zeta_\alpha} c_{j\zeta_\alpha}, \tag{1}$$

where hopping with amplitude $-t$ between sites i and j occurs only for the directional orbitals $|\zeta_\alpha\rangle$ oriented along the bond $\langle ij \rangle$, i.e., $|\zeta_\alpha\rangle \propto 3x^2 - r^2$, $3y^2 - r^2$, and $3z^2 - r^2$, when $\langle ij \rangle$ is along the cubic axis $\alpha = a, b$, and c, respectively. To describe the local electron interactions one needs to choose an orthogonal basis for the two orbital flavors. One possibility is $|x\rangle \equiv x^2 - y^2$ and $|z\rangle \equiv (3z^2 - r^2)/\sqrt{3}$, called *real orbitals*, whereupon the local Coulomb interaction becomes $U n_{ix} n_{iz}$. The disadvantage of this choice is that the expression for the kinetic energy then takes a different form for each axis [9]. We thus use instead the basis of *complex orbitals* $|+\rangle = \frac{1}{\sqrt{2}}(|z\rangle - i|x\rangle)$ and $|-\rangle = \frac{1}{\sqrt{2}}(|z\rangle + i|x\rangle)$, corresponding to "up" and "down" pseudospin flavors, and write the e_g orbital-Hubbard model in the form

$$\mathcal{H} = -\frac{1}{2} \sum_\alpha \sum_{\langle ij \rangle \| \alpha} \left[t \left(c^\dagger_{i+} c_{j+} + c^\dagger_{i-} c_{j-} \right) \right.$$
$$\left. + \gamma t \left(e^{-i\chi_\alpha} c^\dagger_{i+} c_{j-} + e^{+i\chi_\alpha} c^\dagger_{i-} c_{j+} \right) \right] + U \sum_i n_{i+} n_{i-}, \tag{2}$$

with $\chi_{a,b} = \pm 2\pi/3$, $\chi_c = 0$, and $\gamma = 1$. The phase factors $e^{\pm i\chi_\alpha}$ are characteristic of the orbital problem – the orbitals have an actual shape in real space so that each hopping process depends on the bond direction.

The representation in Eq. (2) has several advantages: (*i*) It displays manifestly the cubic symmetry, as the transformation $c^\dagger_{i\pm} \to c^\dagger_{i\pm} e^{\pm 2i\pi/3}$ and simultaneous cyclic permutation of the cubic axes leaves the Hamiltonian Eq. (2) invariant. (*ii*) It exhibits clearly the difference between the spin case and the orbital case, and in fact allowed us to introduce the parameter γ by which one can turn the e_g-band orbital-Hubbard model ($\gamma = 1$) into what looks formally like a spin-Hubbard model ($\gamma = 0$). (*iii*) It shows explicitly that rotational SU(2) symmetry for the pseudospins is absent [2], since although the diagonal hopping $\propto c^\dagger_{i\pm} c_{j\pm}$ is pseudospin-conserving, the off-diagonal terms $\propto c^\dagger_{i\pm} c_{j\mp}$

are non-pseudospin-conserving. Thus the total pseudospin operator $\mathcal{T}^z = \sum_i T_i^z$, with $T_i^z = \frac{1}{2}(n_{i+} - n_{i-})$, is conserved only at $\gamma = 0$ (i.e., $[\mathcal{T}^z, \mathcal{H}] = 0$), while the terms $\propto \gamma$ commute instead with the staggered pseudospin operator $\mathcal{T}_{\mathbf{Q}}^z = \sum_i \exp(i\mathbf{Q} \cdot \mathbf{R}_i) T_i^z$, where $\mathbf{Q} = (\pi, \pi, \pi)$.

3. ORDERED STATES

Because of the local Coulomb interaction U, there is a tendency towards orbital order, analogous to the tendency towards spin order, i.e. magnetism, in the spin case [10]. At half-filling ($n = 1$) the simplest possibility to reduce the interaction energy $\propto U$ would be to polarize the system completely into *ferro orbital* (FO) states, with uniform order [11], $|\Phi_{FO}\rangle = \prod_i c_i^\dagger(\psi, \theta)|0\rangle$. As in the spin case, another possibility is AO order, $|\Phi_{AO}\rangle = \prod_{i \in A} c_i^\dagger(\psi_A, \theta_A) \prod_{j \in B} c_j^\dagger(\psi_B, \theta_B)|0\rangle$, i.e., with orbitals alternating between two sublattices A and B. If the band is partly filled ($n < 1$), such states must be modified to involve a coherent mixture of occupied and empty sites.

It is an important feature of the $|+\rangle$-polarized (FO+) and $|+\rangle/|-\rangle$-staggered (AO±) *complex states*, that they retain cubic symmetry [12]. By contrast, the FO and AO *real states*, such as $|x\rangle$-polarized (FOx), $|z\rangle$-polarized (FOz), or $(|x\rangle + |z\rangle)/(|x\rangle - |z\rangle)$-staggered (AOxz), with either quasi-one-dimensional (FOz and AOxz) or two-dimensional (FOx) dispersion, break cubic symmetry (and are thus favored in lower dimensional systems, e.g. FOx in a 2D square lattice [13]). This nonequivalence between real and complex states is a manifestation of the broken SU(2) symmetry. We focus here on the orbital-ordered states with complex orbitals [14], which occur in Hartree-Fock (HF) approximation for large enough U [12].

In the case of the FO+ state, the electron bands split above a critical value of U, and lead [15] to a finite order parameter $T^z = \langle T_i^z \rangle \neq 0$, but at finite γ the mechanism of the instability is different from that in the spin case. At $\gamma = 0$ one recovers the Stoner criterion $U_0 N(E_F) = 1$ for the existence of FS order, with the FS states becoming

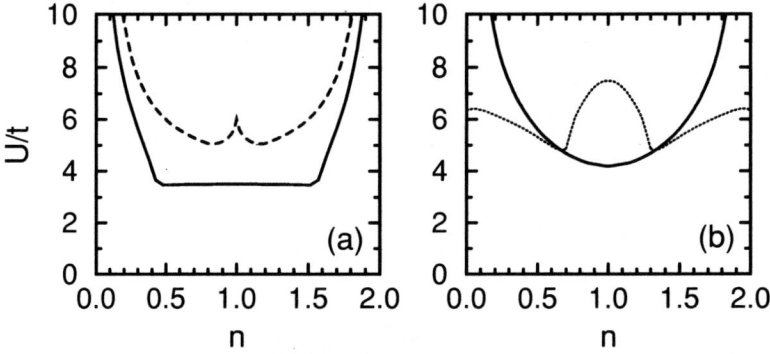

FIGURE 1. HF instabilities towards FO states (full lines) for: (a) $\gamma = 0$, and (b) $\gamma = 1$ [dotted line shows $N^{-1}(\omega)$]. Saturated FS states occur only in (a) above the dashed line.

saturated only at larger U [see Fig. 1(a)]. By contrast, at $\gamma > 0$ the FO states appear as a global property of the band rather than as a Fermi surface instability [Fig. 1(b)]. For large U ($> 6t$) a gap opens, only the lower band $\varepsilon_\mathbf{k}^{FO} = -tA_\mathbf{k} + U(\frac{1}{2}n - T^z E_\mathbf{k})$ is occupied, and

$$T^z = \frac{1}{2N}\sum_\mathbf{k} \frac{1}{E_\mathbf{k}}, \quad E_\mathbf{k} = \left[1 + \left(\frac{\gamma t}{UT^z}\right)^2 B_\mathbf{k}^2\right]^{1/2}, \tag{3}$$

where $A_\mathbf{k} = c_x + c_y + c_z$, $B_\mathbf{k}^2 = c_x^2 + c_y^2 + c_z^2 - (c_x c_y + c_y c_z + c_z c_x)$, with $c_x = \cos k_x$, etcetera, and the sum is over the occupied part of the Brillouin zone (BZ). Unlike in the spin case, $T^z = n/2$ only at $U = \infty$, since the saturated FO+ state is not an eigenstate of $\mathcal{H}$. Thus the FO+ state resembles the AS phase in the spin-Hubbard model.

In the case of the AO± state one has four subbands in the reduced BZ. For large U only the lowest two with dispersion $\varepsilon_{\mathbf{k},\pm}^{AO} = \pm\gamma t B_\mathbf{k} + U(\frac{1}{2}n - T^z F_\mathbf{k})$ are occupied, and the order parameter, $T^z = \langle T_{i \in A}^z \rangle = -\langle T_{j \in B}^z \rangle$, is given by

$$T^z = \frac{1}{2N}\sum_\mathbf{k} \frac{1}{F_\mathbf{k}}, \quad F_\mathbf{k} = \left[1 + \left(\frac{t}{UT^z}\right)^2 A_\mathbf{k}^2\right]^{1/2}, \tag{4}$$

rather similar to the FO case Eqs. (3), but with the interchange $A_\mathbf{k} \leftrightarrow \pm\gamma B_\mathbf{k}$. The reason for this interchange is readily recognized from Eq. (2): for FO order, the diagonal hopping $\propto c_{i\pm}^\dagger c_{j\pm}$ that gives $A_\mathbf{k}$ is order-preserving, while the off-diagonal terms $\propto \gamma$ are order-perturbing and reduce T^z. For AO order this is reversed: the off-diagonal hopping $\propto c_{i\pm}^\dagger c_{j\mp}$ that gives $B_\mathbf{k}$ is compatible with the order, while the diagonal one disturbs it. The similarity between the FO and AO states at $\gamma \simeq 1$ becomes even more transparent at large U, where, at $x = 1 - n > 0$,

$$T_{FO}^z = \frac{1}{2}\left\{(1-x) - \frac{3}{(1-x)^2}\left(\frac{\gamma t}{U}\right)^2\right\}, \tag{5}$$

$$T_{AO}^z = \frac{1}{2}\left\{(1-x) - \frac{3-2x}{(1-x)^2}\left(\frac{t}{U}\right)^2\right\}. \tag{6}$$

Note that a SE contribution $\propto (\gamma t)^2/U$ appears also in the FO+ state, because the off-diagonal hopping permits virtual charge fluctuations. In the genuine orbital case ($\gamma = 1$) the reduction of the order parameter by SE is the same for FO and AO at $x = 0$, but at $x > 0$ it is slightly larger for the FO phase, and surprisingly the energy per site of the FO phase is *lower* than that of the AO phase. Thus near half-filling FO+ order is more stable than AO± order at any U, because the FO phase not only gains more kinetic energy $\propto -3tx$ than the AO phase $\propto -2t\gamma x$ due to the difference in band edge between $-tA_\mathbf{k}$ and $-\gamma t B_\mathbf{k}$, but also has lower SE energy [16]. Instead, AO± order dominates at larger doping $x > 0.27$ [12, 17], as a consequence of its peculiar density of states with large weight close to the band edges. Note that this is *opposite* to the spin case ($\gamma = 0$), where the Néel (AS) state has lower energy near $n = 1$ and the FS state takes over above a critical doping $x_c \simeq t/2U$.

4. ORDER VERSUS DISORDER

Another striking difference is that the Nagaoka theorem, stating that at $U = \infty$ a single hole in the half-filled Hubbard model gives a spin-polarized (FS) ground state [18], *does not hold* for the orbital-Hubbard model at $\gamma > 0$, because it requires conservation of $\mathcal{T}^z$ [see, e.g., the proof in Ref. [18]]. This signals that the tendency towards polarization is weaker for orbitals than for spins, as is illustrated explicitly by an exact calculation for a four-site plaquette with three electrons. Whereas at $\gamma = 0$ the ground state is fourfold degenerate, corresponding to maximum spin $S = \frac{3}{2}$ as required by the Nagaoka theorem, at $\gamma > 0$ it splits into a nondegenerate ground state and three excited states (Fig. 2), none of which can be classified by a pseudospin quantum number. One notes the large energy gain in the ground state (from $E_{\text{Plaq}} = -0.25t$ at $\gamma = 0$ down to $E_{\text{Plaq}} \simeq -0.44t$ at $\gamma = 1$) when the orbitals get disordered in order to take full advantage of the non-pseudospin-conserving hopping.

We will now argue that indeed orbital (FO or AO) order is not robust at $\gamma > 0$ and gets replaced by a disordered (OL) phase, as soon as one goes *beyond* the HF approximation and includes electron correlation effects in the disordered phase as well. We consider specifically the $U = \infty$ limit, where the OL competes with fully saturated FO (5) and AO (6) states. Needing a reliable variational method to calculate the correlation energy, we have adapted the slave-boson approach introduced by Kotliar and Ruckenstein (KR) for the spin-Hubbard model [20] to the orbital case. The introduction of slave bosons ($b_{i+}^\dagger$ and $b_{i-}^\dagger$ for occupied, $e_i^\dagger$ for empty sites) and pseudofermions ($f_{i+}^\dagger$ and $f_{i-}^\dagger$) was done for the *complex* $\{|+\rangle, |-\rangle\}$ orbitals, which is essential to arrive at a gauge (cubic) invariant formulation [21]. The local constraints [20] that guarantee single-occupancy and correct correspondence between bosons and pseudofermions were implemented by means of

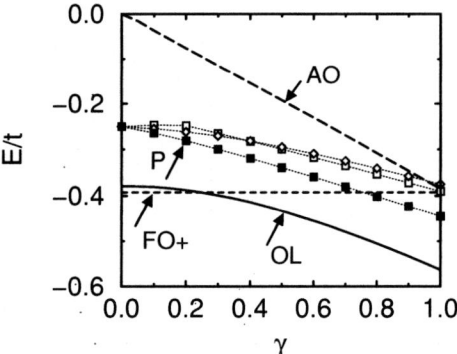

FIGURE 2. Energies per site at $n = 0.75$ and $U = \infty$: E in the KR approach for OL (solid line), FO+ (dashed line), AO± state (long-dashed line), and E_{Plaq} for: the ground state (filled squares) and excited states (empty symbols) of a single plaquette (P).

Lagrange multipliers $\{\mu_{i+}, \mu_{i-}\}$. Treating the bosons in mean-field approximation yields

$$\mathcal{H}_{U=\infty} = -\sum_{i\lambda}\mu_{i\lambda}n_{i\lambda} - \frac{1}{2}t\sum_{\langle ij\rangle}\left[q_+ f_{i+}^\dagger f_{j+} + q_- f_{i-}^\dagger f_{j-}\right.$$
$$\left. + \gamma\sqrt{q_+ q_-}\left(e^{+i\chi_\alpha}f_{i+}^\dagger f_{j-} + e^{-i\chi_\alpha}f_{i-}^\dagger f_{j+}\right)\right], \quad (7)$$

with $n_{i\lambda} = f_{i\lambda}^\dagger f_{i\lambda}$, $\sqrt{q_\pm} = [x/(1 - \langle n_{i\pm}\rangle)]^{1/2}$. The self-consistent solution corresponds to an *OL state* with $\langle n_{i+}\rangle = \langle n_{i-}\rangle\frac{1}{2}(1-x)$ and Gutzwiller renormalization factor $q(x) = q_\pm(x) = 2x/(1+x)$ [22]. The fermion bands, $\varepsilon_{\mathbf{k},\pm}^{OL} = -tq(x)[A_\mathbf{k} \pm \gamma B_\mathbf{k}]$, interpolate correctly between the uncorrelated ($x \simeq 1$) and Mott insulator ($x = 0$) limits. Since the OL state is incoherent (it is described by the density matrix $\hat{\rho}_i = \frac{1}{2}\mathbf{1}_i$ at every site), it is SU(2) symmetric: random complex or random real orbitals are equivalent, and indeed the same *correlated disordered* OL state is obtained using real orbitals [23].

The ordered states can be obtained too within the present KR slave-boson formalism by a suitable choice of the Lagrange multipliers, e.g. $\mu_+ = 0$, $\mu_- = -\infty$ in Eq. (7) gives the FO+ state. Such states do not experience any band narrowing, as double occupancy is eliminated at $U = \infty$, and the correlation energy vanishes [24]. As a result, only the $\varepsilon_\mathbf{k}^{FO} = -tA_\mathbf{k}$ band ($\varepsilon_{\mathbf{k},\pm}^{AO} = \pm\gamma tB_\mathbf{k}$ bands) is (are) partly filled in the FO+ (AO±) state. Therefore, the bands in the OL state represent formally a (renormalized) superposition of the FO and AO bands.

It is instructive to consider the variation with γ of the total energy E of ordered and disordered states at fixed doping not too far from half-filling, for which we take $x = 0.25$ (Fig. 2). In the spin model ($\gamma = 0$) the FS phase then has somewhat lower energy than the disordered state [25]. When γ is increased, E_{FO} does not change, whereas $E_{AO\pm}$ decreases from zero $\propto \gamma$, but still does not surpass the FO+ state at $\gamma = 1$. However, in spite of the band narrowing $\propto q(x)$, considerably more (kinetic) energy is gained in the OL state [26], essentially because both hopping channels contribute, as is obvious from the expression for $\varepsilon_{\mathbf{k},\pm}^{OL}$. The presence of the additional non-pseudospin-conserving hopping channel $\propto \gamma$, associated with the absence of SU(2) symmetry, implies that more kinetic energy can be gained by paying correlation energy than in the spin case. We thus may interpret the stronger tendency towards disorder in the orbital case as compared to the spin case as being due to the availability of more kinetic energy because of the lower symmetry.

Finally, we compare at $U = \infty$ the energies of all states, complex as well as real, varying both n and γ. One finds that AO states are never stable, while FO states are stable only at small γ (Fig. 3). At $\gamma = 0$ the FO+ and FOx(z) states are degenerate, but at any $\gamma > 0$ the real phases have lower energy, with FOz (FOx) being more stable at $n < 0.71$ ($n > 0.71$). The range of FO order shrinks gradually with increasing γ, and above $\gamma \simeq 0.94$ *the OL phase has lowest energy in the entire range of n*. At finite U the kinetic energy will become more dominant and favor disorder even more, except near $n \simeq 1$ where SE stabilizes real-orbital AO order [2, 3, 4, 6, 17]. We thus argue that for the e_g orbital-Hubbard model ($\gamma = 1$) doping induces a crossover to the OL state *at any U*, supporting earlier conjectures that such a disordered state is realized [22, 27].

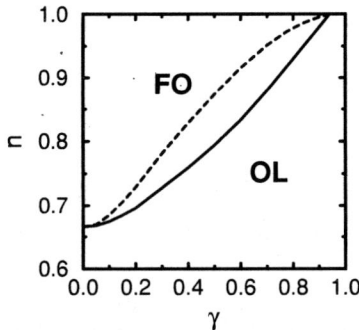

FIGURE 3. Phase diagram as function of γ: OL versus real FOx(z) (full line) and complex FO+ (dashed line) states at $U = \infty$.

Indeed, the disordered OL state provides a natural explanation why the magnons in the ferromagnetic metallic phase of the manganites are isotropic. As we have shown elsewhere [28], the stiffness constant (determined mainly by the double exchange) is proportional to the Gutzwiller band narrowing factor $q(x)$ and increases with hole doping x — thus it measures the kinetic energy of strongly correlated e_g electrons that is released by doping. Remarkably, even when the orbital degeneracy is slightly lifted, as in the bilayer manganite $La_{2-2x}Sr_{1+2x}Mn_2O_7$, at sufficiently large doping the tendency towards orbital disorder prevails, and an anisotropic orbital liquid is realized, which explains the doping dependence of the exchange interactions [29].

5. CONCLUSIONS

In conclusion, the Nagaoka theorem does not apply to the orbital-Hubbard model of correlated e_g electrons, and ordered states are harder to realize than in the spin case. This is manifested by the inverted stability of the ordered phases with complex orbitals, with ferro (staggered) orbital order favored at small (large) doping. More significantly, at finite doping such complex-orbital ordered states are not stable with respect to an *orbital liquid* state because of the *inherent tendency of e_g systems towards orbital disorder*, due to the enhancement of the kinetic energy associated with the absence of SU(2) symmetry.

ACKNOWLEDGMENTS

We thank P. Horsch, G. Khaliullin, D. I. Khomskii, P. Wölfle, and particularly K. Rościszewski for valuable discussions. This work was supported by the Committee of Scientific Research (KBN) Project No. 5 P03B 055 20.

REFERENCES

1. Tokura, Y., and Nagaosa, N., *Science* **288**, 462 (2000).
2. Kugel, K.I., and Khomskii, D.I., *Usp. Fiz. Nauk* **136**, 621 (1982) [*Sov. Phys. Usp.* **25**, 231 (1982)]; Oleś, A.M., Feiner, L.F., and Zaanen, J., *Phys. Rev. B* **61**, 6257 (2000).
3. Oleś, A.M., *Acta Phys. Polon. B* **32**, 3303 (2001).
4. Feiner, L.F., Oleś, A.M., and Zaanen, J., *Phys. Rev. Lett.* **78**, 2799 (1997).
5. Khaliullin, G., and Maekawa, S., *Phys. Rev. Lett.* **85**, 3950 (2000); Khaliullin, G., Horsch, P., and Oleś, A.M., *ibid.* **86**, 3879 (2001).
6. Feiner, L.F., and Oleś, A.M., *Phys. Rev. B* **59**, 3295 (1999).
7. Ramirez, A.P., *J. Phys.: Condens. Matter* **9**, 8171 (1997).
8. The present discussion is based upon: Feiner, L.F., and Oleś, A.M., "Orbital physics versus spin physics: the orbital-Hubbard model," in *Concepts in Electron Correlation*, edited by A. Hewson and V. Zlatic, NATO Science series II, Vol. **110**, Kluwer, Dordrecht, 2003, pp. 123–132; with kind permission of Kluwer Academic Publishers.
9. Takahashi, A., and Shiba, H., *Eur. Phys. J. B* **5**, 413 (1998); Van den Brink, J., and Khomskii, D.I., *Phys. Rev. Lett.* **82**, 1016 (1999).
10. Fazekas, P., *Lecture Notes on Electron Correlation and Magnetism*, World Scientific, Singapore, 1999.
11. Electron creation operators $c_i^\dagger(\psi_i, \theta_i)$ depend on two angles $\{\psi_i, \theta_i\}$ and create e_g electrons in orbital coherent states $|\Omega_i\rangle = \cos(\psi_i/2)e^{-i\theta_i}|i+\rangle + \sin(\psi_i/2)e^{+i\theta_i}|i-\rangle$.
12. Takahashi, A., and Shiba, H., *J. Phys. Soc. Jpn.* **69**, 3328 (2000); Van den Brink, J., and Khomskii, D.I., *Phys. Rev. B* **63**, 140416 (2001).
13. Mack, F., and Horsch, P., *Phys. Rev. Lett.* **82**, 3160 (1999).
14. Such states would however cost Jahn-Teller energy which we neglect here, see: Englman, R., *The Jahn-Teller Effect in Molecules and Crystals*, Wiley, London, 1972; Motome, Y., and Imada, M., *Phys. Rev. B* **60**, 7921 (1999).
15. In the absence of SU(2) symmetry only the decoupling $Un_{i+}n_{i-} \simeq U(\langle n_{i+}\rangle n_{i-} + n_{i+}\langle n_{i-}\rangle - \langle n_{i+}\rangle\langle n_{i-}\rangle)$ gives complex states with finite $T = \langle T_i^z\rangle$.
16. The reduction of T in the FOx (FOz) state is only half of that in the FO+ state. As the SE energy gain is halved as well, these states are unstable at finite U for $n \simeq 1$. For the real states in HF: $Un_{i+}n_{i-} \simeq -U(\langle T_i^+\rangle c_{i-}^\dagger c_{i+} + c_{i+}^\dagger c_{i-}\langle T_i^-\rangle - \langle T_i^+\rangle\langle T_i^-\rangle)$.
17. Maezono, S., and Nagaosa, N., *Phys. Rev. B* **62**, 11 576 (2000).
18. Nagaoka, Y., *Phys. Rev.* **147**, 392 (1966).
19. Yuan, Q., Yamamoto, T., and Thalmeier, P., *Phys. Rev. B* **62**, 12 696 (2000).
20. Kotliar, G., and Ruckenstein, A.E., *Phys. Rev. Lett.* **57**, 1362 (1986).
21. Like the SU(2) invariant formulation for spins: Frésard, R., and Wölfle, P., *Int. J. Mod. Phys. B* **6**, 685 (1992).
22. The nonvariational slave-fermion approximation gives a different band renormalization $\propto x$ [Ishihara, S., Yamanaka, M., and Nagaosa, N., *Phys. Rev. B* **56**, 686 (1997)], and underestimates the stability of the OL phase.
23. The cubic invariant KR approach for real states consists in substituting $b_{i\pm}^\dagger \mapsto (b_{iz}^\dagger \pm ib_{ix}^\dagger)/\sqrt{2}$, and treating the amplitudes $\langle b_{iz}\rangle$ and $\langle b_{ix}\rangle$ in mean field.
24. Equivalent results are therefore obtained for the orbital ordered states by a single slave-fermion approach.
25. At $\gamma = 0$ the present method is believed to give an upper bound for the stability of FS states;[10] they are stable below $x \simeq 0.33$ in the cubic lattice, very close indeed to $x \simeq 0.32$ found for a single spin-flip in the Gutzwiller wave function [Shastry, B.S., Krishnamurthy, H.R., and Anderson, P.W., *Phys. Rev. B* **41**, 2375 (1990)].
26. At $\gamma = 1$ and $n = 0.75$ the energy gain in the OL state comes close to that in the exact ground state for a plaquette (Fig. 2).
27. Kilian, R., and Khaliullin, G., *Phys. Rev. B* **58**, R11 841 (1998).
28. Oleś, A.M., and Feiner, L.F., *Phys. Rev. B* **65**, 052414 (2002).
29. Oleś, A.M., and Feiner, L.F., *Phys. Rev. B* **67**, 092407 (2003).

Local moment systems: magnetism and electronic correlations

W. Nolting*, W. Müller*, C. Santos* and P. Sinjukow*

*Lehrstuhl Festkörpertheorie, Institut für Physik,
Humboldt-Universität zu Berlin, Newtonstraße 15, 12489 Berlin, Germany*

Abstract. We describe local-moment systems by the (multiband) s-f model (ferromagnetic Kondo-lattice model) which traces back the characteristic properties of such materials to an interband exchange coupling between itinerant conduction electrons and localized magnetic moments. We first present a many-body approach to the electronic and magnetic properties of the single-band model. The exchange coupling leads, on the one hand, to a distinct temperature-dependence of the electronic quasiparticle spectrum and, on the other hand, to magnetic properties, as e. g. the Curie temperature T_C or the magnon dispersion, which are strongly influenced by the band electron selfenergy and therewith in particular by the carrier density. Results for the electronic part are given in terms of quasiparticle densities of states and quasiparticle band structures and for the magnetic part in terms of the selfconsistently derived Curie temperature and spin wave spectra. The transition from weak-coupling (RKKY) to strong-coupling (double exchange) behaviour is worked out.

The multiband model is combined with an ab-initio bandstructure calculation to describe real magnetic materials. The proposed method avoids the double counting of relevant interactions and takes into account the correct symmetry of atomic orbitals. For the ferromagnetic metal Gd we get a selfconsistently derived Curie temperature of 301.5 K and a $T = 0$-moment of $7.81\mu_B$, very close to the experimental values. Furthermore a striking induced temperature-dependence of the 5d conduction bands explains respective photoemission data. For the ferromagnetic semiconductors EuO and EuS we present results for electronic and magnetic bulk properties as well as for thin films.

1. INTRODUCTION

The Kondo-lattice model (KLM), being also denoted as *s-f* or *s-d model* or, in its strong-coupling regime, as *double exchange model*, is today surely one of the most prominent models in solid state theory, and that mainly because of its great variety of important applications to rather hot topics in the wide field of collective magnetism. It refers to magnetic materials which get their magnetic properties from a system of localized magnetic moments being indirectly coupled via an interband exchange to itinerant conduction electrons. Many characteristic properties of such materials can be traced back to this interband exchange.

The classical local-moment systems are magnetic semiconductors such as EuO and EuS [1] and magnetic metals like Gd [2] which possess localized moments due to the half-filled $4f$ shell of the rare earth ion (Eu^{2+}, Gd^{3+}) while the conductivity properties are determined by respectively empty and partially occupied $5d$ states. A striking temperature dependence of the unoccupied conduction band states in EuX was experimentally first observed as *red shift* of the optical absorption edge for the electronic $4f - 5d$ transition when cooling the sample below T_C [1]. Another dramatic consequence of the mentioned temperature dependence is the insulator-metal transition in Eu-rich EuO with the biggest jump in resistivity ever observed in solid state physics [3]. The temperature dependence of the more or less uncorrelated bandstates is explainable only by an interband exchange coupling between localized $4f$ and extendend $5d$ states. The same mechanism is needed to understand the ferromagnetic moment coupling in Gd via a conduction electron spin polarization (RKKY) [2]. Other local-moment systems are the intensively discussed diluted magnetic semiconductors [4], as e. g. $Ga_{1-x}Mn_xAs$, and the colossal magnetoresistance (CMR) materials [5] such as $La_{1-x}(Ca,Sr)_xMnO_3$. Both classes appear attractive because of their technological potential. It can be shown that in both cases the main properties are likely to be caused by an interband exchange between localized moments and itinerant charge carriers as in EuX and Gd.

A proper model for describing the reported situation is the s-f model or Kondo-lattice model [6–8], the multiband version of which is introduced in the next section. In section 3 we inspect a rather simple, but nevertheless very instructive exactly solvable limiting case, which refers to a ferromagnetically saturated semiconductor. Section 4 presents a compact review of the many-body approach which we have used to solve the Kondo-lattice problem. Section 5 is then devoted to the electronic and magnetic model properties of the single-band case for arbitrary temperatures and band occupations in order to prepare the discussion of the real magnetic materials, Gd in section 6 and EuO, EuS in section 7.

2. MULTIBAND KONDO-LATTICE MODEL

The system under consideration consists of quasi-free electrons in rather broad conduction bands (d bands) and localized electrons with extremely flat dispersions (f levels). While there is no contribution of the f electrons to the kinetic energy, the part of the d electrons reads

$$H_d = \sum_{ijmm'\sigma} T_{ij}^{mm'} c_{im\sigma}^+ c_{jm'\sigma} \tag{1}$$

The hopping process from site $\mathbf{R}_i$ to site $\mathbf{R}_j$ may be accompanied by an orbital change $(m \to m')$. $T_{ij}^{mm'}$ are the respective hopping integrals. $c_{jm\sigma}^+$ ($c_{jm\sigma}$) is the creation (annihilation) operator for a Wannier electron at site $\mathbf{R}_j$ in the orbital m with spin σ ($\sigma =\uparrow, \downarrow$).

When applying the model study to real local-moment materials, intended as final goal, we shall require that the single-electron energies do not only account for the kinetic energy and the influence of the lattice potential, but also for all those interactions which are not explicitly covered by the model Hamiltonian. That means that the hopping integrals are to be taken from a proper "ab initio" band calculation. We exemplify the procedure in Sect. 6 for the special case of Gd. In particular we show how to avoid the well-known double-counting problem of important interactions.

The Coulomb interaction is restricted to intraatomic terms. Furthermore, we assume that only two subbands are involved in the scattering process. Then the interaction part can be written as [8]

$$H_C = H_{dd} + H_{ff} + H_{df} \tag{2}$$

In an obvious manner the Coulomb interaction may be split into three different parts depending on whether both interacting particles stem from a conduction band H_{dd}, or both from a flat band H_{ff}, or one from a flat band and the other from a conduction band H_{df}. The first term refers to electron correlations in the conduction bands, not explicitly taken into account by the KLM. As explained below our procedure for describing real materials regards such correlations by a proper renormalization of the single particle energies. The term H_{ff} describes interactions between electrons from the flat bands which interest us only with respect to the fact that they form permanent magnetic moments (spins):

$$\mathbf{S}_i = \sum_f \boldsymbol{\sigma}_f \qquad (3)$$

$\boldsymbol{\sigma}_f$ is the spin operator of an electron in subband f. If it is necessary (insulators !) to consider a (super)exchange interaction between the local spins then H_{ff} is chosen to be a Heisenberg Hamiltonian possibly with a symmetry-breaking single-ion anisotropy D:

$$H_{ff} = -\sum_{ij} J_{ij} \mathbf{S}_i \cdot \mathbf{S}_j - D \sum_i (S_i^z)^2 \qquad (4)$$

The third term in (2) H_{df} refers to the interaction between a localized and an itinerant electron. Neglecting unimportant spin-independent contributions it can be written as an intraatomic exchange, i. e. a local interaction between the conduction electron spin $\boldsymbol{\sigma}_{jm}$ and the local moment spin $\mathbf{S}_j$ [8]:

$$H_{df} = -J \sum_{jm} \boldsymbol{\sigma}_{jm} \cdot \mathbf{S}_j \qquad (5)$$

Here m denotes the conduction electron orbital. J is the exchange coupling constant being assumed to be identical for all df pairs. Using second quantization for the itinerant electron spin ($n_{jm\sigma} = c^+_{jm\sigma} c_{jm\sigma}$), this interaction reads:

$$H_{df} = -\frac{1}{2} J \sum_{jm} \left(S_j^z (n_{jm\uparrow} - n_{jm\downarrow}) + S_j^+ c^+_{jm\downarrow} c_{jm\uparrow} + S_j^- c^+_{jm\uparrow} c_{jm\downarrow} \right) \qquad (6)$$

We see that the first term describes an Ising-like interaction of the two spin operators while the other two provide spin exchange processes between localized moment and itinerant electron. Spin exchange may happen in three different elementary processes: Magnon emission by an itinerant $\downarrow$ electron, magnon absorption by an $\uparrow$ electron and formation of a quasiparticle (*magnetic polaron*). The latter can be understood as a propagating electron dressed by a virtual cloud of repeatedly emitted and reabsorbed magnons corresponding to a polarization of the immediate localized spin neighbourhood.

The total Hamiltonian of the multiband Kondo lattice is composed of (1), (2) and (6)

$$H = H_d + H_{ff} + H_{df} \qquad (7)$$

An important model parameter is of course the effective coupling constant $\frac{JS}{W}$ where W is the Bloch-bandwidth and S the local spin. Especially the sign of J is decisive. Other parameters are the lattice structure and above all the band occupation $n = \sum_\sigma \langle n_\sigma \rangle$. In case of a nondegenerate band n is a number in between 0 and 2.

3. FERROMAGNETICALLY SATURATED SEMICONDUCTOR

To get a first insight into the physics of the Kondo-lattice model we start our investigations with the single-band Hamiltonian:

$$H = \sum_{ij\sigma} T_{ij} c_{i\sigma}^{+} c_{j\sigma} - J \sum_j \mathbf{S}_j \cdot \boldsymbol{\sigma}_j = H_0 + H_1 \qquad (8)$$

The many-body problem of the KLM is not exactly solvable for the general case, even for the simplified version (8). Fortunately, however, there exists a non-trivial, very illustrative limiting case which is rigorously tractable and nevertheless exhibits all the above-mentioned elementary excitation processes [9–12]. It refers to a single electron in an otherwise empty conduction band being coupled to a ferromagnetically saturated moment system, e. g. EuO at $T = 0$. In this case the $\uparrow$ spectrum is simple, because the $\uparrow$ electron cannot exchange its spin with the parallely aligned spin system. Only the

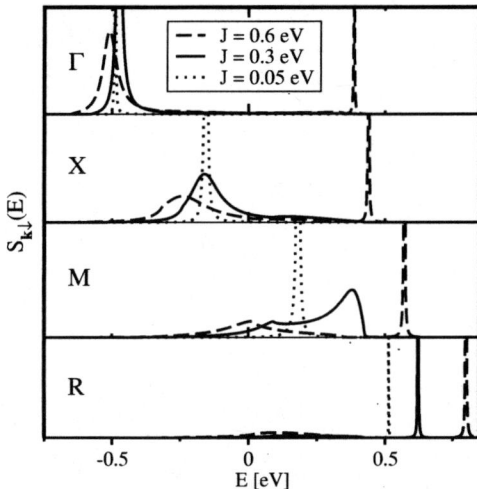

FIGURE 1. $\downarrow$-spectral density as function of energy for several symmetry points in the first Brillouin zone and for different exchange couplings J. Parameters: $S = \frac{1}{2}, W = 1\text{eV}$, sc lattice.

Ising-type interaction in (6) takes care for a rigid shift of the total spectrum by $-\frac{1}{2}JS$. The spectral density is a δ-function at $\varepsilon(\mathbf{k}) - \frac{1}{2}JS$ where ε is the free Bloch energy, the Fourier transform of the hopping integral T_{ij}. Real correlation effects appear, however, in the $\downarrow$ spectrum. Fig. 1 shows the energy dependence of the $\downarrow$-spectral density $S_{\mathbf{k}\downarrow}(E)$ for some symmetry points. For weak couplings the spectral density consists of a single pronounced peak. The finite width points to a finite quasiparticle lifetime due to first spinflip processes, but the sharpness of the peak points to a long-living quasiparticle.

This changes drastically even for rather moderate effective exchange couplings JS/W. One observes in certain parts of the Brillouin zone, for strongly coupled system in the whole Brillouin zone, that the excitation energy splits into two parts. The sharp high-energy peak belongs to the magnetic polaron while the broader low-energy part consists of scattering states due to magnon emission. As long as the polaron peak is above

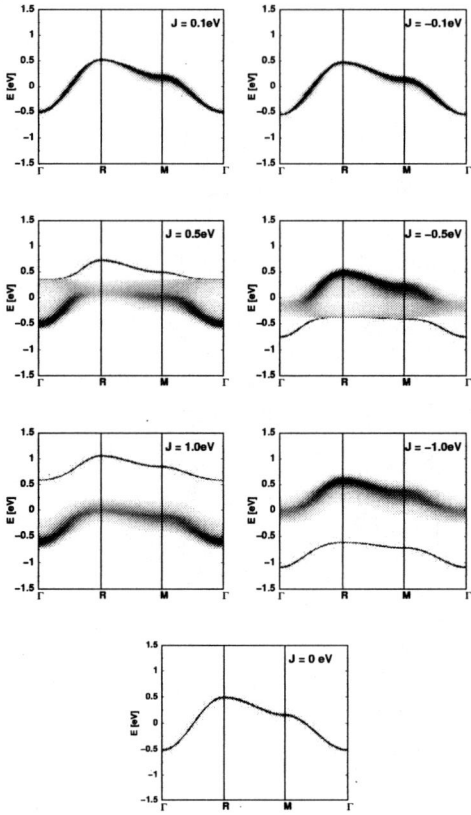

FIGURE 2. ↓ quasiparticle dispersion of a ferromagnetically saturated semiconductor for different exchange couplings J; left column for ferromagnetic exchange, right column for antiferromagnetic exchange. Parameters: $S = \frac{1}{2}$, $W = 1\text{eV}$, sc lattice

the scattering spectrum the quasiparticle has even an infinite lifetime. The scattering spectrum is in general rather broad because the emitted magnon can carry away any wave-vector from the first Brillouin zone. Because of the concomitant spinflip magnon emission can happen only if there are ↑ states within reach. Therefore, the scattering part extends just over that energy region where $\rho_\uparrow \neq 0$ (ρ_σ: quasiparticle density of states (Q-DOS)). Sometimes, as e. g. for $J = 0.6$ eV at the Γ point (Fig. 1), the scattering part is surprisingly bunched together to a prominent peak.

Fig. 2 shows the quasiparticle dispersion, derived as density plot from the spectral density. The sharp polaron part splits off from the scattering spectrum already for rather moderate exchange couplings. In case of antiferromagnetic coupling ($J < 0$) it even represents the lowest-lying excitation. For interaction strengths of about $|JS| = 0.5$ eV the scattering states give rise for a rather pronounced peak certainly visible in a respective (inverse) photoemission experiment. Note that the results in Fig.1 and 2 are exact and free of any uncontrollable approximation. They represent typical correlation effects which are by no means reproducible by a single-electron theory.

4. MANY-BODY APPROACH TO THE KONDO-LATTICE

For the general case (finite temperature, finite band occupation) the many-body problem of the KLM cannot be solved exactly. Approximations must be tolerated. The central quantity is the selfenergy $\Sigma_{\mathbf{k}\sigma}(E)$, which appears in the single-electron Green function:

$$\langle\langle c_{\mathbf{k}\sigma}; c_{\mathbf{k}\sigma}^+\rangle\rangle = \frac{\hbar}{E+\mu-\varepsilon(\mathbf{k})-\Sigma_{\mathbf{k}\sigma}(E)} \qquad (9)$$

To get the selfenergy we have applied several approaches [6, 13, 14], which qualitatively lead to the same results. The figures in the next section are from a Green function procedure developed in [13]. This non-perturbational theory can be considered a *"moment conserving decoupling approximation"* (MCDA) which interpolates between exact limiting cases as, e. g., the above discussed example. The method decouples a "higher" Green function of the type $\langle\langle A_i [c_{m\sigma}, H_{df}]_-; c_{j\sigma}^+\rangle\rangle$, where A_i is any combination of local-moment and band operators, in the following way: The off-diagonal terms $i \neq m$ are simplified by use of the selfenergy equation:

$$\langle\langle A_i [c_{m\sigma}, H_{df}]_-; c_{j\sigma}^+\rangle\rangle \Longrightarrow \sum_r \Sigma_{mr\sigma}(E) \langle\langle A_i c_{r\sigma}; c_{j\sigma}^+\rangle\rangle \qquad (10)$$

The right-hand side is a linear combination of "lower" Green functions with the selfenergy elements as coefficients. The latter are to be determined selfconsistently. To account for the strong local correlations the diagonal $i = m$ are handled with special care:

$$\langle\langle A_i [c_{i\sigma}, H_{df}]_-; c_{j\sigma}^+\rangle\rangle \Longrightarrow \sum_n \gamma_n \langle\langle a_n; c_{j\sigma}^+\rangle\rangle \qquad (11)$$

The right-hand side contains "lower" Green functions being already involved in the hierarchy of equations of motion. The choice of these functions is done in such a way that all known exact limiting cases (atomic limit, ferromagnetically semiconductor, local spin $S = \frac{1}{2}$, $n = 0$, $n = 2$, ...) are reproducable. The coefficients γ_n are eventually found by exact high-energy expansions (spectral moments) of the Green functions. As the interpolating selfenergy approach (ISA) in [6, 14], too, the described MCDA leads to the following structure of the selfenergy:

$$\Sigma_{\mathbf{k}\sigma}(E) = -\frac{1}{2}Jz_\sigma\langle S^z\rangle + J^2 D_{\mathbf{k}\sigma}(E,J) \qquad (12)$$

Restriction to the first term, only, yields the mean-field approach of the KLM, which is correct for sufficiently weak couplings J. It is mainly due to the Ising-part in eq. (6) ($z_\sigma = \delta_{\sigma\uparrow} - \delta_{\sigma\downarrow}$). Without the second part of the selfenergy it would give rise to a spin-polarized splitting of the conduction band. The term $D_{\mathbf{k}\sigma}(E,J)$ is more complicated being predominantly determined by the spin exchange processes. It is a complicated functional of the selfenergy itself, i. e. (12) is an implicit equation for $\Sigma_{k\sigma}(E)$ and not at all an analytical solution. $D_{k\sigma}(E)$ depends, furtheron, on mixed spin correlations such as $\langle S_i^z n_{i\sigma}\rangle$, $\langle S_i^+ c_{i\downarrow}^+ c_{i\uparrow}\rangle$, ..., built up by combinations of localized-spin and itinerant-electron operators. Fortunately, all these mixed correlations can rigorously be expressed via the spectral theorem by any of the Green functions involved in the hierarchy of the MCDA.

However, there are also pure local-moment correlations of the form $\langle S_i^z\rangle$, $\langle S_i^\pm S_i^\mp\rangle$, $\langle (S_i^z)^3\rangle$, ..., which also have to be expressed by the selfenergy (12). For this purpose we use the *modified RKKY* theory of [7] which exploits a mapping of the interband exchange (5) to an effective Heisenberg model,

$$H_f = -\sum_{ij} \hat{J}_{ij} \mathbf{S}_i \cdot \mathbf{S}_j \tag{13}$$

by averaging out the conduction electron degrees of freedom:

$$-J\sum_j \mathbf{S}_j \cdot \boldsymbol{\sigma}_j \longrightarrow -J\sum_j \mathbf{S}_j \cdot \langle \boldsymbol{\sigma}_j\rangle^{(c)} \longrightarrow H_f \tag{14}$$

In the last analysis this means to determine the expectation value $\langle c_{\mathbf{k}+\mathbf{q}\sigma}^+ c_{\mathbf{k}\sigma'}\rangle^{(c)}$. We use again a Green-function procedure [7] by defining a proper function in the conduction electron subspace which allows for a determination of the above expectation value via the spectral theorem. Since the local-spin operators act in the conduction electron subspace as c-numbers the equation of motion hierarchy results in an exact series expansion which, however, has to be truncated accordingly [7]. The lowest non-trivial weak-coupling order arrives at the conventional RKKY result. In the last analysis the *modified* RKKY theory leads to effective exchange integrals in (13) of the following form:

$$\hat{J}_{ij} = \frac{J^2}{4\pi N^2} \sum_{\mathbf{kq}\sigma} e^{i\mathbf{q}\cdot(\mathbf{R}_i - \mathbf{R}_j)} \int_{-\infty}^{+\infty} dE f_-(E) \mathrm{Im}\left[(E - \varepsilon(\mathbf{k}) + i0^+)\right.$$
$$\left.(E - \varepsilon(\mathbf{k}+\mathbf{q}) - \Sigma_{\mathbf{k}+\mathbf{q}\sigma}(E))\right]^{-1} \tag{15}$$

$f_-(E)$ denotes the Fermi function. The effective exchange integrals are decisively influenced by the conduction electron selfenergy Σ_σ which brings into play a distinct band occupation and temperature dependence of the $\hat{J}_{ij}$. Neglecting Σ_σ leads to the "conventional" RKKY formula with $J_{ij} \propto J^2$ as a result of second order perturbation theory. Via $\Sigma_\sigma(E)$ higher order terms of the electron spin polarization enter the *modified* RKKY being therefore not restricted to weak couplings, only.

To get from the effective operator (13) the magnetic properties of the KLM we apply the standard Tyablikow-approximation which is known to yield convincing results in the low as well as in the high temperature region [7]. All the above-mentioned local-moment correlations are then expressed by the electronic selfenergy. We therefore end up with a closed system of equations that can be solved self-consistently for all entities of interest.

5. MAGNETIC AND ELECTRONIC MODEL PROPERTIES

We discuss some typical magnetic and electronic model results. In Fig. 3 we have plotted the quasiparticle bandstructure (Q-BS) and the quasiparticle density of states (Q-DOS) for the moderately coupled sc Kondo-lattice model (effective coupling $\frac{JS}{W} = 0.7$) with a finite band occupation ($n = 0.2$). The Q-BS is derived from the spectral density as density plot, i. e. the degree of blackening is a measure of the magnitude of the respective spectral density peak. The selfconsistenly calculated Curie temperature amounts to $T_C = 238$ K. Note that we did not use a direct exchange interaction between the moments ($J_{ij} \equiv 0$). At $T = 37$ K the local-moment magnetization is 3.4 and therewith very close to saturation. The theory reproduces, in spite of the finite carrier concentration, the same features as for the exact ($n = 0, T = 0$)-case exhibited in Figs. 1 and 2. In the ↓ spectrum

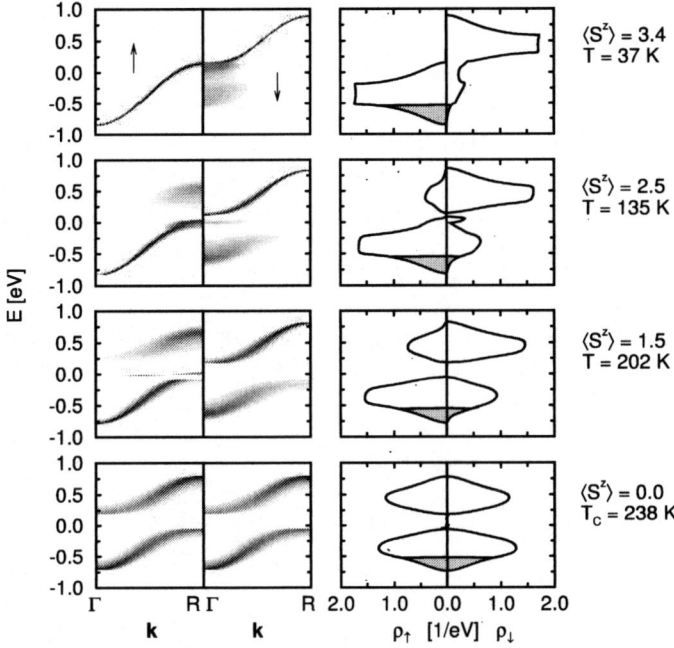

FIGURE 3. Spin dependent quasiparticle bandstructure (left column) as a function of wave vector and quasiparticle density of states (right column) as a function of energy for the Kondo lattice for four different temperatures. Parameters: $J = 0.2$ eV, $W = 1$ eV, $n = 0.2$, $S = \frac{7}{2}$, $J_{ij} \equiv 0$, sc lattice

the scattering states take away a substantial part of the spectral weight near the Γ point while near the R point the polaron state clearly dominates. On the other hand, the ↑ spectrum is for low temperatures only rigidly shifted compared to the *free* case. This holds in particular for the Q-DOS ρ_σ. With increasing temperature (decreasing $\langle S^z \rangle$) a finite magnon density appears to allow for scattering states in the ↑ spectrum, too, because of magnon absorption and simultaneous spinflip by the ↑ electron. Furthermore, polaron and scattering states separate, where, surprisingly, the rather broad scattering spectrum is in wide wave-vector regions bunched to a prominent peak. At $T = T_C$

the spin asymmetry is removed but there remains a correlation-caused splitting of the spectrum into two branches due to the interband exchange coupling J. Because of the possibility of mutual spin exchange $\rho_\uparrow$ and $\rho_\downarrow$ occupy for finite temperatures always the same energy regions.

The key-quantity of ferromagnetism is the Curie temperature T_C. A finite T_C comes out as a consequence of an indirect coupling between the local moments, mediated by a polarization of the conduction electron spins. Therefore, T_C exhibits a strong carrier-

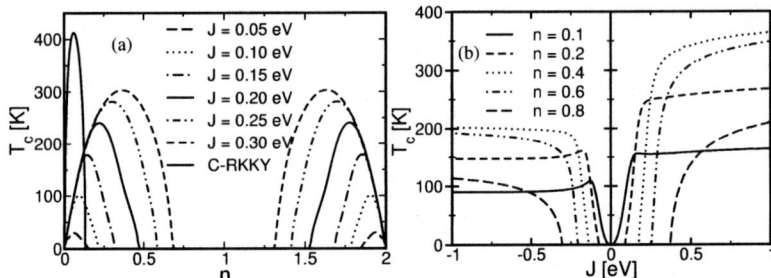

FIGURE 4. Curie temperature as a function of (a) the bandoccupation n for various J, (b) the interband exchange coupling J for various n, calculated by use of the *modified* RKKY.

concentration dependence which is plotted in Fig. 4. Ferromagnetism appears for low electron (hole) concentrations, while being excluded in the region around half-filling ($n = 1$). A similar n-dependence of T_C has been found in ref. [15]. It is interesting to compare the results with those of the *conventional* RKKY, given by $\Sigma_\sigma \equiv 0$ in eq. (15). A corresponding curve is inserted in Fig. 4. The maximal T_C values are higher, but ferromagnetism exists only in a very narrow region of low electron (hole) concentrations where this region turns out to be independent of J.

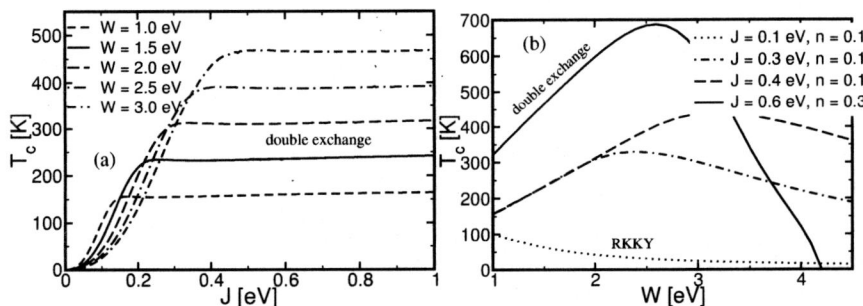

FIGURE 5. Curie temperature as a function of (a) the interband exchange J for various W and $n = 0.1$, (b) the Bloch bandwidth W for various J and n, calculated by use of the *modified* RKKY.

Two features dominate the J-dependence of the Curie temperature plotted in part (b) of Fig. 4. The first is the appearance of a critical J for bandoccupations for which the *conventional* RKKY does not allow ferromagnetism ($n \geq 0.13$). This is in accordance with the fact that for $J \to 0$ the *modified* RKKY reproduces the *conventional* RKKY. The second fact is that T_C runs into a saturation for strong couplings J, where the saturation value depends on the band occupation n. These results are far beyond *conventional RKKY* which can work of course only in the weak coupling regime.

It is interesting to observe (see Fig. 5) that in the weak coupling *RKKY region* T_C scales with the effective coupling constant $\frac{J}{W}$ and in the strong coupling *double exchange region* with the kinetic energy, i. e. $\propto W$. In part (b) of Fig. 5 the change from RKKY behaviour (small $\frac{J}{W}$) to double exchange behaviour (large $\frac{J}{W}$) is clearly to be seen.

6. FERROMAGNETIC METAL GADOLINIUM

In this section we apply the general theory of the KLM to a special system, namely to the ferromagnetic 4f metal Gd. It crystallizes in the hcp structure with a lattice constant $a = 3.629$ Å, $c/a = 1.597$ Å. Magnetic properties result from the half-filled 4f shell ($L = 0, J = S = \frac{7}{2}$) which gives rise to strictly localized magnetic moments. Conductivity properties are due to partially filled $5d/6s$ conduction bands. As to the purely magnetic properties Gd is considered an almost ideal Heisenberg ferromagnet with a Curie temperature of $T_C = 293.2$ K and a $T = 0$ moment of $\mu(T = 0) = 7.63\mu_B$.

Two main extensions of the model treatment in the last section are necessary. We have to account for the 5d band degeneracy and the proper single-particle spectrum of Gd. The first point means that we have to exploit the multiband version of the KLM developed in section 2. Fortunately all the approaches used for the single-band case work for the multiband situation, too.

As mentioned after equation (1) the hopping integrals $T_{ij}^{mm'}$ shall be taken from a bandstructure calculation in order to guarantee that the influences of all the other interactions, which are not directly covered by our model Hamiltonian, are accounted for by a proper renormalization of the single-electron energies. For this purpose we have used an ASW code (*augmented sperical wave*) [16]. LDA-typical difficulties arise with the strongly localized character of the 4f levels. A conventional LDA calculation for Gd results in an antiferromagnetic metal. To circumvent the problem we followed ref. [17] considering the 4f electrons as core electrons, since in our study the 4f levels appear only as localized spins. The result of our bandstructure calculation for Gd is plotted in Fig. 6.

We have to choose the proper single-particle input for our many-body evaluation in such a way, that a double counting of just the decisive interband exchange, explicitly by the model Hamiltonian (6, 7) and implicitly by the ASW input, is avoided. The most direct solution of this problem would be to switch off the interband exchange H_{df} in the LDA code, what turns out to be impossible. However, we can exploit the exact limiting case discussed in section 3. There we have seen that for the ($n = 0$, $T = 0$) limiting case (see Fig. 1) the ↑ spectrum is identical to the *free* Bloch spectrum except for an unimportant rigid shift by $\frac{1}{2}JS$. This cannot be proven exactly for finite band occupations. However, our model investigations clearly indicate (see Fig. 3) that this is also true, at least to a very good approximation, for finite and less than half-filled energy bands. Thus we can use without any manipulation the ↑ dispersions of the ASW calculation ($T = 0$!) for the Gd-5d conduction bands as input for the single-electron Hamiltonian (1). There is no need to switch off H_{df} because in this special case it leads

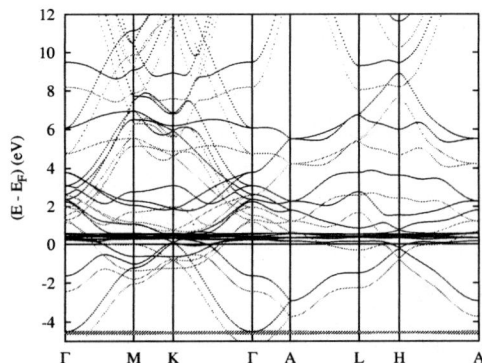

FIGURE 6. ASW result for the Gd bandstructure: Energy versus wave-vector for different symmetry directions. The energy zero is defined by the Fermi edge. The flat dispersions are the 4f levels. $\sigma=\uparrow$: green lines; $\sigma=\downarrow$: black lines.

only to an unimportant rigid shift. It should be stressed, however, that the $\downarrow$ spectrum, on the contrary, is strongly influenced by the interband exchange, even at $T=0$ (see Fig. 1).

After defining the single-particle input there remains only one parameter, namely the exchange coupling J. It is not considered as a free parameter, but also taken from the bandstructure calculation. It is commonly accepted that an LDA treatment of ferromagnetism is quite compatible with a mean-field picture, so that the $T=0$ shift of the $\uparrow$ and $\downarrow$ dispersions in Fig. 6 should amount to $\Delta=JS$. However, it can be seen that the assumption of a rigid splitting is too simple. A certain energy and wave-vector dependence of the exchange splitting is found by LDA, too. We have therefore averaged the $T=0$ exchange splitting over prominent features in the Q-DOS of Gd arriving at $J=0.34$ eV. This is of the same order of magnitude as found for the Eu chalcogenides [18, 19]. There are no further free parameters in our theory.

Figure 7 shows the temperature-dependence of the selfconsistently calculated spontaneous magnetization of Gd. First of all, our theory predicts correctly ferromagnetism with astonishing accurate magnetic key-data, the magnetic $T=0$ moment and the very sensitive Curie temperature.

$$\mu(T=0) = 7.81\mu_B; \quad T_C = 301.5\text{K} \tag{16}$$

The whole magnetization curve fits excellently the experimental data [20]. Since our model does not contain a direct exchange between the moments, the ferromagnetism is excludingly due to an indirect coupling via an induced spin polarization of the conduction electrons. (H_{ff} (4) is used only in the case of insulators (see section 7)). The calculated magnetization curve shows the same characteristic deviations from the $S=\frac{7}{2}$ Brillouin function as the data points, being below for low temperatures and above the Brillouin function for temperatures near T_C. The stronger demagnetization at low temperatures is certainly due to magnon excitations.

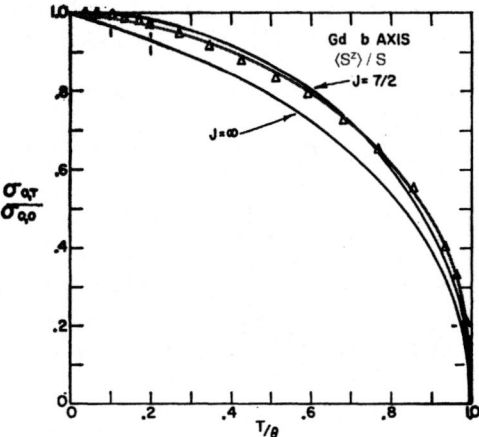

FIGURE 7. Calculated magnetization of Gd as function of the temperature (red line) compared to a $S = J = \frac{7}{2}$ Brillouin function (black line). Data points are from [20].

As mentioned the moment coupling is indirect via conduction electron spin polarization. The respective Q-DOS shows (Fig. 8) indeed a spectacular temperature-dependence which arises from the interband exchange coupling of the "a priori" non-magnetic 5d electrons to the localized 4f moments. The lower edge of the ↑ part exhibits a distinct red shift upon cooling below T_C as it is known since long for ferromagnetic semicon-

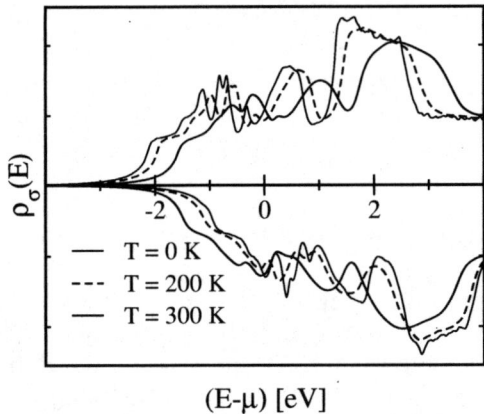

FIGURE 8. Quasiparticle density of states of the Gd-conduction band as function of energy for three different temperatures. The energy zero coincides with the chemical potential.

ductors such as EuO and EuS [21]. The induced spin-polarized splitting is not at all rigid (*"Stoner-like"*) but shows strong deformations mainly because of spin exchange processes according to the partial Hamiltonian (6). The induced spin polarization at the Fermi edge is responsible for the excess moment of $0.81\mu_B$ at $T = 0$, which is only slightly higher than the experimental value of $0.63\mu_B$.

A spin and angle resolved photoemission experiment refers to the single particle spectral density $S_{\mathbf{k}\sigma}(E)$ which is plotted in Figure 9 for the Γ point. Lifetime broadened

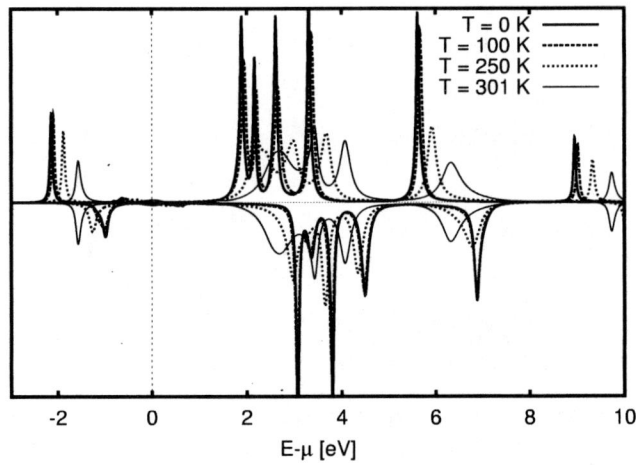

FIGURE 9. Single electron spectral density at the Γ point for Gadolinium as function of energy for four different temperatures.

but well-defined peaks are observed below the chemical potential μ with a temperature dependent spin splitting in the ferromagnetic phase. The spin splitting collapses (*Stoner-like*) when T approaches T_C. This is a long-standing controversy whether or not the induced exchange splitting of the Gd-conduction bands collapses or persists (*spin mixing*) above T_C [2]. At least at the Γ point and below μ our theory clearly shows that it collapses. The situation above μ is not so clear because of stronger lifetime effects and partial overlapping of different spectral density peaks. Figures 8 and 9 present the first finite temperature electronic structure calculation for the ferromagnetic 4f metal Gadolinium [22].

7. FERROMAGNETIC SEMICONDUCTORS EUO AND EUS

EuO and EuS are archetypal ferromagnetic semiconductors which crystallize in the rocksal structure with lattice constants $a = 5.14$ Å (EuO) and $a = 5.95$ Å (EuS), respectively. As in the case of Gadolinium the magnetic properties are mainly determined by localized and half-filled 4f levels, while the conductivity properties are due to the (empty) 5d conduction band. The magnetic properties of the insulators EuO and EuS are exclusively due to the partial operator H_{ff} (4). Because of the empty conduction bands the RKKY mechanism does not work. The superexchange integrals J_{ij} are known from the experiment [23]. Only nearest ($J_1/k_B = 0.625(0.221)$ K) and next-nearest neighbour interaction ($J_2/k_B = 0.125(-0.100)$) have to be taken into account in the case of EuO (EuS). The single-ion anisotropy constant D is unimportant as long as the bulk band structure and its temperature dependence are aimed at. However, when treating systems of lower dimensionality (films, surfaces), then a finite D is necessary to overcome the

Mermin-Wagner theorem [24, 25]. D has been chosen in such a way that the experimental bulk T_C value is correctly reproduced (see figure 3 in [19]). The evaluation of the effectiove Heisenberg model (4) is done by an RPA decoupling of a properly chosen spin Green function (*"Tyablikov decoupling"*) which is known to yield reasonable results over the whole temperature range.

We performed a band structure calculation using the Anderson scheme [26, 27] of a *tight binding linear muffin-tin orbital* (TB-LMTO) ansatz. As in the case of Gd we treated the f states as core states, a *normal* LDA calculation for EuO would produce a metal with the 4f levels lying well within the conduction band. The interband exchange coupling constant is determined in the same manner as above described for Gd. We find $J = 0.25(0.23)$ eV for EuO and EuS, respectively. Our many-body approach to the multiband KLM, explained in section 4, is then used to determine, e. g., the temperature-dependent Q-DOS of EuO as exhibited in Fig. 10. The exchange coupling of the (empty) 5d conduction bands to the localized 4f moments leads to a remarkable temperature dependence which does not comply with the simple Stoner picture of an energy-independent induced exchange splitting of the spectra. The reason is that in our theory correlation is treated in a way distinctly beyond mean-field. In particular, we recognize a shift of the lower ↑ band edge to lower energies with decreasing temperature below T_C. This explains the famous red shift effect of the optical absorption edge for the electronic $4f - 5d_{t_{2g}}$ transition, first experimentally observed some fourty years ago for EuO [28], in the meantime found for all ferromagnetic local-moment systems [21] (see also Fig. 8). Since we have taken into account the full band structure of the Eu-5d con-

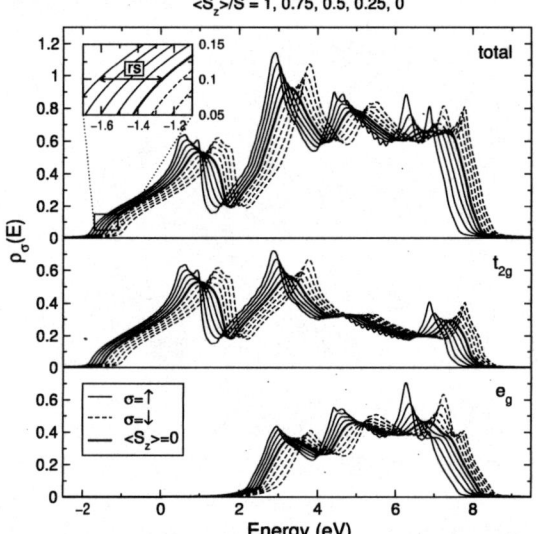

FIGURE 10. Quasiparticle density of states for the (empty) 5d bands of EuO as function of the energy for five different temperatures, i. e. five different 4f magnetizations. ↑ spectrum: solid lines, ↓ spectrum: broken lines. The thick middle line belongs to $T = T_C$, the outermost curves to $T = 0$. The curves in between are for the other temperatures.

duction bands, the symmetry of the 5d orbitals is preserved. The 5d bands can therefore be decomposed into t_{2g} and e_g subbands where the t_{2g} part is much broader ($\simeq 9$ eV for EuO, $\simeq 7$ eV for EuS) than the e_g part ($\simeq 6$ eV for EuO, $\simeq 4$ eV for EuS).

The mentioned red shift is responsable for the striking temperature-driven insulator-metal transition observed in Eu-rich EuO [3]. The Eu-richness manifests itself in an oxygen-deficiency giving rise to twofold positively charged vacancies and two electrons per vacancy which are no longer needed for the binding. One of the electrons will be tightly bound to the vacancy. If the second electron is also trapped at the same place the Coulomb repulsion between the two charge carriers will result in a substantially higher impurity level, possibly located slightly below the lower conduction band edge. Due to the red shift of the ↑ band states upon cooling below T_C the excess electrons can be freed into the conduction band. This scenario has been modelled by us [29] leading to a rather accurate description of the phase transition (Fig. 11). Even details are nicely reproduced. The order of magnitude of the resistivity jump is strongly dependent on the vacancy concentration [30]. Furthermore, for some high-resistivity samples an

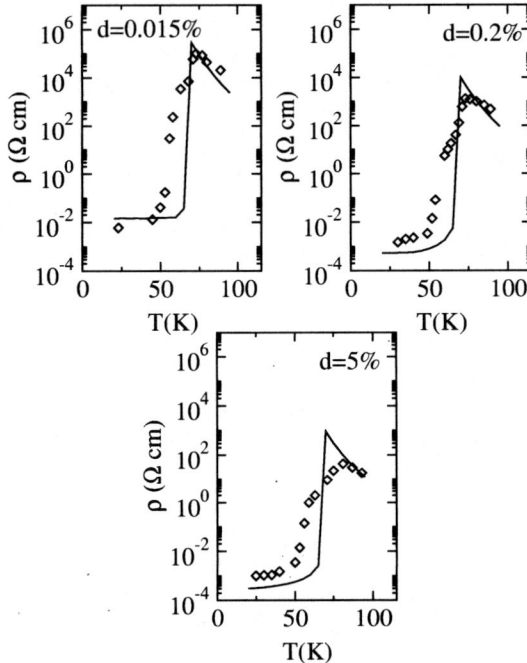

FIGURE 11. Comparison of measured (diamonds [30]) and calculated (solid lines, [29]) resistivity of Eu-rich EuO in dependence on temperature for three different impurity concentrations d.

interesting low-temperature minimum is observed which can be understood by a close approaching the impurity states by the band states (red shift!) without a real crossing. The decrease of resistivity is first due to thermally excited electrons while at very low temperatures the thermal energy is is no longer sufficient for an excitation out of the impurity level into the conduction band. That leads again to an increase of the resistivity for very low temperatures. For details the reader is referred to [29].

The last point, we want to focus on, is the influence of a dimension reduction on magnetic and electronic properties of a local-moment system like EuO [8, 31] or EuS [32]. We have calculated the thicknes and layer-dependence of a thin film consisting of n monolayers with translational symmetry in the film plane. The influence of the two surfaces consists in a forbidden electron hopping in the third space direction.

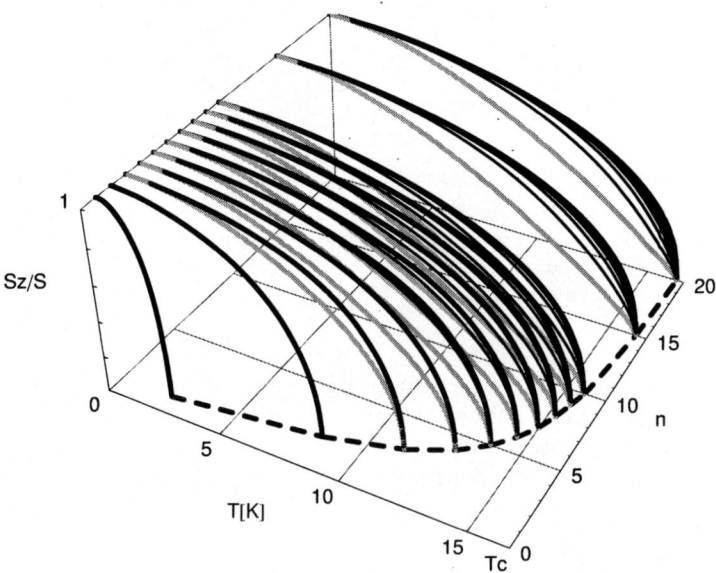

FIGURE 12. Layer-dependent magnetization of EuS(100) films as function of the temperature for various numbers n of monolayers. Gray line: surface layer; black line: central layer; thin black lines: middle layers. Dashed line in $S_z/S = 0$-plane: Curie temperature T_C as function of film-thickness.

Fig. 12 shows the temperature- and layer-dependent magnetizations of EuS(100) films for thicknesses of $n = 1$ to $n = 20$ monolayers. The magnetization of a film increases monotonously from the surface to the film center. The surface always tries to realize a lower T_C being, however, forced by the other layers into a common critical temperature. The Curie temperature shows up a remarkable thickness dependence. Starting at about 2 K for the monolayer, T_C steadily grows up with n reaching the bulk value of 16.33K [1] not before $n \simeq 30$. The 20-layer film has $T_C = 16.2$ K. The results are in excellent agreement with the experimental data of ref. [33].

Fig. 13 shows as a typical electronic example the projected ↑ spin band structure (5d part) of a 20 layer EuS(100) film compared to the respective bulk spectrum. This figure clearly indicates the existence of surface states which appear in energy regions where no bulk spectrum occurs. An additional important feature of a surface state is the exponential decay of its spectral weight with increasing distance from the surface. This is

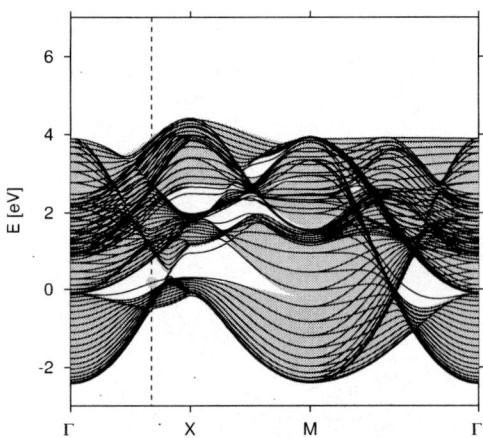

FIGURE 13. LMTO-↑ bandstructure of a EuS(100)20-layer film at $T = 0K$ for some symmetry directions. The gray background represents the respective projected bulk band structure. The dashed line shows the point in k-space used for Fig. 14, the gray circle on it shows the position in energy for which the layer-dependent spectral weight in Fig. 14.

exemplified in Fig. 14, where the spectral weights are plotted for excitations along $\frac{2}{3}\Gamma X$. The prominent peak slightly below the zero point (shaded region) is obviously a surface state. The electronic structure of the film shows up a temperature dependence similar to

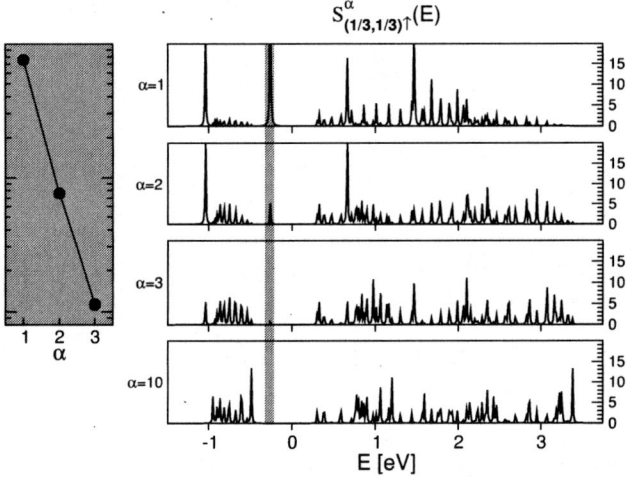

FIGURE 14. Spectral weight of the excitations for the $T = 0$-↑-spectrum at $\frac{2}{3}\overline{\Gamma X}$ (dashed line in Fig. 13) as function of energy for different layers ($\alpha = 1, 2, 3, 10$) of an EuS(100) 20-layer film. The left inset shows the exponential decrease with increasing distance from the surface of the spectral weight of a state within the gray marked energy region. $\alpha = 1$ means the surface layer, $\alpha = 10$ the middle layer.

that of the bulk spectrum (see e. g. Fig. 10 in the case of EuO). The surface states carry the same temperature behaviour, starting at $T = 0$ from an induced spin splitting which collapses *Stoner like* when T approaches T_C. Contrary to EuO films we do not find for EuS a surface state below the 5d spectrum. In ref. [31] it is speculated that such a surface state may lead to a remarkable metal-insulator transition at the surface of an EuO film. The positions of the surface states in EuS, however, are surprisingly different from those in EuO. There is hardly any surface state below the bulk spectrum which could lead to a respective speculation of a surface-metal insulator transition.

8. CONCLUSIONS

We have demonstrated how the multiband Kondo-lattice model can be used for a realistic description of the strikingly temperature-dependent electronic and magnetic properties of local-moment ferromagnets. For this purpose we combined a many-body evaluation of the KLM with a *first principles* band structure calculation (ASW for Gd, TB-LMTO for EuS, EuO)

The first step was to inspect an exactly solvable, non-trivial limiting case of the KLM to learn about the fundamental excitations (spinflip processes, magnetic polaron formation). In addition, this didactic limiting case provides a strong testing criterion for unavoidable approximations for the general, not rigorously tractable case.

By using a self-consistent approach we found out the electronic as well as magnetic properties of the KLM as functions of decisive model parameters such as band occupation and interband exchange coupling. In connection with the Curie temperature we could demonstrate the transition from the weak-coupling RKKY behaviour to a strong-coupling double exchange scenario. That both features are simultaneously involved in our theory means a remarkable support of the model concept.

By the mentioned combination of our model study with an ASW band structure calculation we could reproduce the magnetic key-quantities of the ferromagnetic metal Gd in excellent agreement with the experiment, even with respect to the extremely sensitive Curie temperature. The *a priori* uncorrelated 5d conduction states show up a distinct and unconventional temperature behaviour as a consequence of the exchange coupling to the magnetic 4f moments.

Similar effects are derived for the ferromagnetic semiconductors EuS and EuO. The empty (!) 5d conduction bands exhibit striking temperature dependence in the ferromagnetic phase. As a special detail our calculations reproduce the famous red-shift effect of the optical $4f-5d$ absorption edge. Besides that we could demonstrate how a film geometry influences magnetic and electronic properties of the ferromagnetic semiconductors.

Further goals of our investigation will concern the magnetism of the diluted magnetic semiconductors such as the prototype $Gd_{1-x}Mn_xAs$. Here the main new aspect will be to incorporate the disorder of the localized moments into the theory. Another topic are the colossal magnetoresistance materials such as $La_{1-x}Ca_xMnO_3$ which are believed to be reasonably modelled by an extended KLM. The extension concerns, e. g., the Coulomb interaction of the conduction electrons or the Jahn-Teller effect.

ACKNOWLEDGMENTS

Financial support by the SFB 290 of the *"Deutsche Forschungsgemeinschaft"* is gratefully acknowledged.

REFERENCES

1. P. Wachter, *Handbook of the Physics and Chemistry of Rare Earth* (Amsterdam, North Holland, 1979), vol. 1, chap. 19.
2. M. Donath, P. Dowben, and W. Nolting, eds., *Magnetism and Electronic Correlations in Local-Moment Systems:Rare-Earth Elements and Compounds* (World Scientific, Singapore, 1998).
3. M. R. Oliver, J. A. Kafalas, J. O. Dimmock, and T. B. Reed, Phys. Rev. B **24**, 1064 (1970).
4. F. Matsukara, H. Ohno, A. Shen, and Y. Sugawara, Phys. Rev. B **57**, R2037 (1998).
5. A. P. Ramirez, J. Phys.: Condens. Matter **9**, 8171 (1997).
6. W. Nolting, G. G. Reddy, A. Ramakanth, D. Meyer, and J. Kienert, Phys. Rev. B **67**, 024426 (2003).
7. C. Santos and W. Nolting, Phys. Rev. B **65**, 144419 (2002).
8. R. Schiller, W. Müller, and W. Nolting, Phys. Rev. B **64**, 134409 (2001).
9. D. Meyer, C. Santos, and W. Nolting, J. Phys.: Condens. Matter **13**, 2531 (2001).
10. W. Nolting and U. Dubil, phys. stat. sol. (b) **130**, 561 (1985).
11. W. Nolting, U. Dubil, and M. Matlak, J. Phys.: Condens. Matter **18**, 3687 (1985).
12. B. S. Shastry and D. C. Mattis, Phys. Rev. B **24**, 5340 (1981).
13. W. Nolting, S. Rex, and S. Mathi Jaya, J. Phys.: Condens. Matter **9**, 1301 (1997).
14. W. Nolting, G. G. Reddy, A. Ramakanth, and D. Meyer, Phys. Rev. B **64**, 155109 (2001).
15. A. Chattopadhyay and A. J. Millis, Phys. Rev. B **64**, 024424 (2001).
16. J. Sticht and J. Kübler, Solid State Commun. **72**, 529 (1985).
17. P. Kurz, G. Bihlmayer, and S.Blügel, J. Phys.: Condens. Matter **14**, 6353 (2002).
18. R. Schiller and W. Nolting, Solid State Commun. **118**, 173 (2001).
19. W. Müller and W. Nolting, Phys. Rev. B **66**, 085205 (2002).
20. H. E. Nigh, S. Legvold, and F. H. Spedding, Phys. Rev. **132**, 1092 (1963).
21. B. Batlogg, E. Kaldis, A. Schlegel, and P. Wachter, Phys. Rev. B **12**, 3940 (1975).
22. C. Santos, thesis, Humboldt University at Berlin (2003).
23. H. G. Bohn, W. Zinn, B. Dorner, and A. Kollmar, Phys. Rev. B **22**, 5447 (1980).
24. N. M. Mermin and H. Wagner, Phys. Rev. Lett. **17**, 1133 (1966).
25. A. Gelfert and W. Nolting, J. Phys.: Condens. Matter **13**, R505 (2001).
26. O. K. Andersen, Phys. Rev. B **12**, 3060 (1975).
27. O. K. Andersen and O. Jepsen, Phys. Rev. Lett. **53**, 2571 (1984).
28. G. Busch, J. Junod, and P. Wachter, Phys. Lett. **12**, 11 (1965).
29. P. Sinjukow and W. Nolting, Phys. Rev. B p. submitted (2003).
30. M. R. Oliver, J. O. Dimmock, A. L-McWhorter, and T. B. Reed, Phys. Rev. B **5**, 1078 (1972).
31. R. Schiller and W. Nolting, Phys. Rev. Lett. **86**, 3847 (2001).
32. W. Müller, thesis, Humboldt University at Berlin (2003).
33. A. Stachow-Wójcik, T. Story, W. Dobrowolski, M. Arciszewska, R. R. Gałązka, M. W. Kreijveld, C. H. W. Swüste, H. J. M. Swagten, W. J. M. de Jonge, A. T. Twardowski, Phys. Rev. B **60**, 15220 (1999).

Superconductivity and ferromagnetism: how to find a compromise

Mario Cuoco*[†], Paola Gentile[†] and Canio Noce[†]**

*Centre de Recherches sur les Très Basses Températures associé à l'Université Joseph Fourier,
C.N.R.S., BP 166, 38042 Grenoble-Cédex 9, France
[†]Unità I.N.F.M. di Salerno, Dipartimento di Fisica "E. R. Caianiello", Università di Salerno,
I-84081 Baronissi (Salerno), Italy
**Unità I.N.F.M. di Salerno-Coherentia

Abstract. We analyze the problem of the interplay of superconductivity and ferromagnetism, by focusing on their coexistence. The interest in this field has been renewed after the recent discovery of coexistence of superconductivity and ferromagnetism in UGe_2, $URhGe$, $ZrZn_2$, and in rutheno-cuprate $RuSr_2RECu_2O_8$ compounds, with RE= Eu or Gd. In this paper we concentrate on the competition between itinerant ferromagnetism and singlet superconductivity looking at the case in which the same electrons participate in both the ordered phases. By using a mean field approximation, we have studied the stability of the coexistence phase when the ferromagnetic instability is described through the Stoner model, and when it can be ascribed to a kinetic exchange mechanism.

INTRODUCTION

The problem of the interplay of superconductivity (SC) and ferromagnetism (FM) has been studied since the sixties. The first theoretical inquiry into this problem was made by Abrikosov and Gorkov [1], who provided an accurate description of the action of magnetic impurities on superconductivity, showing that if introduced in a superconductor they lower the superconducting critical temperature significantly, and totally destroy superconductivity, acting as an internal magnetic field.

There are two different mechanisms of pair breaking due to the action of a magnetic field: the spin pair breaking and the orbital pair breaking. The first one is due to the coupling of the magnetic field to the electron spins, which leads to the Zeeman effect, lifting the degeneracy of spin-up and spin-down electron energies. By increasing the strength of the magnetic field, the system passes, through a first-order phase transition, from the BCS state to the normal state. The critical field above which the superconducting phase is destroyed is obtained by equating the magnetic energy the system gains as a consequence of the pair breaking, with the condensation energy. In this way one can obtain the so-called Pauli limiting field. As derived by Chandrasekhar and Clogston, for singlet s-wave superconductivity, this limit is $H_p = \Delta_0/(\sqrt{2}\mu_B)$, where μ_B is the Bohr magneton and Δ_0 is the superconducting order parameter.

On the other hand, the orbital pair breaking is due to the electromagnetic interaction between the potential vector associated with the magnetic field and the linear momentum $\vec{p}$ of the superconducting electrons. This interaction leads to a shift in the kinetic energies of the electrons forming the Cooper pairs, thus breaking the time reversal symmetry of the pair members. Moreover, the orbital coupling leads the superconducting order parameter to become frustrated and this raises the free energy of the superconducting state, leading, as the field is increased, to a transition back to the normal state. Due to the orbital pair breaking, two different behaviors are experimentally observed, which distinguish between I-type and II-type superconductors. The I-type superconductors exhibit a complete Meissner effect until the field reaches the critical value H_c: then there is a second order phase transition to the normal state. On the contrary, the II-type superconductors are characterized by a complete Meissner effect until $H = H_{c1}$; for field strength higher than H_{c1} there is a partial penetration of the magnetic field through the formation of vortexes (Shubnikov phase) until the field strength reaches a critical value $H = H_{c2}$. When $H > H_{c2}$ the system passes into the normal state.

Nevertheless, a nonuniform superconducting state in the Chandrasekhar-Clogston limit in a strong magnetic field has been predicted by Fulde and Ferrell and independently by Larkin and Ovchinnikov (FFLO state)[2]. They noted that the destructive influence of Pauli paramagnetism on superconductivity can be reduced by pairing spin-up and spin-down electrons with a non-zero total momentum, which depends on the magnetic field. In this way the pairing condition, which requires that opposite spin electrons with equal energy and a given total momentum should be paired, can be fulfilled with improved accuracy over some parts of the Fermi surface. On the other parts of the Fermi surface it may then be not possible to pair electrons at all, so that these regions of the k-space correspond to the formation of depaired electrons. The fact that the superconducting pairs have non zero total momentum leads the phase of the gap function to vary spatially with the wave vector of the total momentum of the pairs, and the formation of depaired electrons causes the pair amplitude to be lowered with respect to the BCS case.

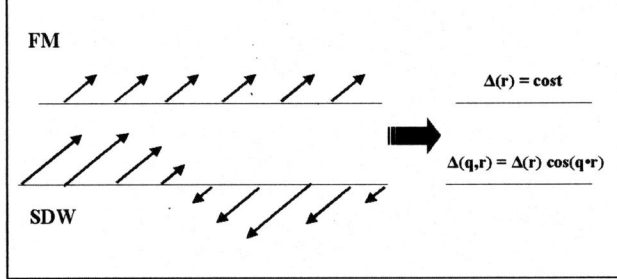

FIGURE 1. Illustration of the analogy between superconducting BCS and Fulde-Ferrell states with ferromagnetic and spin density wave states.

To visualize the differences between the FFLO state and the BCS one, we can consider the analogy with the magnetic case, illustrated in Fig.1. The magnetic analogous for the BCS state is the ferromagnetic phase, since they are both characterized by constant order parameters: in the BCS state the energy gap assumes a constant value in the real space; in the ferromagnetic state the magnetization does not vary in the spin space since the electron spins have equal projection along the direction of total magnetization. On the other side, the superconducting FFLO state corresponds, in the magnetic domain, to the spin density wave state (SDW): like in the FFLO state the energy gap is a periodic function in the real space, in the same way in the SDW the electron spin projection along the magnetization direction varies periodically in the spin space, so that the magnetization has a modulate behavior.

Even if the pair amplitude of the FFLO state is lower than that of the BCS phase, the FFLO state can nonetheless be more stable than the uniform solution, since the presence of depaired electrons allows the system to gain magnetic energy in such a way that in the presence of magnetic field higher than the Pauli limit, the energy of the FFLO phase can become lower than the energy of the paramagnetic state. This condition is verified for temperature values that are lower than a critical one T^*, as one can conclude by looking at the phase diagram [3]. If one considers the effects of the application of a magnetic field on the system, the analysis of the phase diagram shows that by increasing the applied magnetic field at a fixed temperature value $T < T^*$, through a first order phase transition the system passes from the BCS state to the FFLO one before the Pauli limit is reached, and the inhomogeneous superconducting state persists even in the presence of magnetic fields higher than the Pauli one.

Despite many experimental efforts, there has been up to now no unequivocal experimental evidence of the existence of the FFLO state. Indeed, when the orbital effects of the field are considered in a type-II superconductor, the FFLO state can only exist if the diamagnetic effect is weak enough compared to the paramagnetic effect. Moreover, a FFLO state is easily destroyed by paramagnetic impurities so that the FFLO state can be observed only in very clean superconductors and this condition is very hardly verified in ordinary metals. A possible formation of a nonuniform superconducting state has been found in the heavy-fermion compound UPd_2Al_3[4]. In this case, thermal expansion of magnetostriction measurements below the superconducting critical temperature have been utilized to establish a first order phase transition into a inhomogeneous superconducting state. This result has been possible because UPd_2Al_3 is strongly Pauli limited compound and represents an extreme clean limit of a type-II superconductor.

The recent discovery of the coexistence of FM and SC in UGe_2, $URhGe$, $ZrZn_2$, and in rutheno-cuprate $RuSr_2RECu_2O_8$ compounds, with RE= Eu or Gd, has revealed that FM and SC are not incompatible phenomena, as previously outlined. For UGe_2 [5] and $ZrZn_2$ [6] compounds there is experimental evidence that the ferromagnetic order and the superconducting one are due to the same carriers, the $5f$ electrons belonging to the U atoms in UGe_2, and the $4d$ electrons of the Zr atoms in $ZrZn_2$. On the other hand, in the rutheno-cuprate compound $GdSr_2RuCu_2O_8$ it is possible to distinguish among ferromagnetic electrons and superconducting ones. This compound has a crystal structure characterized by CuO_2 layers alternating to RuO_2 ones and experimental results indicate that the ruthenate layers are responsible for magnetic properties while the cuprate layers for the superconductivity[7].

In this paper we will analyze the interplay of itinerant ferromagnetism and singlet superconductivity in systems where the same electrons participate in both the FM and SC order. Therefore the model we propose will be called one-carrier model. We will look at two possible mechanisms responsible for the magnetic instability. Firstly in next Section, we have assumed that the itinerant ferromagnetism can be described by the Stoner model, in which the instability towards the ferromagnetic state is due to a rigid shift in the positions of the majority and minority spin bands, as a consequence of the Coulomb interaction. From this assumption it follows that it is not possible to stabilize a state in which superconductivity and ferromagnetism coexist, if the depaired electrons forming within the superconducting phase are ferromagnetically correlated. We will show hereafter how the depaired electrons that are ferromagnetically correlated modify the interplay between superconductivity and ferromagnetism. In Sect. II we will analyze the kinetic exchange mechanism. In this case the ferromagnetism comes in as a consequence of the spontaneous change in the relative bandwidth of electrons with opposite spin polarization. A detailed comparison between the two situations is discussed in the last Section, containing a discussion and the summary of the results presented in this paper.

STONER EXCHANGE IN THE SUPERCONDUCTING-FERROMAGNETIC PHASE

In this Section we will study the one-carrier model by assuming that the itinerant ferromagnetism rises as a consequence of the Coulomb interaction, which causes the majority and minority spin bands to be rigidly shifted with respect to each other, as illustrated in Fig.2 (Stoner model).

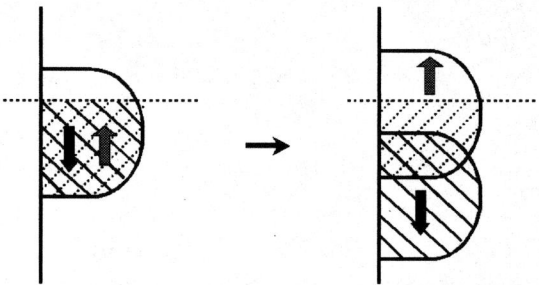

FIGURE 2. Schematic representation of density of states for up and down electrons in the unpolarized (left) and polarized (right) states, within the Stoner model. The dashed line indicates the position of the Fermi level.

An Hamiltonian containing these ingredients is [8]:

$$H = -t \sum_{<ij>\sigma} \left(c^\dagger_{i\sigma} c_{j\sigma} + h.c \right) + U \sum_{i=1}^{N} n_{i\uparrow} n_{i\downarrow} - g \sum_{i=1}^{N} c^\dagger_{i\uparrow} c^\dagger_{i\downarrow} c_{i\downarrow} c_{i\uparrow}$$
$$- \mu \sum_{i=1}^{N} \sum_{\sigma} c^\dagger_{i\sigma} c_{i\sigma} \quad , \tag{1}$$

where the notation $\sum_{<ij>}$ indicates a sum over nearest-neighbor sites; N is the number of sites in the system; t is the hopping amplitude; U is the on-site Coulomb repulsion μ is the chemical potential; g is the superconducting pairing coupling being effective only in a shell of amplitude $2\hbar\omega_D$ around the Fermi surface (as in the usual BCS theory) and μ is the chemical potential. The operators $c_{i\sigma}$ and $c^\dagger_{i\sigma}$ denote the annihilation and creation operators for electrons with spin σ on the i-th site; $n_{i\sigma}$ is the number operator for the electrons with spin σ on the i-th site.

After the mean-field decoupling, the effective Hamiltonian in the k-space is:

$$H_{tot} = \sum_{k\sigma} E_{k\sigma} c^{\dagger}_{k\sigma} c_{k\sigma} + \sum_{\vec{k}} \left[\Delta^* \, c_{-\vec{k}\downarrow} c_{\vec{k}\uparrow} + h.c. \right] + N \left[\frac{|\Delta|^2}{g} + \frac{Um^2}{4} - \frac{Un^2}{4} \right], \quad (2)$$

where

$$E_{k\sigma} = \varepsilon_k - \tilde{\mu} - \frac{1}{2} U m \sigma \ .$$

We have indicated with $\varepsilon_k = -t \sum_{\delta} \exp(ik \cdot \delta)$ the bare dispersion (δ is a vector connecting a site to its nearest neighbors) and $\tilde{\mu} = \mu - (Un)/2$. $\Delta = -g \langle c_{i\uparrow} c_{i\downarrow} \rangle$ is the pairing amplitude, $m = (1/N) \sum_i < n_{i\uparrow} - n_{i\downarrow} >$ and $n = (1/N) \sum_i < n_{i\uparrow} + n_{i\downarrow} >$ are the average magnetization and the occupation number per site, respectively. The magnetization of the system, indicated with M, is linked to m through the relation: $M = (\mu_B/2) N m$, where μ_B is the Bohr magneton.

At this stage, one can diagonalize the above Hamiltonian through a standard Bogoliubov transformation and the quasi-particle Hamiltonian is obtained as

$$\hat{H}_{tot} = \sum_{\vec{k}} \left[\xi^{\alpha}_{\vec{k}} \hat{n}^{\alpha}_{\vec{k}} + \xi^{\beta}_{\vec{k}} \hat{n}^{\beta}_{\vec{k}} \right] + E_0 \ .$$

where

$$\xi^{\alpha,\beta}_{\vec{k}} = \pm I + \sqrt{\Delta^2 + (\varepsilon_{\vec{k}} - \tilde{\mu})^2} \qquad (3)$$

$$E_0 = N \left[\frac{|\Delta|^2}{g} + \frac{Um^2}{4} - \frac{Un^2}{4} \right] + \sum_{\vec{k}} (\varepsilon_{\vec{k}} - E_{\vec{k}}) \ . \qquad (4)$$

The superscripts α and β used above denote the two different Bogoliubov fermions in the coexistence phase and $I = (Um)/2$.

The self-consistent mean-field equations for the two order parameters are derived as:

$$M = \frac{\mu_B}{2} \sum_{\vec{k}} \left(n^{\alpha}_{\vec{k}} - n^{\beta}_{\vec{k}} \right) \ , \qquad (5)$$

$$\Delta = \frac{1}{N} \sum_{\vec{k}} \left[\left(\frac{g\Delta}{2} \right) \frac{1}{\sqrt{(\varepsilon_{\vec{k}} - \tilde{\mu})^2 + \Delta^2}} (1 - n^{\alpha}_{\vec{k}} - n^{\beta}_{\vec{k}}) \right] \ . \qquad (6)$$

We notice that, at $T = 0$, in order to have non vanishing magnetization (that means ferromagnetic order) it is necessary that at least one of the two occupation numbers $n^{\alpha,\beta}_{\vec{k}}$ is non zero, i.e. some depaired electrons are present. This condition can be fulfilled for the values of the electron wave vector $\vec{k}$ that make negative the corresponding quasi-particle energy. By looking at the equation for the energy gap, we realize that the presence of depaired electrons leads to a reduction of the superconducting order parameter with respect to the BCS one.

In the limit $I \simeq \Delta$, by assuming $\varepsilon_{\vec{k}} = \hbar^2 k^2 / 2m^*$, where m^* indicates the effective electron mass, we analytically solve the Eqs. (5)-(6) obtaining the following relations

$$M = \frac{N\mu_B}{U} \frac{r\Delta}{\sqrt{r^2 - 1}} \tag{7}$$

$$\Delta = \Delta_0 \sqrt{\frac{r-1}{r+1}} \tag{8}$$

where $r = (m^* k_F U)/(4\pi^2)$ [8].

Since we have found that M is proportional to Δ, we can conclude that, for weak magnetic exchange, a coexistence solution ($M \neq 0, \Delta \neq 0$) exists. However, it is necessary to establish if this solution is energetically more stable than the other possible ones, in particular with respect to the non magnetic BCS superconducting phase.

By assuming that the density of state can be approximated to the constant value $N(0)$ assumed at the Fermi level, it follows that the free energy of the coexistence state at zero temperature is

$$E_{FS} = E_N(H = 0) - \frac{1}{2} N(0) \Delta^2 \quad . \tag{9}$$

where $E_N(H=0)$ is the energy of the normal state when no magnetic field is applied. Comparing E_{FS} with the zero temperature free energy of the BCS state, whose expression is given by

$$E_S = E_N(H = 0) - \frac{1}{2} N(0) \Delta_0^2 \quad , \tag{10}$$

where Δ_0 is the energy gap of the non magnetic superconducting state (BCS state), we conclude that the superconducting ferromagnetic state has an energy which is always higher than the energy of the non-magnetic superconducting state, due to the fact that $\Delta < \Delta_0$. Thus, the coexistence state is thermodynamically unstable with respect to the superconducting one. This instability can be attributed to the fact that the Zeeman energy gained by the unpaired electrons can not compensate the loss of the superconducting energy due to depaired pairs. Therefore, a small fraction of depaired electrons, if ferromagnetically correlated, destabilize the coexistence phase with respect to the normal BCS state.

For completeness, we have also analyzed the possibility of a coexistence phase when the superconducting state is characterized by Cooper pairs with finite momentum (FFLO superconducting state). The corresponding Hamiltonian, within the mean field approximation, is:

$$\hat{H}_{tot} = \sum_{k\sigma} E_{k\sigma} c^\dagger_{k\sigma} c_{k\sigma} + \sum_{\vec{k}} [\Delta^*_{\vec{q}} \, c_{-\vec{k}+\frac{\vec{q}}{2},\downarrow} c_{\vec{k}+\frac{\vec{q}}{2},\uparrow} + h.c.] + N \left[\frac{|\Delta_{\vec{q}}|^2}{g} + \frac{Um^2}{4} - \frac{Un^2}{4} \right] ,$$

where

$$\Delta_{\vec{q}} = -g \frac{1}{N} \sum_{\vec{k}} \langle c_{-\vec{k}+\frac{\vec{q}}{2}\downarrow} c_{\vec{k}+\frac{\vec{q}}{2}\uparrow} \rangle \quad .$$

Diagonalizing this Hamiltonian by applying a suitable Bogoliubov transformation [3], one obtains two self-consistent equations for the magnetization and the energy gap that are formally identical to Eqs.(5)-(6), where $\varepsilon_{\vec{k}} = (\hbar^2 k^2)/2m^*$ is replaced with $\varepsilon_{\vec{k},\vec{q}} = \hbar^2(k^2+q^2)/2m^*$. The expression of the free energy at $T=0$ for the coexistence state is the same we found in the case of homogeneous superconductivity, even if the energy gap is different, being determined by different equations. However, in both cases the energy gap is less than the BCS state energy gap, so that even in the FFLO phase, the coexistence of superconductivity and ferromagnetism is not energetically favorite. As a consequence, we conclude that, in presence of ferromagnetic correlations between the electrons carrying superconductivity, the s-wave non magnetic state is always more stable than the s-wave superconducting ferromagnetic state, whatever the pair momentum is.

The previous discussion could be applied also to a different pair symmetry such as for instance d-wave singlet superconductivity. If we assume that the superconducting phase of the system is described by means of a $d_{x^2-y^2}$ order parameter, the superconducting order parameter corresponding to the zero momentum pairing state depends on the angular variable θ, and has the following form: $\Delta_\theta = \Delta_d \cos(2\theta)$. Using the same procedure as in the s-wave case, we have obtained that the self-consistent equation for the magnetization is identical to Eq.(5), while the equation for the energy gap is:

$$\Delta_d = \Delta_d \left(-\frac{g}{2N}\right) \sum_{\vec{k}'} \frac{1}{\sqrt{\varepsilon_k^2 + \Delta_d^2 \cos^2(2\theta)}} (1 - n_{\vec{k}}^\alpha - n_{\vec{k}}^\beta) \cos^2(2\theta) \quad .$$

The $T=0$ free energies associated with the superconducting non magnetic state and with the coexistence state are:

$$E_S = E_N(H=0) - \frac{1}{4} N(0) \Delta_{d0}^2, \quad (11)$$

$$E_{FS} = E_N(H=0) - \frac{1}{4} N(0) \Delta_d^2 \quad , \quad (12)$$

respectively, therefore E_{FS} is always larger than E_S, being $\Delta_d < \Delta_{d0}$.

By considering also for the d-wave case the finite momentum pairing state, it is straightforward to deduce that Δ_d is less than Δ_{d0}. The free energy of the coexistence state has the same form we found for the zero momentum state, and thus it is larger than the energy of the superconducting non magnetic state.

Like in the s-wave case, also in the d-wave one, whatever the pair momentum is, the coexistence phase at $T=0$ is thermodynamically unstable with respect to the superconducting no magnetic one, according to the results obtained by Shen et al. [9].

In our previous discussion, we have considered the approximation $N(E) \simeq N(0)$. This assumption does not reproduce correctly the ferromagnetic instability within the Stoner model, since the occurrence of the ferromagnetism is a consequence of the curvature of the density of states close to the Fermi level. For this reason, we have investigated the role played by a non-constant density of state on the coexistence state within the same model. By assuming a density of state having the following semi-circular form

$$N(\varepsilon) = \frac{2}{\pi \Gamma} \sqrt{\Gamma^2 - \varepsilon^2} \quad , \quad (13)$$

where Γ is the electron bandwidth, we have calculated the energies of (i) the ferromagnetic superconducting state ($\Delta \neq 0, M \neq 0$), (ii) the non magnetic superconducting state ($\Delta \neq 0, M = 0$), (iii) the non superconducting ferromagnetic state ($\Delta = 0, M \neq 0$). By varying U, for fixed value of the bandwidth, we have found that if U is lower than a critical value U^*, which is larger than the Stoner limiting value $U_{St} = 1/N(0)$, the energetically favorable phase is the superconducting one. If $U > U^*$ the system exhibits a first order phase transition from the superconducting non magnetic state to the ferromagnetic non superconducting one.

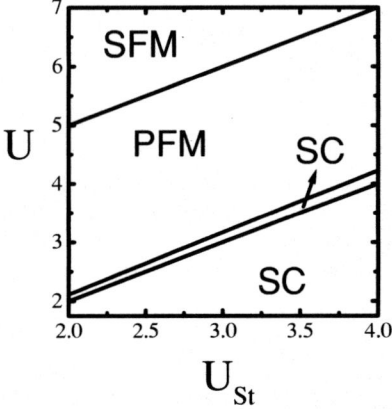

FIGURE 3. Phase diagram at $T = 0$ for a system characterized by one type of carrier, having at the same time superconducting and ferromagnetic interactions. The three curves (from the bottom to the top), represent the behavior of U_{St}, U^*, and U_{sat}, as functions of U_{St}, respectively. U_{sat} represents the value of U in correspondence of which the magnetization reaches its maximum value M_{sat}. PFM stands for partial ferromagnetic state which represents the ferromagnetic phase characterized by a value of the magnetization lower than M_{sat}; on the other side, SFM indicates the saturate ferromagnetic state, where $M = M_{sat}$. SC indicates the superconducting non-magnetic state.

In Fig.3 it is reported the phase diagram obtained varying U and Γ.
Besides, by taking into account that $N(0)$ can be expressed in the following form:

$$N(0) = \frac{2}{\pi \Gamma},$$

it is straightforward to conclude that U_{St} varies proportionally with Γ. The obtained results show that as Γ increases (and thus as U_{St} increases), the range of values of U above U_{St}, corresponding to the superconducting phase, is wider. This circumstance can be ascribed to the assumptions we have made. Indeed, we have assumed that the Debye energy $\hbar \omega_D$, which represents the cut-off used in the integrals involving the superconducting coupling constant g, varies proportionally to Γ, so that as Γ increases, $\hbar \omega_D$ increases too and consequently the range of electron energies producing a non vanishing superconducting coupling is wider. Then, although the exchange energy increases, favoring the ferromagnetic phase, also the superconducting interaction increases. By solving the equation for the magnetization, we find that if $U > U^*$, M increases from a non vanishing value lower than the saturation one $M_{sat} = \mu_B N/2$ up to M_{sat}. Finally, we notice that U_{sat} increases with increasing U_{St}, with a nearly linear law.

From this analysis we can conclude that if the ferromagnetic instability is due to the Stoner exchange splitting, the competition between singlet superconductivity and ferromagnetism is always detrimental to the formation of the coexistence phase, since the magnetic energy gained by the depaired electrons deriving from the breaking of superconducting pairs cannot compensate the loss in condensation energy due to the pair breaking.

KINETIC EXCHANGE IN THE SUPERCONDUCTING-FERROMAGNETIC PHASE

We have extended the one carrier model by proposing a novel mechanism for the coexistence of metallic ferromagnetism and singlet superconductivity, assuming that the magnetic instability is due to kinetic exchange. In this case the metallic ferromagnetism is considered due to a change in the relative bandwidth of electrons with up and down spin polarization (Fig.4) rather than, as in the previous section, to a rigid shift in the positions of the majority and minority spin bands (i.e. Stoner model).

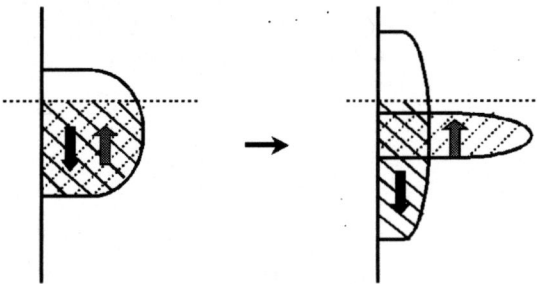

FIGURE 4. Schematic representation of the density of states for up and down electrons in the unpolarized (left) and polarized (right) states, when the ferromagnetic instability is attributed to the kinetic exchange. The dashed line indicates the position of the Fermi level.

Indeed, the mass of the majority spins becomes undressed, thus inducing a bandwidth enlargement which lowers the kinetic energy. This energy gain leads to the ferromagnetic instability.

The model Hamiltonian we have used has the following form [10]:

$$H = -t \sum_{<ij>\sigma} \left(c_{i\sigma}^\dagger c_{j\sigma} + h.c \right) + U \sum_{i=1}^{N} n_{i\uparrow} n_{i\downarrow} + V \sum_{<ij>} n_i n_j + J \sum_{<ij>\sigma\sigma'} c_{i\sigma}^\dagger c_{j\sigma'}^\dagger c_{i\sigma'} c_{j\sigma}$$

$$+ J' \sum_{<ij>} \left(c_{i\uparrow}^\dagger c_{i\downarrow}^\dagger c_{j\downarrow} c_{j\uparrow} + h.c \right) - g \sum_{i=1}^{N} c_{i\uparrow}^\dagger c_{i\downarrow}^\dagger c_{i\downarrow} c_{i\uparrow} - \mu \sum_{i=1}^{N} \sum_\sigma c_{i\sigma}^\dagger c_{i\sigma} \quad . \tag{14}$$

The notations are the same used in Eq.1. Moreover, V represents the nearest-neighbor Coulomb repulsion and the parameters J and J' describe nearest-neighbor exchange and pair hopping processes. n_i is the electron total number operator on the i-th site: $n_i = n_{i\uparrow} + n_{i\downarrow}$.

By applying the mean-field decoupling we can rewrite the model Hamiltonian in terms of a bond charge quantity $I_\sigma = \sum_{<ij>} \langle c_{i\sigma}^\dagger c_{j\sigma} \rangle$ that in the momentum space is given by

$$I_\sigma = \frac{1}{N} \sum_k n_{k\sigma} \left(\frac{-\varepsilon_k}{\Gamma/2} \right) , \qquad (15)$$

where $n_{k\sigma} = \langle n_{k\sigma} \rangle$ is the Fermi distribution of electrons with spin σ and wave vector $\vec{k}$ and Γ is the electron bandwidth. The mean field Hamiltonian has the following expression:

$$H_{mf} = \sum_{k\sigma} E_\sigma(\varepsilon_k) c_{k\sigma}^\dagger c_{k\sigma} + \sum_{\vec{k}} \left[\Delta^* c_{-\vec{k}\downarrow} c_{\vec{k}\uparrow} + h.c. \right] + f(I_\uparrow, I_\downarrow) , \qquad (16)$$

where

$$E_\sigma(\varepsilon_k) = (1 - 2j_1 I_\sigma - 2j_2 I_{-\sigma}) \varepsilon_k - \mu - \frac{1}{2} m\Gamma k\sigma + \frac{1}{2} \Gamma(n\, k - 2n j_1 - j_1)$$

$$f(I_\uparrow, I_\downarrow) = -\frac{1}{4} N \Gamma \left[n^2(k - 2j_1) - m^2 k + 2j_1(I_\uparrow^2 + I_\downarrow^2) + 4 j_2 I_\uparrow I_\downarrow \right] + \frac{|\Delta|^2}{g} N .$$

The values of j_1, j_2, and k are linked to U, V, J and J' by means of the relations

$$j_1 = (J - V)/\Gamma$$
$$j_2 = (J + J')/\Gamma$$
$$k = (U + zJ)/\Gamma .$$

By indicating with $2\omega_\sigma = 1 - 2j_1 I_\sigma - 2j_2 I_{-\sigma}$ the spin mass renormalization, the electron energy can be written in the following form:

$$E_\sigma(\varepsilon_k) = 2\omega_\sigma \varepsilon_k - \tilde{\mu} - \frac{1}{2} m\Gamma k\sigma .$$

The exchange interaction k gives an overall rigid shift in the band energies, as in the Stoner model, so that in the following analysis will be neglected.
By applying the standard Bogoliubov transformations we have diagonalized the Hamiltonian (16), obtaining:

$$H = \sum_k \left(E_k^\alpha \alpha_k^\dagger \alpha_k + E_k^\beta \beta_k^\dagger \beta_k \right) + \sum_{\vec{k}} \left(E_\downarrow(\varepsilon_{\vec{k}}) - E_{\vec{k}}^\beta \right) + f(I_\uparrow, I_\downarrow) .$$

The field operators α_k, β_k correspond to fermionic excitations with quasi-particle dispersion

$$E_k^{\alpha,\beta} = \pm E_-(\varepsilon_k) + \sqrt{E_+^2(\varepsilon_k) + \Delta^2} , \qquad (17)$$

with

$$E_\pm(\varepsilon_k) = \frac{1}{2} \left[E_\uparrow(\varepsilon_k) \pm E_\downarrow(\varepsilon_k) \right] .$$

If we introduce two parameters a and b, defined as $a = w_\uparrow - w_\downarrow$ and $b = w_\uparrow + w_\downarrow$, it is possible to write down the quasi-particle energies (17) in the form

$$E_k^{\alpha,\beta} = \pm a\,\varepsilon_k + \sqrt{(b\,\varepsilon_k - \mu)^2 + \Delta^2} \ . \tag{18}$$

By using the explicit form of $w_\uparrow$ and $w_\downarrow$, it follows that:

$$\begin{aligned} b &= 1 - (j_1 + j_2)(I_\uparrow + I_\downarrow) \\ a &= (j_2 - j_1)(I_\uparrow - I_\downarrow) \ . \end{aligned} \tag{19}$$

From these relations, we notice that a and b are determined by $j_- = (j_2 - j_1)$ and $j_+ = (j_1 + j_2)$, respectively. Being $w_\uparrow$, $w_\downarrow$ the electron mass renormalization factors, they are positive, and consequently b is positive, whatever values j_1 and j_2 assume. Moreover, the sign of j_+ controls whether the starting value in the SC state for b is larger or smaller than 1. Hereafter, we will assume that $j_\pm$ are positive, which is consistent, in the weak coupling regime, with realistic values of the microscopic couplings (J, J', V) above introduced.

In terms of the α- and β- band occupation numbers, the equations for the pairing amplitude, the magnetization and the chemical potential can be rewritten in the following form:

$$\Delta = \frac{1}{N} \sum_k \left[\frac{g\,\Delta}{2\sqrt{(b\,\varepsilon_k - \mu)^2 + \Delta^2}} (1 - n_k^\alpha - n_k^\beta) \right] \tag{20}$$

$$M = \frac{\mu_B}{2} \sum_k \left(n_k^\beta - n_k^\alpha \right) \tag{21}$$

$$n = \frac{1}{N} \sum_k \left[1 - \frac{(b\,\varepsilon_k - \mu)}{\sqrt{(b\,\varepsilon_k - \mu)^2 + \Delta^2}} (1 - n_k^\alpha - n_k^\beta) \right] \ , \tag{22}$$

where $n_k^{\alpha,\beta}$ are the Fermi distributions in the momentum space of the fermionic fields having $E_k^{\alpha,\beta}$ as energy dispersion relations, respectively.
Still I_σ, depends on the occupation number of the α (β) band, via the following relations:

$$\begin{aligned} n_{k\uparrow} &= v_k^2 + u_k^2(n_k^\alpha + n_k^\beta) - n_k^\beta \\ n_{k\downarrow} &= v_k^2 + u_k^2(n_k^\alpha + n_k^\beta) - n_k^\alpha \ , \end{aligned}$$

where v_k and u_k represent the coefficients of the Bogoliubov rotations we used to diagonalize the mean-field Hamiltonian:

$$\alpha_k = u_k\,c_{k\uparrow} + v_k\,c^\dagger_{-k\downarrow} \tag{23}$$

$$\beta_k^\dagger = -v_k\,c_{k\uparrow} + u_k\,c^\dagger_{-k\downarrow} \ . \tag{24}$$

They depends on a and b through the following relations:

$$u_k^2 = \frac{1}{2}\left[1 + \frac{(b\varepsilon_k - \mu)}{\sqrt{(b\varepsilon_k - \mu)^2 + \Delta^2}}\right] \quad (25)$$

$$v_k^2 = \frac{1}{2}\left[1 - \frac{(b\varepsilon_k - \mu)}{\sqrt{(b\varepsilon_k - \mu)^2 + \Delta^2}}\right]. \quad (26)$$

Looking at Eq.18 we notice that the quasi-particle energies could be negative and consequently the quasi-particle occupation numbers could be non-vanishing, for suitable values of the bare energy ε_k. In particular, if we assume that the $b > a > 0$ and $\mu > 0$, it follows that only the occupation number of β quasi-particles can be non-vanishing, and this happens when the following two conditions are fulfilled:

$$\frac{a}{b} > \frac{\Delta}{\sqrt{\Delta^2 + \mu^2}};$$

$$\varepsilon_k \in (\lambda_-, \lambda_+) \quad ,$$

with

$$\lambda_{k,\pm} = \frac{b\mu \pm \sqrt{(b\mu)^2 - (b^2 - a^2)(\mu^2 + \Delta^2)}}{b^2 - a^2}.$$

On the contrary, when $b > a$ and $\mu > 0$, but a is negative, only the occupation number of α quasi-particles is non-vanishing, and it happens when

$$\frac{|a|}{b} > \frac{\Delta}{\sqrt{\Delta^2 + \mu^2}},$$

$$\varepsilon_k \in (\lambda_-, \lambda_+) \quad .$$

These results can be symmetrically extended within the same procedure to the negative μ case. In the limit of $\mu \to 0$ ($n \to 1$) or $a \to 0$ ($M \to 0$), there are no real solutions for $\lambda_{k,\pm}$, so that the only possibility for the system is the superconducting non-magnetic state.

By solving the right side of Eq.(21) in the thermodynamical limit, using a model density of states given by $N(\varepsilon) = 1/\Gamma$, if $-(\Gamma/2) \leq \varepsilon \leq (\Gamma/2)$, we get:

$$M = -\left[\frac{a}{b}(b^2 - a^2)\right]\frac{\mu_B N}{4\mu(j_2 - j_1)}. \quad (27)$$

Being b always positive, the sign of the total magnetization M is the same of a. Therefore, without any loss of generality, one can focus only on the case with $a > 0$, the other one being just derived by inverting the direction of the magnetization. Moreover, this expression shows that the magnetization is completely determined by calculating a, b, and μ. Thus, the solution of the problem can be completely obtained by solving self-consistently the equations for a, b, μ and Δ.

In a preliminary analysis, we have fixed the value of b and n and we have then studied how the pairing amplitude, the energy and the magnetization are modified by changing *a*. The formation of depaired electrons within the superconducting phase leads *b* to assume different values (b_{sf}) with respect to the superconducting non-magnetic state (b_{sc}). Hence, two different situations might occur: a) $b_{sf} > b_{sc}$, so that the average effective mass becomes undressed; b) $b_{sf} < b_{sc}$, which leads the average effective mass to be dressed. Our analysis shows that only when $b_{sf} > b_{sc}$, and for suitable value of *a* the coexistence phase is the more stable one. Indeed, when the average mass becomes dressed, we found that the energy gap grows, thus reducing the condensation energy, but at the same time, the loss in kinetic energy of the superconducting electrons cannot be compensated to stabilize the superconducting-ferromagnetic phase. As a consequence, when the self-consistent equations are solved, it is crucial to establish whether the formation of depaired electrons in the coexistence phase causes the increase or the decrease of the average effective mass with respect to the non magnetic case. By assuming $b > a > 0$ and $\mu > 0$, we found that b_{sf} becomes undressed in the coexistence phase, so that this phase can be stabilized. This circumstance can be easily deduced by looking at the equation for *b* in the thermodynamical limit:

$$b = 1 - \frac{(j_1 + j_2)}{\Gamma/2} \left[\int_{-\Gamma/2}^{\Gamma/2} d\varepsilon\, N(\varepsilon) \frac{(b\varepsilon - \mu)\varepsilon}{\sqrt{(b\varepsilon - \mu)^2 + \Delta^2}} - \int_{\varepsilon^-}^{\varepsilon^+} d\varepsilon\, N(\varepsilon) \frac{(b\varepsilon - \mu)\varepsilon}{\sqrt{(b\varepsilon - \mu)^2 + \Delta^2}} \right].$$

From this equation it is straightforward to conclude that the formation of depaired electrons gives rise, in the right side, to a positive contribution that is absent in the superconducting non ferromagnetic case, so that in the coexistence state the effective electron mass becomes undressed. By comparing the ground state energy relative to

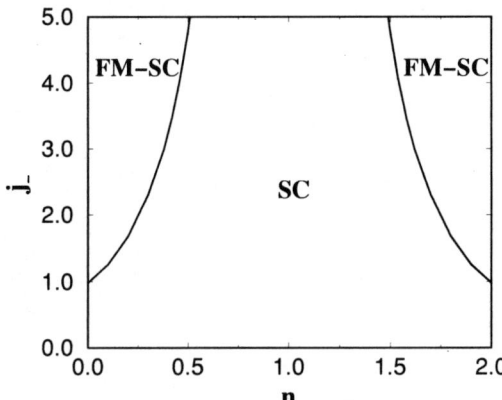

FIGURE 5. Phase diagram relative to the model in which the ferromagnetic instability is due to kinetic exchange. SC indicates the superconducting non-magnetic state, and FM+SC stands for ferromagnetic-superconducting phase.

the superconducting non-magnetic state and to the ferromagnetic non-superconducting one with the energy of the coexistence state, we have derived the phase diagram for the system. It has been constructed by varying the average number of electrons per site with respect j_-, for fixed values of j_+. As it is possible to observe in Fig.5 (where we have reported the phase diagram for $j_+ = 0.5, g/t = 0.25$), only when the density of electrons per site is far from the half-filling limit it is possible to stabilize the coexistence phase with not very large couplings. By changing j_+, the critical line is shifted, but the qualitative behavior of the phase diagram does not change. In particular the character of the phase transition from the superconducting non-magnetic state to the coexistence state remains the same: first-order like with respect to magnetization, and second-order like with respect to the pair amplitude. Indeed, going through the critical line, from the superconducting non-magnetic state to the coexistence state, the magnetization has a jump to its possible maximum value which depends on Δ and μ, while the energy gap grows continuously from zero.

SUMMARY AND DISCUSSION

In this paper we have studied the interplay between singlet superconductivity and itinerant ferromagnetism by considering a single band model. Firstly, we have described the ferromagnetic interaction through the Stoner model, in which the instability towards the ferromagnetic state is attributed to a spontaneous shift of spin-up and spin-down electrons bands. On the other side, the superconducting pairing has been described by using the usual BCS theory, by taking into account two possible pair symmetries: the s-wave symmetry and the d-wave one. We have also considered the possibility of inhomogeneous superconductivity, by assuming that the superconducting state is the Fulde-Ferrel-Larkin-Ovchinnikov one (FFLO), since in this state the singlet superconductivity is allowed to survive in presence of spin-exchange field higher than the critical one predicted by Chandrasekhar and Clogston as a consequence of Pauli paramagnetic effect. In both cases (zero momentum and non-zero momentum for the superconducting pairs), and independently of the symmetry of the superconducting order parameter, the formation of depaired electrons that are ferromagnetically correlated within the superconducting phase leads the coexistence phase to be instable with respect to the superconducting non-magnetic one. By comparing these results with those obtained by analyzing a model where the same mechanisms responsible for the itinerant ferromagnetism and the singlet superconductivity are assumed, but where the superconducting electrons and the ferromagnetic ones are supposed to be different (two carrier-model) [11], we conclude that it is crucial for having a state in which superconductivity and ferromagnetism coexist that the depaired electrons are not ferromagnetically correlated, so that their density can be tuned in an unconstrained way by changing the amplitude of the superconducting modulation.

Different results are obtained within the one-carrier model by assuming that the mechanism giving rise to the ferromagnetic instability is the kinetic exchange. In this case the ferromagnetism is driven by the gain in kinetic energy resulting from the spontaneous change in the relative bandwidth of majority and minority spin electrons upon spin polarization. Within this assumption it is possible to get a coexistence of ferromagnetism and superconductivity only when the formation of depaired electrons causes the undressing of the effective mass of the carriers which participate into the pairing. This circumstance is fulfilled for realistic values of the coupling constants when the system is far from the limit of exact particle-hole symmetry, as it follows from the comparison of the ground state energy associated with the possible states for the system. It worth pointing out that, contrary to the case of the Stoner model, when ferromagnetism is due to kinetic exchange, a non trivial ferromagnetic solution can be obtained without the need of a curvature in the density of states close to the Fermi level.

The obtained results, showing that the mechanism of kinetic exchange allows for the stability of the coexistence phase, could be relevant to analyze the interplay of ferromagnetism and superconductivity in system like UGe_2 or in systems like manganese oxides, rare-earth hesaborides and magnetic semiconductors, where the ferromagnetism is due to the kinetic exchange mechanism, and the superconducting pairing could be induced artificially.

ACKNOWLEDGMENTS

The interplay of superconductivity and (ferro)magnetism has been one of the research fields of interest of Prof. F. Mancini. Thus, it is a great pleasure for us to dedicate this work to him in the occasion of his sixty anniversary.

REFERENCES

1. A. A. Abrikosov and L. P. Gorkov, *Sov.Phys. JETP* **19**, 1243 (1963)
2. P. Fulde and R. A. Ferrell, Phys. Rev. **135**, A550 (1964); A. I. Larkin and Yu. N. Ovchinnikov, Zh. Eksp. Teor. Fiz. **47**, 1136 (1964) [Sov. Phys. JETP **20**, 762 (1965).
3. H. Shimahara, Phys. Rev. B **50**, 12760 (1994)
4. K. Gloos, R. Modler, H. Schimanski, C. D. Bredl, C. Geibel, F. Steglich, A. I. Buzdin, N. Sato, and T. Komatsubara, Phy. Rev. Lett. **70**, 501 (1993).
5. S. Saxena *et al.*, Nature **406**, 587 (2000)
6. C. Pfleiderer, M. Uhlarz, S. M. Hayden, R. Vollmer, H. v. Löhneysen, N. R. Bernhoeft, and G. G. Lonzarich, Nature **412**, 58 (2001)
7. See for instance: C. Noce, A. Vecchione, M. Cuoco and A. Romano: *Ruthenate and rutheno-cuprate materials: Unconventional Superconductivity, Magnetism and Quantum Phase Transitions* Springer Verlag, Lecture Notes in Physics (2002); C. Bernhard, J.L.Tallon, Ch. Niedermayer, Th. Blasius, A. Golnik, B. Btucher, R.K. Kremer, D.R. Noakes, C.E. Stronach, and E.J. Ansaldo, Phys. Rev. B **59**, 14099 (1999); F. Tallon, C. Bernhard, M. Bowden, P. Gilberd, T. Stoto, D. Pringle IEEE Trans. Appl. Supercond. **9**, 1696 (1999); L. Baurfeind, W. Widder, H. F. Braun, Physica C **254**, 151 (1995); L. Baurfeind, W. Widder, H. F. Braun, J. Low Temp. Phys. **105**, 1605 (1996); I. Felner, U. Asaf, Y.Levi, Physica C **311**, 163 (1999).
8. N. I. Karchev, K. B. Blagoev, K. S. Bedell and P. B. Littlewood, Phys. Rev. Lett. **86**, 846 (2001).
9. R. Shen, Z. M. Zheng, S. Liu and D. Y. Xing, Phys. Rev. B **67**, 024514 (2003).
10. J. E. Hirsch, Phys. Rev. B **40**, 2354 (1989);**40**, 9061 (1989);**59**, 6256 (1999).
11. M. Cuoco, P. Gentile, C. Noce, Phys. Rev. B **68**, 054521 (2003)

Superfluid properties of the Boson-Fermion model [1]

R. Micnas*, S. Robaszkiewicz* and A. Bussmann-Holder[†]

Institute of Physics, A. Mickiewicz University, Umultowska 85, 61-614 Poznań, Poland
[†]*Max-Planck Institut für Festkörperforschung, Heisenbergstrasse 1, D-70569 Stuttgart, Germany*

Abstract. Superconductivity in the two component model of coexisting local electron pairs (charged bosons) and itinerant fermions coupled via a charge exchange mechanism is discussed. The cases of isotropic s-wave and anisotropic pairing of extended s-wave and $d_{x^2-y^2}$ symmetries are analyzed for a 2D square lattice within the BCS-mean field approximation and the Kosterlitz-Thouless theory. The phase diagrams and superconducting characteristics of the model as a function of the position of the local pair (LP) level and the total carrier concentration, are determined. The model exhibits interesting crossovers from the BCS like behavior to that of LP's. Finally, we analyze the pairing fluctuation effects (in 3D) within a generalized T-matrix approach and determine the transition temperatures from the pseudogap state.

1. INTRODUCTION

The model of coexisting local electron pairs (with $q = 2e$) and itinerant fermions coupled via a charge exchange mechanism, which mutually induces superconductivity in both subsystems was proposed some time ago and it has been studied in a large number of recent papers. Such a two-component (boson-fermion) model can show features intermediate between those of the local pair (bipolaronic) superconductors and those of classical BCS systems. It is of relevance for high temperature superconductors (HTS) and other exotic superconductors [1–16]. A related model has also been adopted for the description of a resonance s-wave superfluidity in Fermi atomic gases with a Feshbach resonance [17, 18].
Recently, we have studied a generalization of this model to the case of anisotropic pairing [14–16]. Here, we briefly outline the study and present some further results concerning the phase diagrams and superconducting properties of such a system in the case of isotropic s-wave and anisotropic d-wave pairing, for a 2D square lattice (Sec.3), as well as for s-wave pairing for a 3D simple cubic (sc) lattice (Sec.4).

---

[1] This paper is dedicated to Professor Ferdinando Mancini on the occasion of his 60th birthday.

2. THE MODEL

We consider the model of coexisting electron pairs (hard-core bosons "b") and itinerant "c" electrons defined by the following effective Hamiltonian

$$H = \sum_{k\sigma}(\varepsilon_k - \mu)c^\dagger_{k\sigma}c_{k\sigma} + 2\sum_i (\Delta_0 - \mu)b^\dagger_i b_i - \sum_{ij} J_{ij} b^\dagger_i b_j + \sum_{k,q}\left[V_q(k)c^\dagger_{k+q/2,\uparrow}c^\dagger_{-k+q/2,\downarrow}b_q + h.c.\right], \quad (1)$$

where ε_k refers to the band energy of the c-electrons, Δ_0 measures the relative position of the LP level with respect to the bottom of the c-electron band, μ is the chemical potential which ensures that the total number of particles in the system is constant, i.e. $n = \frac{1}{N}\left(\sum_{k\sigma}\langle c^\dagger_{k\sigma}c_{k\sigma}\rangle + 2\sum_i \langle b^\dagger_i b_i\rangle\right) = n_c + 2n_B$. n_c is the concentration of c-electrons, n_B is the number of local pairs per site. J_{ij} is the pair hopping integral. We will not discuss the density-density interaction terms [3]. The operators for local pairs $b^\dagger_i, b_i$ obey the Pauli spin 1/2 commutation rules: $[b_i, b^\dagger_j] = (1 - 2n_i)\delta_{ij}$, $[b_i, b_j] = 0$, $(b^\dagger_i)^2 = (b_i)^2 = 0$, $b^\dagger_i b_i + b_i b^\dagger_i = 1$, $n_i = b^\dagger_i b_i$. $V_q(k)$ describes the coupling between the two subsystems. We will consider the case $V_q(k) = V_0(k) = I\phi_k/\sqrt{N}$, and neglect its q dependence at small q. Then the interaction term takes the form of a coupling, via the center of mass momenta q, of the singlet pair of c-electrons $B^\dagger_q$ and the hard-core boson b_q:

$$H_1 = \frac{1}{\sqrt{N}}\sum_q I(B^\dagger_q b_q + b^\dagger_q B_q). \quad (2)$$

$B^\dagger_q = \sum_k \phi_k c^\dagger_{k+q/2,\uparrow} c^\dagger_{-k+q/2,\downarrow}$ denotes the singlet pair creation operator of c-electrons and I is the coupling constant. The pairing symmetry is determined by the form of ϕ_k, which is constant (1) for on-site pairing (s), and in the case of a 2D square lattice $\phi_k = \gamma_k = \cos(k_x) + \cos(k_y)$ for extended s-wave (s^*) and $\phi_k = \eta_k = \cos(k_x) - \cos(k_y)$ for $d_{x^2-y^2}$-wave pairing (d).

In the absence of interactions, depending on the relative concentration of "c" electrons and LP's we distinguish three essentially different physical situations. For $n \leq 2$ it will be: (i) $\Delta_0 < 0$ such that at $T = 0K$ all the available electrons form local pairs ($2n_B \gg n_c$) (LP); (ii) $\Delta_0 > 0$ such that the "c" electron band is filled up to the Fermi level $\mu = \Delta_0$ and the remaining electrons are in the form of local pairs (the "c+b" or Mixed regime, $0 < 2n_B, n_c < 2$) (LP+E); (iii) $\Delta_0 > 0$ such that the Fermi level $\mu < \Delta_0$ and consequently at $T = 0K$ all the available electrons occupy the "c" electron states (the c-regime or "BCS", $n_c \gg 2n_B$) (E).

For $|I_0| \neq 0$, in the case (ii) the superconductivity is due to the interchange between local pairs and pairs of "c" electrons. In this process "c" electrons become "polarized" into Cooper pairs and local pairs increase their mobility by decaying into "c" electron pairs. In this intermediate case neither the standard BCS picture nor the picture of local pairs applies and the superconductivity has a "mixed" character. The system shows features which are intermediate between those of the BCS and those of the preformed local pair regime. This concerns the energy gap in the single-electron excitation spectrum ($E_g(0)$), the $k_B T_c/E_g(0)$ ratio, the critical fields, the Ginzburg ratio κ, the width of the critical regime as well as the normal state properties. In the case (i) the local pairs can move via a mechanism of virtual excitations into empty c-electrons states. Such a mechanism gives rise to the long range hopping of LP's (in analogy to the RKKY interaction for s-d mechanism in the magnetic equivalent). The superconducting properties are analogous to those of a pure local pair (bipolaronic) superconductor [1, 8, 20, 21]. In the case (iii), on the contrary, we find a situation which is similar to the BCS case: Cooper pairs of "c" electrons are exchanged via virtual transitions into local pair states.

The superconducting state of the model is characterized by two order parameters: $x_0 = \frac{1}{N}\sum_k \phi_k \langle c^\dagger_{k\uparrow} c^\dagger_{-k\downarrow}\rangle$ and $\rho_0^x = \frac{1}{2N}\sum_i \langle b^\dagger_i + b_i\rangle$. In the BCS-mean-field approximation (MFA) the free energy of the system is evaluated to be:

$$F/N = -\frac{2}{\beta N}\sum_k \ln\left[2\cosh(\beta E_k/2)\right] - \frac{1}{\beta}\ln\left[2\cosh(\beta\Delta)\right] + C, \tag{3}$$

$$C = -\varepsilon_b + \Delta_0 + \mu(n_c + 2n_B) - 2\mu - 2I|x_0|\rho_0^x + J_0(\rho_0^x)^2, \tag{4}$$

where the quasiparticle energy of the c-electron subsystem is given by $E_k = \sqrt{\bar{\varepsilon}_k^2 + \bar{\Delta}_k^2}$ and $\bar{\varepsilon}_k = \varepsilon_k - \mu$, $\bar{\Delta}_k^2 = I^2 \phi_k^2 (\rho_0^x)^2$, $\Delta = \sqrt{(\Delta_0 - \mu)^2 + (-I|x_0| + J_0\rho_0^x)^2}$. $J_0 = \sum_{i\neq j} J_{ij}$. $\beta = 1/k_B T$. It should be noted that the energy gap in the c-band is due to nonzero Bose condensate amplitude ($|\langle b\rangle| \neq 0$), and well defined Bogoliubov quasiparticles can exist in the superconducting phase. The order parameters and the chemical potential are given by

$$\frac{\partial F}{\partial x_0} = 0, \quad \frac{\partial F}{\partial \rho_0^x} = 0, \quad \frac{\partial F}{\partial \mu} = 0. \tag{5}$$

The superfluid stiffness derived within the linear response method and BCS theory, for the case $J_{ij} = 0$, is of the form:

$$\rho_s = \frac{1}{2N}\sum_k \left\{\left(\frac{\partial \varepsilon_k}{\partial k_x}\right)^2 \frac{\partial f(E_k)}{\partial E_k} + \frac{1}{2}\frac{\partial^2 \varepsilon_k}{\partial k_x^2}\left[1 - \frac{\bar{\varepsilon}_k}{E_k}\tanh\left(\frac{\beta E_k}{2}\right)\right]\right\}, \tag{6}$$

where $f(E_k) = 1/\left[\exp(\beta E_k) + 1\right]$ is the Fermi-Dirac distribution function. In the local limit: $\lambda^{-2} \propto (16\pi e^2/\hbar^2 c^2)\rho_s$, where λ is the London penetration depth.

The mean-field transition temperature (T_c^{MFA}), at which the gap amplitude vanishes, yields an estimation of the c-electron pair formation temperature [15, 16] and is given by

$$1 = \left[J_0 + \frac{I^2}{N}\sum_k \phi_k^2 \frac{\tanh\left(\beta_c^{MFA}\bar{\varepsilon}_k/2\right)}{2\bar{\varepsilon}_k}\right]\frac{\tanh\left[\beta_c^{MFA}(\Delta_0 - \mu)\right]}{2(\Delta_0 - \mu)}. \tag{7}$$

Due to the fluctuation effects the superconducting phase transition will occur at a critical temperature lower than that given by the BCS-MFA theory.

In 2D, T_c can be derived within the Kosterlitz-Thouless (KT) theory for 2D superfluids [19], which describes the transition in terms of vortex-antivortex pair unbinding. We evaluate T_c using the KT relation for the universal jump of the (in-plane) superfluid density ρ_s at T_c [19]:

$$\frac{2}{\pi}k_B T_c = \rho_s(T_c), \qquad (8)$$

where $\rho_s(T)$ is given by Eq.(6) and $x_0(T), \rho_0^x(T), \mu(T)$ are given by Eqs.(5). Thus, the critical temperature denoted further by T_c^{KT} is determined from the set of four self-consistent equations. In the weak coupling limit ($|I_0|/2D \ll 1, J_0 = 0, D$-the half-bandwidth, $I = -|I_0|$), $T_c^{KT}/T_c^{MFA} \to 1$ if $|I_0|/2D \to 0$.

3. RESULTS FOR 2D ELECTRONIC DISPERSION

A detailed study of the phase diagrams and superfluid properties of the model Eq.(1) for different pairing symmetries including s, the extended s (s^*) and $d_{x^2-y^2}$-wave symmetries for 2D and quasi2D lattices was performed in Refs.[14–16]. For the 2D square lattice the c-electron dispersion is $\varepsilon_k = \tilde{\varepsilon}_k - \varepsilon_b = -2t\left[\cos(k_x) + \cos(k_y)\right] - 4t_2\cos(k_x)\cos(k_y) - \varepsilon_b$, with the nn and nnn hopping parameters t and t_2, respectively, $\varepsilon_b = min\tilde{\varepsilon}_k$.
The typical phase diagrams for s-wave pairing symmetry plotted as a function of the position of the LP level Δ_0 at fixed n and $J_{ij} = 0$ are shown in Fig.1. For the $d_{x^2-y^2}$-symmetry the transition temperatures are given in Fig.2. The evolution of the order parameters, the chemical potential, n_c and n_B with Δ_0 at fixed n and $T = 0$ is shown in Fig.3a, whereas the corresponding plots of the superfluid stiffness $\rho_s(0)$ and the fermionic gap $\Delta_F(0)$ are presented in Fig.3b.

For all the pairing symmetries one observes a drop in the superfluid stiffness (and in the KT transition temperature) when the bosonic level reaches the bottom of the c-electron band and the system approaches the LP limit. In the opposite, BCS like limit, T_c^{KT} approaches asymptotically T_c^{MFA}, with a narrow fluctuation regime. Between the KT and MFA temperatures, *phase fluctuation effects* are important. In this regime a pseudogap in the c-electron spectrum will develop and the normal state of LP's and itinerant fermions will exhibit non-Fermi liquid properties [6].

A closer inspection of the Mixed-LP crossover indicates that when the LP level is lowered and reaches the bottom of the fermionic band an effective attraction between fermions becomes strong, since it varies as $I^2/(2\Delta_0 - 2\mu)$ and $\mu \approx \Delta_0$ [15, 16]. In this regime the density of c electrons is low and formation of bound c-electron pairs occurs. It gives rise to an energy gap in the single-electron spectrum *independently of the pairing symmetry*. We have calculated the binding energies of c- electron pairs and have found that T_c^{MFA} essentially scales with the half of their binding energy for $\Delta_0 < 0$. The superconducting transition temperature is here always much lower than the c-pair formation temperature (T_c^{MFA}) and decreases rapidly with $|\Delta_0/D|$. In such a case, the superconducting state can be formed by two types of coexisting bosons: preformed c-electron pairs and LP's [15, 18].

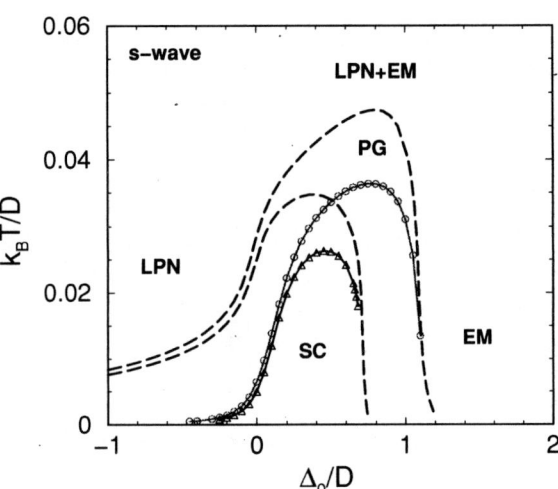

FIGURE 1. Phase diagrams of the hard-core boson-fermion model as a function of Δ_0/D at fixed n derived for s-wave symmetry and 2D square lattice. $J_0 = 0$ and $t_2 = 0$. $|I_0|/D = 0.25$, D=4t. The dashed lines show the BCS-MFA transition temperature (upper for $n = 1$ and lower for $n = 0.5$), while the lines with circles and triangles show the KT transition temperatures calculated for $n = 1$ and $n = 0.5$, respectively. LPN–normal state of predominantly LP's, EM–electronic metal, SC–superconducting (LPS+ES) state, PG – pseudogap region. A weak interplanar coupling stabilizes the SC state.

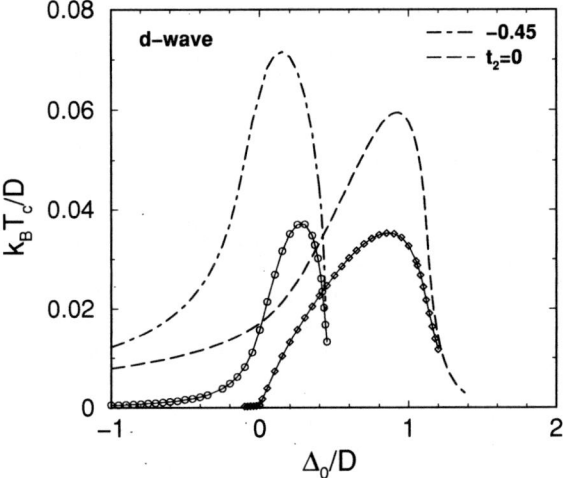

FIGURE 2. MFA and KT transition temperatures as a function of Δ_0/D at fixed $n=1$ derived for $d_{x^2-y^2}$ - pairing symmetry. The dashed and dot-dashed lines show the BCS-MFA transition temperatures for $t_2 = 0$ and for $t_2/t = -0.45$, respectively. The line with diamonds shows the corresponding KT transition temperature calculated for $t_2 = 0$ and the line with circles for $t_2/t = -0.45$. $|I_0|/D = 0.25$, $J_0 = 0$, D=4t.

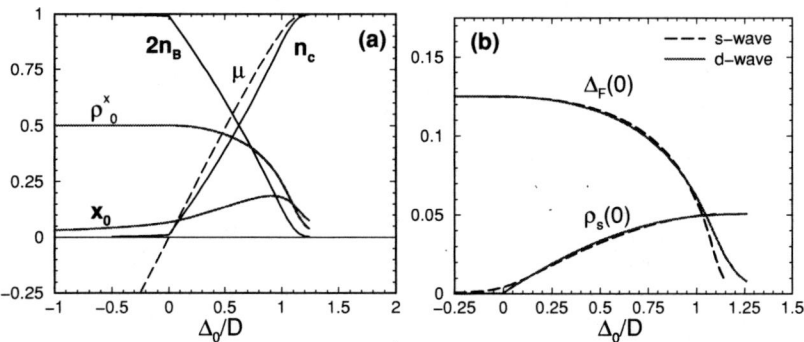

FIGURE 3. (a). Variation of n_c (concentration of c-electrons), n_B (concentration of LP's), superconducting order parameters ρ_0^x, x_0 and the chemical potential μ/D at $T = 0$ as a function of Δ_0/D, for $n = 1, |I_0|/D = 0.25, t_2 = 0. J_0 = 0$. d-wave pairing. (b) $\Delta_F(0)/D$ and $\rho_s(0)/D$ versus Δ_0/D for same parameters. $\Delta_F = |I_0|\rho_0^x$.

Comparing T_c vs Δ_0 plots for various pairing symmetries one finds that in the case of nn hopping only, the d and s -wave pairings are favored for higher concentration of c-electrons, while the s^*-wave can be stable at low n_c (Fig.3 and Fig.5 in Ref. [15]).

The nnn hopping t_2 (with opposite sign to t) can strongly enhance T_c for d-wave symmetry, moreover it favors the d and s-wave pairings for lower values of n_c [15].

With increasing n at fixed Δ_0 one finds three possible types of density driven change-overs [11, 15]: (i) for $2 \geq \Delta_0/D \geq 0$, "BCS"$\rightarrow$ Mixed $\rightarrow$ "BCS"; (ii) for $\Delta_0/D > 2$: "BCS" $\rightarrow$ "LP" and (iii) for $\Delta_0/D < 0$: "LP" $\rightarrow$ "BCS". Only if the LP level is deeply located below the bottom of the c-band, the system remains in the LP regime for any $n \leq 2$. As for the evolution of the superconducting properties with the increase of Δ_0/D there are two sequences of change-overs possible: "LP $\rightarrow$ "Mixed" $\rightarrow$ "BCS" (if $n < 2$) and "BCS" $\rightarrow$ "Mixed" $\rightarrow$ "LP" (if $n > 2$). In particular, Fig.3b shows that the superfluid stiffness $\rho_s(0)$ is smaller than the fermionic gap $\Delta_F(0)$ in the mixed and LP regimes, in contrast to the "BCS" limit.

The region between T_c^{MFA} and T_c^{KT}, where the system can exhibit a pseudogap, expands with increasing intersubsystem coupling $|I_0|$. As we have found [16], except for $|I_0|/D \ll 1$, the coupling dependences of T_c^{MFA} and T_c^{KT} are qualitatively different. T_c^{MFA} is an increasing function of $|I_0|$ for all the pairing symmetries. On the other hand, T_c^{KT} vs $|I_0|$ increases first, goes through a round maximum and then decreases (similarly as it is observed in the attractive Hubbard model). The position of the maximum corresponds to the intermediate values of $|I_0|/D$ and it depends on the pairing symmetry as well as the values of Δ_0/D and n. For large $|I_0|$, the T_c^{KT} are close to the upper bound for the phase ordering temperature which is given by $\pi \rho_s(0)/2$.

Let us also comment on the effects of a weak interlayer coupling on the calculated transition temperatures [12, 22]. In the KT theory the 2D correlation length behaves as follows for $T > T^{KT}$: $\xi(T) = a\exp\left(b/\sqrt{T/T^{KT} - 1}\right)$, where $b \approx 1.5$ and a is the size of the vortex core. If U_c is the coupling energy per unit length between the planes and $U_c \ll T^{KT}$, then the actual T_c can be estimated by calculating the energy needed to destroy phase coherence between two regions of size $\sim \xi^2$ in different planes i.e. $T_c \sim \bar{c} U_c (\xi(T_c)/a)^2$, where $\bar{c}$ is the interplanar distance. The resulting equation for T_c can be solved asymptotically

$$T_c = T^{KT}\left(1 + \frac{4b^2}{\ln^2(T^{KT}/\bar{c}U_c)}\right). \tag{9}$$

Therefore T_c is only weakly dependent on the interplanar distance $\bar{c}$ and is close to T^{KT}, if $U_c \ll T_{KT}$. In the presence of the interplanar coupling there is no discontinuous jump in ρ_s but a crossover from 2D like to 3D like (XY) behavior occurs.

4. SUPERFLUID TRANSITION FROM THE PSEUDOGAP STATE IN 3D

Let us now consider the pseudogap behavior and present the recent evaluation of the superconducting transition temperature from a pseudogap state by going beyond the BCS-MFA. In our analysis we have applied a generalized T-matrix approach adapted to a two-component boson-fermion model[25]. It includes pairing fluctuations and the bosonic self-energy effect. Our approach is an extension of the pairing fluctuation theory of the BCS-Bose-Einstein crossover [23, 24] developed previously for a one-component fermion systems with attractive interaction. The numerical results presented in Fig.4 are for a 3D sc lattice assuming the tight-binding dispersion for fermions and bosons of the following form: $\varepsilon_{\mathbf{k}} = D(1 - \tilde{\gamma}_{\mathbf{k}})$, $D = zt$; $J_{\mathbf{q}} = J_0 \tilde{\gamma}_{\mathbf{q}}$, $J_0 = zJ$, $\tilde{\gamma}_{\mathbf{k}} = [\cos(k_x) + \cos(k_y) + \cos(k_z)]/3$, $z = 6$.

The results are shown for both cases with and without the direct hopping of LP's J_{ij}. Except for the c-regime, the calculated T_c's are much lower as compared to BCS-MFA results (these are given by Eq.(7)), and if $J = 0$, T_c is strongly depressed as soon as the LP level is close to the bottom of the electronic band. In the pseudogap region the electronic spectrum is gapped, and the pseudogap parameter at T_c for $\Delta_0 > 0$ essentially measures a mean square amplitude of the pairing field (of the "c" electrons). The values of pseudogap parameter at T_c are comparable to the zero temperature gap values in the fermionic spectrum ($\Delta_F(0)$), beyond the c-regime.

With the direct LP hopping $J_0/D = 0.1$, which corresponds to $m_B = 10 m_F$, the hard-core bosons can undergo a superfluid transition even without the intersubsystem coupling $|I_0|$. For the case $|I_0| = 0$ our approach reduces to the RPA for hard-core bosons and the critical temperature is determined by [20]:

$$(1 - 2n_B)^{-1} = \frac{1}{N} \sum_{\mathbf{q}} \coth(\beta_c E_{\mathbf{q}}^0/2), \tag{10}$$

where $E_{\mathbf{q}}^0 = J_0(1 - \tilde{\gamma}_{\mathbf{q}})(1 - 2n_B)$, together with the constraint $n = n_c + 2n_B$, $n_c = 1 - (1/N)\sum_{\mathbf{k}} \tanh(\beta_c \bar{\varepsilon}_{\mathbf{k}}/2)$, $2\mu = 2\Delta_0 - J_0(1 - 2n_B)$. The solution (for parabolic dispersion of bosons) is plotted in Fig.4 by the dotted line.

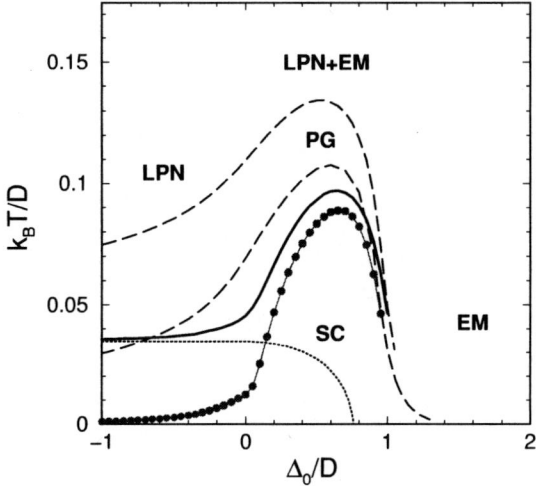

FIGURE 4. Phase diagrams of the hard-core boson-fermion model as a function of Δ_0/D for s-wave pairing and sc lattice. $n = 0.5, |I_0|/D = 0.5, D = 6t$. The transition temperatures derived with a T-matrix approach are shown by the solid line ($J_0/D = 0.1$) and the line with symbols ($J_0 = 0$), respectively. The dashed lines indicate the BCS-MFA transition temperatures (upper for $J_0/D = 0.1$, lower for $J_0 = 0$). The dotted line is the RPA result for $J_0/D = 0.1$, $|I_0| = 0$. LPN–normal state of predominantly LP's, EM–electronic metal, (SC)–superconducting state(LPS+ES), PG – pseudogap region.

From Fig.4 we see that in the presence of the boson-fermion coupling $|I_0|$ the transition temperature is much enhanced in the mixed regime.

In the self-consistent T-matrix approach the fluctuations of the order parameter are included at the Gaussian level. Nevertheless, it is interesting to observe that the phase diagram for $J_0 = 0$ shown in Fig.4 displays similar regimes as that of Fig.1 determined in Sec.3 from BCS and KT theories in 2D. We also note that for $J_0 = 0$ in both cases the shapes of T_c vs Δ_0/D are qualitatively similar. As we proceed from the regime of predominantly c-electrons to that of predominantly LP's with decreasing Δ_0, T_c first sharply increases then it goes through a maximum inside the mixed regime and is suppressed when the LP level reaches the bottom of the c-band and the system enters the LP regime.

5. FINAL REMARKS

In conclusion, we summarize the important features of the model considered [6, 14–16].

1. Well defined Bogoliubov quasiparticles can exist in the superconducting ground state. However, above T_c (in a mixed regime) local pairs coexist with itinerant fermions and the normal state properties deviate from Fermi liquid behavior.
2. The T_c's calculated beyond the BCS-MFA in a T-matrix approach in 3D and in a KT

scenario for 2D show crucial effects of pair fluctuations (and phase fluctuations) in the mixed and LP regimes.

3. In the mixed regime, for $T_c^{MFA} < T < T_c$, the system will exhibit a pseudogap in the c-electron spectrum, which will evolve into a real gap as one moves to the LP regime. For $\Delta_0 < 0$, LP's coexist with preformed c-electron pairs, which have the binding energy $E_b^c/2 \propto T_c^{MFA}$.

4. The Uemura-type plots i.e., the T_c vs zero-temperature phase stiffness $\rho_s(0)$, are obtained for d, s^* and s-wave symmetry in the KT scenario [15]. The reason for Uemura scaling $T_c \sim \rho_s(0)$ is the separation of the energy scales for the pairing and for the phase coherence [14, 15].

5. For d-wave pairing, the superfluid density exhibits a linear in T behavior at low T due to the presence of nodal quasiparticles (beyond LP regime).

Some of our findings can be qualitatively related to experimental results for the cuprate HTS where a pseudogap exists. It has been suggested by ARPES experiments, that for underdoped cuprates the Fermi surface in the pseudogap phase is truncated around the corners due to the formation of preformed pairs ("bosons") with charge $2e$, whereas the "electrons" on the diagonals remain unpaired [10]. In the two component model such a situation is obtained when LP's and c-electrons coexist in the mixed regime. The linear T-dependence of the superfluid density has been observed experimentally in copper oxides and also in several organic superconductors. This points to an order parameter of $d_{x^2-y^2}$ – symmetry and existence of nodal quasiparticles. In the present model the gap ratio is nonuniversal for all the pairing symmetries and can deviate strongly from BCS predictions (particularly in the d-wave case for which it is always enhanced) [15]. This feature is also found in several exotic superconductors. The Uemura plots and the scaling $T_c \sim \rho_s(0)$ reported for cuprates and organic superconductors can be reproduced within the model for extended s- and d-wave order parameter symmetry [14, 15].

ACKNOWLEDGMENTS

R.M. and S.R. acknowledge partial support from the State Committee for Scientific Research (KBN Poland): Project No: 2 P03B 154 22. R. M. also acknowledges support from the Foundation for Polish Science.

REFERENCES

1. R. Micnas, J. Ranninger, and S. Robaszkiewicz, Rev. Mod. Phys. **62**, 113 (1990) and Refs. therein.
2. L. P. Gor'kov, J. Supercond. **13**, 765 (2000).
3. S. Robaszkiewicz, R. Micnas, and J. Ranninger, Phys. Rev. **B 36**, 180 (1987).
4. R. Friedberg, T.D. Lee, and H.C. Ren, Phys. Rev. **B 42**, 4122 (1990); *ibid.* **B 45**, 10732 (1992).
5. Y. Bar-Yam, Phys. Rev. **B43**, 359 (1991); *ibid.* **B43**, 2601 (1991).
6. J. Ranninger and J.M. Robin Sol. State Comm. **98**, 559(1996); Phys. Rev. **B 53**, R11961 (1996); P. Devillard and J. Ranninger, Phys. Rev. Lett. **84**, 5200 (2000); T. Domanski and J. Ranninger, Phys. Rev. **B 63**, 134505 (2001).
7. T. Kostyrko and J. Ranninger, Phys. Rev. **B54**, 13105 (1996).

8. R. Micnas and S. Robaszkiewicz *"High-Tc Superconductivity 1996: Ten Years after the Discovery"*, (E. Kaldis, E. Liarokapis and K.A. Müller, eds.), NATO ASI Series E vol.343, p.31, (1997). (Kluwer Academic Publishers, The Netherlands) and Refs. therein.
9. J.A. Wilson and A. Zahir, Rep. Prog. Phys. **60**, 941 (1997).
10. V.B. Geshkenbein, L.B. Ioffe, and A.I. Larkin, Phys. Rev. **B 55**, 3173 (1997).
11. W. Czart and S. Robaszkiewicz, Int. J. Mod. Phys. **B15**, 3125 (2001).
12. A.H. Castro Neto, Phys. Rev. **B 64**, 104509 (2001).
13. E. Altman and A. Auerbach, Phys. Rev. **B65**, 104508 (2002).
14. R. Micnas, S. Robaszkiewicz, and B. Tobijaszewska, Physica B 312-313C, 49 (2002).
15. R. Micnas, S. Robaszkiewicz, and A. Bussmann-Holder, Phys. Rev. **B 66**, 104516 (2002).
16. R. Micnas, S. Robaszkiewicz, and A. Bussmann-Holder, Physica C **387**, 58 (2003).
17. M. J. Holland et al., Phys. Rev. Lett. **87**, 120406 (2001); M. L. Chiofalo et al., Phys. Rev. Lett. **88**, 090402 (2002).
18. Y. Ohashi and A. Griffin, Phys. Rev. Lett. **89**, 130402 (2002); Phys. Rev. **A 67**, 033603 (2003).
19. J.M. Kosterlitz and D.J. Thouless, J. Phys. C: Solid State Phys. **6**, 1181 (1973); D.R. Nelson and J.M. Kosterlitz, Phys. Rev. Lett. **39**, 1201 (1977).
20. R. Micnas and S. Robaszkiewicz, Phys. Rev. **B 45**, 9900 (1992); R. Micnas, S. Robaszkiewicz, and T. Kostyrko, *ibid.* **B52**, 6863 (1995).
21. A.S. Alexandrov and N.F. Mott, Rep. Prog. Phys. **57**, 1197 (1994).
22. S. Hikami and T. Tsuneto, Prog. Theor. Phys. **63**, 387 (1980); V.M. Loktev and V. Turkovskii, Phys. Rev. **67**, 214510 (2003).
23. R. Micnas et al, Phys. Rev. **B52**, 16 223 (1995).
24. Q. Chen, I.Kosztin, B. Janko, and K. Levin, Phys. Rev. **B 59**, 7083 (1999); I. Kosztin, Q. Chen, Y.-J. Kao, and K. Levin, Phys. Rev. **B 61**, 11 662 (2000).
25. R. Micnas (unpublished).

Composite operators and algebra constraints: a formalism for highly interacting systems

Ferdinando Mancini

Dipartimento di Fisica "E.R. Caianiello" - Unità di Ricerca INFM di Salerno
Università degli Studi di Salerno, I-84081 Baronissi (SA), Italy

Abstract. A formalism for the study of highly interacting electronic systems is presented. The proposed scheme is based on two key concepts: composite operators and algebra constraints. Composite field operators, that naturally appear as a consequence of interaction, are promoted to the rank of basic fields in terms of which a perturbation formulation is set up. The formalism is based on the use of Green's function and equation of motion method. The use of composite operators requires a revisitation of the Green's function formulation, where the representation is determined by means of algebra constraints which are a manifestation at macroscopic level of the algebra rules and symmetry properties obeyed at microscopic level.

1. INTRODUCTION

Since the discovery of superconductivity it has become more and more clear that the physics of many-particle interacting systems is very rich and complex. The development of technology, the possibility of changing the external thermodynamical parameters up to very extreme regions, the discovery of new materials, have led to an enormous progress in Condensed Matter Physics. New and unsuspected properties have been discovered and we are faced with a revolution whose influence is not limited to the scientific world but is involving all fields. Still, we are touching the top of an iceberg whose dimensions are not clear. The progress in technology and in experimental science has been accompanied by a parallel progress in the development of new theories and new schemes of calculations.

In the last twenty years most of the progresses have been made just in the discovery of new materials with unusual properties. It is believed that the origin of such anomalous behaviors is generally due to strong electronic correlations in narrow conduction bands [1]. In this line of thinking many analytical methods have been developed for the study of highly correlated electron systems [2], such as the Hubbard approximation [3], the spectral density approach [4,5], the non crossing approximation [6-8], the slave boson method [9-11], the d_∞-method [12-16], the projection operator method [17-22], the composite operator method [23-29]. The main difficulties are connected with the absence of any obvious small parameter in the strong coupling regime and with the simultaneous presence of itinerant (spatial correlations) and atomic (pronounced on-site quantum fluctuations) features. According to this, it is extremely difficult to judge the reliability of the results obtained by the various approximation methods. The comparison with the recently accumulated results of numerical simulations, although severely restricted in cluster size and temperature and, therefore, with generally poor momentum and energy resolutions [30], is a unavoidable basic step. The numerical results are certainly a guide for the construction of any microscopic theory and, in any case, the different theoretical formulations should refer to them as experimental results obtained on model Hamiltonians instead of materials.

2. COMPOSITE OPERATORS

The most important characteristic of these new materials is a strong correlation among the electrons that makes inapplicable classical schemes based on the band picture. It is necessary to pass from a "single-electron" physics to a "many-electron" physics, where the dominant part will be the correlations among the electrons. Usual perturbation schemes are inadequate and new concepts must be introduced.

The "classical" techniques are based on the hypothesis that the interaction among the electrons is weak and can be treated in the framework of some perturbation scheme. However, as many and many experimental and theoretical studies of highly correlated electron systems have shown with more and more convincing evidence, all these methods are not adequate any more and different approaches must be considered. The main concept that breaks down is the existence of the electrons as particles with some well-defined and intrinsic properties. The presence of interaction modifies the properties of the particles and at a macroscopic level, the level of observation, what are observed are new particles with new peculiar properties entirely determined by the dynamics and by the boundary conditions (i.e. all elements characterizing the physical situation under study). These new objects appear as the final result of the modifications imposed by the interactions on the original particles and contain, by the very beginning, the effects of correlations. Collective behaviors in forms of bound states, resonances, diffused modes and so on emerge as the physical fields. Although some of them are not stable excitations, they give considerable contributions in physical processes, and therefore it is sometimes necessary to promote them to the role of well-defined quasi-particle excitations. The choice of new fundamental particles, whose properties have to be self-consistently determined by dynamics, symmetries and boundary conditions, becomes relevant.

As a simple example, let us consider an atomic system described by the Hamiltonian

$$H = -\mu \sum_\sigma \varphi_\sigma^\dagger \varphi_\sigma + V \varphi_\uparrow^\dagger \varphi_\downarrow^\dagger \varphi_\downarrow \varphi_\uparrow \tag{2.1}$$

φ_σ denotes an Heisenberg electronic field with spin $\sigma = \uparrow, \downarrow$, satisfying canonical anticommutation relations; μ is the chemical potential and V is the strength of the interaction. This model is exactly solvable in terms of the operators

$$\xi_\sigma = \varphi_\sigma \varphi_{-\sigma} \varphi_{-\sigma}^\dagger \qquad \eta_\sigma = \varphi_\sigma \varphi_{-\sigma}^\dagger \varphi_{-\sigma} \tag{2.2}$$

which are eigenoperators of the Hamiltonian

$$i\frac{\partial}{\partial t}\xi = [\xi, H] = -\mu\xi \qquad i\frac{\partial}{\partial t}\eta = [\eta, H] = -(\mu - V)\eta \tag{2.3}$$

Due to the presence of the interaction, the original electrons φ_σ are no more observables and new stable elementary excitations, described by the field operators ξ and η, appear. Due to the V-interaction, two sharp features develop in the band structure: the energy level $E = -\mu$ of the bare electron splits in the two levels $E_1 = -\mu$ and $E_2 = V - \mu$. The bare electron reveals itself to be precisely the wrong place to start. A perturbative solution will never give the band splitting.

On the basis of this evidence one can be induced to move the attention from the original fields to the new fields generated by the interaction. The operators describing these excitations, once they have been found, can be written in terms of the original ones and are known as composite operators.

The convenience of developing a formulation to treat composite excitations as fundamental objects has been noticed for the many-body problem of condensed matter physics since long time. Recent years have seen remarkable developments in many-body theory in the form of an assortment of techniques that may be termed composite particle methods. The beginnings of these types of techniques may be traced back to the work of Bogolubov [31], Dancoff [32], Zwanzig [17], Mori [18], Umezawa [33]. The slave boson method, the spectral density approach and the composite operator method (COM) are also along similar lines. This large class of theories is founded on the conviction that an analysis in terms of elementary fields might be inadequate for a system dominated by strong interactions.

All these approaches are very promising because all the different approximation schemes are constructed on the basis of interacting particles: some amount of the interaction is already present in the chosen basis and permits to overcome the problem of finding an appropriate expansion parameter. However, one price must be paid. In general, the composite fields are neither Fermi nor Bose operators, since they do not satisfy canonical (anti)commutation relations, and their properties, because of the inherent definition, must be self-consistently determined. They can only be recognized as fermionic or bosonic operators according to the number, odd or even, of the constituting original electronic fields. This fact makes a tremendous difference with respect to the case of the original electronic operators φ, which satisfy a canonical algebra.

New techniques of calculus have to be developed in order to treat with composite fields. In developing perturbation calculations where the building blocks are now the propagators of composite fields one cannot use the consolidated scheme: diagrammatic expansions, Wick's theorem and many other techniques are no more valid. The formulation of the Green's function method must be revisited and new frameworks of calculations have to be formulated.

3 GREEN'S FUNCTION AND EQUATION OF MOTION FORMALISM

Let us consider a system of N_e interacting Wannier-electrons residing on a Bravais lattice of N sites, spanned by the vectors $\mathbf{R}_i = \mathbf{i}$. For the sake of simplicity, we ignore the presence of magnetic impurities and restrict the analysis to single-band electron models. The generalization of the formalism to more complex systems is straightforward (see for example [34]). In a second quantization scheme this system is described by a certain Hamiltonian

$$H = H[\varphi(i)] \tag{3.1}$$

describing, in complete generality, the free propagation of the electrons and all the interactions among them and with external fields. $\varphi(i)$ denotes an Heisenberg electronic field [$i = (\mathbf{i},t)$] satisfying canonical anticommutation relations.

Given the hypothesis that the original fields are not a good basis, we choose a set of composite fields $\{\psi(i)\}$ in terms of which a perturbation scheme will be constructed. Firstly, we choose the set $\psi(i)$ according to the physical properties we want to study. Roughly, the properties of electronic systems can be classified in two large classes: single particle properties, described in terms of fermionic propagators, and response functions, described in terms of bosonic propagators. These two sectors, fermionic and bosonic, are not independent but interplay with each other, and a fully self-consistent solution usually requires that both sectors are simultaneously solved. Once the sector, fermionic or bosonic, has been fixed, we have several criteria for the choice of the new basis. In constructing the composite fields no recipe can be given without thinking to its drawbacks, but many recipes can assure a correct and controlled description of relevant aspects of the dynamics. One can choose: the higher order fields emerging from the equations of motion (in this case the conservation of some spectral moments is assured [35]), the eigenoperators of some relevant interacting terms [36], the eigenoperators of the problem reduced to a small cluster [37].

Let $\psi(i)$ be a n-component field

$$\psi(i) = \begin{pmatrix} \psi_1(i) \\ \vdots \\ \psi_n(i) \end{pmatrix} \tag{3.2}$$

We do not specify the nature, fermionic or bosonic, of the set $\{\psi(i)\}$. In the case of fermionic operators it is intended that we use the spinorial representation

$$\psi_m(i) = \begin{pmatrix} \psi_{\uparrow m}(i) \\ \psi_{\downarrow m}(i) \end{pmatrix} \quad \psi_m^\dagger(i) = \begin{pmatrix} \psi_{\uparrow m}^\dagger(i) & \psi_{\downarrow m}^\dagger(i) \end{pmatrix} \tag{3.3}$$

The dynamics of these operators is governed by the given Hamiltonian $H = H[\{\varphi\}]$ and can be written as

$$i\frac{\partial}{\partial t}\psi(i) = [\psi(i), H] = J(i) \tag{3.4}$$

In general, this equation cannot be exactly solved and some approximations are necessary. In order to construct approximate solutions one procedure is the following. Let us rewrite the equation of motion as

$$i\frac{\partial}{\partial t}\psi(\mathbf{i},t) = \sum_{\mathbf{j}} \varepsilon(\mathbf{i},\mathbf{j})\psi(\mathbf{j},t) + \delta J(\mathbf{i},t) \tag{3.5}$$

where

$$\varepsilon(\mathbf{i},\mathbf{j}) = \sum_{\mathbf{l}} m(\mathbf{i},\mathbf{l}) I(\mathbf{l},\mathbf{j})^{-1} \quad \begin{array}{l} I(\mathbf{i},\mathbf{j}) =< [\psi(\mathbf{i},t),\psi^{\dagger}(\mathbf{j},t)]_{\eta} > \\ m(\mathbf{i},\mathbf{j}) =< [J(\mathbf{i},t),\psi^{\dagger}(\mathbf{j},t)]_{\eta} > \end{array} \quad (3.6)$$

Here $\eta = \pm 1$; usually, it is convenient to take $\eta = 1$ ($\eta = -1$) for a fermionic (bosonic) set $\psi(i)$ (i.e., for a composite field constituted of an odd (even) number of original fields) in order to exploit the canonical anticommutation relations of $\{\psi(i)\}$; but, in principle, both choices are possible. Accordingly, we define

$$[A,B]_{\eta} = \begin{cases} \{A,B\} = AB + BA & \text{for } \eta = 1 \\ [A,B] = AB - BA & \text{for } \eta = -1 \end{cases} \quad (3.7)$$

$<\cdots>$ denotes the quantum statistical average over the grand canonical ensemble. Since the components of $\psi(i)$ contain composite operators, the normalization matrix $I(\mathbf{k})$ is not the identity matrix and defines the spectral content of the excitations.

Let us consider the two-time thermodynamic Green's functions (GF) [38]

$$G^Q(i,j) =< Q[\psi(i)\psi^{\dagger}(j)] > \quad (3.8)$$

where $Q = C$ (causal), R (retarded), A (advanced). By means of the equation of motion (3.4) we can derive a Dyson equation for composite fields

$$G^Q(\mathbf{k},\omega) = G_0^Q(\mathbf{k},\omega) + G_0^Q(\mathbf{k},\omega)\Sigma^Q(\mathbf{k},\omega)G^Q(\mathbf{k},\omega) \quad (3.9)$$

where $G_0^Q(\mathbf{k},\omega)$ is the free propagator for composite fields, satisfying the equation

$$[\omega - \varepsilon(\mathbf{k})]G_0^Q(\mathbf{k},\omega) = I(\mathbf{k}) \quad (3.10)$$

and $\Sigma^Q(\mathbf{k},\omega)$ is the self energy

$$\Sigma^Q(\mathbf{k},\omega) = \Gamma^{-1}(\mathbf{k})B_{irr}^Q(\mathbf{k},\omega)\Gamma^{-1}(\mathbf{k}) \quad (3.11)$$

$B_{irr}^Q(\mathbf{k},\omega)$ is the irreducible part of the propagator $B^Q(\mathbf{k},\omega) = F.T. <Q[\delta J(i)\delta J^{\dagger}(j)]>$.

We have constructed a generalized perturbative approach designed for formulations using composite fields. Equation (3.9) is a Dyson-like equation and may represents the starting point for a perturbative calculation in terms of the propagator $G_0^Q(\mathbf{k},\omega)$. Contrarily to the usual perturbation schemes, the calculation of the "free propagator" $G_0^Q(\mathbf{k},\omega)$ is not an easy task and the next Sections will be dedicated to this problem. Then, the attention will be given to the calculation of the self-energy $\Sigma^Q(\mathbf{k},\omega)$, and some approximate methods will be presented. It should be noted that the computation of the two quantities $G_0^Q(\mathbf{k},\omega)$ and $\Sigma^Q(\mathbf{k},\omega)$ are intimately related. The total weight of the self-energy corrections is bounded by the weight of the residual source operator $\delta J(i)$. According to this, it can be made smaller and smaller by increasing the components of the basis $\psi(i)$ [e.g. by including higher-order composite operators appearing in $\delta J(i)$]. The result of such a procedure will be the inclusion in the energy matrix of part of the self-energy as an expansion in terms of coupling constants multiplied by the weights of the newly includes basis operators. In general, the enlargement of the basis leads to a new self-energy with a smaller total weight. However, it is necessary pointing out that this process can be quite cumbersome and the inclusion of fully momentum and frequency dependent self-energy corrections can be necessary to effectively take into account low-energy and virtual processes. According to this, one can chose a reasonable number of components for the basic set and then use another approximation method to evaluate the residual dynamical corrections.

4. THE FREE PROPAGATOR $G_0^Q(\mathbf{k},\omega)$

In this Section we concentrate on the calculation of the Green's functions $G_0^Q(\mathbf{k},\omega)$ which constitute the building blocks of the perturbation scheme we are trying to formulate. To keep the notation as simple as possible, we will drop the sub index 0 in the definition of $G_0^Q(\mathbf{k},\omega)$.

One fundamental aspect in a Green's function formulation is the choice of the representation. The knowledge of the Hamiltonian and of the operatorial algebra is not sufficient to completely specify the GF. The GF refer to a specific representation (i.e., to a specific choice of the Hilbert space) and this information must be supplied as a boundary condition to the equations of motion that alone are not sufficient to completely determine the GF. The use of composite operators leads to an enlargement of the Hilbert space by the inclusion of some unphysical states. As a consequence of this, it is difficult to satisfy a priori all the sum rules and, in general, the symmetry properties enjoined by the system under study. In addition, since the representation where the operators are realized has to be dynamically determined, the method clearly requires a process of self-consistency.

From this discussion it is clear that fixing the representation is not an easy task and requires special attention. In the literature the properties of the GF are usually determined by starting from the knowledge of the representation. Owing to the difficulties discussed above we cannot proceed in this way. Therefore, we will derive the general properties of the GF on the basis of the two elements we have: the dynamics, fixed by the choice of the Hamiltonian (3.1), and the algebra, fixed by the choice of the basic set (3.2). The problem of fixing the representation will be considered in Section 7.

Let $\psi(i)$ be a n-component field satisfying linear equations of motion

$$i\frac{\partial}{\partial t}\psi_m(\mathbf{i},t) = \sum_{\mathbf{j}} \sum_{l=1}^{n} \varepsilon_{ml}(\mathbf{i},\mathbf{j})\psi_l(\mathbf{j},t) \quad (4.1)$$

with the energy matrix $\varepsilon(\mathbf{i},\mathbf{j})$ defined by (3.6). If the fields $\psi(i)$ are eigenoperators of the total Hamiltonian, the equations of motion (4.1) are exact. If the fields $\psi(i)$ are not eigenoperators of H, the equations are approximated; they correspond to neglecting the residual source operator $\delta J(i)$ in the full equation of motion (3.5) and all the formalism is developed with the intention of using the propagators of these fields as a basis to set up a perturbative scheme of calculations on the ground of the Dyson equation (3.9) derived in the previous Section.

By means of the field equation (4.1), the Fourier transforms of the various Green's functions and of the correlation function $C(i,j) = <\psi(i)\psi^\dagger(j)>$ satisfy the following equations

$$[\omega - \varepsilon(\mathbf{k})]G^{Q(\eta)}(\mathbf{k},\omega) = I^{(\eta)}(\mathbf{k})$$
$$[\omega - \varepsilon(\mathbf{k})]C(\mathbf{k},\omega) = 0 \quad (4.2)$$

where the dependence on the parameter η has been explicitly introduced. It can be shown that the energy matrix $\varepsilon(\mathbf{k})$ does not depend on the choice of η. As mentioned in Section 3, the set $\{\psi(i)\}$ can be fermionic or bosonic and the parameter η generally takes the value $\eta = 1$ ($\eta = -1$) for a fermionic (bosonic) set $\psi(i)$. The three Green's functions G^C, G^R and G^A satisfy the same equation of motion which alone is not sufficient and must be supplemented by other equations. Indeed, the GF are determined by solving a first order differential equation of motion, thereby the GF are given only within an arbitrary constant of integration. The retarded and advanced GF can be completely determined because the factor $\theta[\pm(t_i - t_j)]$ provides the boundary condition: $G^{R,A}(i,j) = 0$ for $t_i = t_j \mp \delta$. The determination of the causal GF is not so immediate. In the following we consider the case of finite temperature. For T=0 see Ref. [29].

The most general solution of equation (4.2) is

$$G^{C,R,A,(\eta)}(\mathbf{k},\omega) = \sum_{l=1}^{n} \left\{ P\left(\frac{\sigma^{(l,\eta)}(\mathbf{k})}{\omega - \omega_l(\mathbf{k})}\right) - i\pi\delta[\omega - \omega_l(\mathbf{k})]g^{(l,\eta)C,R,A}(\mathbf{k}) \right\}$$

$$C(\mathbf{k},\omega) = \sum_{l=1}^{n} \delta[\omega - \omega_l(\mathbf{k})]c^{(l)}(\mathbf{k}) \quad (4.3)$$

$g^{(l,\eta)C,R,A}(\mathbf{k})$ and $c^{(l)}(\mathbf{k})$ are momentum functions, not fixed by the equations of motion, to be determined by means of the boundary conditions. $\omega_l(\mathbf{k})$ are the eigenvalues of the matrix $\varepsilon(\mathbf{k})$; $\sigma^{(l,\eta)}(\mathbf{k})$ are the spectral density functions, completely determined by the matrices $\varepsilon(\mathbf{k})$ and $I^{(\eta)}(\mathbf{k})$ as

$$\sigma^{(l,\eta)}_{\alpha\beta}(\mathbf{k}) = \Omega_{\alpha l}(\mathbf{k}) \sum_{\delta} \Omega^{-1}_{l\delta}(\mathbf{k}) I^{(\eta)}_{\delta\beta}(\mathbf{k}) \quad (4.4)$$

where $\Omega(\mathbf{k})$ is the $n \times n$ matrix whose columns are the eigenvectors of the matrix $\varepsilon(\mathbf{k})$. By calculations we obtain [29]:

Fermionic fields (i.e., $\eta = 1$)

$$G^{R,A,(+1)}(\mathbf{k},\omega) = \sum_{l=1}^{n} \frac{\sigma^{(l,+1)}(\mathbf{k})}{\omega - \omega_l(\mathbf{k}) \pm i\delta}$$

$$G^{C,(+1)}(\mathbf{k},\omega) = \sum_{l=1}^{n} \sigma^{(l,+1)}(\mathbf{k}) \left[\frac{1 - f_F(\omega)}{\omega - \omega_l(\mathbf{k}) + i\delta} + \frac{f_F(\omega)}{\omega - \omega_l(\mathbf{k}) - i\delta}\right]$$

$$C(\mathbf{k},\omega) = 2\pi \sum_{l=1}^{n} \delta[\omega - \omega_l(\mathbf{k})]\{1 - f_F[\omega_l(\mathbf{k})]\}\sigma^{(l,+1)}(\mathbf{k}) \quad (4.5)$$

where $f_F(\omega)$ is the Fermi distribution function.

Bosonic fields (i.e., $\eta = -1$)

For any given momentum $\mathbf{k}$ we can always write

$$\omega_l(\mathbf{k}) = \begin{cases} = 0 & \text{for } l \in A(\mathbf{k}) \subseteq N = \{1,\ldots n\} \\ \neq 0 & \text{for } l \in B(\mathbf{k}) = N - A(\mathbf{k}) \end{cases} \quad (4.6)$$

Obviously, $A(\mathbf{k})$ can also be the empty set (i.e., $A(\mathbf{k}) = \emptyset$ and $B(\mathbf{k}) = N$).

$$G^{R,A,(-1)}(\mathbf{k},\omega) = \sum_{l=1}^{n} \frac{\sigma^{(l,-1)}(\mathbf{k})}{\omega - \omega_l(\mathbf{k}) \pm i\delta}$$

$$G^{C,(-1)}(\mathbf{k},\omega) = -2i\pi\Gamma(\mathbf{k})\delta(\omega)$$

$$+ \sum_{l \in B(\mathbf{k})} \sigma^{(l,-1)}(\mathbf{k}) \left[\frac{1 + f_B(\omega)}{\omega - \omega_l(\mathbf{k}) + i\delta} - \frac{f_B(\omega)}{\omega - \omega_l(\mathbf{k}) - i\delta} \right]$$

$$C(\mathbf{k},\omega) = 2\pi\Gamma(\mathbf{k})\delta(\omega) + 2\pi \sum_{l \in B(\mathbf{k})} \delta[\omega - \omega_l(\mathbf{k})]\{1 + f_B[\omega_l(\mathbf{k})]\}\sigma^{(l,-1)}(\mathbf{k})$$
(4.7)

where $f_B(\omega)$ is the Bose distribution function. The zero-frequency function (ZFF) $\Gamma(\mathbf{k})$ has been defined as

$$\Gamma(\mathbf{k}) = \frac{1}{2\pi} \sum_{l \in A(\mathbf{k})} c^{(l)}(\mathbf{k}) = \frac{1}{2} \sum_{l \in A(\mathbf{k})} g^{(l,-1)C}(\mathbf{k})$$
(4.8)

and it is left undetermined within the bosonic sector. $\Gamma(\mathbf{k})$ could be computed by considering an anticommutating algebra: remaining in the bosonic sector we make the choice $\eta = 1$ and $\Gamma(\mathbf{k})$ can be calculated by means of the following relations

$$\Gamma(\mathbf{k}) = \frac{1}{2} \sum_{l \in A(\mathbf{k})} \sigma^{(l,+1)}(\mathbf{k}) = \frac{1}{2} \lim_{\omega \to 0} \omega G^{C,(+1)}(\mathbf{k},\omega)$$
(4.9)

However, the calculation of the $\sigma^{(l,+1)}(\mathbf{k})$ requires the calculation of the normalization matrix $I^{(+1)}(\mathbf{k})$ that, for bosonic fields, generates unknown momentum dependent correlation functions whose determination can be very cumbersome as requires, at least in principle, the self-consistent solution of the integral equations connecting them to the corresponding Green's functions. In practice, also for simple, but anyway composite, bosonic fields the $\Gamma(\mathbf{k})$ remains undetermined and other methods rather than equation (4.9) should be used. Similar methods, like the use of the relaxation function [39], would lead to the same problem.

It is worth pointing out that in the bosonic sector we generally have

$$\sum_{l \in A(\mathbf{k})} \sigma^{(l,-1)}(\mathbf{k}) = 0$$
(4.10)

A situation where $\sum_{l \in A(\mathbf{k})} \sigma^{(l,-1)}(\mathbf{k}) \neq 0$ would lead to the fact that for $l \in A(\mathbf{k})$ the Fourier coefficients $c^{(l)}(\mathbf{k})$ diverge as $[\beta\omega_l(\mathbf{k})]^{-1}$. Since the correlation function in direct space must be finite, at finite temperature this is admissible only in the thermodynamic limit and if the dispersion relation $\omega_l(\mathbf{k})$ is such that the divergence in momentum space is integrable and the corresponding correlation function in real space remains finite. For finite systems and for infinite systems where the divergence is not integrable we must have $\sum_{l \in A(\mathbf{k})} \sigma^{(l,-1)}(\mathbf{k}) = 0$. The calculation of the spectral density matrices $\sigma^{(l,-1)}(\mathbf{k})$ it not a simple dynamical problem, but requires the self-consistent calculation of some expectation values, where the boundary conditions and the choice of the representation play a crucial role. A finite value of $\sum_{l \in A(\mathbf{k})} \sigma^{(l,-1)}(\mathbf{k})$ is generally related to the presence of long-range order and the previous statement is nothing but the Mermin-Wagner theorem [40].

We see that the general structure of the GF is remarkably different according to the statistics. For fermionic composite fields all the Green's functions and correlation functions are completely determined. The zero-frequency function $\Gamma(\mathbf{k})$, defined on the Fermi surface $\omega_l(\mathbf{k}) = \mu$, contributes to the spectral function, is directly related to the spectral density functions $\sigma^{(l,+1)}(\mathbf{k})$ by means of equation (4.9), and its calculation does not require more information. Also, it does not contribute to the imaginary part of the causal GF. For bosonic composite fields the retarded and advanced GF are completely determined, but the causal GF and the correlation function depend on the zero-frequency function $\Gamma(\mathbf{k})$, defined on the surface $\omega_l(\mathbf{k}) = 0$. It is now clear that the causal and retarded (advanced) GF contain different information and that the right procedure of calculation is controlled by the statistics. In particular, in the case of bosonic fields one must start from the causal function and compute the other GF by means of the expressions

$$\text{Re}[G^{R,A(-1)}(\mathbf{k},\omega)] = \text{Re}[G^{C(-1)}(\mathbf{k},\omega)]$$

$$\text{Im}[G^{R,A(-1)}(\mathbf{k},\omega)] = \pm \tanh\left(\frac{\beta\omega}{2}\right) \text{Im}[G^{C(-1)}(\mathbf{k},\omega)]$$

$$C(\mathbf{k},\omega) = -\left[1 + \tanh\left(\frac{\beta\omega}{2}\right)\right] \text{Im}[G^{C(-1)}(\mathbf{k},\omega)]$$

(4.11)

On the contrary, for fermionic fields the right procedure requires first the calculation of the retarded (advanced) function and then computing the other GF by means of the expressions

$$\text{Re}[G^{C(+1)}(\mathbf{k},\omega)] = \text{Re}[G^{R,A(+1)}(\mathbf{k},\omega)]$$

$$\text{Im}[G^{C(+1)}(\mathbf{k},\omega)] = \pm \tanh\left(\frac{\beta\omega}{2}\right) \text{Im}[G^{R,A(+1)}(\mathbf{k},\omega)]$$

$$C(\mathbf{k},\omega) = \mp\left[1 + \tanh\left(\frac{\beta\omega}{2}\right)\right] \text{Im}[G^{R,A(+1)}(\mathbf{k},\omega)]$$

(4.12)

5. THE ZERO-FREQUENCY PROBLEM

Given two appropriate operators A and B one can define physical response functions χ_{AB}, called generalized susceptibilities. It was noticed by Kubo [39] that the isolated susceptibility $\chi^I_{AB}(\omega)$, defined for a situation where the system is isolated and the external force is turned on adiabatically, in the limit of zero frequency is in general different from the isothermal susceptibility χ^T_{AB}, defined for a situation where the system is in thermal equilibrium in the presence of a time-independent force. Several years later it was shown [41] that the difference between the two susceptibilities is related to the zero-frequency anomaly exhibited by the bosonic correlation functions, as discussed in the previous Section. Indeed, it can be shown that

$$\chi_{AB}^T - \chi_{AB}^I(\omega = 0) = \beta \frac{1}{N} \sum_{\mathbf{k}} e^{i\mathbf{k}\cdot(\mathbf{R}_i-\mathbf{R}_j)} \Gamma_{AB}(\mathbf{k}) - \beta <A><B> \qquad (5.1)$$

Kubo pointed out that the problem of the difference between the two susceptibilities is related to the ergodic property of the system. If the operator $\psi(i)$ has an ergodic dynamics with respect to the Hamiltonian H, then the zero-frequency function $\Gamma_{\psi\psi^\dagger}(\mathbf{k})$ must satisfies the following equation

$$\lim_{T\to\infty}\frac{1}{T}\int_0^T <\psi(\mathbf{j},0)\psi^\dagger(\mathbf{j},t)> dt = \frac{1}{N}\sum_{\mathbf{k}} e^{i\mathbf{k}\cdot(\mathbf{R}_i-\mathbf{R}_j)} \Gamma_{\psi\psi^\dagger}(\mathbf{k}) = <\psi(\mathbf{i})><\psi^\dagger(\mathbf{j})> \qquad (5.2)$$

If this is true, then the problem of calculating the zero-frequency function is solved and from (5.1) the two generalized susceptibilities χ_{AB}^T and $\chi_{AB}^I(0)$ are equal. However, we have not to forget that the condition (5.2) is the same as the standard ergodic requirement only for statistical averages computed in the microcanonical ensemble [39, 42]; in other ensemble it holds only in the thermodynamic limit. Moreover, the condition (5.2) is not satisfied by any integral of motion and, more generally, by any operator that has a diagonal part with respect to the Hamiltonian [43]. This latter consideration clarifies why the ergodic nature of the dynamics of an operator mainly depends on the Hamiltonian which is subject to. It is really remarkable that the zero-frequency constants (ZFC), which are the values of the zero-frequency function $\Gamma(\mathbf{k})$ over the moments for which $A(\mathbf{k}) \neq \emptyset$, are directly related to relevant measurable quantities such as the compressibility, the specific heat, the magnetic susceptibility. According to this, in the case of infinite systems too the correct determination of the zero-frequency constants cannot be considered as an irrelevant issue. In conclusion, Eq. (5.2) generally cannot be used to compute the ZFC and $\Gamma(\mathbf{k})$ has to be computed case by case according to the dynamics and boundary conditions.

The presence of undetermined constants in the bosonic correlation functions is some time known as the zero-frequency anomaly problem. It was first put in evidence in Ref. [44] and then studied by several authors [41, 43, 45-49]. There is a general belief that this problem is of academic interest and in the last years no much attention has been dedicated to it. The main reason is that the response functions, the experimentally observed quantities, are given by retarded bosonic GF which, as we have shown, formally do not depend on the zero-frequency constants, which are, therefore, considered of no physical interest. The general attitude [39, 45] is to believe that in macroscopic real systems at equilibrium at temperature T, the fluctuations are very small and the interaction between the system and the reservoir would introduce an irreversible relaxation and decouple the correlation functions. Then, as suggested in Ref. [45], the zero-frequency constants should be always determined by requiring the ergodicity and therefore fixed by means of Eq. (5.2). This procedure is some how an artifice and may lead to serious problems because it might break the internal self-consistency of the entire formulation. The fact that the retarded GF do not depend on the zero-frequency constant is true only for noninteracting systems. In general, for interacting systems the retarded GF do depend on the ZFC. To understand this we must recall that in the equations of motion of all the GF appears an inhomogeneous term, the normalization matrix $I(\mathbf{i},\mathbf{j}) = <[\psi(\mathbf{i},t), \psi^{\dagger}(\mathbf{j},t)]_\eta >$. This quantity is expressed in terms of various correlation functions, depending on the algebra of the set $\{\psi(i)\}$, of fermionic and bosonic nature, to be determined in a self-consistent way. Since the bosonic correlation functions depend on the ZFC, the normalization function and therefore all the GF do depend on the ZFC. These quantities cannot be fixed in an arbitrary way, but they must be calculated in order to keep the internal self-consistency of the global formulation.

6. ARE THE GREEN'S FUNCTIONS FULLY DETERMINED?

By means of the equations of motion and by using the boundary conditions related to the definitions of the various Green's functions we have been able to derive explicit expressions for these latter [cfr. (4.5) and (4.7)]. However, these expressions can only determine the functional dependence; the knowledge of the GF is not fully achieved yet. The reason is that the algebra of the field $\psi(i)$ is not canonical. As a consequence, the inhomogeneous terms $I^{(\eta)}(\mathbf{k})$ in the equations of motion (4.2) and the energy matrix $\varepsilon(\mathbf{k})$ contain some unknown static correlation functions, correlators, that have to be self-consistently calculated. These functions can be both of fermionic and bosonic nature and usually one needs to study more sectors at the same time. Furthermore, these correlation functions are expectation values of higher-order operators not belonging to the chosen basis $\{\psi(i)\}$. This is the most serious problem! In order to calculate these correlators one should enlarge the basis by including the new higher-order operators and repeat the scheme of calculation. It is clear that the calculation of the new matrices $I^{(\eta)}(\mathbf{k})$ and $\varepsilon(\mathbf{k})$ will lead to new correlators and new higher-order field operators will appear. In general the process might not converge, or a huge number of basic operators will be needed. In addiction to this problem, in the case of bosonic fields, there is the presence of the zero-frequency functions $\Gamma(\mathbf{k})$ whose determination is not easy at all. The self-consistent calculation of the unknown correlators and zero-frequency functions must be performed in order to completely determine the GF. It is important to remark that the entire process of self-consistency will affect all the GF at the same time and, therefore, all the physical properties of the system. For instance, as noticed in the previous Section, although the retarded GF do not explicitly depend on the ZFC, there is an implicit dependence through the internal self-consistent parameters. A self-consistent scheme of calculations for the various GF will be given in the next Section.

7. A SELF-CONSISTENT SCHEME

In the approximation scheme we are proposing, an essential element is the knowledge of the free propagators $G_0^Q(\mathbf{k},\omega)$. These quantities have been largely studied in Section 5 and the explicit expressions have been obtained. However, three serious problems arise with the study of these functions:

(a) the calculation of some parameters expressed as correlation functions of field operators not belonging the chosen basis;

(b) the appearance of some zero-frequency constants (ZFC) and their determination;

(c) the problem of fixing the representation where the Green's functions are formulated.

In most of the approaches found in the literature the solution of the previous problems is the following.

(a) In order to determine the unknown parameters several methods (arbitrary ansatz, decoupling schemes, use of the equation of motion) have been considered in the context of different approaches (Hubbard I approximation [3], Roth's method [50], projection method [2], spectral density approach [4, 5]). As shown in Ref. 28 in the context of the Hubbard model, these procedures lead to a series of unpleasant results: several sum rules and the particle-hole symmetry are violated, there is no presence of a Mott transition, all local quantities strongly disagree with the results of the numerical simulation.

(b) The ZFC are usually fixed by requiring the ergodicity of the dynamics of the relative operators. This is clearly a very strong assumption. There are many examples where the zero-frequency constants do not assume their ergodic value: if we would force the ZFC to assume it, this choice leads to wrong results. In general, these quantities must be calculated case by case.

(c) The knowledge of the Hamiltonian and of the operatorial algebra is not sufficient to completely specify the GF. The GF refer to a specific representation (i.e., to a specific choice of the Hilbert space) and this information must be supplied to the equations of motion that alone are not sufficient to completely determine the GF. The construction of the Hilbert space where the GF are realized is not an easy task and is usually ignored. The use of composite operators leads to an enlargement of the Hilbert space by the inclusion of some unphysical states. As a consequence of this, it is difficult to satisfy a priori all the sum rules and, in general, the symmetry properties enjoyed by the system under study.

In the composite operator method (COM) the three problems are not considered separately but they are all connected in one self-consistent scheme. The main idea is that fixing the values of the unknown parameters and of the ZFC implies to put some constraints on the representation where the GF are realized. As the determination of this representation is not arbitrary, it is clear that there is no freedom in fixing these quantities. They must assume values compatible with the dynamics and with the right representation. Which is the right representation? This is a very hard question to answer.

From the algebra it is possible to derive several relations among the operators. We will call Pauli constraints (PC) all possible relations among the operators dictated by the algebra. This set of relations valid at microscopic level must be satisfied also at macroscopic level, when expectations values are considered. In general, the correlation functions calculated by means of the equation of motion, as shown in Section 4, without having specified the representation, do not satisfy the relations called by the algebra. To see this, let us consider as an example the correlation function $C_{\xi\eta^\dagger}(i,j) = <\xi(i)\eta^\dagger(j)>$, where $\xi(i)$ and $\eta(i)$ are the Hubbard operators defined by Eq. (2.2). Owing to the fact that the algebra of these operators is not canonical, the correlation function $C_{\xi\eta^\dagger}(i,j)$ will depend on a set of parameters $\{p_1, p_2, ... p_m\}$, not known a priori, which must be calculated by some appropriate methods. By means of the Pauli principle, the operators $\xi(i)$ and $\eta(i)$ satisfy the relation $\xi(i)\eta^\dagger(i) = 0$. However, when we consider the expectation value it is clear that the relation

$$< \xi(i)\eta^\dagger(i) > = 0 \qquad (7.1)$$

will be satisfied only when the parameters $\{p_1, p_2, ... p_m\}$ will take appropriate values. For any other values the relation (7.1) will be violated. It is then evident that there is no freedom in determining the parameters $\{p_1, p_2, ... p_m\}$. If (7.1) is not satisfied, it is clear that in the Hilbert space we are picking up states of the type $|i(\uparrow), i(\uparrow)\rangle$, which are incompatible with the Pauli principle and must be eliminated.

We also note that, in general, the Hamiltonian has some symmetry properties (i.e. rotational invariance in coordinate and spin space, phase invariance, gauge invariance,......). These symmetries generate a set of relations among the matrix elements: the Ward-Takahashi identities [51] (WT).

Now, certainly the right representation must be the one where all relations among the operators satisfy the conservation laws present in the theory when expectation values are taken (i.e., where all the PC and WT are preserved). Then, we impose these conditions and obtain a set of self-consistent equations that will fix the unknown correlators, the ZFC and the right representation at the same time. Several equations can be written down, according to the different symmetries we want to preserve. A large class of self-consistent equations is given by the following equation

$$< \psi(i)\psi^\dagger(i) > = \frac{1}{N}\sum_{\mathbf{k}} \frac{1}{2\pi} \int_{-\infty}^{+\infty} d\omega\, C_{\psi\psi^\dagger}(\mathbf{k}, \omega) \qquad (7.2)$$

where the l.h.s. is fixed by the PC, the WT and the boundary conditions compatible with the phase under investigation and in the r.h.s. the correlation function $C_{\psi\psi^\dagger}(\mathbf{k}, \omega)$ is computed by means of the equation of motion, as illustrated in Section 4. Equations (7.2) generate a set of self-consistent equations which determine the unknown parameters (i.e., ZFC and unknown correlators) and, consequently, the proper representation, avoiding the problem of uncontrolled and uncontrollable decoupling. Condition (7.2) can be considered as a generalization, to the case of composite fields, of the equation that, in the non-interacting case, fixes the way of counting the particles per site, according to the algebra, by determining the chemical potential.

Another important relation, that will be largely used in the applications, is the requirement of time translational invariance which leads to the condition that the m-matrix, defined by Eq. (1.3.14), must satisfy the following relation:

$$m_{ab}(\mathbf{k}) = (m_{ba}(\mathbf{k}))^* \qquad (7.3)$$

This is a particular case of a more general condition on the spectral moments [35]. It can be shown that if (7.3) is violated, then states with a negative norm are included in the Hilbert space.

It should be noted that the number of constraints generated by Eqs. (7.2) and (7.3) can be different from the number of unknown parameters. Generally, the coincidence of these two numbers signals that the chosen basic set gives a reasonable description of the dynamics.

It is worth noting that by means of Eqs. (7.2) is often possible to close one sector (i.e., fermionic, spin, charge, pair, ...) at a time without resorting to the opening of all or many of them simultaneously. Obviously, this occurrence enormously facilitates the calculations.

8. THE DYSON EQUATION

The generalized Dyson equation (3.9) is an exact equation and permits, in principle, once the normalization matrix $I(\mathbf{i},\mathbf{j})$, the m-matrix $m(\mathbf{i},\mathbf{j})$ and the propagator $B(i,j)$ are known, in the framework of the self-consistent scheme outlined in Section 7, the calculation of the various Green's functions. However, for most of the physical systems of interest the calculation of the propagator $B(i,j)$ is a very difficult task and some approximations are needed. Various approximate schemes have been proposed.

The simplest approximation is based on completely neglecting the dynamical part $\Sigma(\mathbf{k},\omega)$. This approximation is largely used in the literature [2-5, 18-21, 23-29, 50, 52-58] and is called pole approximation. This approximation consists in retaining that one can neglect finite life-time effects paying attention to the choice of a proper extended operatorial basis, with respect to which the self-energy corrections have a small total weight. Indeed, the total weight of the corrections is bounded by the thermal average involving the residual source $\delta J(i)$. It is worth noting [35] that the n-pole structure of the various GF corresponds to a Dyson-like equation

$$G_{ab}^Q(\mathbf{k},\omega) = \frac{I_{ab}(\mathbf{k})1}{\omega - \Sigma_{ab}^Q(\mathbf{k},\omega)} \quad (8.1)$$

where the self-energy components $\Sigma_{ab}^Q(\mathbf{k},\omega)$ have a (n-1)-pole structure. A theorem concerning the conservation of the spectral moments $M^{(p)}(\mathbf{k}) = F.T.\left\langle\left[(i\partial/\partial t)^p \psi(\mathbf{i},t), \psi^\dagger(\mathbf{j},t)\right]_\eta\right\rangle$ can be assessed [35].

Theorem: Consider a Hamiltonian H and choose a set of composite fields $\{\psi_l, l = 1, \cdots n\}$. If the subset $\{\psi_l, l = 1, \cdots n-1\}$ is chosen so that

$$i\frac{\partial}{\partial t}\psi_l(i) = [\psi_l(i), H] = \sum_{p=1}^{l+1} \gamma_{lp}(-i\nabla)\psi_p(i) \qquad \text{[for } 1 \leq l \leq n-1] \quad (8.2)$$

then the first $2(n-l+1)$ spectral moments for the field $\psi_l(i)$ ($1 \leq l \leq n-1$) are conserved.

In other words, the conservation of the first $2(n-l+1)$ spectral moments is automatically assured if we construct a multiplet whereby, at any stage, the sources rule what should enter as a new operator.

As a corollary, this theorem shows that the n-pole approximation is equivalent to the spectral density approach [4, 5] when the specific choice (8.2) for the basis is considered. However, it is important to remark that the choice (8.2) suffers from severe limitations. For several systems, for example for a multi-orbital model, a basis diagonalizing the atomic problem could be more appropriate than the one coming from the equations of motion [23, 34]. In some other cases, by choosing the appropriate field it is possible to catch the low-energy physics of the system [59]. This is unfeasible through a finite sum of spectral moments as we would need an increasingly large number of them to describe lower and lower energy scales.

In order to go beyond the n-pole approximation one needs to take into account self-energy corrections by developing some methods to calculate the effects of $\Sigma(\mathbf{k}, \omega)$. Various approximate schemes have been proposed. We mention some of them.

Born approximation

In the self-consistent Born approximation (SCBA), or non-crossing approximation, the many-particle Green's functions, appearing in the expression of $\Sigma(\mathbf{k}, \omega)$ [see (3.11)], are calculated by assuming that the fermionic and bosonic modes propagate independently. In order to illustrate the approximation, let us consider the case where the basic set $\{\psi(i)\}$ is of a fermionic type. Then, typically we have to calculate GF of the form

$$H^R(i,j) = <R[B(i)F(i)F^\dagger(j)B^\dagger(j)]> \qquad (8.3)$$

where $F(i)$ and $B(i)$ are fermionic and bosonic fields, respectively. By means of (4.12) we can write

$$H^R(\mathbf{k},\omega) = -\frac{1}{\pi}\int_{-\infty}^{+\infty} d\omega' \frac{1}{\omega-\omega'+i\varepsilon} \coth\frac{\beta\omega'}{2} \operatorname{Im}[H^c(\mathbf{k},\omega')] \qquad (8.4)$$

where $H^c(i,j) = <T[B(i)F(i)F^\dagger(j)B^\dagger(j)]>$ is the causal function. In the SCBA we approximate

$$H^c(i,j) \approx f^c(i,j)b^c(i,j) \qquad (8.5)$$

where $f^c(i,j) = <T[F(i)F^\dagger(j)]>$ and $b^c(i,j) = <T[B(i)B^\dagger(j)]>$. Approximation (8.5) has been used in many works (as an example see 60-62). By assuming that the system is ergodic we can use the spectral representation to obtain

$$H^R(\mathbf{k},\omega) = \frac{1}{\pi}\int_{-\infty}^{+\infty} d\omega' \frac{1}{\omega - \omega' + i\delta} \frac{a^d}{(2\pi)^{d+1}} \int_{\Omega_B} d^d p\, d\Omega \operatorname{Im}[f^R(\mathbf{p},\Omega)] \\ \operatorname{Im}[b^R(\mathbf{k}-\mathbf{p}, \omega'-\Omega)][\tanh\frac{\beta\Omega}{2} + \coth\frac{\beta(\omega'-\Omega)}{2}] \qquad (8.6)$$

Two-site resolvent approach

In this scheme [26, 27] the dynamical part $\Sigma(\mathbf{k},\omega)$ of the self-energy is estimated by a two-site approximation in combined use with the resolvent method [6]. Let us approximate the higher order propagator as

$$B^Q(\mathbf{k},\omega) = F.T. < Q[\delta J(i)\delta J^\dagger(j)] > \approx B_0^Q(\omega) + \alpha(\mathbf{k})B_1^Q(\omega) \qquad (8.7)$$

where $B_0^Q(\omega)$ is related to level transitions on equal site, while $B_1^Q(\omega)$ is related to transitions across the two sites. The Green's function (3.8) takes the form

$$G^Q(\mathbf{k},\omega) = \frac{1}{\omega - \varepsilon(\mathbf{k}) + t^2 V(\omega)\alpha(\mathbf{k})} I(\mathbf{k}) \qquad (8.8)$$

where $V(\omega)$ has to be calculated from the definition (3.11) by making use of (8.7). This approximation has been applied to the study of the t-J [26] and Hubbard [27, 63] models. It has been shown that the approach produces most of the features seen in the numerical simulation as well as the features of spectral distributions near the metal-insulator transition.

9. CONCLUSIONS

I have illustrated a formalism for the study of highly correlated electronic systems, based on two main concepts: propagators of composite operators as building blocks of a perturbation calculation; use of algebra constraints to fix the representation of the GF in order to maintain the algebraic and symmetry properties. The outline of the method can be so schematized:

(i) Given a certain Hamiltonian expressed in terms of electronic fields, one chooses a set $\{\psi(i)\}$ of composite operators.

(ii) A generalized Dyson equation is derived

$$G(\mathbf{k},\omega) = G^{(0)}(\mathbf{k},\omega) + G^{(0)}(\mathbf{k},\omega)\Sigma(\mathbf{k},\omega)G(\mathbf{k},\omega)$$

where $G(\mathbf{k},\omega)$ is the complete GF and $G^{(0)}(\mathbf{k},\omega)$ is the "free" propagator obtained by linearizing the dynamics of the composite fields through a projection on the basis itself.

(iii) The functional dependence of $G^{(0)}(\mathbf{k},\omega)$ in terms of internal parameters (ZFC and correlators) is determined.

(iv) The internal parameters are determined by a set of self-consistent equations which restore the algebra constraints and the symmetry properties of the Hamiltonian.

(v) An approximation is chosen for the determination of the dynamical self-energy $\Sigma(\mathbf{k},\omega)$.

During the last years this formulation has been applied to the study of several systems: Hubbard, t-t'-U, t-J, p-d, double exchange, Kondo, Anderson, Heisenberg models. A systematic comparison with the results of numerical simulation has been carried out. The interested reader may refer to the works cited in the bibliography and references therein.

ACKNOWLEDGMENTS

This manuscript summarizes the research work we have done at the University of Salerno in the last eighth years. During this period, the collaborations with Hideki Matsumoto and Adolfo Avella have been of essential importance. They have brought decisive and vital contributions and I take this opportunity to express to them my deepest gratitude. I also wish to mention the numerous graduate students and post-doc fellows that during these years took part in various phases of the work giving significative contributions: A.M. Allega, M. Bak, T. Di Matteo, S.-S. Feng, V. Fiorentino, S. Krivenko, S. Marra, R. Münzner, S. Odashima, V. Oudovenko, N. Perkins, T. Saikawa, M.M. Sanchez, V. Turkowski, D. Villani, E. Zasinas. I also wish to acknowledge several international collaborations: A.F. Barabanov, F.D. Buzatu, R. Hayn, N.M. Plakida, L. Siurakshina, R. Sridhar, M. Tachiki, V.Yu. Yushankhai.

REFERENCES

1. Anderson, P. W., *Science* **235**, 1196 (1987).
2. Fulde, P., *Electron Correlations in Molecules and Solids*. Springer-Verlag, Berlin Heidelberg New York, 3rd edn,. 1995.
3. Hubbard, J., *Proc. Roy. Soc. A* **276**, 238 (1963); ibid. A **277**, 237 (1964); ibid. A **281**, 401 (1964).
4. Kalashnikov, O. K., and Fradkin, E. S., *Sov. Phys. JETP* **28**, 317 (1969).
5. Nolting, W., *Z. Phys.* **255**, 25 (1972).
6. Kuramoto, Y., *Z. Phys. B* **53**, 37 (1983).
7. Grewe, N., *Z. Phys. B* **53**, 271 (1983).
8. Pruschke, T., *Z. Phys. B* **81**, 319 (1990).
9. Barnes, S. E., *J. Phys. F* **6**, 1375 (1976).
10. Coleman, S. P., *Phys. Rev. B* **29**, 3035 (1984).
11. Kotliar, G., and Ruckenstein, A. E., *Phys. Rev. Lett.* **57**, 1362 (1986).
12. Metzner, W., and Vollhardt, D., *Phys. Rev. Lett.* **62**, 325 (1989).
13. Georges, A., and Kotliar, G., *Phys. Rev. B* **45**, 6479 (1992).
14. Georges, A., and Krauth, W., *Phys. Rev. B* **48**, 7167 (1993).
15. Rozenberg, M. J., and Kotliar, G., and X.Y. Zhang, Phys. Rev. B **49**, 10181 (1994).
16. Georges, A., Kotliar, G., Krauth, W., and Rozenberg, M. J., *Rev. Mod. Phys.* **68**, 13 (1996).
17. Zwanzig, R., in *Lectures in Theoretical Physics*, Interscience, New York (1961).
18. Mori, H., *Progr. Theor. Phys.* **33**, 423 (1965); ibid. **34**, 399 (1965).
19. Becker, K. W., Brenig, W., and Fulde, P., *Z. Phys. B* **81**, 165 (1990).
20. Fedro, A. J., Zhou, Y., Leung, T. C., Harmon, B. N., and Sinha, S. K., *Phys. Rev. B* **46**, 14785 (1992).
21. Mehlig, B., Eskes, H., Hayn, R., and Meinders, M. B. J., *Phys. Rev. B* **52**, 2463 (1995).
22. Onoda, S. and Imada, M. *J. Phys. Soc. Jap.*, **70**, 3398 (2001); *J. Phys.Chem. Solids* **63**, 2225 (2002); *Phys. Rev. B* **67**, 161102 (2003).
23. Ishihara, S, Matsumoto, H., Odashima, S., Tachiki, M., and Mancini, F., *Phys. Rev. B* **49**, 1350 (1994).
24. Mancini, F., Marra, S., Allega, A. M., and Matsumoto, H., *Physica C* **235-240**, 2253 (1994).
25. Mancini, F., Marra, S., and Matsumoto, H., *Physica C* **244**, 49 (1995), ibid. **250**, 184 (1995); ibid. **252**, 361 (1995).
26. Matsumoto, H., Saikawa, T., and Mancini, F., *Phys. Rev. B* **54**, 14445 (1996).
27. Matsumoto, H., and Mancini, F. *Phys. Rev. B* **55**, 2095 (1997).
28. Avella, A., Mancini, F., Villani, D., Siurakshina, L., and Yushankhai, V. Y., *Int. J. Mod. Phys. B* **12**, 81(1998).
29. Avella, A. and Mancini, F., Eur. Phys. J. B (in press) - cond-mat/0006377
30. Dagotto, E., Rev. Mod. Phys. **66**, 763 (1994), and references therein.
31. Bogolubov, N. N., *J. Phys. USSR* **11**, 23 (1947).
32. Dancoff, S. M., *Phys. Rev.* **78**, 382 (1950).
33. Umezawa, H., *Acta Phys. Hung.* **19**, 9 (1965); Umezawa, H., *Nuovo Cimento* **38**, 1415 (1965); H. Umezawa, *Suppl. of Progr. Theor. Phys.* **37** & **38**, 585 (1966); Leplae, L., Mancini, F., and Umezawa, H., *Physics Reports* **10** C, 151 (1974).
34. Villani, D., Lange, E., Avella, A., and Kotliar, G., *Phys. Rev. Lett.* **85**, 804 (2000); Fiorentino, V., Mancini, F., Zasinas, E., and Barabanov, A., *Phys. Rev. B* **64**, 214515 (2001); Mancini, F., Perkins, N., and Plakida, N., *Phys. Lett. A* **284**, 286 (2001); Bak, M., and Mancini, F., *Physica B* **312**, 732 (2002); Avella, A., Mancini, F. and Hayn, R., *Acta Physica Polonica* **34**, 1345 (2003); Bak, M., Avella, A., and Mancini, F., *Phys. Stat. Sol.***236**, 396 (2003)
35. Mancini, F., *Physics Letters A* **249**, 231 (1998).
36. Avella, A., Mancini, F., and Odashima, S., *Physica C* **388**, 76 (2003)
37. Avella, A., Mancini, F., and Odashima, S., to be published in *Journal of Magnetism and Magnetic Materials*
38. Bogolubov, N. N., and Tyablikov, S. V., *Dokl. Akad. Nauk SSSR* **126**, 53 (1959); Zubarev, D. N., *Sov. Phys. Uspekhi* **3**, 320 (1960); Zubarev, D. N., *Non Equilibrium Statistical Thermodynamics* (Consultant Bureau, New York, 1974).

39. Kubo, R., *J. Phys. Soc. Japan* **12**, 570 (1957).
40. Mermin, N. D., and Wagner, H., *Phys. Rev. Lett.* **17**, 1133 (1966).
41. Morita, T., and Katsura, S., *J. Phys. C* **2**, 1030 (1969); Kwork, P. C., and Schultz, T. D., *J. Phys. C* **2**, 1196 (1969).
42. Khintchin, A. I., *Mathematical Foundations of Statistical Mechanics*, Dover Publ. Inc. (1949).
43. Suzuki, M., *Physica* **51**, 277 (1971).
44. Stevens, K. W. H., and Toombs, G. A., *Proc. Phys. Soc.* **85**, 1307 (1965).
45. Callen, H., Swendsen, R. H., and Tahir-Kheli, R., *Phys. Lett. A* **25**, 505 (1967).
46. Fernandez, J. F., and Gersch, H. A., *Proc. Phys. Soc.* **91**, 505 (1967).
47. Ramos, J. G., and Gomes, A. A., *Il Nuovo Cimento* **3** A, 441 (1971).
48. Huber, D. L., *Physica* **87** A, 199 (1977).
49. Aksionov, V., Konvent, H., and Schreiber, J., *Phys. Stat. Sol. (b)* **88**, K43 (1978); Aksionov, V., and Schreiber, J., *Phys. Lett. A* **69**, 56 (1978); Aksionov, V., Bobeth, M., Plakida, N., and Schreiber, J., *Phys. C* **20**, 375 (1987).
50. Roth, L. M., *Phys. Rev.* **184**, 451 (1969).
51. Ward, J. C., *Phys. Rev.* **78**, 182 (1950); Takahashi, Y., *Nuovo Cimento* **6**, 370 (1957).
52. Plakida, N. M., Yushankhai, V. Y., and Stasyuk, I. V., *Physica C* **162-164**, 787 (1989).
53. Rowe, D. J., *Rev. Mod. Phys.* **40**, 153 (1968).
54. Beenen, J., and Edwards, D.M., *Phys. Rev. B* **52**, 13636 (1995).
55. Geipel, G., and Nolting, W., *Phys. Rev. B* **38**, 2608 (1988).
56. Tserkovnikov, Y. A., *Teor. Mat. Fiz.* **49**, 219 (1981); ibid. **50**, 261 (1981).
57. Kruger, P., and Schuck, P., *Eur. Phys. Lett.* **27**, 395 (1994).
58. Kondo, J., and Yamaji, K., *Progr. Theor. Phys.* **47**, 807 (1972).
59. See papers 1 and 5 in Ref. 34.
60. Prelovsek, P., *Z. Phys. B* **103**, 363 (1997).
61. Plakida, N. M., and Oudovenko, V. S., *Phys. Rev. B* **59**, 11949 (1999).
62. Plakida, N. M., Anton, L., Adam, S., and Adam, Gh., cond-mat/0104234.
63. Stanescu, T. D., Martin, I., and Phillips, P., *Phys. Rev. B* **62**, 4300 (2000); Stanescu, T. D., and Phillips, P., *Phys. Rev. B* **64**, 235117 (2001).

Self-energy corrections within the Composite Operator Method

Adolfo Avella

Dipartimento di Fisica "E.R. Caianiello" - Unità di Ricerca INFM di Salerno
Università degli Studi di Salerno, I-84081 Baronissi (SA), Italy

Abstract. The possibility to match the projection technique at the basis of the Composite Operator Method and the determination of the residual self-energy through the mode-coupling approximation is explored. The natural irreducibility given by the projection procedure is fully preserved and exploited. The expressions of the higher order propagators, within the mode-coupling approximation, in terms of the basic fermionic and bosonic propagators is given. The self-energy for the spin and charge propagators is computed in terms of convolutions of electronic propagators.

1. INTRODUCTION

In the last decade, the group of Professor Ferdinando Mancini at the University of Salerno, in tight collaboration with the one of Professor Hideki Matsumoto (Tsukuba University), has been developing a formulation to study highly interacting systems: the Composite Operator Method [1, 2]. This latter has two main features.

First, strong correlations, arising by complex interactions among charge, spin and orbital degrees of freedom, completely destroy the individuality of the electrons. New composite particles (collective modes, extended excitations, resonances, ...) appear in the system and the correct description of the dynamics requires their individuation (i.e., the definition of the corresponding composite operators) and use as a starting point for any approximate treatment [2]. This kind of conviction may be traced back to the work of Bogoliubov [3] and later to that of Dancoff [4]. The work of Zwanzig [5], Mori [6] and Umezawa [7] has to be mentioned. Closely related to this work is that of Hubbard [8, 9, 10], Rowe [11], Roth [12] and Tserkovnikov [13, 14]. The slave boson method [15, 16, 17], the spectral density approach [18, 19] are also along similar lines.

Second, the use of composite operators requires a complex procedure to fix the representation in which the Green's functions are defined. In this case, it is not sufficient to self-consistently fix the chemical potential, as we usually do for weak- and not- interacting systems, but more parameters are needed and naturally appear. The determination of these latter should be seen as an opportunity to fix the representation and, at the same time, impose relevant symmetry constraints to the solution [2]. In particular, the constraints derived by the local algebra satisfied by the composite operators revealed as very powerful tools in order to obtain the solution of many complex strongly correlated systems: Kondo and Anderson models, Hubbard, t-J and p-d models, double-exchange model, ... [20, 21, 22, 23, 24, 25, 26, 27, 28, 29, 30, 31, 32].

When the number of degrees of freedom of the system is extremely large (or just infinite), the application of the Composite Operator Method requires the choice of an approximation treatment for the otherwise intractable hierarchy of equations of motion obeyed by the propagators of the "ad hoc" chosen composite operators. In the great majority of cases, we have used the N-pole approximation having in mind that the chosen operators described well-defined medium/high-energy excitations and that the finite lifetime effects were marginal with respect to the investigated properties of the system. On the other hand, the description of low-energy excitations, which naturally requires to take into account finite life-time effects, calls for an approximate treatment which models the self-energy not only in terms of coherent excitations, but that also gives an incoherent part. This latter will take into account the natural inclination of collective bosonic (spin, charge, orbital, pair) low-energy excitations to decompose in terms of electron(hole)-electron(hole) excitations. In this short manuscript, we indicate a way to conciliate the use of the projection technique (which results in a inherent polar structure) with the computation of the residual self-energy (generated by the source counter-terms) by means of a diagrammatic approximation. The two-dimensional Hubbard model will be used as example in order to describe the procedure in detail.

2. THE TWO-DIMENSIONAL HUBBARD MODEL

The Hamiltonian of the two-dimensional Hubbard model [8, 9, 10, 33] reads as

$$H = \sum_{ij} (t_{ij} - \mu \delta_{ij}) c^\dagger(i) \cdot c(j) + U \sum_i n_\uparrow(i) n_\downarrow(i) \tag{1}$$

where μ is the chemical potential, $c^\dagger(i) = \left(c_\uparrow^\dagger(i), c_\downarrow^\dagger(i)\right)$ is the electronic creation operator at the site $\mathbf{i}$ in spinorial notation, $i = (\mathbf{i}, t)$, U is the on-site Coulomb interaction strength, $n_\sigma(i) = c_\sigma^\dagger(i) c_\sigma(i)$ is the charge density operator for spin σ at the site $\mathbf{i}$ and $t_{ij} = -4t\alpha_{ij}$, where t is the hopping integral, $\alpha_{ij} = \frac{1}{4}\delta_{\langle ij \rangle}$ is the projection operator on the nearest-neighbor sites,

$$\alpha(\mathbf{k}) = \mathscr{F}[\alpha_{ij}] = \frac{1}{2} \sum_{n=1}^{2} \cos(k_n a_n) \tag{2}$$

and a_n is the lattice constant along the n-th direction (it will be set to 1). We will extensively use the following definition ($\Psi(i)$ is a generic operator)

$$\Psi^\alpha(\mathbf{i}, t) = \sum_{\mathbf{j}} \alpha_{ij} \Psi(\mathbf{j}, t) \tag{3}$$

3. THE EQUATIONS OF MOTION AND THE BASIS

After the Hubbard Hamiltonian (1), the electronic field $c(i)$ satisfies the following equation of motion

$$i\frac{\partial}{\partial t} c(i) = -\mu c(i) - 4t c^\alpha(i) + U \eta(i) \tag{4}$$

with $\eta(i) = n(i)c(i)$. According to this, we can decompose $c(i)$ as $c(i) = \xi(i) + \eta(i)$, where $\xi(i) = [1-n(i)]c(i)$ and $\eta(i)$ are the Hubbard operators and describe the transitions $n=0 \leftrightarrow n=1$ and $n=1 \leftrightarrow n=2$, respectively. Moreover, they are the local eigenoperators of the atomic term of the Hubbard Hamiltonian and describe the original electron operator dressed by the on-site charge and spin excitations. They satisfy the following equations of motion

$$i\frac{\partial}{\partial t}\xi(i) = \varepsilon_\xi \xi(i) - 4t\left(1-\frac{n}{2}\right)c^\alpha(i) - 4t\pi(i) \tag{5a}$$

$$i\frac{\partial}{\partial t}\eta(i) = \varepsilon_\eta \eta(i) - 4t\frac{n}{2}c^\alpha(i) + 4t\pi(i) \tag{5b}$$

with $\pi(i) = \frac{1}{2}\sigma^\mu \delta n_\mu(i) c^\alpha(i) + c^{\dagger\alpha}(i) \cdot c(i) \otimes c(i)$. $\varepsilon_\xi = -\mu$ and $\varepsilon_\eta = U - \mu$. $n_\mu(i) = c^\dagger(i)\sigma_\mu c(i)$ is the total charge ($\mu=0$) and spin ($\mu=1,2,3$) density operator at the site $\mathbf{i}$ and $\delta n_\mu(i) = n_\mu(i) - \langle n_\mu(i)\rangle$. $\sigma_\mu = (1,\vec{\sigma})$, $\sigma^\mu = (-1,\vec{\sigma})$ and $\vec{\sigma}$ are the Pauli matrices. In the paramagnetic phase, we have $\langle n_\mu(i)\rangle = n$.

We choose $\xi(i)$ and $\eta(i)$ as components of the basic field [1]

$$\psi(i) = \begin{pmatrix} \xi(i) \\ \eta(i) \end{pmatrix} \tag{6}$$

Under particle-hole symmetry, we have

$$\xi(i) \mapsto (-)^{\mathbf{i}} \eta^\dagger(i) \tag{7a}$$

$$\eta(i) \mapsto (-)^{\mathbf{i}} \xi^\dagger(i) \tag{7b}$$

$$\pi(i) \mapsto -(-)^{\mathbf{i}} \pi^\dagger(i) \tag{7c}$$

4. THE ENERGY AND THE NORMALIZATION MATRICES

We have the following entries for the normalization matrix [1] $I = \mathcal{F}\langle\{\psi(\mathbf{i}),\psi^\dagger(\mathbf{j})\}\rangle$

$$I_{11} = \mathcal{F}\langle\{\xi(\mathbf{i}),\xi^\dagger(\mathbf{j})\}\rangle = 1 - \frac{n}{2} \tag{8a}$$

$$I_{12} = I_{21} = \mathcal{F}\langle\{\xi(\mathbf{i}),\eta^\dagger(\mathbf{j})\}\rangle = 0 \tag{8b}$$

$$I_{22} = \mathcal{F}\langle\{\eta(\mathbf{i}),\eta^\dagger(\mathbf{j})\}\rangle = \frac{n}{2} \tag{8c}$$

where $\langle\ldots\rangle$ indicates the thermal average in the grand-canonical ensemble and $\mathcal{F}$ stands for the Fourier transform.

The equation of motion of $\psi(i)$ (5) can be rewritten, by splitting the current $J(i)$ into two terms [1]: the linear and the residual ones, as

$$i\frac{\partial}{\partial t}\psi(\mathbf{i},t) = J(\mathbf{i},t) = \sum_\mathbf{j} \varepsilon(\mathbf{i}-\mathbf{j})\psi(\mathbf{j},t) + \delta J(\mathbf{i},t) \tag{9}$$

where the residual current $\delta J(i)$ is defined through the condition

$$\left\langle \left\{ \delta J(i), \psi^\dagger(j) \right\} \right\rangle = 0 \tag{10}$$

which gives for the energy matrix [1] $\varepsilon(i-j)$

$$\varepsilon(i-j) = m(i-j) I^{-1} \tag{11}$$

The matrix $m(\mathbf{k}) = \mathscr{F}\left\langle \left\{ i\frac{\partial}{\partial t}\psi(i), \psi^\dagger(j) \right\} \right\rangle$ has the following entries [1]

$$m_{11}(\mathbf{k}) = \varepsilon_\xi I_{11} - 4t\left[\alpha(\mathbf{k})\left(I_{11}^2 + p\right) + \Delta\right] \tag{12a}$$
$$m_{12}(\mathbf{k}) = m_{21}(\mathbf{k}) = -4t\left[\alpha(\mathbf{k})\left(I_{11} I_{22} - p\right) - \Delta\right] \tag{12b}$$
$$m_{22}(\mathbf{k}) = \varepsilon_\eta I_{22} - 4t\left[\alpha(\mathbf{k})\left(I_{22}^2 + p\right) + \Delta\right] \tag{12c}$$

where

$$\Delta = \left\langle \xi^\alpha(i) \xi^\dagger(i) \right\rangle - \left\langle \eta^\alpha(i) \eta^\dagger(i) \right\rangle \tag{13a}$$
$$p = \frac{1}{4}\left\langle \delta n_\mu^\alpha(i) \delta n_\mu(i) \right\rangle - \left\langle [c_\uparrow(i) c_\downarrow(i)]^\alpha c_\downarrow^\dagger(i) c_\uparrow^\dagger(i) \right\rangle \tag{13b}$$

Then,

$$\varepsilon_{11}(\mathbf{k}) = m_{11}(\mathbf{k}) I_{11}^{-1} = \varepsilon_\xi - 4t\left[\alpha(\mathbf{k})\left(I_{11}^2 + p\right) + \Delta\right] I_{11}^{-1} \tag{14a}$$
$$\varepsilon_{12}(\mathbf{k}) = m_{12}(\mathbf{k}) I_{22}^{-1} = -4t\left[\alpha(\mathbf{k})\left(I_{11} I_{22} - p\right) - \Delta\right] I_{22}^{-1} \tag{14b}$$
$$\varepsilon_{21}(\mathbf{k}) = m_{21}(\mathbf{k}) I_{11}^{-1} = -4t\left[\alpha(\mathbf{k})\left(I_{11} I_{22} - p\right) - \Delta\right] I_{11}^{-1} \tag{14c}$$
$$\varepsilon_{22}(\mathbf{k}) = m_{22}(\mathbf{k}) I_{22}^{-1} = \varepsilon_\eta - 4t\left[\alpha(\mathbf{k})\left(I_{22}^2 + p\right) + \Delta\right] I_{22}^{-1} \tag{14d}$$

5. THE GREEN'S FUNCTION

Let us now compute the thermal retarded Green's function $G(\mathbf{k}, \omega) = \mathscr{F}\left\langle \mathscr{R}\left[\psi(i) \psi^\dagger(j)\right] \right\rangle$ that, after the equation of motion of $\psi(i)$ (9), satisfies the following equation [2]

$$\omega G(\mathbf{k}, \omega) = I + \varepsilon(\mathbf{k}) G(\mathbf{k}, \omega) + A(\mathbf{k}, \omega) \tag{15}$$

where $A(\mathbf{k}, \omega) = \mathscr{F}\left\langle \mathscr{R}\left[\delta J(i) \psi^\dagger(j)\right] \right\rangle$, $\mathscr{R}[\ldots]$ stands for the usual retarded operator. If we neglect $\delta J(i)$, the solution of Eq. 15 is the following

$$G_0(\mathbf{k}, \omega) = \frac{1}{\omega - \varepsilon(\mathbf{k}) + i\delta} I = \sum_{l=1}^{2} \frac{\sigma^{(l)}(\mathbf{k})}{\omega - E_l(\mathbf{k}) + i\delta} \tag{16}$$

The energies $E_l(\mathbf{k})$ are the eigenvalues of the energy matrix $\varepsilon(\mathbf{k})$ and the spectral weights $\sigma^{(l)}(\mathbf{k})$ can be computed by means of the following expression

$$\sigma_{ab}^{(l)}(\mathbf{k}) = \Lambda_{al}(\mathbf{k}) \sum_c \Lambda_{lc}^{-1}(\mathbf{k}) I_{cb} \tag{17}$$

where the matrix $\Lambda(\mathbf{k})$ has the eigenvectors of the energy matrix $\varepsilon(\mathbf{k})$ as columns.
Now, we also have

$$\omega A(\mathbf{k},\omega) = A(\mathbf{k},\omega)\varepsilon^T(\mathbf{k}) + B(\mathbf{k},\omega) \tag{18}$$

where $B(\mathbf{k},\omega) = \mathscr{F}\langle\mathscr{R}[\delta J(i)\delta J^\dagger(j)]\rangle$.
Then, we have

$$A(\mathbf{k},\omega) = B(\mathbf{k},\omega) I^{-1} G_0(\mathbf{k},\omega) \tag{19}$$

and

$$G(\mathbf{k},\omega) = G_0(\mathbf{k},\omega) + G_0(\mathbf{k},\omega) T(\mathbf{k},\omega) G_0(\mathbf{k},\omega) \tag{20}$$

where $T(\mathbf{k},\omega) = I^{-1} B(\mathbf{k},\omega) I^{-1}$ is the scattering matrix.
To introduce the self-energy $\Sigma(\mathbf{k},\omega)$ we need to define

$$A(\mathbf{k},\omega) = \Sigma(\mathbf{k},\omega) G(\mathbf{k},\omega) \tag{21}$$

then

$$G(\mathbf{k},\omega) = \frac{1}{\omega - \varepsilon(\mathbf{k}) - \Sigma(\mathbf{k},\omega) + i\delta} I \tag{22}$$

and

$$\Sigma(\mathbf{k},\omega) = \frac{1}{1 + B(\mathbf{k},\omega) I^{-1} G_0(\mathbf{k},\omega) I^{-1}} B(\mathbf{k},\omega) I^{-1} = B^{irr}(\mathbf{k},\omega) I^{-1} \tag{23}$$

where we have also introduced the irreducible part of $B(\mathbf{k},\omega)$.

6. THE SELF-ENERGY

We have to compute $B^{irr}(\mathbf{k},\omega) = \mathscr{F}\langle\mathscr{R}[\delta J(i)\delta J^\dagger(j)]\rangle^{irr}$ in order to obtain the self-energy $\Sigma(\mathbf{k},\omega) = B^{irr}(\mathbf{k},\omega) I^{-1}$ [2]. According to the definition of $\delta J(i)$ as residual counter-term of the source

$$\delta J(i) = J(i) - \sum_{\mathbf{j}}\varepsilon(\mathbf{i}-\mathbf{j})\psi(\mathbf{j},t) = 4t\begin{pmatrix}1\\-1\end{pmatrix}\Delta J(i) \tag{24}$$

where

$$\Delta J(i) = \Delta[I_{11}^{-1}\xi(i) - I_{22}^{-1}\eta(i)] + p[I_{11}^{-1}\xi^\alpha(i) - I_{22}^{-1}\eta^\alpha(i)] - \pi(i) \tag{25}$$

we can consider the irreducibility requirement already satisfied by the full propagator B: the projection procedure systematically eliminate terms proportional to G_0. Then, we have

$$B^{irr}(\mathbf{k},\omega) = 16t^2(1-\sigma_1)b(\mathbf{k},\omega) \tag{26}$$

with

$$\begin{aligned}b(\mathbf{k},\omega) &= \mathscr{F}\langle\mathscr{R}[\Delta J(i)\Delta J^\dagger(j)]\rangle\\ &= [\Delta + \alpha(\mathbf{k})p]^2[I_{11}^{-2}G_{11}(\mathbf{k},\omega) - 2I_{11}^{-1}I_{22}^{-1}G_{12}(\mathbf{k},\omega) + I_{22}^{-2}G_{22}(\mathbf{k},\omega)]\\ &\quad - 2[\Delta + \alpha(\mathbf{k})p][I_{11}^{-1}c(\mathbf{k},\omega) - I_{22}^{-1}d(\mathbf{k},\omega)] + g(\mathbf{k},\omega)\end{aligned} \tag{27}$$

where

$$c(\mathbf{k}, \omega) = \mathscr{F}\left\langle \mathscr{R}\left[\xi(i)\pi^\dagger(j)\right]\right\rangle \tag{28a}$$

$$d(\mathbf{k}, \omega) = \mathscr{F}\left\langle \mathscr{R}\left[\eta(i)\pi^\dagger(j)\right]\right\rangle \tag{28b}$$

$$g(\mathbf{k}, \omega) = \mathscr{F}\left\langle \mathscr{R}\left[\pi(i)\pi^\dagger(j)\right]\right\rangle \tag{28c}$$

We could compute $c(\mathbf{k}, \omega)$ and $d(\mathbf{k}, \omega)$ in an approximate way by truncating the equation of motion of $\psi(i)$ (9), i.e., neglecting $\delta J(i)$. We would get

$$c(\mathbf{k}, \omega) = \frac{I_{\xi\pi}(\mathbf{k})}{I_{11}} G_{110}(\mathbf{k}, \omega) + \frac{I_{\eta\pi}(\mathbf{k})}{I_{22}} G_{120}(\mathbf{k}, \omega) \tag{29a}$$

$$d(\mathbf{k}, \omega) = \frac{I_{\xi\pi}(\mathbf{k})}{I_{11}} G_{120}(\mathbf{k}, \omega) + \frac{I_{\eta\pi}(\mathbf{k})}{I_{22}} G_{220}(\mathbf{k}, \omega) \tag{29b}$$

where

$$I_{\xi\pi}(\mathbf{k}) = \mathscr{F}\left\langle \left\{\xi(i), \pi^\dagger(j)\right\}\right\rangle = \Delta + \alpha(\mathbf{k})(p - I_{11}I_{22}) \tag{30a}$$

$$I_{\eta\pi}(\mathbf{k}) = \mathscr{F}\left\langle \left\{\eta(i), \pi^\dagger(j)\right\}\right\rangle = -\Delta - \alpha(\mathbf{k})(p + I_{22}^2) \tag{30b}$$

In this case, we have no contribution from $c(\mathbf{k}, \omega)$ and $d(\mathbf{k}, \omega)$ as they result linear combinations of entries of $G_0(\mathbf{k}, \omega)$ and have no irreducible component.

Another way to compute $c(\mathbf{k}, \omega)$ and $d(\mathbf{k}, \omega)$ in an approximate way is through the mode coupling approximation [34]. This latter treats propagators of composite particles decoupling the bosonic and fermionic modes. For instance, we can give the general formulation for basic fields ψ made up of:

(a) two fermionic modes: $\psi = f_1^\dagger(i)\sigma_\mu f_2(i)$ and $\varphi = f_3^\dagger(i)\sigma_\mu f_4(i)$

$$\begin{aligned}\mathscr{F}\left\langle \mathscr{T}[\psi(i)\varphi(j)]\right\rangle &= \mathscr{F}\left\langle \mathscr{T}\left[f_1^\dagger(i)\sigma_\mu f_2(i) f_3^\dagger(j)\sigma_\mu f_4(j)\right]\right\rangle \\ &\approx -2\pi i \delta(\omega)\delta(\mathbf{k})\left\langle f_1^\dagger(i)\sigma_\mu f_2(i)\right\rangle\left\langle f_3^\dagger(i)\sigma_\mu f_4(i)\right\rangle \\ &- 2\mathscr{D}\left[\mathscr{F}\left\langle \mathscr{T}\left[f_2(i)f_3^\dagger(j)\right]\right\rangle \mathscr{F}\left\langle \mathscr{T}\left[f_4(j)f_1^\dagger(i)\right]\right\rangle\right]\end{aligned} \tag{31a}$$

(b) a fermionic mode and a bosonic one: $\psi = b_1(i)f_1(i)$ and $\varphi = b_2(i)f_2(i)$

$$\begin{aligned}\mathscr{F}\left\langle \mathscr{T}[\psi(i)\varphi(j)]\right\rangle &= \mathscr{F}\left\langle \mathscr{T}\left[b_1(i)f_1(i)f_2^\dagger(j)b_2^\dagger(j)\right]\right\rangle \\ &\approx \mathscr{C}\left[\mathscr{F}\left\langle \mathscr{T}\left[b_1(i)b_2^\dagger(j)\right]\right\rangle \mathscr{F}\left\langle \mathscr{T}\left[f_1(i)f_2^\dagger(j)\right]\right\rangle\right]\end{aligned} \tag{31b}$$

where $\mathcal{T}[\ldots]$ stands for the usual causal operator and $\mathcal{C}$ and $\mathcal{D}$ for the usual convolution operators

$$\mathcal{C}[A(\mathbf{k},\omega)B(\mathbf{k},\omega)] = \frac{1}{(2\pi)^3}\int dq d\varpi A(\mathbf{k}-\mathbf{q},\omega-\varpi)B(\mathbf{q},\varpi) \quad (32a)$$

$$\mathcal{D}[A(\mathbf{k},\omega)B(\mathbf{k},\omega)] = \frac{1}{(2\pi)^3}\int dq d\varpi A(\mathbf{k}+\mathbf{q},\omega+\varpi)B(\mathbf{q},\varpi) \quad (32b)$$

Then, taking into account that $\eta(i) = -\frac{1}{4}\sigma^\mu n_\mu(i) c(i)$ and $\xi(i) = \left[1+\frac{1}{4}\sigma^\mu n_\mu(i)\right]c(i)$ and neglecting the pair propagator, we get

$$c^C(\mathbf{k},\omega) \approx \frac{1}{8}\mathcal{C}\left[S_{\mu 11}(\mathbf{k},\omega)\left[\alpha(\mathbf{k}) G_{cc}^C(\mathbf{k},\omega)\right]\right] \quad (33a)$$

$$d^C(\mathbf{k},\omega) \approx -\frac{1}{8}\mathcal{C}\left[S_{\mu 11}(\mathbf{k},\omega)\left[\alpha(\mathbf{k}) G_{cc}^C(\mathbf{k},\omega)\right]\right] \quad (33b)$$

where

$$G_{cc}(\mathbf{k},\omega) = \sum_{\alpha,\beta=1}^{2} G_{\alpha\beta}(\mathbf{k},\omega) \quad (34)$$

$$S_{\mu 11}(\mathbf{k},\omega) = \mathcal{F}\langle \mathcal{T}[\delta n_\mu(i) \delta n_\mu(j)]\rangle \quad (35)$$

and the superscript ^{C} stays for the causal propagator. We can compute $g(\mathbf{k},\omega)$ in the same approximation and get

$$g^C(\mathbf{k},\omega) \approx \frac{1}{4}\mathcal{C}\left[S_{\mu 11}(\mathbf{k},\omega)\left[\alpha^2(\mathbf{k}) G_{cc}^C(\mathbf{k},\omega)\right]\right] \quad (36)$$

According to this, we have

$$b^C(\mathbf{k},\omega) \approx [\Delta+\alpha(\mathbf{k})p]^2 \left[I_{11}^{-2}G_{11}^C(\mathbf{k},\omega) - 2I_{11}^{-1}I_{22}^{-1}G_{12}^C(\mathbf{k},\omega) + I_{22}^{-2}G_{22}^C(\mathbf{k},\omega)\right]$$
$$+ \frac{1}{4}\mathcal{C}\left[S_{\mu 11}(\mathbf{k},\omega)\left[\alpha^2(\mathbf{k}) G_{cc}^C(\mathbf{k},\omega)\right]\right]$$
$$- \frac{1}{4}I_{11}^{-1}I_{22}^{-1}[\Delta+\alpha(\mathbf{k})p]\mathcal{C}\left[S_{\mu 11}(\mathbf{k},\omega)\left[\alpha(\mathbf{k}) G_{cc}^C(\mathbf{k},\omega)\right]\right] \quad (37)$$

7. THE BOSONIC GREEN'S FUNCTION

We are now left with the problem of computing the bosonic propagator $S_{\mu 11}(\mathbf{k},\omega) = \mathcal{F}\langle \mathcal{T}[\delta n_\mu(i) \delta n_\mu(j)]\rangle$. In order to do that, we need to open the bosonic sector [27]. The equation of motion of $n_\mu(i)$ reads as follows

$$i\frac{\partial}{\partial t}n_\mu(i) = -4t\rho_\mu(i) \quad (38)$$

where $\rho_\mu(i) = c^\dagger(i)\sigma_\mu c^\alpha(i) - c^{\dagger\alpha}(i)\sigma_\mu c(i)$.

According to this, we choose the following basis

$$N_\mu(i) = \begin{pmatrix} \delta n_\mu(i) \\ \rho_\mu(i) \end{pmatrix} \tag{39}$$

and study the following propagator

$$S_\mu(\mathbf{k},\omega) = \mathscr{F}\left\langle \mathscr{T}\left[N_\mu(i)N_\mu^\dagger(j)\right]\right\rangle \tag{40}$$

The equation of motion of $\rho_\mu(i)$ reads as follows

$$i\frac{\partial}{\partial t}\rho_\mu(i) = -4tl_\mu(i) + U\kappa_\mu(i) \tag{41}$$

where

$$l_\mu(i) = c^\dagger(i)\sigma_\mu c^{\alpha^2}(i) + c^{\dagger\alpha^2}(i)\sigma_\mu c(i) - 2c^{\dagger\alpha}(i)\sigma_\mu c^\alpha(i) \tag{42a}$$

$$\kappa_\mu(i) = c^\dagger(i)\sigma_\mu\eta^\alpha(i) - \eta^\dagger(i)\sigma_\mu c^\alpha(i) + \eta^{\dagger\alpha}(i)\sigma_\mu c(i) - c^{\dagger\alpha}(i)\sigma_\mu\eta(i) \tag{42b}$$

Then, the equation of motion of $N_\mu(i)$ could be cast as follows

$$i\frac{\partial}{\partial t}N_\mu(\mathbf{k},t) = \Upsilon_\mu(\mathbf{k},t) = \varepsilon_\mu(\mathbf{k})N_\mu(\mathbf{k},t) + \delta\Upsilon_\mu(\mathbf{k},t) \tag{43}$$

The current expression of $\varepsilon_\mu(\mathbf{k})$ as well as of the related normalization matrix $I_\mu(\mathbf{k}) = \mathscr{F}\left\langle\left[N_\mu(\mathbf{i}),N_\mu^\dagger(\mathbf{j})\right]\right\rangle$ are not reported here for the sake of brevity and can be found in Ref. [27], where a comprehensive and detailed account of the analysis of the bosonic sector of the Hubbard model in the two-pole approximation is given.

In complete analogy with what presented for the fermionic propagator, we have

$$S_\mu(\mathbf{k},\omega) = \frac{1}{\omega - \varepsilon_\mu(\mathbf{k}) - \Sigma_\mu(\mathbf{k},\omega) + i\delta}I_\mu(\mathbf{k}) \tag{44}$$

where $\Sigma_\mu(\mathbf{k},\omega) = B_\mu^{irr}(\mathbf{k},\omega)I_\mu^{-1}(\mathbf{k})$ and $B_\mu(\mathbf{k},\omega) = \mathscr{F}\left\langle\mathscr{T}\left[\delta\Upsilon_\mu(i)\delta\Upsilon_\mu^\dagger(j)\right]\right\rangle$.

According to (43), the residual current $\delta\Upsilon_\mu(i)$ has the following expression

$$\delta\Upsilon_\mu(i) = \begin{pmatrix} 0 \\ 1 \end{pmatrix}\Delta\Upsilon_\mu(i) \tag{45}$$

$$\Delta\Upsilon_\mu(i) = -4tl_\mu(i) + U\kappa_\mu(i) - \sum_\mathbf{j}\varepsilon_{\mu 21}(\mathbf{i}-\mathbf{j})\delta n_\mu(\mathbf{j},t) \tag{46}$$

and the irreducible propagator can be computed as (following the same line of thinking reported for the fermionic sector)

$$B_\mu^{irr}(\mathbf{k},\omega) = \frac{1}{2}(1-\sigma_3)b_\mu(\mathbf{k},\omega) \tag{47}$$

Then, we have the following expression for $b_\mu(\mathbf{k}, \omega)$

$$b_\mu(\mathbf{k}, \omega) = \mathscr{F} \left\langle \mathscr{T} \left[\Delta \Upsilon_\mu(i) \Delta \Upsilon_\mu^\dagger(j) \right] \right\rangle = 16 t^2 \mathscr{F} \left\langle \mathscr{T} \left[l_\mu(i) l_\mu^\dagger(j) \right] \right\rangle$$
$$- 8tU \left\langle \mathscr{T} \left[l_\mu(i) \kappa_\mu^\dagger(j) \right] \right\rangle + U^2 \mathscr{F} \left\langle \mathscr{T} \left[\kappa_\mu(i) \kappa_\mu^\dagger(j) \right] \right\rangle$$
$$+ 8t \varepsilon_{\mu 21}(\mathbf{k}) \mathscr{F} \left\langle \mathscr{T} \left[\delta n_\mu(i) l_\mu^\dagger(j) \right] \right\rangle - 2U \varepsilon_{\mu 21}(\mathbf{k}) \mathscr{F} \left\langle \mathscr{T} \left[\delta n_\mu(i) \kappa_\mu^\dagger(j) \right] \right\rangle$$
$$+ \varepsilon_{\mu 21}^2(\mathbf{k}) \mathscr{F} \left\langle \mathscr{T} \left[\delta n_\mu(i) \delta n_\mu(j) \right] \right\rangle \tag{48}$$

Finally, within the mode-coupling approximation, we have

$$\mathscr{F} \left\langle \mathscr{T} \left[l_\mu(i) l_\mu^\dagger(j) \right] \right\rangle \approx$$
$$- 4\mathscr{D} \left[G_{cc}^C(\mathbf{k}, \omega) \left[\alpha^4(\mathbf{k}) G_{cc}^C(\mathbf{k}, \omega) \right] \right] + 16 \mathscr{D} \left[\left[\alpha(\mathbf{k}) G_{cc}^C(\mathbf{k}, \omega) \right] \left[\alpha^3(\mathbf{k}) G_{cc}^C(\mathbf{k}, \omega) \right] \right]$$
$$- 12 \mathscr{D} \left[\left[\alpha^2(\mathbf{k}) G_{cc}^C(\mathbf{k}, \omega) \right] \left[\alpha^2(\mathbf{k}) G_{cc}^C(\mathbf{k}, \omega) \right] \right] \tag{49a}$$

$$\mathscr{F} \left\langle \mathscr{T} \left[l_\mu(i) \kappa_\mu^\dagger(j) \right] \right\rangle$$
$$\approx 4 \mathscr{D} \left[\left[\alpha^3(\mathbf{k}) G_{cc}^C(\mathbf{k}, \omega) \right] G_{c\eta}^C(\mathbf{k}, \omega) \right] - 12 \mathscr{D} \left[\left[\alpha^2(\mathbf{k}) G_{cc}^C(\mathbf{k}, \omega) \right] \left[\alpha(\mathbf{k}) G_{c\eta}^C(\mathbf{k}, \omega) \right] \right]$$
$$+ 12 \mathscr{D} \left[\left[\alpha(\mathbf{k}) G_{cc}^C(\mathbf{k}, \omega) \right] \left[\alpha^2(\mathbf{k}) G_{c\eta}^C(\mathbf{k}, \omega) \right] \right] - 4 \mathscr{D} \left[G_{cc}^C(\mathbf{k}, \omega) \left[\alpha^3(\mathbf{k}) G_{c\eta}^C(\mathbf{k}, \omega) \right] \right] \tag{49b}$$

$$\mathscr{F} \left\langle \mathscr{T} \left[\kappa_\mu(i) \kappa_\mu^\dagger(j) \right] \right\rangle \approx 8 \mathscr{D} \left[G_{c\eta}^C(\mathbf{k}, \omega) \left[\alpha^2(\mathbf{k}) G_{c\eta}^C(\mathbf{k}, \omega) \right] \right]$$
$$- 8 \mathscr{D} \left[\left[\alpha(\mathbf{k}) G_{c\eta}^C(\mathbf{k}, \omega) \right] \left[\alpha(\mathbf{k}) G_{c\eta}^C(\mathbf{k}, \omega) \right] \right] - 4 \mathscr{D} \left[G_{cc}^C(\mathbf{k}, \omega) \left[\alpha^2(\mathbf{k}) G_{\eta\eta}^C(\mathbf{k}, \omega) \right] \right]$$
$$+ 8 \mathscr{D} \left[\left[\alpha(\mathbf{k}) G_{cc}^C(\mathbf{k}, \omega) \right] \left[\alpha(\mathbf{k}) G_{\eta\eta}^C(\mathbf{k}, \omega) \right] \right] - 4 \mathscr{D} \left[\left[\alpha^2(\mathbf{k}) G_{cc}^C(\mathbf{k}, \omega) \right] G_{\eta\eta}^C(\mathbf{k}, \omega) \right] \tag{49c}$$

$$\mathscr{F} \left\langle \mathscr{T} \left[\delta n_\mu(i) l_\mu^\dagger(j) \right] \right\rangle \approx -4 \mathscr{D} \left[G_{cc}^C(\mathbf{k}, \omega) \left[\alpha^2(\mathbf{k}) G_{cc}^C(\mathbf{k}, \omega) \right] \right]$$
$$+ 4 \mathscr{D} \left[\left[\alpha(\mathbf{k}) G_{cc}^C(\mathbf{k}, \omega) \right] \left[\alpha(\mathbf{k}) G_{cc}^C(\mathbf{k}, \omega) \right] \right] \tag{49d}$$

$$\mathscr{F} \left\langle \mathscr{T} \left[\delta n_\mu(i) \kappa_\mu^\dagger(j) \right] \right\rangle \approx -4 \mathscr{D} \left[G_{cc}^C(\mathbf{k}, \omega) \left[\alpha(\mathbf{k}) G_{c\eta}^C(\mathbf{k}, \omega) \right] \right]$$
$$+ 4 \mathscr{D} \left[\left[\alpha(\mathbf{k}) G_{cc}^C(\mathbf{k}, \omega) \right] G_{c\eta}^C(\mathbf{k}, \omega) \right] \tag{49e}$$

8. THE INTERNAL PARAMETERS

The last problem to be solved in order to get a fully self-consistent scheme is the computation of the internal parameters like p and Δ and those appearing in $\varepsilon_\mu(\mathbf{k})$ [27]. According to the procedure explained in detail in the Introduction, the Composite Operator Method exploits the local algebra constraints [2] in order to accomplish this task (for a detailed analysis see Refs. [1, 27]) and, at the same time, fix the correct representation and impose fundamental symmetry constraints (Pauli principle, particle-hole symmetry, correct hydrodynamic behavior).

ACKNOWLEDGMENTS

This manuscript is dedicated to my master: Professor Ferdinando Mancini. In the last ten years he has been (and I hope he wants to be for much longer!) a guide and a fellow. My best wishes to him on this occasion and many many thanks for all he has done and is still doing for me.

REFERENCES

1. Mancini, F., Marra, S., Allega, A. M., and Matsumoto, H., *Physica C*, **235**, 2253 (1994).
2. Mancini, F., and Avella, A., Equation of motion method for composite field operators (2000), cond-mat/0006377; to be published in Eur. Phys. J. B.
3. Bogoliubov, N., *J. Phys. USSR*, **11**, 23 (1947).
4. Dancoff, S. M., *Phys. Rev.*, **78**, 382 (1950).
5. Zwanzig, R., "Statistical Mechanics of Irreversibility," in *Lectures in Theoretical Physics*, edited by W. Britton, B. Downs, and J. Downs, Interscience, New York, 1961, vol. 3, p. 106.
6. Mori, H., *Progr. Theor. Phys.*, **33**, 423 (1965).
7. Umezawa, H., *Advanced Field Theory: Micro, Macro and Thermal Physics*, A.I.P., New York, 1993, and references therein.
8. Hubbard, J., *Proc. Roy. Soc. A*, **276**, 238 (1963).
9. Hubbard, J., *Proc. Roy. Soc. A*, **277**, 237 (1964).
10. Hubbard, J., *Proc. Roy. Soc. A*, **281**, 401 (1964).
11. Rowe, D. J., *Rev. Mod. Phys.*, **40**, 153 (1968).
12. Roth, L. M., *Phys. Rev.*, **184**, 451 (1969).
13. Tserkovnikov, Y. A., *Teor. Mat. Fiz.*, **49**, 219 (1981).
14. Tserkovnikov, Y. A., *Teor. Mat. Fiz.*, **50**, 261 (1981).
15. Barnes, S. E., *J. Phys. F*, **6**, 1375 (1976).
16. Coleman, P., *Phys. Rev. B*, **29**, 3035 (1984).
17. Kotliar, G., and Ruckenstein, A. E., *Phys. Rev. Lett.*, **57**, 1362 (1986).
18. Kalashnikov, O. K., and Fradkin, E. S., *Sov. Phys. JETP*, **28**, 317 (1969).
19. Nolting, W., *Z. Phys.*, **255**, 25 (1972).
20. Villani, D., Lange, E., Avella, A., and Kotliar, G., *Phys. Rev. Lett.*, **85**, 804 (2000).
21. Avella, A., Hayn, R., and Mancini, F., Energy-scale-dependent analysis of the single-impurity anderson model (2002), preprint University of Salerno.
22. Mancini, F., Matsumoto, H., and Villani, D., *J. Phys. Studies*, **4**, 474 (1999).
23. Mancini, F., *Europhys. Lett.*, **50**, 229 (2000).
24. Avella, A., Mancini, F., and Münzner, R., *Phys. Rev. B*, **63**, 245117 (2001).
25. Avella, A., and Mancini, F., *Int. J. Mod. Phys. B*, **17**, 554 (2003).
26. Avella, A., Mancini, F., Villani, D., and Matsumoto, H., *Eur. Phys. J. B*, **20**, 303 (2001).
27. Avella, A., Mancini, F., and Turkowski, V., *Phys. Rev. B*, **67**, 115123 (2003).
28. Avella, A., Mancini, F., and Sànchez-Lopez, M., *Eur. Phys. J. B*, **29**, 399 (2002).
29. Avella, A., Feng, S.-S., and Mancini, F., *Physica B*, **312**, 537 (2002).
30. Fiorentino, V., Mancini, F., Zasinas, E., and Barabanov, A., *Phys. Rev. B*, **64**, 214515 (2001).
31. Mancini, F., Perkins, N., and Plakida, N., *Phys. Lett. A*, **284**, 286 (2001).
32. Bak, M., and Mancini, F., *Physica B*, **312**, 732 (2002).
33. Hubbard, J., *Proc. Roy. Soc. A*, **285**, 542 (1965).
34. Bosse, J., Götze, W., and Lücke, M., *Phys. Rev. A*, **17**, 434 (1978).

The cumulant expansion approach for strongly correlated electron models

R. Citro and M. Marinaro

Dipartimento di Fisica "E. R. Caianiello", Università di Salerno
via S. Allende – I-84081 Baronissi (Sa), Italy and Unità I.N.F.M. di Salerno

Abstract. In a large variety of systems, like the high-temperature superconductors, the dominant parameter of the system is the on-site Coulomb repulsion that becomes comparable with the kinetic energy in the vicinity of the metal-insulator transition. This implies that perturbation expansions must include this interaction in the zeroth order Hamiltonian and treat the hopping term as a perturbation. Thus, one gets a situation radically different from the usual one in which the particles are initially free and needs to develop a nonstandard perturbation theory based on a cumulant expansion. In this paper, we revisit the general formalism of the cumulant expansion approach for the single band and two-band Hubbard model, being interested in introducing a systematic way to perform approximations which take into account cooperative phenomena. As an example we analyze the effect of the charge fluctuations on the single-particle spectral properties.

INTRODUCTION

This article is dedicated to Ferdinando Mancini in occasion of his 60th birthday. In the article we review the perturbation expansion around the atomic limit to illustrate a scheme to find approximate solutions for strongly correlated fermion systems, an old exciting problem to which Ferdinando Mancini has devoted a long period of his more recent scientific activity, obtaining important results and stimulating the interest on it of many collaborators.

A proper treatment of the correlation effects induced by strong short-ranged interactions in Fermi systems remains one of the most challenging problems in the condensed matter theory. Even for the simplest models of strongly correlated electrons, like the one-band Hubbard model, exact solutions exist in a few special cases, such as in one spatial dimension, while in the general one only numerical methods give exact results on finite cluster systems. Since the early work of Hubbard[1, 2, 3] much effort has been invested in developing a useful method to study the Hubbard Hamiltonian, and related models, using perturbation theory around the solution of the atomic limit[1, 2, 3, 4, 5, 6]. The advantage of such approach lies on the nonperturbative treatment of the on-site Coulomb interaction U, which is a peculiar feature in the case of strongly correlated systems. However the price to pay for the inclusion of the interaction in the unperturbed Hamiltonian is the breakdown of the Wick's theorem, preventing utilization of the well-known many-body techniques. Besides that, the atomic limit destroys the translational invariance which must be recovered by summing infinite subsets of diagrams of the perturbation series. The simplest approximation satisfying this requirement is the well known Hubbard I approximation which qualitative reproduce some expected behavior of the strongly correlated limit, but it tends to retain a strong memory of the atomic limit favoring an insulating state even in the case of half-filled band. The summation of infinite subsets of diagrams is also required to introduce cooperative phenomena, like magnetic, metal-insulating, superconducting transitions. To this end a diagram technique has been introduced which follow the standard field theoretical scheme as close as possible[4, 5]. In this work we review the general formalism of the cumulant expansion, and exemplify its applicability to the Hubbard model. In Section I we introduce the basic notions of the diagram method for the one-band Hubbard model and extend it to the two-band Hubbard model (or $p-d$ model). In both cases we obtain a Dyson equation which can be written in terms of a matrix $Z_{\sigma\sigma'}(\mathbf{k},\omega)$ which is the sum of all the two roots irreducible diagrams containing irreducible many-particle Green's function (GF) or cumulants. The relation between the matrix $Z_{\sigma\sigma'}$ and the self-energy is also determined. As an example, in Section II we discuss the effect of the charge fluctuations on the spectral function starting from the equation which expresses $Z_{\sigma\sigma'}(\mathbf{k},\omega)$ in terms of the two-particle irreducible charge vertex. Concluding remarks are presented in Section III. Other applications of the method within the periodic Anderson model have been discussed elsewhere in literature[7, 8, 9, 10] and we do not report them here.

I THE FORMALISM

The one-band Hubbard model

We write the Hamiltonian of the one-band Hubbard model as follows:

$$H = H_0 + H_{int} \; ; \; H_0 = \sum_{i=1}^{N} H_i^0$$

$$H_i^0 = -\mu \sum_\sigma n_{i\sigma} + U n_{i\uparrow} n_{i\downarrow},$$

$$H_{int} = \sum_{<i,j>} t_{ij} c_{i\sigma}^\dagger c_{j\sigma} + h.c. \tag{1}$$

where $c_{i\sigma}^\dagger (c_{i\sigma})$ is the creation(annihilation) operator of an electron with spin σ on the lattice site i, t_{ij} is the intersite hopping matrix element up to next nearest neighbors, μ is the chemical potential, $n_{i\sigma} = c_{i\sigma}^\dagger c_{i\sigma}$, U is the Coulomb repulsion.

We will be interested in the temperature dependent m-particle Green's function (GF) defined as:

$$G_m(i_1,\sigma_1,\tau_1,\ldots i_m,\sigma_m,\tau_m|i'_1,\sigma'_1,\tau'_1,\ldots i'_m,\sigma'_m,\tau'_m) =$$
$$= (-1)^m \frac{\langle T_\tau c_{i_1\sigma_1}(\tau_1)\ldots c_{i'_m\sigma'_m}^\dagger(\tau')U(\beta)\rangle_0}{\langle U(\beta)\rangle_0},$$
$$c_{i\sigma}(\tau) = e^{H_0\tau} c_{i\sigma} e^{-H_0\tau}; c_{i\sigma}^\dagger(\tau) = e^{H_0\tau} c_{i\sigma}^\dagger e^{-H_0\tau},$$

$$U(\beta) = T_\tau \exp\left(-\int_0^\beta H_{int}(\tau)d\tau\right) =$$
$$= \sum_k \frac{(-1)^k}{k!} \int_0^\beta d\tau_1 \ldots d\tau_k T_\tau(H_{int}(\tau_1) \ldots H_{int}(\tau_k)). \quad (2)$$

where T_τ is the time-order operator, $U(\beta)$ is the evolution operator in the interaction picture. Finally, we have introduced $\langle\ldots\rangle_0$ to indicate the statistical average:

$$\langle X \rangle_0 = Tr(\rho_0 X)/Tr(\rho_0), \quad (3)$$

where ρ_0 the statistical operator related to H_0. From (2) the following perturbative expansion in terms of the intersite hopping matrix is obtained:

$$G_m(l_1, l_2, \ldots, l_m | l'_1, l'_2, \ldots, l'_m) = \frac{1}{\langle U(\beta)\rangle_0} \sum_k \frac{1}{k!} \sum_{i_1, j_1} \cdots \sum_{i_k, j_k}$$
$$G^0_{m+k}(l_1, \ldots, l_m, i_1, \ldots i_k | l'_1, \ldots, l'_m, j_1, \ldots, j_k) t_{i_1 j_1} \cdots t_{i_k j_k}, \quad (4)$$

where, to simplify the notation, we have put:

$$l_k \equiv (x_k, \tau_k, \sigma_k); l'_k \equiv (x'_k, \tau'_k, \sigma'_k); i_s \equiv (i_s, \tau_s, \sigma_s); j_s \equiv (j_s, \tau_s, \sigma_s), \quad (5)$$

where i_s and j_s are at equal times and G^0_m is the unperturbed atomic Green's function. By taking advantage of the factorization of the statistical operator over the lattice sites, $\rho_0 = \exp(-\beta H_0) = \Pi_i e^{-\beta H^0_i}$, it is immediate to see that the multi-particle Green's function G^0_m factorizes in products of *single-site* multiparticles GF $G^0_m(\tau_1\sigma_1, \ldots, \tau_m\sigma_m | \tau'_1\sigma'_1, \ldots, \tau'_m\sigma'_m)$. For example, the two-particle GF factorizes like:

$$G^0_2(l_1, l_2 | l'_1, l'_2) = \delta_{x_1 x_2} \delta_{x_1 x'_1} \delta_{x_1 x'_2} G^0_2(\tau_1\sigma_1, \tau_2\sigma_2 | \tau'_1\sigma'_1, \tau'_2\sigma'_2)$$
$$+ \delta_{x_1 x'_1} \delta_{x_2 x'_2} (1 - \delta_{x_1 x_2}) G^0_1(\tau_1\sigma_1 | \tau'_1\sigma'_1) G^0_1(\tau_2\sigma_2 | \tau'_2\sigma'_2)$$
$$- \delta_{x_1 x'_2} \delta_{x_2 x'_1} (1 - \delta_{x_1 x_2}) G^0_1(\tau_1\sigma_1 | \tau'_2\sigma'_2) G^0_1(\tau_2\sigma_2 | \tau'_1\sigma'_1). \quad (6)$$

Equation (6) is usually written in terms of the *irreducible* single-site GF (or *Kubo cumulants*):

$$G^0_2(l_1, l_2 | l'_1, l'_2) = \delta_{x_1 x_2} \delta_{x_1 x'_1} \delta_{x_1 x'_2} G^{0irr}_2(\tau_1\sigma_1, \tau_2\sigma_2 | \tau'_1\sigma'_1, \tau'_2\sigma'_2)$$
$$+ \delta_{x_1 x'_1} \delta_{x_2 x'_2} G^0_1(\tau_1\sigma_1 | \tau'_1\sigma'_1) G^0_1(\tau_2\sigma_2 | \tau'_2\sigma'_2)$$
$$- \delta_{x_1 x'_2} \delta_{x_2 x'_1} G^0_1(\tau_1\sigma_1 | \tau'_2\sigma'_2) G^0_1(\tau_2\sigma_2 | \tau'_1\sigma'_1), \quad (7)$$

where

$$G^{0irr}_2(\tau_1\sigma_1, \tau_2\sigma_2 | \tau'_1\sigma'_1, \tau'_2\sigma'_2) = G^0_2(\tau_1\sigma_1, \tau_2\sigma_2 | \tau'_1\sigma'_1, \tau'_2\sigma'_2)$$
$$- G^0_1(\tau_1\sigma_1 | \tau'_1\sigma'_1) G^0_1(\tau_2\sigma_2 | \tau'_2\sigma'_2) + G^0_1(\tau_1\sigma_1 | \tau'_2\sigma'_2) G^0_1(\tau_2\sigma_2 | \tau'_1\sigma'_1). \quad (8)$$

This procedure allows us to write a generalized Wick's theorem for the multiparticle GF:

$$G_n^0(l_1, l_2, \ldots, l_n | l'_1, l'_2, \ldots, l'_n) \equiv (-1)^n \langle T_\tau(c(l_1) \ldots c^\dagger(l'_n)) \rangle_0 =$$

$$\sum_{\{p\}} (-1)^p \delta_{x_1 x'_{p_1}} \delta_{x_2 x'_{p_2}} \ldots \delta_{x_n x'_{p_n}} G_1^0(\tau_1 \sigma_1 | \tau'_{p_1} \sigma'_{p_1}) \ldots G_1^0(\tau_n \sigma_n | \tau'_{p_n} \sigma'_{p_n}) +$$

$$\sum_{m>1} \sum_{\{p\},\{s\}} (-1)^{p+s} \delta_{x_{s_1} x'_{p_1}} \ldots \delta_{x_{s_{n-m}} x'_{p_{n-m}}} G_1^0(\tau_{s_1} \sigma_{s_1} | \tau'_{p_1} \sigma'_{p_1}) \ldots G_1^0(\tau_{s_{n-m}} \sigma_{s_{n-m}} | \tau'_{p_{n-m}} \sigma'_{p_{n-m}})$$

$$\times \delta_{x_{s_{n-m+1}} \ldots x'_{p_n}} G_m^{0irr}(\tau_{s_{n-m+1}} \sigma_{s_{n-m+1}}, \ldots, \tau_{s_n} \sigma_{s_n} | \tau'_{p_{n-m+1}} \sigma'_{p_{n-m+1}}, \ldots, \tau'_{p_n} \sigma'_{p_n}) +$$

$$\sum_{m>2} \sum_{\{p\},\{s\}} (-1)^{p+s} \delta_{x_{s_1} \ldots x'_{p_m}} \delta_{x_{s_{m+1}} \ldots x'_{p_n}} G_m^{0irr}(\tau_{s_1} \sigma_{s_1}, \ldots, \tau_{s_m} \sigma_{s_m} | \tau'_{p_1} \sigma'_{p_1}, \ldots, \tau'_{p_m} \sigma'_{p_m})$$

$$\times G_{n-m}^{0irr}(\tau_{s_{m+1}} \sigma_{s_{m+1}}, \ldots, \tau_{s_n} \sigma_{s_n} | \tau'_{p_{m+1}} \sigma'_{p_{m+1}}, \ldots, \tau'_{p_n} \sigma'_{p_n}) \tag{9}$$

where we have introduced the symbol $\delta_{x_1 \ldots x_n} = \delta_{x_1 x_2} \delta_{x_1 x_3} \ldots \delta_{x_1 x_n}$, $\{s\}, \{p\}$ represents all the possible permutations of the indices and the factors $(-1)^{s(p)}, (-1)^{s+p}$ depend on the number of permutations. Finally, G_n^{0irr} is the cumulant of order n:

$$G_n^{0irr}(\{\tau\sigma\}_n | \{\tau'\sigma'\}_n) = G_n^0(\{\tau\sigma\}_n | \{\tau'\sigma'\}_n)$$

$$- \sum' (-1)^p G_{m_1}^{0irr}(\{\tau\sigma\}_{m_1} | \{\tau'\sigma'\}_{m_1}) \ldots G_{m_k}^{0irr}(\{\tau\sigma\}_{m_k} | \{\tau'\sigma'\}_{m_k}), \tag{10}$$

where $\{\ldots\}_k$ indicates a set of k elements, $\sum'$ is the summation on all the partition $m_1, \ldots, m_k$ of m such that $m_i < n$ and $\sum_i m_i = n$. Eq.(10) is the generalization of the formula (6) to n-particles. A diagrammatic method to calculate the single-site cumulants of order n can be found in Ref.[18]. The first sum in (9) represents the usual Wick's terms, followed by those containing single-site irreducible many-particle GF.

A diagrammatic representation of the perturbation expansion (4) is obtained by associating to the single-site GF $G_k^{0irr}(\tau_1 \sigma_1, \ldots \tau_k \sigma_k | \tau'_1 \sigma'_1, \ldots, \tau'_k \sigma'_k)$ a polygon with $2k$ vertices, and a dashed line to the hopping, as illustrated in Fig.1.

In Fig.2 several diagrams of the series expansion for the one-particle GF are plotted following the previous prescriptions. The correspondence between diagrams and analytic expansion is immediate. As an example we report the expression of the first term in the second line:

$$\sum_{i_1, \sigma_1} \int d\tau_1 d\tau_2 G_2^{0irr}(\tau, \sigma, \tau_1, \sigma_1 | \tau', \sigma', \tau_2, \sigma_1) t_{i i_1} t_{i_1 i} G_1^0(\tau_2, \sigma_1 | \tau_1, \sigma_1). \tag{11}$$

The theorem for linked diagrams remains valid for this approach also, and therefore only linked diagrams are considered. The graphs in the first line of Fig.2 are the only ones that survive if cumulants of order higher than one are neglected. They can be summed to reproduce the Dyson equation with t_{ij} as self-energy. The remaining diagrams describe renormalization processes from collective phenomena as charge and spin fluctuations. These effects can be taken into considerations by introducing the two-roots diagrams. We call $Z(l_1 | l_2)$ the sum of all two-roots irreducible diagrams which cannot be broken into two parts by cutting a single propagator or hopping line. Examples of

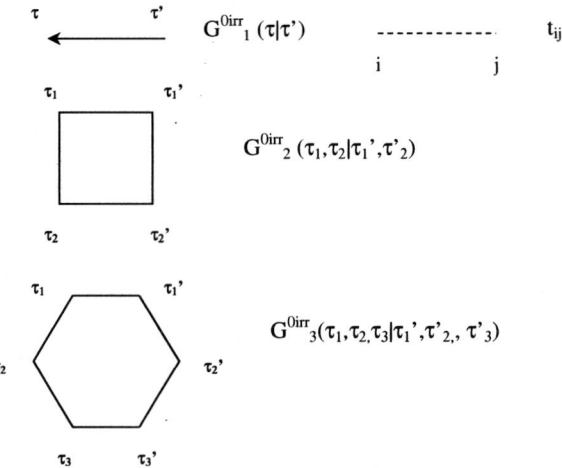

FIGURE 1. Elements of the diagram representation. The single line represents a cumulant of order one. The polygons (square, etc...) represent the two, three-particle irreducible GF. The spin indices have been omitted for brevity.

contributions to Z are given by the graphs in the second line of Fig.2. By using this function we write the one-particle GF as

$$G_1(l|l') = \sum_{i,j} \int_0^\beta d\tau_i [G_1^0(\tau, \sigma | \tau_i, \sigma_i) + Z(l|i)](\delta_{\sigma_i \sigma'} + t_{ij} G_1(j|l')). \quad (12)$$

Eq.(12) can be written in more compact form in the $(\mathbf{k}, \omega)$-space. By introducing the Fourier transforms:

$$c_{i\sigma} = \frac{1}{\sqrt{N}} \sum_{\mathbf{k}} c_{\mathbf{k}} e^{i\mathbf{k}\mathbf{x}_i}, \quad c_{\mathbf{k},\sigma} = \frac{1}{\sqrt{N}} \sum_i c_{i,\sigma} e^{-i\mathbf{k}\mathbf{x}_i}; \quad t(\mathbf{k}) = \frac{1}{N} \sum_{i,j} t_{i,j} e^{-i\mathbf{k}(\mathbf{x}_i - \mathbf{x}_j)}, \quad (13)$$

$$G(i, \tau, \sigma | i', \tau', \sigma') = \frac{1}{N} \sum_{\mathbf{k}} \frac{1}{\beta} \sum_\omega G_{\sigma\sigma'}(\mathbf{k}, i\omega) \exp(i\mathbf{k}(\mathbf{x}_i - \mathbf{x}'_i) - i\omega(\tau - \tau')), \quad (14)$$

where $\omega \equiv \omega_n = \frac{2\pi}{\beta}(n + \frac{1}{2})$ are the fermion Matsubara frequencies, we get the Dyson equation for the one-particle GF.

$$G_{\sigma\sigma'}(\mathbf{k}, i\omega)) = \sum_{\sigma_1} [Z_{\sigma\sigma_1}(\mathbf{k}, i\omega)) + G^0(i\omega) \delta_{\sigma\sigma_1}]\{\delta_{\sigma_1\sigma'} + t(\mathbf{k}) G_{\sigma_1\sigma'}(\mathbf{k}, i\omega)\}, \quad (15)$$

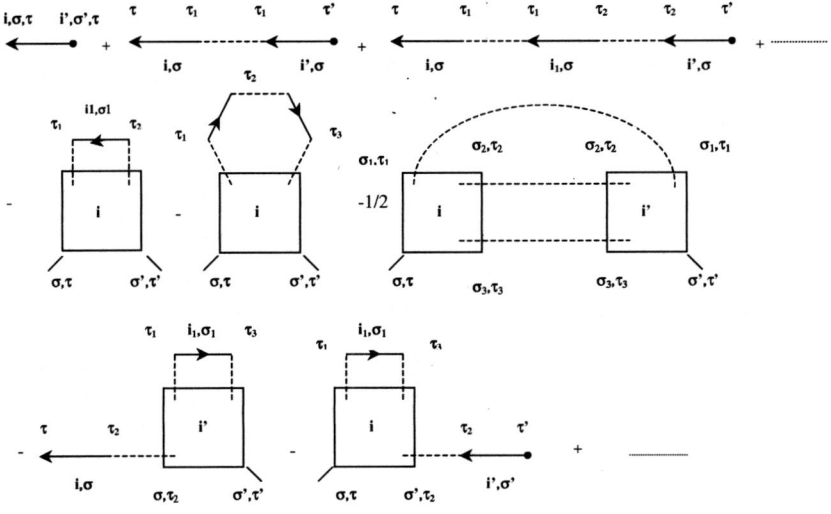

FIGURE 2. Diagram expansion for the one-particle GF: $i, j, \ldots$ are direct lattices's points. The dashed line represents the hopping between two sites i and j and the $2n$-sided polygon represent the unperturbed irreducible n-particle GF. The diagrams in the first line (so called chain-like diagrams) can be summed by a Dyson-like equation.

where $G^0(i\omega)$ is the atomic GF, given by

$$G^0(i\omega) = \frac{1 - \langle n_{-\sigma} \rangle}{i\omega + \mu} + \frac{\langle n_{-\sigma} \rangle}{i\omega - U + \mu} \tag{16}$$

and $\langle n_{-\sigma} \rangle$ is the mean-number of electrons with spin $-\sigma$.

The Dyson equation (15) can be also written in terms of the *mass operator*:

$$\begin{aligned}G_{\sigma\sigma'}(\mathbf{k}, i\omega)) &= G^0(i\omega)(\delta_{\sigma\sigma'} + \sum_{\sigma_1} M_{\sigma\sigma_1}(\mathbf{k}, i\omega) G_{\sigma_1\sigma'}(\mathbf{k}, i\omega)) \\ \hat{M}(\mathbf{k}, i\omega)) &= [t(\mathbf{k}) + [\hat{G}^0(i\omega)]^{-1}]\hat{I} - [\hat{Z}(\mathbf{k}, i\omega)) + \hat{G}^0(i\omega)]^{-1}.\end{aligned} \tag{17}$$

where the symbol $\hat{A}$ indicates a matrix in the spin space. The real problem of the cumulant expansion is to calculate the matrix $Z_{\sigma\sigma'}(\mathbf{k}, i\omega)$. Since the computation of the whole series of Z diagrams is an unsolvable problem, one takes approximate expressions for Z corresponding to a summation of a subset of diagrams. The choice of this subset corresponds usually to the inclusion of spin, charge and/or pair fluctuations, depending on the physical problem under consideration, as will be discussed in Sec.II.

The p-d model

The formalism introduced in the previous Section can be extended to obtain the cumulant expansion for the $p-d$-model[13] (or two-band Hubbard model) which has been extensively studied in connection with the physics of high-temperature superconductors. We write the Hamiltonian as

$$H = H_0 + H_{int}, \tag{18}$$

$$H_0 = (E_d - \mu)\sum_{j,\sigma} d^\dagger_{j,\sigma} d_{j,\sigma} + (E_p - \mu)\sum_{i,\sigma} p^\dagger_{i,\sigma} p_{i,\sigma}$$

$$+ \frac{U_d}{2} \sum_{j,\sigma} d^\dagger_{j,\sigma} d_{j,\sigma} d^\dagger_{j,-\sigma} d_{j,-\sigma} \tag{19}$$

$$H_{int} = - \sum_{\langle ij \rangle,\sigma} V_{ij}(p^\dagger_{i\sigma} d_{j\sigma} + d^\dagger_{j\sigma} p_{i\sigma}), \tag{20}$$

where the labels d and p stand for the copper and oxygen fermion operators in the CuO_2 plane[13] and V_{ij} represents the hybridization term. As before, we are interested in the calculation of the n-particle GF:

$$G_n(l_1,\ldots,l_{2n}) = (-1)^n \langle T_\tau A(l_1)\ldots A^\dagger(l_{2n}) U(\beta)\rangle_0 / \langle U(\beta)\rangle_0, \tag{21}$$

where the average $\langle\ldots\rangle_0$ is performed with respect to the unperturbed Hamiltonian H_0 and $A = p$ or d. Using the series expansion for $U(\beta)$ in H_{int}, we obtain the following expansion for the n-particle GF:

$$G_n(l_1,\ldots,l_{2n}) = \frac{1}{\langle U(\beta)\rangle_0} \sum_k \frac{(-1)^{2k}}{2k!} \binom{2k}{k} \sum_{i_1,j_1,\sigma_1}\ldots \sum_{i_{2k},j_{2k},\sigma_{2k}} V_{i_1 j_1}\ldots V_{i_{2k} j_{2k}}$$

$$\times \int_0^\beta d\tau_1 \ldots \int_0^\beta d\tau_{2k} \langle T_\tau(A(l_1)\ldots A^\dagger(l_{2n}) B_{i_1,j_1}(\tau_1) B^\dagger_{i_2,j_2}(\tau_2)\ldots B^\dagger_{i_{2k},j_{2k}}(\tau_{2k})\rangle_0 \tag{22}$$

where $B_{i,j}(\tau) \equiv p^\dagger_{i\sigma}(\tau) d_{j\sigma}(\tau)$. From Eq.(22) we can immediately deduce the expression for the d one-particle GF:

$$G^d_1(l_1,l'_1) \equiv -\langle T_\tau d(l_1) d^\dagger(l'_1)\rangle = -\sum_k \frac{1}{k!}\frac{1}{k!} \sum_{i_1,j_1,\sigma_1}\ldots \sum_{i_{2k},j_{2k},\sigma_{2k}} V_{i_1 j_1}\ldots V_{i_{2k} j_{2k}}$$

$$\times \int_0^\beta d\tau_1 .. \int_0^\beta d\tau_{2k} \langle T_\tau(d(l_1) d^\dagger(l'_1) p^\dagger_{i_1\sigma_1}(\tau_1) d_{j_1\sigma_1}(\tau_1) d^\dagger_{j_2\sigma_2}(\tau_2) p_{i_2\sigma_2}(\tau_2)..)\rangle_0 \tag{23}$$

where only connected terms are taken into account for the linked cluster theorem. Since the Hilbert space is the direct product:

$$|\ldots\rangle_0 = |\ldots\rangle_d \otimes |\ldots\rangle_p, \tag{24}$$

we can separate the averages of the p and d operators in the expression (23) as follows

$$G_1^d(l_1,l_1') = -\sum_k \frac{1}{k!}\frac{1}{k!}\int_0^\beta d\tau_1 \ldots \int_0^\beta d\tau_{2k} \sum_{i_1,j_1,\sigma_1} \ldots \sum_{i_{2k},j_{2k},\sigma_{2k}} V_{i_1 j_1}\ldots V_{i_{2k}j_{2k}} \times$$
$$\langle T_\tau(d(l_1)d^\dagger(l_1')d_{j_1\sigma_1}(\tau_1)\ldots d^\dagger_{j_{2k}\sigma_{2k}}(\tau_{2k}))\rangle_0 \times \langle T_\tau(p^\dagger_{i_1\sigma_1}(\tau_1)\ldots p_{i_{2k}\sigma_{2k}}(\tau_{2k}))\rangle_0, \quad (25)$$

The two averages that appear in (25) are the unperturbed multi-particle GF for p and d particles, respectively:

$$\langle T_\tau(d(l_1)d^\dagger(l_1')d_{j_1\sigma_1}(\tau_1)\ldots d^\dagger_{j_{2k}\sigma_{2k}}(\tau_{2k}))\rangle_0 = G^{0d}_{2k}(l_1,j_1\ldots,j_{2k-1}|l_1',j_2\ldots,j_{2k})$$
$$\langle T_\tau(p^\dagger_{i_1\sigma_1}(\tau_1)\ldots p_{i_{2k}\sigma_{2k}}(\tau_{2k}))\rangle_0 = G^{0p}_{2k}(i_2,\ldots,i_{2k}|i_1,\ldots,i_{2k-1}). \quad (26)$$

The p-particles are free and, by using the standard Wick's theorem, we have:

$$G^{0p}_{2k}(i_2,\ldots,i_{2k}|i_1,\ldots,i_{2k-1}) = \sum_{\{p\}}(-1)^P G^{0p}(i_2|i_1)\ldots G^{0p}(i_{2k}|i_{2k-1}), \quad (27)$$

while the averages for the d-particles must be treated by the generalized Wick's theorem (9). Introducing the same diagrammatic elements as before, with the addition of an extra line for the p-propagator, we can build up the series reported in Fig.3. The graphs in the first line of Fig.3 represent an iterative process leading to the Dyson equation with $V_{ij}^2 G^{0p}$ as self-energy instead of t_{ij} as for the single-band Hubbard model. These diagrams are the only ones that survive if we neglect cumulants of order greater than one. All the other diagrams contain at least one cumulant of order larger than one. For example the first one in the second line is given by:

$$\int_0^\beta\int_0^\beta d\tau_1 d\tau_2 \sum_{\sigma_1}[G^{0d}_2]^{irr}(\sigma,\tau,\sigma_1,\tau_1|\sigma_1,\tau_2,\sigma',\tau')G^{0p}(\tau_2-\tau_1)\sum_i V_{ij}^2, \quad (28)$$

where the irreducible two-particle d propagators appear (second order cumulant).

As before, the complete one-particle GF for the d particle can be obtained in the k-space by introducing the matrix $Z_{\sigma\sigma'}(\mathbf{k},i\omega)$ which is the sum of all the irreducible diagrams with two-roots:

$$G^d_{\sigma\sigma'}(\mathbf{k},i\omega)) =$$
$$\sum_{\sigma_1}[Z_{\sigma\sigma_1}(\mathbf{k},i\omega)) + G^{0d}(i\omega)\delta_{\sigma\sigma_1}]\{\delta_{\sigma_1\sigma'}+V^2(\mathbf{k})G^{0p}(i\omega)G^d_{\sigma_1\sigma'}(\mathbf{k},i\omega)\}. \quad (29)$$

where $V(\mathbf{k}) = V\sqrt{2(2-\cos k_x -\cos k_y)}$ for the CuO_2 planes, and $G^{0p}(i\omega) = 1/(i\omega - \varepsilon_p + \mu)$ is the atomic GF of p-type and $G^{0d}(i\omega)$ has been defined before for the Hubbard model.

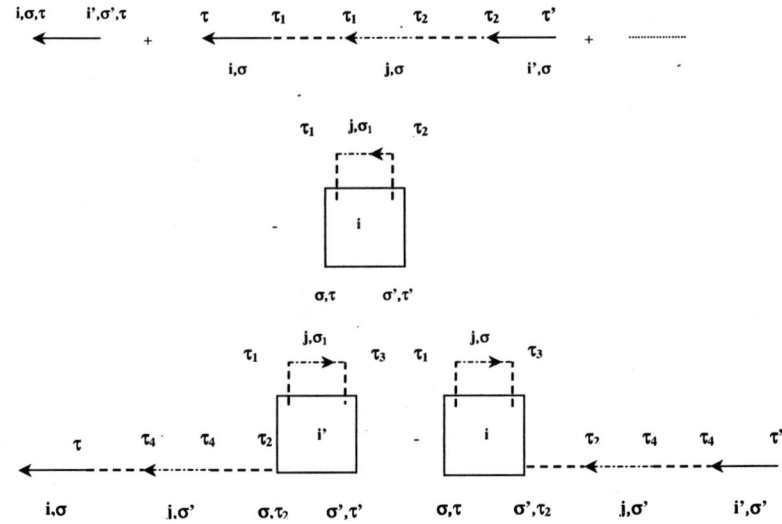

FIGURE 3. Diagram expansion for the one-particle GF of type d. The same notation as in Fig.2 has been used with the extra dashed-dotted line that represents the unperturbed one-particle GF of type p.

II APPROXIMATE SOLUTIONS OF THE ONE-BAND HUBBARD MODEL

Hubbard I approximation and the metal-insulator transition

When in Eq.(15) we neglect the cumulants of order greater than one, i.e. $Z_{\sigma\sigma'}$ vanishes, we fall in the Hubbard I approximation[1]. In this approximation the one-particle GF reads:

$$G_\sigma^1(\mathbf{k}, i\omega) = \frac{G_\sigma^0(i\omega)}{1 - t(\mathbf{k}) G_\sigma^0(i\omega)} = \sum_{i=1,2} \frac{A_{i\sigma}(\mathbf{k})}{i\omega - \tilde{\varepsilon}_{i\sigma}(\mathbf{k})}, \tag{30}$$

where $\tilde{\varepsilon}_{i\sigma} = (\varepsilon_{i\sigma} - \mu)$, and $\varepsilon_i(\mathbf{k})$ (i=1,2) are the energy spectrum of Hubbard subbands:

$$\varepsilon_{1,2\,\sigma}(\mathbf{k}) = \frac{1}{2}[U + t(\mathbf{k}) \mp \sqrt{(U - t(\mathbf{k}))^2 + 4t(\mathbf{k})\langle n_{-\sigma}\rangle U}], \tag{31}$$

$A_{i\sigma}(\mathbf{k})$ (i=1,2) are the residues at each pole:

$$A_{1\sigma}(\mathbf{k}) = \frac{\varepsilon_{1\sigma}(\mathbf{k}) - U(1 - \langle n_{-\sigma}\rangle)}{\varepsilon_{1\sigma}(\mathbf{k}) - \varepsilon_{2\sigma}(\mathbf{k})} = 1 - A_{2\sigma}(\mathbf{k}). \tag{32}$$

This approximation is equivalent to take into account only chain-like diagrams, as discussed previously. It gives qualitatively the expected behavior in the strong correlation limit $U \gg t$, where a lower and upper Hubbard subbands are separated by a correlation gap in the single-particle density of states (DOS). However, this structure appears for any nonzero value of U, with the gap collapsing only strictly at $U = 0$, where the tight-binding band is recovered. At half-filling the expected metal-insulator (MI) transition at some finite U is not present[12]. In the next Section we introduce a renormalization of the one-particle GF taking into account charge fluctuations and will recover a MI transition at a finite value of U. In Ref.[12] a MI has been obtained by introducing a self-consistent solution for the irreducible part of the one-particle GF. There was found a MI for a critical value of the Coulomb interaction $U \simeq 1.5t$ and a non-Fermi liquid behavior characterized by the absence of the peak in the density of states at $\omega = 0$ was observed.

Effects of charge fluctuations and phase-separation in single-band Hubbard model

Appropriate corrections to the Hubbard I approximation can be obtained by introducing suitable approximation of the Z matrix computed as the sum of an infinite subset of diagrams. This allows us to take into account the collective excitations present in the physical system as the charge, spin and pair excitations. In the following, to illustrate our approach, we consider the effect of the charge fluctuations on the single-particle spectral properties. It has been shown[14] that in correspondence of specific values of the parameters charge fluctuations become critical (i.e. density-density correlation function diverges) and a phase-separation takes place. Our aim is to consider the influence of this phenomenon on the spectral function. By taking the analytic continuation $(i\omega_n \to \omega + i\eta)$, this quantity (per fixed spin direction) is given by the imaginary part of the one-particle GF $A(\mathbf{k},\tilde{\omega}) = -\frac{1}{\pi}Im G_1(\mathbf{k},\tilde{\omega})$ where $\tilde{\omega} = \omega - \mu$. To calculate it, we need first to evaluate the function Z. By keeping in mind the diagrammatic expansion introduced in the previous Sections, we write $Z(l_1,l_2)$ in terms of the four-roots irreducible charge vertex $\Gamma^{irr}(l_1,l_2|l_1',l_2')$:

$$Z(l_1,l_2) = \sum_{i_s,j_s,\sigma_s} \sum_{i_k,j_k,\sigma_k} \int d\tau_s d\tau_k (\Gamma^{irr}(l_1,i_s|l_2,j_k) - \Gamma^{irr}(i_s,l_1|l_2,j_k)) t_{i_s j_s} t_{i_k j_k} G_1(j_s|i_k), \quad (33)$$

where the usual relation between the two-particle GF and the vertex Γ has been used:

$$G_2(l_1,l_2|l_1',l_2') = \sum_{\sigma_{m_1},\sigma_{m_2},\sigma_{m_1'},\sigma_{m_2'}} \int d\tau_{m_1} d\tau_{m_2} d\tau_{m_1'} d\tau_{m_2'}$$
$$G_1(l_1|m_1')G_1(l_2|m_2')\Gamma(m_1,m_2|m_1',m_2')G_1(m_1|l_1')G_1(m_2|l_2'). \quad (34)$$

An approximate expression for Z is obtained by substituting in (33) G_1 with the one-particle GF in the Hubbard I approximation G_1^1, and calculating the charge vertex function within a generalized random phase approximation (RPA)[14]. Within this approximation the matrix Z becomes diagonal $Z = \delta_{\sigma\sigma'}Z_{\sigma\sigma'}$ and the Fourier transform of Eq.(33) is

$$Z(\mathbf{k},i\omega_n) = 2\beta^{-1}\sum_{\omega_{n'}}\sum_{\mathbf{k'}}\Gamma^c(\mathbf{k'},i\omega_{n'})G^1(\mathbf{k'}-\mathbf{k},i\omega_{n'}-i\omega_n)t_{\mathbf{k'}-\mathbf{k}}^2. \quad (35)$$

Γ^c is the charge vertex $\Gamma^c = (\Gamma_{\sigma\sigma} + \Gamma_{\sigma-\sigma})$, given by a Bethe-Salpeter equation[16]

$$\Gamma^c(\mathbf{k},i\omega_n) = \Gamma^{c0}(i\omega_n) + \Gamma^{c0}(i\omega_n)[\tilde{t}_\mathbf{k} - 2\beta^{-1}\sum_{\mathbf{q},i\omega'_n}t_{\mathbf{k+q}}^2 t_\mathbf{q}^2 G^1(\mathbf{k+q},i\omega_n+i\omega'_n)G^1(\mathbf{q},i\omega'_n)]\Gamma(\mathbf{k},i\omega_n), \quad (36)$$

where $\tilde{t}_\mathbf{k} = 2(\cos k_x + \cos k_y) + 4t'^2 \cos k_x \cos k_y$ comes from the polarization graph made up of two hopping lines only. The difference between the classical RPA and our generalized RPA consists in the character of the bare vertex Γ^{c0}, that in our case is a two-particle cumulant instead of a spatially local non-retarded interaction, whose expression is:

$$\Gamma^{c0}(i\omega_n) = \langle n_{-\sigma}\rangle^2 [\frac{1}{i\omega_n + (U-\mu)} - \frac{1}{i\omega_n - (U-\mu)}]. \quad (37)$$

In a previous work[14] we have shown that the charge vertex becomes singular and a phase separation takes place when $\tilde{\omega} \to 0$. The low-energy behavior of the charge vertex has been determined as:

$$\Gamma^c(\mathbf{k'},\tilde{\omega}' \to 0) \simeq -\frac{1}{\Omega(\mathbf{k'}) - i\gamma_{\mathbf{k'}}\tilde{\omega}'}, \quad (38)$$

where $\Omega(\mathbf{k'})$ is an anisotropic function in the $\mathbf{k'}$ space, $\gamma_{\mathbf{k'}}$ is the inverse relaxation time of charge fluctuations. Both these function can be determined analitically[14]. At $\omega' = 0$, the singular behavior of the vertex function is determined by the zeros of the function $\Omega(\mathbf{k'})$. In particular, we have found a singularity of Γ^c consistent with a quantum critical behavior of a gaussian type[17]: $\Omega(\mathbf{k'}) \simeq M(\delta) + \alpha k^2$, where $M(\delta)$ is the mass term, that we found to be a linearly vanishing function of the doping, $M(\delta) \propto (\delta - \delta_c)$, similarly to what found in Ref.[15]. This relation shows that the doping determines the distance from the criticality and drives the system towards a phase-separation. Concerning phase separation (PS) in the Hubbard model, it is well known that phase separation occurs in models with short range interactions, provided the strong local e-e repulsion inhibits the stabilizing role of the kinetic energy. One evidence for PS in the Hubbard model with local Coulomb interaction comes from the Fixed Node Quantum Monte Carlo study[19] where at $U \gg t$ a PS is found near half-filling. From our numerical results of the critical doping we have obtained that PS occurs at doping lower than $\delta_c = 0.15$ at $U/t = 10$, $\delta_c = 0.06$ at $U/t = 15$, in good agreement with findings of Ref.[19].

To determine the influence of the singularity in the charge vertex on the one-particle GF at low-energy, we use the Dyson equation (15) that permits us to write the spectral function in terms of Z and, in our approximation, we obtain:

$$A(\mathbf{k},\tilde{\omega}) = -\frac{1}{\pi}G_1(\mathbf{k},\tilde{\omega}) = -\frac{1}{\pi}\frac{ImZ(\mathbf{k},\tilde{\omega})}{[1 - t_\mathbf{k}(G^0(\tilde{\omega}) + ReZ(\mathbf{k},\tilde{\omega}))]^2 + (t_\mathbf{k}ImZ(\mathbf{k},\tilde{\omega}))^2}, \quad (39)$$

where, using the equation (35) for Z, the low-energy expression of the charge vertex (38) and the approximate expression of the Green's function (30), ImZ is given:

$$ImZ(\mathbf{k},\tilde{\omega}) = -sign(\tilde{\omega}) \int_{\Omega_B} \frac{d^2\mathbf{k}}{(2\pi)^2} \frac{\gamma_{\mathbf{k}'}[\tilde{\omega}+\tilde{\varepsilon}_{\mathbf{k}'-\mathbf{k}}][f(\tilde{\varepsilon}_{\mathbf{k}'-\mathbf{k}})+b(\tilde{\varepsilon}_{\mathbf{k}'-\mathbf{k}}+\tilde{\omega})]}{\Omega^2(\mathbf{k}') + \gamma_{\mathbf{k}'}^2[\tilde{\omega}+\tilde{\varepsilon}_{\mathbf{k}'-\mathbf{k}}]^2} A_{\mathbf{k}'-\mathbf{k}} t_{\mathbf{k}'-\mathbf{k}}^2,$$

(40)

(a summation over the two Hubbard subbands is intended and the integral is over the first Brillouin zone). The expression for $ImZ(\mathbf{k},\tilde{\omega})$, and that for the real part, obtained by the Kramers-Kronig relation, permits us to calculate the spectral function (39). Numerical results show that in our approximation we recover the MI transition at half-filling at a critical value of the Coulomb interaction.

III CONCLUSIONS

We have presented the general formalism of the cumulant expansion approach to treat some models of strongly correlated systems, where the Coulomb interaction is the largest parameter in the Hamiltonian and must be included in the unperturbed Hamiltonian. We have shown that the cumulant expansion formalism allows to write the Dyson equation in terms of a two-root Z matrix and to connect this function with the two-particle irreducible vertex Γ. By using suitable approximations for Z and Γ, the single-particle spectral function and other physical quantities can be computed. Diagrammatically, the approximations are obtained by summing infinite subsets of diagrams of the perturbation expansion, selected in such a way to take into account collective phenomena. An application of this formalism to study the effect of charge fluctuations on the spectral function of the single-band Hubbard model has been discussed as an example. Another application regards the inclusion of long-range Coulomb interactions in the model. In this case the phase-separation is replaced by an instability towards an incommensurate charge density wave and results for this case could be found in the works [14, 20, 21].

REFERENCES

1. J. Hubbard, *Proc. R. Soc. London Ser. A*, **276**, 238 (1963).
2. J. Hubbard, *Proc. R. Soc. London Ser. A*, **277**, 237 (1964).
3. J. Hubbard, *Proc. R. Soc. London Ser. A*, **281**, 401 (1964).
4. W. Metzner, *Phys. Rev. B*, **43**, 8549 (1991).
5. V.A. Moskalenko, L.Z. Kon, *Cond. Matt. Phys.*, **1**, 23 (1998).
6. S. Pairault, D. Senechal and A.-M. S. Tremblay, *Phys. Rev. Lett.*, **80**, 5389 (1998).
7. A. Romano, PhD Thesis, University of Salerno (1992) and references therein.
8. M.S. Gigueira, M.E. Foglio and G.G. Martinez, *Phys. Rev. B*, **50**, 17933 (1994).
9. M.E. Foglio and M.S. Finguera, *J. Phys. A: Math. Gen.*, **30**, 7879 (1997) and references therein.
10. V. A. Moskalenko, P. Entel, M. Marinaro, N. B. Perkins, and C. Holtfort *Phys. Rev. B*, **63**, 245119 (2001).
11. R. Kubo, *J. Phys. Soc. Jpn.*, **17**, 1100 (1962).
12. L. Craco and M.A. Gusmao, *Phys. Rev. B*, **52**, 17135 (1995).

13. V.J. Emery, *Phys. Rev. Lett.*, **58**, 2794 (1987).
14. R. Citro and M. Marinaro, *Eur. Phys. J. B*, **20**, 343 (2001).
15. C. Castellani, C. Di Castro, and M. Grilli, *Phys. Rev. Lett.*, **75**, 4650 (1995); C. Castellani, C. Di Castro, and M. Grilli, *Z. Phys. B*, **103**, 137 (1997).
16. V.M. Galitskii, *JETP*, **34**, 279 (1958).
17. S. Sachdev and J. Ye, *Phys. Rev. Lett.*, **69**, 2411 (1992).
18. F. Mancini, M. Marinaro and Y. Nakano, *Physica B*, **159**, 330 (1989).
19. A.C. Cosentini, M. Capone, L. Guidoni and G.B. Bachelet, *Phys. Rev. B* **58**, R14685 (1998). 25)
20. R. Citro and M. Marinaro, *Eur. Phys. J. B* **22**, 343-349 (2001).
21. R. Citro and M. Marinaro, *Eur. Phys. J. B*, **28**, 55 (2002).

Pseudospin-electron Model for Strongly Correlated Electron Systems (Thermodynamics and Dynamics)

Ihor V. Stasyuk

Institute for Condensed Matter Physics of the Nat. Acad. Sci. Ukr., 1 Svientsitskii Str., Lviv, UA-79011, Ukraine

Abstract. The subject of consideration is the pseudospin-electron model (PEM) which appears in the description of locally anharmonic crystalline and molecular systems with strong electron correlations. A review of the main results concerning thermodynamics, energy spectrum and dynamics of the model is given. Thermodynamically stable states and phase transitions in the system are investigated in the regimes of constant chemical potential or the fixed electron concentration both within the framework of the generalized random phase approximation and the dynamical mean field scheme. A comparison is performed with thermodynamics of the Falicov-Kimball model. An attention is called to the possibility of applying of the PEM to the description of high-T_c superconductors as well as to the H-bonded systems with the transition metal ions.

INTRODUCTION

The models used in describing the strongly correlated electron systems with strong short-ranged interactions of particles are based on the Hubbard model and on its generalizations. Among the latter, at the presence of other degrees of freedom (in particular, vibrational), one can mention the pseudospin-electron model (PEM). The model appeared in the recent time in connection with the investigations of the high-T_c superconductors and can be also applied to the description of the molecular and crystalline systems with hydrogen bonds. Electron system in PEM is described by the Hubbard Hamiltonian while the locally anharmonic vibrational modes are treated with the help of the pseudospin formalism. The model Hamiltonian is as follows:

$$H = \sum_i [U n_{i,\uparrow} n_{i,\downarrow} + (g S_i^z - \mu)(n_{i,\uparrow} + n_{i,\downarrow}) - h S_i^z - \Omega S_i^x] + \sum_{i,j,\sigma} t_{ij} c_{i,\sigma}^\dagger c_{j,\sigma}. \quad (1)$$

Here, the single-site part includes, besides the electron correlation (U-term), the interaction with pseudospin (g-term) and the energy of the tunnelling-like splitting of vibrational levels (Ω-term); the field h describes the asymmetry of local potential. The electron transfer (t-term) is included as well.

The pseudospin-electron Hamiltonian (with inclusion only of the g-interaction) was introduced by Müller with the aim of describing the anharmonic vibrations in the oxygen subsystem of the high-T_c superconducting crystals of the $YBa_2Cu_3O_{7-\delta}$ type [1]. A similar Hamiltonian was used by Hirsch and Tang [2] in the study of electron states in these compounds on the basis of the cluster calculations.

At an early stage of the PEM investigations the possible connection between superconducting pairing and the lattice anharmonicity was considered by Frick at al. [3] (the quantum Monte-Carlo calculations) in the context of the idea concerning the influence of anharmonicity on the superconducting transition temperature [4, 5, 6, 7]. By now, the investigations of the PEM were devoted to the analysis of the electron spectrum [8], the pseudospin and collective dynamics [9, 10], the charge and pseudospin pair correlations and the behavior of dielectric susceptibility [11]. An attention was paid to the thermodynamics of the model in special cases and simplifications: (i) a model with the infinitely large correlation, $U \to \infty$; when the double occupation of electron states on the site is excluded; (ii) simplified PEM with $U = 0$ and $\Omega = 0$; (iii) simplified model ($U = 0$) with the tunnelling-like dynamics ($\Omega \neq 0$); (iv) two-sublattice PEM adapted to the description of the layered structures of the YBaCuO-type.

The study of the PEM thermodynamics and dynamics was performed mainly within the generalized random phase approximation (GRPA) [12]. The DMFT method was used in the case of the simplified PEM [13], when the analytic formulation of the theory is possible. The structural instabilities of the dielectric (ferroelectric) type as well as the instabilities with respect to the appearance of spatially modulated phases were revealed and analyzed. Such a consideration was performed both in the limit $U \to \infty$ and for the model with $U = 0$ at the presence or absence of the tunnelling splitting [10, 14].

There was performed a study of the phase transitions between the states with different electron concentrations and with different orientations of pseudospins (in the regime of the fixed chemical potential, $\mu = \text{const}$), of the phase separation effects at the given concentration of electrons ($n = \text{const}$), metastable states as well as bistability phenomena [14, 15]. The possibility of the appearance of a doubly modulated (so-called chessboard) phase or an incommensurate phase (in case of weak coupling) was established.

The PEM is closely related to the Falicov-Kimball (FK) model intensively studied in recent years, in which the interaction between the localized and the moving particles (electrons) is taken into account. The simplified version of PEM corresponds to the FK model in the case when there is no tunnelling-like splitting in the PEM and when the localized and the moving particles in FK model have different chemical potentials. The regimes of thermodynamic averaging are different for both models (a fixed concentration of the localized particles for FK model and a given value of the field h for the PEM).

This paper presents a review of the main results concerning thermodynamics, energy spectrum and dynamics of the PEM in the above mentioned cases and approximations. The case of the simplified PEM ($U = 0$; $\Omega = 0$ or $\Omega \neq 0$) is considered more in detail. An attention is called to the possible application of the PEM to the description of inhomogeneous states and structural instabilities in the high T_c superconductors as well as transitions into the phases with the charge modulation in the H-bonded crystalline systems with the transition metal ions.

SIMPLIFIED PEM ($\Omega = 0$, $U = 0$) IN THE $D \to \infty$ LIMIT

The pseudospin-electron model at $\Omega = 0$, $U = 0$, like the FK model, is analytically exactly solvable in the limit of infinite dimension ($d \to \infty$). The correspondence between

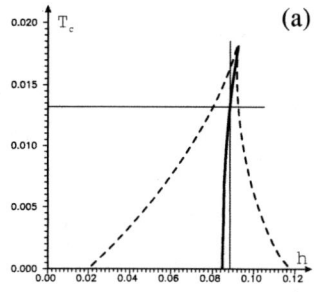 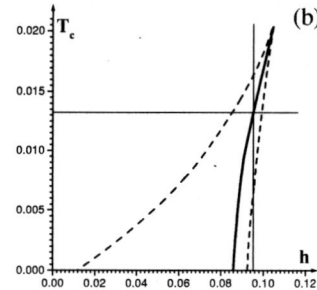

FIGURE 1. $T_c - h$ phase diagram ($g = 1$, $t_{k=0} = 0.2$, $\mu = 0.5$). (a) – DMFT, (b) – GRPA.

the models can be seen when the projective operators $P_i^{\pm} = \frac{1}{2} \pm S_i^z$ are introduced; the Hamiltonian of the FK model is obtained from (1) when the operators $P_i^{\pm}$ are expressed in terms of occupation numbers of localized particles ($P_i^+ = f_i^+ f_i; P_i^- = 1 - f_i^+ f_i^-$). Here, the main difference between the models consists in the above mentioned averaging procedure: the fixation of the field h for the PEM.

Investigations of thermodynamics of the PEM were performed within the framework of DMFT in [13] in the strong coupling case ($g \gg W$) and the transition into another uniform phase without symmetry breaking was considered. In this case, the connection between single-site Green's function and coherent potential $J_\sigma(\omega)$ is as follows:

$$G_{ii(\sigma)} = \frac{\langle P^+ \rangle}{i\omega + \mu - J_\sigma(\omega) - g/2} + \frac{\langle P^- \rangle}{i\omega + \mu - J_\sigma(\omega) + g/2}; \quad (2)$$

$$\langle S^z \rangle = \frac{1}{2} \tanh \frac{1}{2}(\beta h - Q_+ + Q_-),$$

$$Q_{\pm} = \frac{1}{\pi} \int_{-\infty}^{+\infty} \frac{d\omega}{e^{\beta \omega} + 1} \, \text{Im} \ln \left(1 - \frac{J_\sigma(\omega + i0^+)}{\omega + \mu \mp g/2 + i0^+} \right). \quad (3)$$

Numerical solution of the DMFT set of equations was performed in the case of the semi-elliptic DOS for the $d = \infty$ Bethe lattice.

In the $\mu = \text{const}$ regime there exists the first order phase transition with the jumps of the mean value $\langle S^z \rangle$ and electron concentration n. Phase diagrams on the $(T - h)$ plane have the form of a diagram presented in Fig. 1a, where the coexistence curve is bent with respect to vertical. It leads to the phase transition with the change of temperature at certain values of the field h. Such transitions take place in the cases when the μ and h values correspond to the split subbands in an electron spectrum (see Fig. 2).

For the fixed values of n, the regions appear where the derivative $\partial \mu / \partial n$ is negative; it is the evidence of instability with respect to the phase separation. The corresponding phase diagram $(T - n)$ is built (see [13]) which describes the separation on the states with large and small electron concentrations (and with the $\langle S^z \rangle \approx -\frac{1}{2}$ and $\langle S^z \rangle \approx +\frac{1}{2}$ pseudospin averages at low temperatures, respectively).

As a whole, such a picture of phase transitions into an uniform phase is in an agreement with the known picture for the FK model in the case of strong coupling [16].

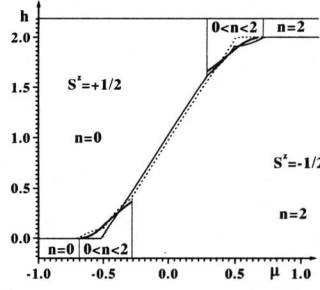 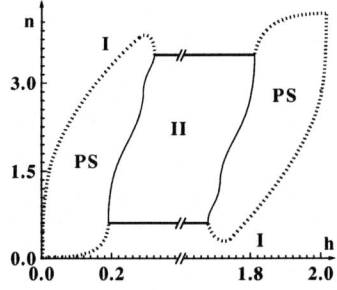

FIGURE 2. left panel: $\mu - h$ phase diagram ($T = 0$, $t_{k=0} = 0.2$, $g = 1$) in GRPA; thick solid lines correspond to the first order phase transitions; right panel: $n - h$ phase diagram ($T = 0.005$, $t_{k=0} = 0.2$, $g = 1$) in GRPA.

SIMPLIFIED PEM IN GRPA

Now, let us compare the results obtained in the DMFT approach with the ones in the generalized random phase approximation (GRPA). Such an approach was formulated by Izyumov and Letfulov [12] for the calculation of the pair correlation functions and magnetic susceptibility of the Hubbard and $t - J$ models. It is based on the expansions in terms of electron transfer and consists in the summation of the diagrams having a structure of sequences of the electron loops (created by the electron Green's functions) joined by vertices of various type, appearing due to short-range interactions [12].

At first we consider the case of strong coupling. In the GRPA scheme the Hubbard-I type approximation for the Green's function for a crystal is used in the loop diagrams

$$G_k(\omega) = \frac{1}{g^{-1}(\omega) - t_k}, \qquad g(\omega) = \frac{\langle P^+ \rangle}{\omega - \varepsilon} + \frac{\langle P^- \rangle}{\omega - \tilde{\varepsilon}} \qquad (4)$$

(here $\varepsilon = g/2 - \mu$; $\tilde{\varepsilon} = -g/2 - \mu$). The corresponding electron spectrum consists of two subbands ($\varepsilon_I(\vec{k})$ and $\varepsilon_{II}(\vec{k})$) and is always split, contrary to the DMFT results (in the latter case the splitting takes place at the values $g > g_c$, where $g_c \sim W$).

In this case, the calculation of the pair correlation function gives the result

$$\langle S^z S^z \rangle_\mathbf{q} = \frac{1/4 - \langle S^z \rangle^2}{1 + \sum_{\alpha\beta}(-1)^{\alpha+\beta}\boxed{\Pi}_\mathbf{q}^{\alpha\beta}(1/4 - \langle S^z \rangle^2)}, \qquad (5)$$

$$\boxed{\Pi}_\mathbf{q}^{\alpha\beta} = \frac{2}{N}\sum_{n,\mathbf{k}} t_\mathbf{k} t_{\mathbf{k}+\mathbf{q}} \Gamma^\alpha(\mathbf{k}, \omega_n) \Gamma^\beta(\mathbf{k}+\mathbf{q}, \omega_n). \qquad (6)$$

Here,

$$\Gamma^\alpha(\mathbf{k}, \omega_n) = (i\omega_n - \varepsilon^\alpha)^{-1}(1 - t_k g(\omega_n))^{-1}; \qquad (7)$$

$\varepsilon^\alpha = \varepsilon, \tilde{\varepsilon}$. Mean values $\langle S^z \rangle$ and $\langle \sum_\sigma n_{i\sigma} = n \rangle$ (as well as cumulant averages which join in sequences the electron loops) are calculated with an inclusion of the mean field contributions which are obtained by incorporation of the loop diagrams into basic semi-invariants. For the $\langle S^z \rangle$ mean value, we have

$$\langle S^z \rangle = \frac{1}{2}\tanh\left\{\frac{\beta}{2}(h+\alpha_2-\alpha_1)+\ln\frac{1+e^{-\beta\varepsilon}}{1+e^{-\beta\tilde{\varepsilon}}}\right\},$$

$$\alpha_2-\alpha_1 = \frac{2}{N}\sum_k t_k \frac{\varepsilon-\tilde{\varepsilon}}{\varepsilon_I(t_k)-\varepsilon_{II}(t_k)}\left[n(\varepsilon_{II}(t_k))-n(\varepsilon_I(t_k))\right]. \tag{8}$$

The expression for grand canonical potential obtained in the same approximation is thermodynamically consistent with the given scheme.

The solutions of the set of equations (6)–(8) and (11) which correspond to the absolute minimum value of Φ were determined and analyzed. The results obtained at the same values of the model parameters as in the case of the DMFT approach are shown in the form of the phase diagrams in Figs.1b and 2. The $\mu-h$ diagrams are practically the same in the cases of DMFT and GRPA approaches (see [13]); the $(T-h)$ diagrams only slightly differ between each other. The difference in the shape of the separation areas on the $(T-n)$ diagrams is more noticeable but not the principal one. Thus, the GRPA can give a satisfactory description of thermodynamics of PEM, at least at the coupling constant values far from the intermediate coupling ($g \sim W$) case.

In addition, the calculations of the pair $\langle SS \rangle$, $\langle nn \rangle$ and $\langle nS \rangle$ correlators and search of their divergence points permit to investigate the stability of the system with respect to the charge and pseudospin orderings. At $g \gg W$, the appearance of the doubly modulated chess-board phase is possible. At the given value of μ such a phase can occur when chemical potential μ is placed between electron subbands or partially enters into one of them. The phase transitions between uniform and modulated phases can be of the first or the second order. The phase separation in the system in the $n = $ const regime is illustrated in Fig. 2, where the separation regions and the region of the modulated phase occurence are shown (depending on the field h value, the separation is possible on the uniform and on the modulated phases or on the two different uniform phases).

The above described picture of phase transitions can be considered as an supplementation of results obtained for the FK model in the case of strong coupling [16, 17, 18] giving a more complete description of the transitions into the modulated phase.

A similar analysis was performed in [10, 11] for the PEM with $U \neq 0$ (when the model is not analogous to the FK model) basing on the analysis of the behaviour of the pair correlators. In the $U \to \infty$ limit, there exists at $h > 0$ and $h > g$ an instability with $\vec{q} = 0$; besides that, at $0 < h < g$, an instability in the M point ($\vec{q} = (\frac{\pi}{a},\frac{\pi}{a})$) appears, which is connected with the charge ordering. The latter dominates at large electron concentrations while the instability in the Γ point has a maximum instability temperature at concentrations near half-filling. The divergences in the temperature dependences of susceptibility are caused by the polarization contributions from the electron transitions between the subbands split due to short-range interactions [10].

In the $U \to \infty$ limit there was also considered the two-sublattice PEM, which appeared as a generalization of the usual model to the case of the YBaCuO type structures in describing the anharmonic subsystem of the apex oxygen ions [19]. In this case, the pseudospin energy in the internal field has the form $h\sum_i(S^z_{i1}-S^z_{i2})$ that is a reflection of the mirror symmetry of the problem. Investigations of thermodynamics of the model performed within self-consistent version of GRPA revealed, besides the transitions similar to the above considered, the possibility of the ferroelectric type instabilities with $\langle S_1^z \rangle + \langle S_2^z \rangle \neq 0$ [19]. Respectively, the separation in the system can take place with the participation of polar phases.

Now let us consider the case of weak coupling. We can base on the perturbation theory approach when the mean field Hamiltonian is the zero-order Hamiltonian [20]. The parameters $\langle n_i \rangle$ and $\langle \eta_i \rangle = \langle S_i^z \rangle$ are determined from the set of equations which is as follows:

$$n = \frac{1}{N}\sum_{k\sigma}(e^{\beta(g\eta+t_k-\mu)} + 1)^{-1}, \qquad (9)$$

$$\eta = \frac{h-gn}{2\lambda}\tanh(\frac{\beta\lambda}{2}); \quad \lambda = \sqrt{(gn-h)^2 + \Omega^2}$$

in the uniform (nonmodulated) structure case.

The GRPA scheme, where the electron Green's functions are calculated in the mean field approximation, is used in to determine the pseudospin, electron and mixed correlation functions. In this approximation, the correlator $\langle TS^zS^z \rangle$ is equal to

$$\langle TS^zS^z \rangle_{q,\omega} = \frac{\Sigma_q(\omega)}{1 - g^2\Sigma_q(\omega)\Pi_q(\omega)}, \qquad (10)$$

where the function

$$\Sigma_q(\omega) = \sin^2\theta\frac{\lambda\langle\sigma_z\rangle}{\omega^2 - \lambda^2} - \beta b'\cos^2\theta\delta(\omega) \qquad (11)$$

has its origin in the boson and in semi-invariant parts, which join the loops ($\Pi_q(\omega) = \frac{2}{N}\sum_k \frac{n(t_k) - n(t_{k-q})}{\omega + t_k - t_{k-q}}$; $b' = \frac{1}{2} - \langle\sigma_z\rangle^2$, $\sin\theta = \Omega/\lambda$; $\langle\sigma^z\rangle = \frac{1}{2}\tanh(\frac{\beta\lambda}{2})$).

Electron band spectrum in this approach remains unsplit in the uniform phase; the shift of the band as a whole takes place, depending on the mean value of pseudospin. In a doubly modulated phase the electron band splits into two subbands. The difference $n_1 - n_2$ (or $\eta_1 - \eta_2$) plays the role of the order parameter. Instability with respect to transition into modulated (or another uniform) phase, which is determined from the condition $\langle TS^zS^z \rangle_{q,\omega=0} \to \infty$, is illustrated in Figs.3a,3b in the form of (T,μ) and $(\vec{q},\mu)$ diagrams. The transition into another uniform phase takes place (in the regime $\mu = $ const) at the decrease of T when the chemical potential is placed near the edge of the band; doubly modulated phase occurs at the localization of μ near the band centre. An intermediate situation corresponds to the appearance of the incommensurate phase.

The given results at $\Omega = 0$ correspond in general to the picture of phase transitions in the FK model obtained in DMFT in the case of weak coupling [17, 21]. In addition, the GRPA gives an explicit dependence of the $\langle TS^zS^z \rangle_{q,\omega}$ susceptibility on the wave vector, while in the DMFT approach at $d \to \infty$ such a dependence enters only through the function $X(\vec{q}) = \frac{1}{d}\sum_{j=1}^d \cos q_j$ that leads to some difficulties at the consideration of incommensurate ordering. Besides, in [17, 21] the regime of the fixed concentration of

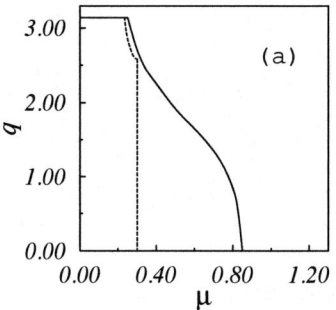 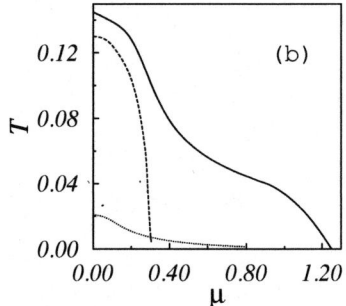

FIGURE 3. (a) the dependences of the modulation wave vector $\vec{q} = (q,q)$ and (b) the temperature of absolute instability of high-temperature phase on the chemical potential, $\Omega = 0$ (solid line), $\Omega = 0.2$ (dashed line), $g = 0.5, W = 1$. Dotted line denotes the transition to the superconducting state ($\Omega = 0.2$).

localized particles was used; in the PEM this corresponds to the regime $\langle S^z \rangle = $ const. In this case the conclusion was made that the transition to the chess-board phase at $g < W$ is always continuous and spinodals are the lines of phase transitions [18]. In such a situation the separation into the uniform and modulated phases would be impossible.

The increase of the tunnelling parameter Ω leads to the narrowing of the region of the μ values at which the phase transitions take place. At the large enough values of Ω only the transition into the chess-board phase remains (Fig. 3b). The corresponding transition temperature has the maximum value at $\mu = 0, h = g$ and is equal to [20]

$$T_c^* \approx 2W \exp\left(-\frac{\pi\sqrt{\Omega W}}{g\sqrt{2}}\right) \qquad (12)$$

in the case of DOS with the logarithmic singularity at the band centre (which corresponds to the square lattice, $d = 2$). When $h \neq g$, T_c^* decreases rapidly. After passing the tricritical points, the order of phase transition changes from the second to the first one (Fig. 4a). In the latter case, a separated state with segregation into phases of different symmetry (uniform phase and doubly modulated phase), can appear (Fig. 4b).

At the presence of intrinsic dynamics PEM, also possesses an instability with respect to the transition into the superconducting state. Calculation the static susceptibility χ^{SC} in the superconducting channel within the GRPA (similarly to (12)) leads to the Bethe-Salpeter equation for the superconducting vertex part $\Gamma_{\omega_1,\omega_2}(k_1,k_2)$ [22]

$$\Gamma = \Gamma^0 + T \sum_{k_3,\omega_3} \Gamma_0 \chi^0 \Gamma, \qquad (13)$$

where $\chi^0_{\omega_1}(k_1) = \frac{1}{N} G^0_{k_1}(\omega_1) G^0_{-k_1}(-\omega_1)$; $\Gamma^0 = -g^2 \langle TS^z S^z \rangle_{k_2-k_1,\omega_1-\omega_2}$ is given by the expression (12). The temperature T_{SC} of the absolute instability with respect to the superconducting transition is determined from the condition of divergence of χ^{SC} (and,

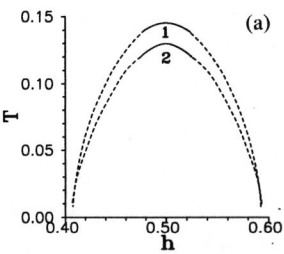 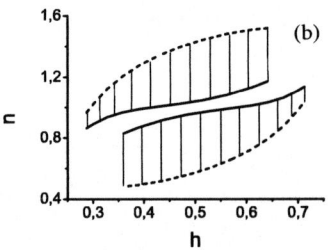

FIGURE 4. (a) – the lines of the phase transitions (solid line denotes the second order phase transition, dotted line denotes the first order one) from the uniform to doubly modulated phase (1 – $\Omega = 0$, 2 – $\Omega = 0.2$; $\mu = 0$, $g = 0.5$, $W = 1$); (b) – (n, h) phase diagram, $\Omega = 0$. Phase separation regions are shown for the case $T = 0.08$. Dashed lines denote the borders of the uniform phase, thick solid lines denote the borders of the phase with doubly modulated lattice period.

respectively, Γ) as the highest one from such temperatures on the (T, h) plane. Numerical calculations performed in [22] show that such a transition is possible at the electron concentrations away from half-filling (outside the region of the above described transitions with the modulation of the electron and pseudospin density). There takes place a competition between transitions to the modulated phase and to the superconducting state depending on the position of the chemical potential (Fig. 3b). It is similar to the results obtained within the DMFT for the Holstein model [23]. But, in the latter case the incommensurate phase does not appear at the intermediate values of μ.

COLLECTIVE DYNAMICS IN PEM

Collective excitations which determine the dynamic response of the PEM are described by dynamic susceptibilities $\chi_d^{SS}(q,\omega)$, $\chi_d^{Sn}(q,\omega)$ and $\chi_d^{nn}(q,\omega)$. They are connected with the pseudospin reorientations and with the accompanying changes in the electron spectrum as well as in the occupation of electron states. For the model with $U \to \infty$ the dynamic susceptibility was considered in [10]. In the limit of the zero transfer

$$\chi_d^{SS}(\omega) = \frac{1}{2} \sum_{r=1}^{4} \frac{\Omega^2}{\sqrt{(n_r g - h)^2 + \Omega^2}} \frac{(b_r + b_{\tilde{r}}) \tanh \frac{\beta}{2}\sqrt{(n_r g - h)^2 + \Omega^2}}{(n_r g - h)^2 + \Omega^2 - \omega^2}, \quad (14)$$

where $n_1 = 0$, $n_2 = 2$, $n_3 = n_4 = 1$, $b_3 + b_{\tilde{3}} = n/2$, $b_1 + b_{\tilde{1}} = 1 - n$ and $b_2 + b_{\tilde{2}} = 0$, at $n < 1$. Here, the single-site electron states $r = |1\rangle, |2\rangle, |\downarrow\rangle, |\uparrow\rangle$ at $S^z = 1/2$ (or $\tilde{r}$ at $S^z = -1/2$) are introduced.

The spectrum and spectral intensities of the pseudospin excitations are obtained from the imaginary part of the susceptibility (14). Corresponding frequencies are equal to $\omega = \sqrt{(n_2 g - h)^2 + \Omega^2}$. There takes place a redistribution of the intensities between different vibrational modes depending on the average electron concentration n value; besides that, the lines shift at the change of the field h, see Fig. 5. Each of these modes describes the reorientation of a pseudospin at given occupation of the site electron states.

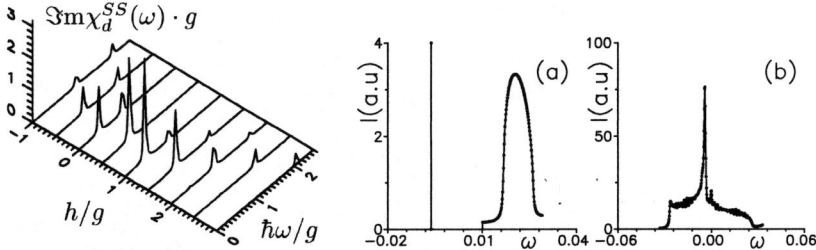

FIGURE 5. Imaginary part of $\chi_d^{SS}(\omega)$ vs h at $n = 0.7$ (left panel). Spectrum $\mathrm{Im}\langle\langle X^{1\bar{1}}|X^{\bar{1}1}\rangle\rangle \equiv I$ $h = 0.4, \Omega = 0.5, W = 0.1, n = 0.9$, (a) $T = 0$, (b) $T = 0.005$; $\omega - \sqrt{h^2 + \Omega^2}/10W \to \omega$ (right panel).

The electron transfer leads to the appearance of collective excitations. In the weak coupling case, the spectrum of the imaginary parts of the $\langle TS^z S^z\rangle$ correlator contains, besides the pseudospin branch, the broad band of electron excitations of the plasmon type. They mutually overlap in a rather broad region of the $\vec{q}$ values near $\vec{q} = 2\vec{k}_F$. At the strong coupling ($g \gg W$) the pseudospin δ-peak and the band of electron excitations can exist separately or superimpose each other even at $\vec{q} \approx 0$ depending on the temperature and, respectively, on the filling of electron states within the band [24] (see Fig. 5, where the part of the collective spectrum connected with the pole $n_r = 0$ of the function (14) is shown). It can have an effect on the frequency dependences of corresponding contributions to the Raman scattering intensity.

CONCLUSIONS

Pseudospin-electron model (PEM) can be considered as a generalization of the Falicov-Kimball (FK) model to the case of different thermodynamic equilibrium regimes as well as an extension of the latter one due to the inclusion of the pseudospin dynamics and the Hubbard type correlations. The PEM possesses a similar variety of phase transitions but there are diferences in the conditions of their realizations and in the criteria of the appearance of different phases .

In the case of the simplified PEM, an approach based on the self-consistent formulation of the GRPA gives the results which are very close to the ones obtained for the thermodynamics of the model in the DMFT scheme. This fact opens a possibility to apply the GRPA in the investigations of the systems described by the PEM with $\Omega \neq 0$ or/and $U \neq 0$.

Within the framework of GRPA approach, the conditions of the appearance of low temperature uniform or modulated phases in the cases of strong ($g \gg W$) and weak ($g < W$) coupling are established (depending on the values of Ω and h parameters as well as on the position of the chemical potential μ). The criteria of phase separation are obtained. The phase transitions and instabilities described within the framework of PEM can be responsible for the appearance of spatial inhomogeneities in the HTSC systems. Such inhomogeneities manifest themselves, for example, in the mesoscopic structure investigations [25], as well as in the phonon Raman scattering intensity profiles [26].

The obtained results can be used also in describing the metal-insulator transitions and systems with the valency change.

Being applied to the systems with the local lattice anharminicity, the PEM supplements the results obtained for the phonon Holstein model. An instability is observed in the PEM at $\Omega \neq 0$ with respect to transition into superconducting state. Such a transition prevails over the tendency to the charge (chess-board or incommensurate) ordering at electron concentrations away from half-filling.

PEM extended to the two-sublattice case is capable of describing the ferroelectric-type anomalies and bistability phenomena in the layered centrosymmetric structures with anharmonic structure units. Such effects, including the manifestations of polarity of structure and dielectric anomalies, are observed, though nonsystematically, in YBaCuO crystals at temperatures above the superconducting transition point [27, 28].

Collective pseudospin dynamics in PEM in the case of $\Omega \neq 0$ is formed by indirect interaction of pseudospins via conducting electrons. Besides that, the dynamical pseudospin susceptibility includes a contribution caused by the electron intraband or interband transitions, which accompany the reorientation of pseudospins. The presence of mixed contributions of this type should be taken into account while interpretting the electron Raman scattering spectra in the HTSC and in the systems with the strongly correlated electrons.

REFERENCES

1. Müller, K. A., *Z. Phys. B: Cond. Matter*, **80**, 193 (1990).
2. Hirsch, J. E., Tahg, S., *Phys. Rev. B*, **40**, 2179 (1989).
3. Frick, M., van der Linden, W., Morgenstern, I., Raedt, H., *Z. Phys. B*, **81**, 327–335 (1990).
4. Hardy, J. R., Flocken, J. W., *Phys. Rev. Lett.*, **60**, 2191 (1988).
5. Bussman-Holder, A., Simon, A., Büttner, H., *Phys. Rev. B*, **39**, 207 (1989).
6. Plakida, N. M., *Physica Scripta*, **29**, 77 (1989).
7. Plakida, N. M., Udovenko, V. S., *Mod. Phys. Lett., B*, **6**, 541 (1992).
8. Stasyuk, I. V., Shvaika, A. M., Schachinger, E., *Physica C*, **213**, 57 (1993).
9. Stasyuk, I. V., Shvaika, A. M., *Fiz. Nizk. Temp.*, **22**, 535 (1996).
10. Stasyuk, I. V., Shvaika, A. M., *Ferroelectrics*, **192**, 1 (1997).
11. Stasyuk, I. V., Shvaika, A. M., *Cond. Matt. Phys.*, **3**, 134 (1994).
12. Izyumov, Yu. A., Letfulov, B. M., *J. Phys.: Cond. Matter*, **2**, 8905 (1990).
13. Stasyuk, I. V., Shvaika, A. M., *Journ. Phys. Studies*, **3**, 177 (1999).
14. Stasyuk, I. V., Shvaika, A. M., Tabunshchyk, K. V., *Ukrainian Journ. of Phys.*, **45**, 520 (2000).
15. Stasyuk, I. V., Velychko, O. V., *Ukrainian Journ. of Phys.*, **44**, 772-781 (1999).
16. Freericks, J. K., Gruber, Ch., Macris, N., *Phys. Rev. B*, **60**, 1617 (1999).
17. Brandt. U, Mielsch, C., *Z. Phys. B*, **82**, 37 (1991); **75**, 365 (1989); **79**, 295 (1989).
18. Freericks, J. K., Lemanski, R., *Phys. Rev. B*, **61**, 13438 (2000).
19. Stasyuk, I. V., Danyliv, O. D., *Phys. Stat. Sol.*, **219**, 299 (2000).
20. Stasyuk, I. V., Mysakovych, T. S., *Cond. Matt. Phys.*, **5**, 473–491 (2002).
21. Freericks, J. K., *Phys. Rev. B*, **47**, 9263 (1993); **48**, 14797 (1993).
22. Mysakovych, T. S., Stasyuk, I. V., *Preprint of the ICMP of the NASU*, ICMP-03-07U, 16 p. (2003).
23. Freericks, J. K., Jarrell, M., Scalapino, D. J., *Phys. Rev. B*, **48**, 6302–6313 (1993).
24. Stasyuk, I. V., Mysakovych, T. S., *Journ. Phys. Studies*, **3**, 344 (1999).
25. Browning, V. M., *Phys. Rev. B*, **56**, 2860 (1997).
26. Iliev, M. N., Hadjiev, V. G., Ivanov, V. G., *Journ. Raman. Spectr.*, **27**, 333 (1996).
27. Mihajlovic, D., Heeger, A. J., *Solid State Commun.*, **75**, 319 (1990).
28. Grachev, A. I., Pleshkov, I. V., *Solid State Commun.*, **101**, 507 (1997).

Electron correlations at nanoscale

J. Spałek*, A. Rycerz*, E. M. Görlich* and R. Zahorbeński*

*Marian Smoluchowski Institute of Physics, Jagiellonian University,
Reymonta 4, 30–059 Kraków, Poland*

Abstract. We briefly summarize the fundamental properties of correlated nanoscopic systems obtained within the EDABI method combining an *exact diagonalization* in the Fock space with an *ab initio* readjustment of the single–particle wave functions in the resultant ground state. Explicitly, we address the following questions: evolution of various systems from a nanometal to a nanoinsulator of the Mott–Hubbard type, the appearance of the Mott–Hubbard gap in molecular and cluster (H_n) systems, as well as the stability of those clusters.

1. INTRODUCTION AND EDABI METHOD

The development of scanning tunnelling microscopy of an atomic resolution from one side and the work on single–electron devices from the other, have lead to a general question of building up theoretically, with the help of the first–principles approach, a small (nanoscopic) system, atom by atom. In this manner, we can address directly the questions: How small a metal (or a quantum wire) can be? What is the nature of such a finite quantum liquid: Tomonaga–Luttinger or Landau–Fermi? Where can we place (if at all) the borderline between the molecular and the solid states of matter? What are the specific features on the nanoscale, which are absent on either atomic or macro scale? The answers to these questions concretize the nanoscience program envisioned by Feynman [1].

Let us provide a few examples of nano–specification. First, it has recently been discovered that the Au nanowires can be transformed into nanochains of single atoms [2], which are surprisingly stable, have nearly ideal values of quantum conductance $\sigma = n(2e^2/h)$, with $n = 1, 2, \ldots$, and can sustain enormous current densities. The same takes place for carbon nanotubes [3]. Second, few–electron quantum dots have specific atomic–like shell properties [4]. Finally, although the formation of freely suspended chains of atoms has been explained within the density functional theory [5], the theoretical understanding of electron–transport is still lacking.

The experimental transport properties were studied for the simplest s–band Na nanowire [6]. These are the systems, where the metallicity should show up in its clearest form. However, when analyzing the evolution of e.g. nanochain from the atomic–like to the extended states from the first principles, one cannot regard the interaction among electrons as weak (e.g. tractable in the Hartree–Fock approximation). This is because the interatomic hopping energy may become comparable or even substantially smaller than the electron–electron interaction. Precisely because of this reason, one has to treat the single–particle and many–body aspects of the electronic states on equal footing.

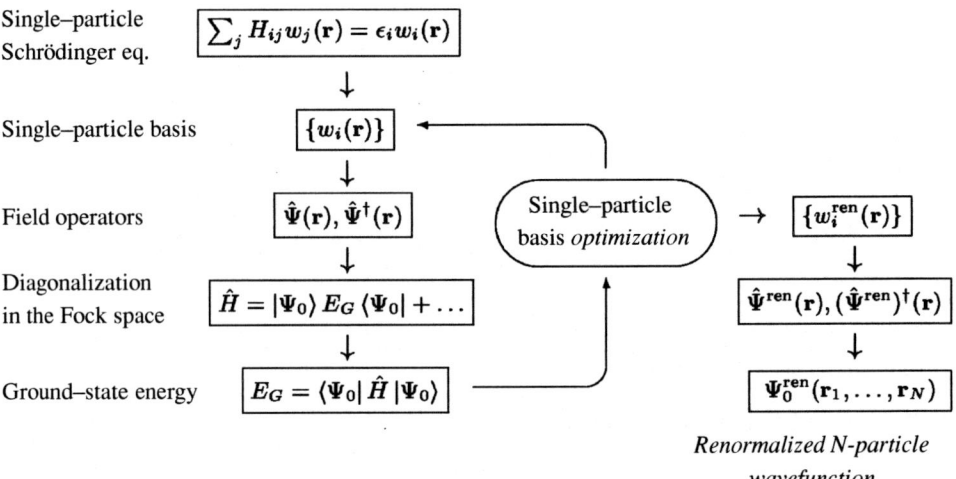

FIGURE 1. Schematic diagram of the *exact diagonalization – ab initio (EDABI) method*. The part on the right characterize *renormalized* quantities: Wannier functions $\{w_i(x)\}$, field operators, and the N-particle wave function.

Having in mind the application to the above systems we have developed a novel method (EDABI) of approach, in which we combine the *Exact Diagonalization*, providing us a rigorous treatment of interparticle interaction, with an *Ab–Initio* optimization of the single–particle wave function in the interacting–system ground state [7]. Earlier results have been reviewed few times as the work progresses [8].

The idea of EDABI is illustrated in Fig.1. We start from selecting the initial orthogonalized atomic or Wannier basis set $\{w_i(\mathbf{r})\}$, composed of atomic–like wave functions $\{\Phi_n(\mathbf{r})\}$ of the radii $\{a_n \equiv \alpha_n^{-1}\}$. Next, we write down the system Hamiltonian in the second–quantized form and determine the ground–state energy E_G and the corresponding state, either by Lanczos or analytic procedure. The ground–state energy is expressed in terms of microscopic parameters $\{t_{ij}, V_{ijkl}\}$ [7] defined, in turn, through the integrals containing the functions $\{w_i(\mathbf{r})\}$. The energy $E_G(t_{ij}\{w_i(\mathbf{r})\}, V_{ijkl}\{w_i(\mathbf{r})\})$ is minimized next with respect to the inverse radii $\{\alpha_n\}$; this represents an iterative procedure entangled with the exact diagonalization, performed at each step and defined by the trial set of $\{\alpha_n\}$ values. The E_G optimization is carried out as a function of interatomic distance for clusters and nanoscopic systems of various geometry and therefore, we have a true microscopic theory. The usual approach to correlated systems terminates with the determination of ground state energy and/or dynamic characteristics of correlated systems as a function of microscopic parameters. Strictly speaking, the procedure outlined above involves a variational determination of the single particle wave function. Before applying this method to concrete examples we describe briefly the concept of *renormalized (self–adjusted) wave equation* for a single particle in milieu of all other particles. The latter concept operates on a more basic level than the variational adjustment of the single–particle wave function.

2. RENORMALIZED (SELF-ADJUSTED) WAVE EQUATION

In the preceding Section we have described a variational procedure of wave–function relaxation to the correlated state. On the fundamental level, one may ask if a renormalized wave equation for a single particle in the environment of all others can be derived. In order to provide the answer in the affirmative, we start with the Hamiltonian of the interacting fermions in the second–quantized form

$$H = \sum_{ij\sigma} t_{ij} a_{i\sigma}^\dagger a_{j\sigma} + \frac{1}{2} \sum_{ijkl\sigma\sigma'} V_{ijkl} a_{i\sigma}^\dagger a_{j\sigma'}^\dagger a_{l\sigma'} a_{k\sigma}. \tag{1}$$

In the above expression $a_{i\sigma}^\dagger$ ($a_{i\sigma}$) is the creation (annihilation) operator of a particle in the single particle state $|i\sigma\rangle \equiv |w_i \chi_\sigma\rangle$ and the microscopic parameters are defined by

$$t_{ij} \equiv \langle w_i | H_1 | w_j \rangle = \int d^3x \, w_i^*(\mathbf{x}) H_1(\mathbf{x}) w_j(\mathbf{x}), \tag{2}$$

and

$$V_{ijkl} \equiv \langle w_i w_j | V | w_k w_l \rangle = \int d^3x \, d^3x' \, w_i^*(\mathbf{x}) w_j^*(\mathbf{x}') V(\mathbf{x} - \mathbf{x}') w_k(\mathbf{x}) w_l(\mathbf{x}'). \tag{3}$$

$H_1(\mathbf{x})$ and $V(\mathbf{x} - \mathbf{x}')$ represent, respectively, the Hamiltonian for a single particle and a single pair of particles in the ordinary (coordinate) representation.

The diagonalization of (1) in the Fock space amounts to determining $E_G = \langle \Psi_G | H | \Psi_G \rangle$ as a function of $\{t_{ij}\}$ and $\{V_{ijkl}\}$. We treat $E_G \equiv E_G\{w_i(\mathbf{x})\}$ as a functional of $\{w_i(\mathbf{x})\}$ under the two constraints: the total particle–number conservation and the fulfilment of the orthogonality condition $\langle w_i | w_j \rangle = \delta_{ij}$. In effect, the *renormalized (self–adjusted) wave equation* takes the form [8]

$$\frac{\delta E_G}{\delta w_i^*(\mathbf{x})} - \nabla \cdot \frac{\delta E_G}{\delta (\nabla w_i^*(\mathbf{x}))} - \sum_{j \geq i, \sigma} (\lambda_{ij} - \mu) w_j(\mathbf{x}) \langle a_{i\sigma}^\dagger a_{j\sigma} \rangle = 0, \tag{4}$$

where λ_{ij} and μ (the chemical potential) are the Lagrange multipliers. If the orbitals are orthogonal then $\lambda_{ij} \equiv 0$. If, additionally, we work with the formalism with fixed number of particles, then also $\mu = 0$, and Eq.(4) reduces to system of *Euler equations* for $\{w_i(\mathbf{x})\}$. This wave equation is extremely difficult to solve at present time. Therefore, in the remaining part we use the variational determination of $\{w_i(\mathbf{x})\}$, as described above.

2.1. Comparison of EDABI with MCI

We compare briefly the proposed EDABI method with the *multi–configurational interaction* (MCI) method developed in quantum chemistry. Both methods belong to the class multi–determinantal expansion of the N–particle wave function. However, the EDABI method is characterized by formal differences in carrying out of the calculations, when compared to MCI. Those differences are threefold:

(i) **Historical.** MCI have evolved from variational methods of quantum physics and chemistry [9] to include the electronic correlations and hence, to obtain lower value of E_G. EDABI represents a procedure of calculating single–particle wave functions concomitant with the exact diagonalization of parametrized models of strongly correlated electrons [7].

(ii) **Technical.** In MCI, we optimize simultaneously the coefficients expressing the weights of different determinants (micro–configurations), as well as the parameters of the trial single–particle basis. In EDABI, we diagonalize Hamiltonian expressed in the Fock space (with the help of analytic or e.g. numerical Lanczos methods), with a simultaneous optimization of the orbital size.

(iii) **Essential.** In the case of exactly soluble models in analytic terms EDABI leads formally to the *renormalized wave equation* (the *nonlinear Schrödinger equation of nonlocal type*). This circumstance opens up a new direction of studies in *mathematical quantum physics*. Additionally, it allows for a direct determination of *dynamical correlation* functions, transport properties, etc., also at temperature $T > 0$.

In studies of the *ground* state of a finite system the differences are mainly technical, as discussed in detail elsewhere [10]. Explicitly, starting from our formulation, one can define the N–particle wave function $\Psi_\alpha(x_1,..,x_N)$ with the help of the field operators $\widehat{\Psi}(x_i)$, as well as the corresponding state $|\Phi_\alpha\rangle$ in the Fock space in the following way

$$\Psi_\alpha(x_1,..,x_N) = \frac{1}{\sqrt{N!}} \langle 0|\widehat{\Psi}(x_1)...\widehat{\Psi}(x_N)|\Phi_\alpha\rangle, \qquad (5)$$

with

$$|\Phi_\alpha\rangle = \frac{1}{\sqrt{N!}} \sum_{j_1...j_N=1}^{M} C_{j_1...j_N} a_{j_1}^\dagger ... a_{j_N}^\dagger |0\rangle, \qquad (6)$$

where $C_{j_1...j_N}$ are the expansion coefficients. A straightforward algebra, combined with the definition of the field operator for the basis containing M states, i.e.

$$\widehat{\Psi}(x) = \sum_{j=1}^{M} w_j(x) a_j, \qquad (7)$$

leads to the expression

$$\Psi_\alpha(x_1,..,x_N) = \frac{1}{\sqrt{N!}} \sum_{i_1...i_N=1}^{M} C_{i_1...i_N} \{A\}[w_{i_1}(x_1)...w_{i_N}(x_N)], \qquad (8)$$

where $\{A\}$ is the symmetrization operations for the case of fermions. This expression supplements the part of Fig.1 in the sense that if we are able to perform the diagonalization (i.e. to find the coefficients $C_{j_1...j_N}$ for the ground state $|\Psi_G\rangle$ for the case of M states composing, usually approximate for $M < \infty$, field operator), then we can write down the corresponding many–particle wave function. As said above, $C_{j_1...j_N}$ are found within EDABI from an exact diagonalization procedure, whereas in MCI approach they are determined variationally.

One methodological remark is in place here. In both EDABI and MCI methods we use a truncated basis containing $M < \infty$ functions $\{w_j(x)\}$ in the expressions (6) or

(7), respectively. This is the reason, why the optimization of any sensible starting basis $\{w_i(\mathbf{x})\}$ is necessary. If the basis was complete, neither variational adjustment of the coefficients C in (6) (in MCI) nor the w_i readjustment (in both methods) were necessary.

2.2. Relation to parametrized models of correlated electrons

Hamiltonian (1) describes interacting fermions. We discuss its explicit form in the single–band (i.e. one orbital per atom) case, where the single–particle basis $|i\rangle \equiv |w_i\rangle$. In that situation, (1) has the form

$$H = \varepsilon_a \sum_{i\sigma} n_{i\sigma} + \sum_{ij\sigma}{}' t_{ij} a_{i\sigma}^\dagger a_{j\sigma} + U \sum_i n_{i\uparrow} n_{i\downarrow} + \frac{1}{2}\left(K_{ij} - \frac{1}{2}J_{ij}\right) \sum_{ij\sigma\sigma'}{}' n_{i\sigma} n_{j\sigma'}$$

$$-\frac{1}{2}\sum_{ij}{}' J_{ij} \mathbf{S}_i \cdot \mathbf{S}_j + J_{ij} \sum_{ij}{}' a_{i\uparrow}^\dagger a_{i\downarrow}^\dagger a_{j\downarrow} a_{j\uparrow} + \sum_{ij}{}' V_{ij} n_{i\bar\sigma} a_{i\sigma}^\dagger a_{j\sigma} + \text{(3– and 4–site terms)}. \quad (9)$$

The first term describes the atomic energy ($\varepsilon_a \equiv \langle i|H_1|i\rangle$), the second is the so-called hoping energy ($t_{ij} \equiv \langle i|H_1|j\rangle$, $i \neq j$), the third and the fourth are, respectively, the intraatomic and interatomic Coulomb interactions ($U \equiv \langle ii|V|ii\rangle$, $K_{ij} \equiv \langle ij|V|ij\rangle$, the fifth and the sixth are the direct exchange and the pair–site hopping $J_{ij} \equiv \langle ij|V|ji\rangle$), and the before last is the so–called correlated hopping ($V_{ij} \equiv \langle ii|V|ij\rangle$). The 3– and 4–site terms are also important for nanoscopic and low–dimensional systems because the electron screening is not as effective, as for 3–dimensional bulk systems.

We diagonalize Hamiltonian (9) for clusters and nanoscopic systems, calculate all the microscopic parameters $\{\varepsilon_a, t_{ij}, U, ...\}$ when the single–particle function is known, all for given lattice parameter R. In this respect, we supplement the usual theoretical analysis of parametrized models with an explicit calculation of the parameters. Therefore, both ground and dynamical characteristics are determined as a function of R.

3. NANOCLUSTERS

As a first application of EDABI we discuss various nanoclusters composed of $N \leq 6$ hydrogen atoms. We take 1s atomic–like orbitals of adjustable size a, composing the Wannier function. In Fig.2 and 3 we provide the ground–state energies (the lowest curves) and those of the excited states for the square and the pentagon configurations, respectively; all curves are drawn as a function of interatomic distance R. A clear evidence for the first three Hubbard subbands can be observed (the horizontal lines represent the energies in the atomic limit with zero, one and two double occupancies, respectively).

So, the principal feature of the correlated systems with narrow bands is preserved in the cluster systems as well. However, for $N = 5$ atoms in the cluster, the Hubbard subbands are not as clearly evidenced when the Lanczos procedure is used for the diagonalization in the Fock space for not too large R.

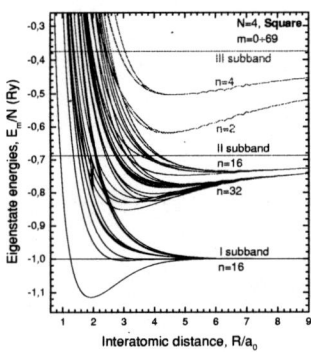

FIGURE 2. Ground– and excited–state energies per atom (E_i/N) versus nearest–neighbor distance for the cluster of $N = 4$ atoms. The horizontal lines mark the positions of the lowest and the two higher Hubbard subbands.

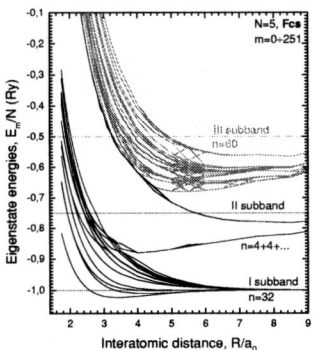

FIGURE 3. Same as in the previous Figure but for $N = 5$ face–centered square (fcs) configuration. The higher subbands are not as clearly defined because the atom in the center has a smaller distance ($\sim 0.7R$) to the corner atoms.

In Fig.4 we display a space profile of the renormalized Wannier function $w_i(\mathbf{x})$ for the cluster of $N = 6$ atoms arranged in a hexagon that is projected onto the cluster plane. The figure is drawn for the interatomic distance $R = 2a_0$ (a_0 is the 1s Bohr radius) and for the atomic 1s wavefunction radius $a_{1s} \simeq 0.89a_0$, for which the energy has its absolute minimum ($E_G = -14.4eV/atom$). Since this is an orthogonalized atomic (Wannier) wave function, its value is negative on the nearest neighboring sites. Additionally, the square or other regular multigon cluster configurations are unstable with respect to dissociation into molecules (with their characteristics at minimum: $E_G = -15.62eV/atom$,

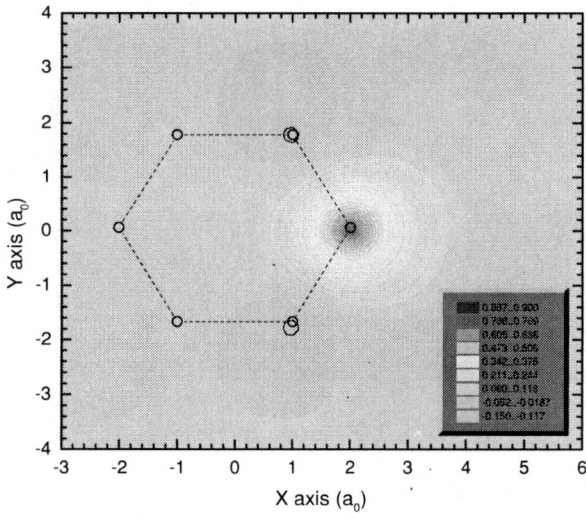

FIGURE 4. Profile of the orthogonalized atomic (Wannier) function $w_i(\mathbf{x})$ for the ground–state configuration. Note the negative values of the function on nearest neighboring sites.

$R/a_0 = 1.45 a_0$, and $a_{1s} = 0.84 a_0$), calculated to the same accuracy level.

The calculations above were performed for Hamiltonian (9) with the 3– and 4–site terms estimated from the condition that their values in the Wannier representation are negligible. The details of that analysis will be reported separately [11].

As an illustration of the method versatility we plot in Fig.5 the electron H_4–cluster density profile, projected onto the cluster plane. To calculate this density $n(\mathbf{x})$ we have used the equivalence between first– and second–quantization representations of the multiparticle wave function [10] to obtain

$$n(\mathbf{x}) = \sum_{i\sigma} |w_i(\mathbf{x})|^2 \langle n_{i\sigma} \rangle + \sum_{ij}{}' w_i^*(\mathbf{x}) w_j(\mathbf{x}) \sum_\sigma \langle a_{i\sigma}^\dagger a_{j\sigma} \rangle. \qquad (10)$$

The first term represent the Hartree–Fock expression, where $\langle n_i \rangle = 1$, whereas the second expresses the multideterminent nature of the N–particle wave function $\Psi_G(\mathbf{x}_1,...,\mathbf{x}_N)$, in which the transitions of particles between the single–particle states $|i\rangle \leftrightarrow |j\rangle$, induced by their mutual interaction, take place. The non–Hartree–Fock–part contribution is essential.

3.1. Stable H_4 clusters

To obtain an energetically stable planar H_4 cluster we should abandon the regular configurations discussed above. Namely, we should distinguish between the lateral distance (say, along the bond length a) and the horizontal intermolecular distance b. In *Table I* we list E_G/N vs. b for the optimal bond length a_{min} and the optimal inverse orbital size

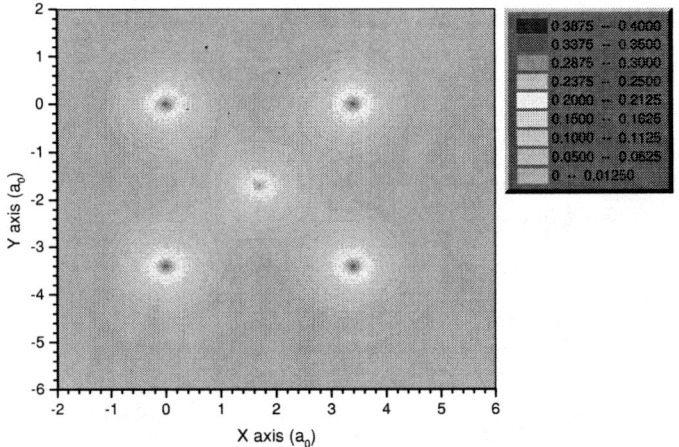

FIGURE 5. Translationally invariant density profile $\rho(\mathbf{x})$ for the cluster containing $N = 4$ atoms.

$\alpha_{min} = 1/a_{min}$. The corresponding energy of two separate H_2 molecules is also shown there. The stable configuration is achieved for $b \simeq 3.2 a_{min}$. We should note that the results listed in *Table I* have been obtained with the help of STO–3G basis, for which the 3– and 4–site interactions in the atomic basis have been taken into account. The simplest version of the Hamiltonian of the type (9), which has been used for these calculations is of the form

$$H = \varepsilon_a^{eff} \sum_{i\sigma} n_{i\sigma} + \sum_{ij\sigma}' t_{ij} a_{i\sigma}^\dagger a_{j\sigma} + U \sum_i n_{i\uparrow} n_{i\downarrow} + \frac{1}{2} \sum_{ij\sigma\sigma'}' K_{ij} \delta n_i \delta n_j, \quad (11)$$

where $\delta n_i = 1 - n_i$ and

$$\varepsilon_a^{eff} = \varepsilon_a + \frac{1}{2N} \sum_{ij}' \left(K_{ij} + \frac{2}{R_{ij}} \right), \quad (12)$$

with the last term in the paranthesis being the Coulomb repulsion of the hydrogen nuclei placed at the distance R_{ij} (is written in atomic units), together with the electron–electron repulsion K_{ij} (such redefinition is needed to achieve the convergence in the atomic limit, $R \to \infty$).

One can extend the present calculation to obtain stable H_{2n} fermionic ladders, both with the planar and the twisted H_2–molecule stackings [12]. The results of this analysis will be reported elsewhere.

4. NANOCHAINS

As a second illustration of EDABI we discuss nanochains of $N \leq 16\, H$ atoms by starting from Hamiltonian (11) and using an adjustable Gaussian (STO–3G) basis represent-

TABLE 1. The optimal bond length a_{min}, inverse orbital size α_{min}, and the ground–state energy E_G per atom for the planar H_4 cluster vs. intermolecular distance b. The corresponding energy of the two molecules $E_G(2H_2)$ is also provided.

b/a_0	a_{min}/a_0	$\alpha_{min}a_0$	E_G/N	$E_G(2H_2)/N$
1.7	1.3627	1.2231	-0.928424	-0.922411
2.0	1.3291	1.2395	-1.019152	-1.016365
2.5	1.3518	1.2304	-1.098551	-1.097314
3.0	1.3829	1.2157	-1.131770	-1.131167
3.5	1.4075	1.2041	-1.144613	-1.144317
4.0	1.4238	1.1969	-1.148598	-1.148454
5.0	1.4373	1.1911	-1.148093	-1.148056
6.0	1.4390	1.1908	-1.145975	-1.145964
8.0	1.4375	1.1924	-1.143651	-1.143649
10.0	1.4366	1.1929	-1.142756	-1.142755
20.0	1.4357	1.1940	-1.141908	-1.141908
∞	1.4356	1.1943	-1.141783	

ing single particle 1s–like states, of which we compose the Wannier functions (in the tight–binding approximation, with up to six atomic Coulomb wells representing the periodic potential). In Fig.6 we plot the ground state energy per atom and in the inset the inverse orbital size α (in units of inverse a_0), for $N = 6 \div 10$ atoms. The continuous lines: INS and M correspond, respectively, the Mott insulating state, representing by $E_G^{INS} = \varepsilon_a^{eff}$, and to the ideal metallic state, for which

$$E_G^M = \varepsilon_a^{eff} - \frac{4|t|}{\pi} + \frac{1}{2N}\sum_{ij}{}'K_{ij}\langle \delta n_i \delta n_j \rangle, \tag{13}$$

where

$$\langle \delta n_i \delta n_j \rangle = -\frac{\sin^2(\pi|i-j|/2)}{\pi|i-j|^2}, \tag{14}$$

corresponds to the value for an ideal fermionic gas.

The crossover from the *metallic* to the *Mott insulating* state can be seen clearly. The dotted line represents the Hartree–Fock estimate of the energy per atom (as the best single–determinant wave function, it provides its upper estimate). The atomic orbital radius, as a rule, is smaller than the corresponding atomic size a_0).

The dynamical spin–spin correlations $\langle \mathbf{S}_i \cdot \mathbf{S}_j \rangle$ have also been determined and shown for the half–filling (i.e. with one electron per atom); they exhibit a typical oscillating feature for the Heisenberg (or Slater) two–sublattice antiferromagnet; these are represented in Fig.7. The density of states shows the Hubbard two–subband structure (cf. Rycerz and Spałek [8]).

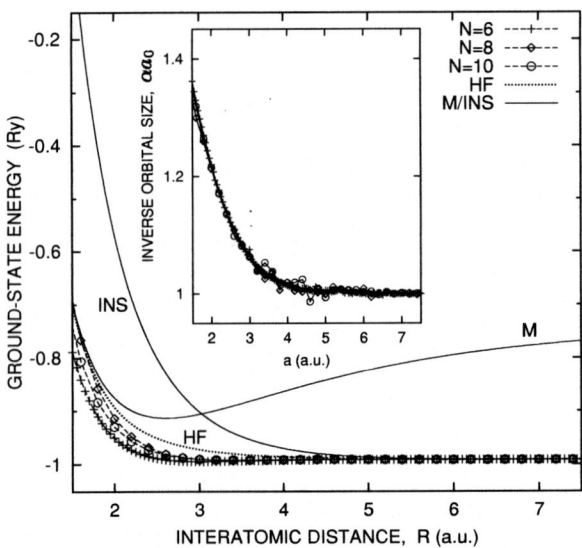

FIGURE 6. Ground state energy obtained from EDABI, as well as from approximate approaches discussed in the text. Inset: the inverse optimal Gaussian orbital size α (in units of a_0).

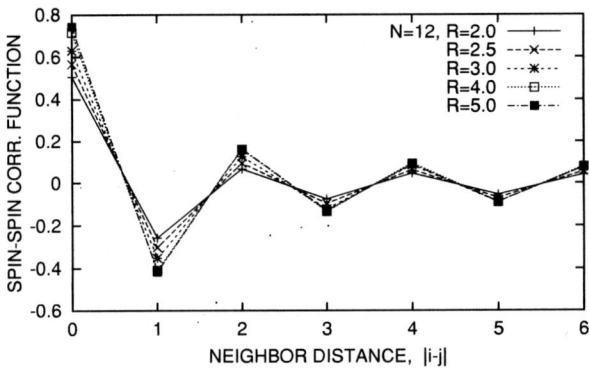

FIGURE 7. Spin–spin correlation function *vs.* neighbor distance $|i - j|$, for different lattice constant R and for $N = 12$ atoms.

4.1. Landau–Fermi or Tomonaga–Luttinger nanoliquid?

One dimensional conductors are very special, since they retain Fermi–surface volume (obey the Luttinger theorem) and, at the same time, there are no *fermionic quasiparticles*, but instead, the collective excitations are combined the *bosonic charge (holons)* and the spin *(spinons)* fluctuations propagating with different velocities. Namely, a propagating electron decomposes into such charge– and spin–excitation branches, which separate spatially with time (exhibit the phenomenon of *charge–spin separation*) [13]. Therefore, it is obviously interesting whether a lattice nanosystem can be represented in these categories. In Fig.8 we plot the statistical distribution function $\langle n_{\mathbf{k}\sigma}\rangle$ for a half–filled system containing up to $N = 14$ electrons. The continuous lines represent the formula

$$\langle n_{\mathbf{k}\sigma}\rangle = \frac{1}{2} + sgn(k - k_F)[\alpha(k - k_F)^2 + \beta|k - k_F|]. \qquad (15)$$

One can say that, to a good approximation, Eq.(15) is the scaling relation, valid even for $N \sim 10$. From this scaling law one can determine the quasiparticle mass enhancement m^*/m_B (where m_B is the band mass), which defines the criterion of Mott–Hubbard localization as $m^* \to \infty$ (cf. Spałek and Rycerz [7]). It is quite suprising that even small system of $N \sim 10$ particles and in the limit of small interatomic distances $R \lesssim 3a_0$, exhibits the Fermi–Dirac behavior, obviously modified by the interparticle interaction (the Fermi–liquid effects). At the critical distance $R \sim 3.4a_0$, the Fermi ridge (the discontinuity at $k_F \simeq \pi/(2R)$ – the Fermi wavevector), vanishes and the system cannot be regarded as metal; this is the exactly the spacing, at which $m^* \to \infty$. So, we have a criterion: only the nanochains with $R < R_c$ can be regarded as nanometals (with a gap for transport properties due to geometrical quantization of k, of course).

This nice Fermi–liquid picture can be supplemented with an alternative Tomonaga–Luttinger picture for which to the leading orders the statistical distribution takes the form

$$\ln|n_F - \langle n_{\mathbf{k}\sigma}\rangle| = -\Theta \ln z + b \ln \ln z + c + o(1/\ln z), \qquad (16)$$

where $z \equiv \pi/|k - k_F|$ and $n_F = 1/2$ is the value of $\langle n_{\mathbf{k}\sigma}\rangle$ at k_F. In Fig.9 we plot the same points as in Fig.8, together with Tomonaga–Luttinger–model (TLM) scaling law (16).

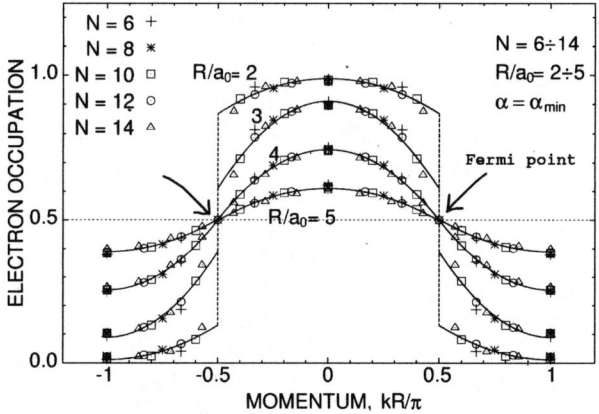

FIGURE 8. Momentum distribution function $\langle n_{\mathbf{k}\sigma}\rangle$ for a half–filled 1D chain and its Fermi–liquid interpretation for R and N specified.

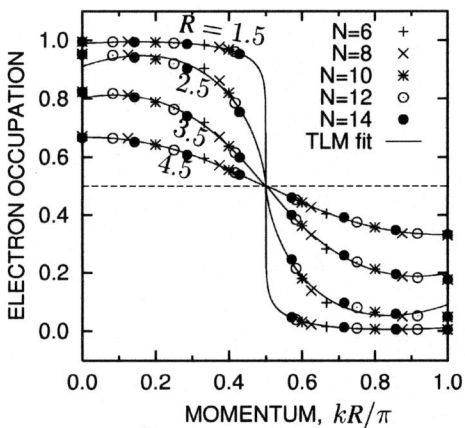

FIGURE 9. Tomonaga–Luttinger scaling of the momentum distribution function for a half–filled 1D chain of $N = 6 - 14$ atoms with long-range Coulomb interaction.

The main drawback of the present fit is that the present curve dives in as one departs below the k_F point. The second, more serious, is the circumstance that we have no points very close to k_F; this follows from the insufficient system size to test reliably TLM. However, one sees clearly that the two viewpoints: the Fermi–liquid and the Tomonaga–Luttinger, can compete at this point, at least at half filling. The lack of clear–cut nature of such small quantum liquid is obviously caused by its nanosize. However, it is suprising that the description, developed for $N \to \infty$ systems, is being observed to some degree, even for $N \sim 10$ atoms. For a quarter filling TLM picture seems more appropriate. A more detailed discussion, together with the spacing dependence of the distribution parameters (α, β, Θ, and b), will be performed separately [14].

5. CONCLUDING REMARKS

In this minireview we have presented the method of *exact diagonalization* combined with *ab initio* approach (EDABI), within which one can calculate the electronic properties of correlated nanoscopic systems as a function of interatomic distance. The same method can be applied to extended systems, for which an exact solution in an analytic form is possible, viz. the Hubbard model in one spatial dimension [15]. Minimally, we supplement the parametrized–model solution with the calculation of the wave–functions contained in the model parameters and thus, complete the solution. The method is particularly suited for strongly correlated systems, for which we cannot start from the one–particle wave equation and after solving it include the electron correlations. In fact, we show how the *renormalized wave equation* can be derived that describes the single–particle wave function readjustment to the correlated state of *all* other particles. This renormalized (self–adjusted) wave equation is extremely difficult to solve even for simple (e.g. H_2)systems and therefore, we resorted here to either numerical and/or variational solutions. The near future should show to applicability of the EDABI approach to nonperturbational problems, the physics of strongly correlated (and other) systems brings about.

ACKNOWLEDGMENTS

The work reviewed here was supported by the KBN of Poland, Grant Nos. 2P03B 050 23 and 2P03B 064 22. Two authors (J.S. and A.R.) acknowledge also the scholarships of the Polish Science Foundation (FNP) awarded to them. All of us would like to thank cordially to Prof. Ferdinando Mancini for the Schools his group organizers, as well as to dedicate him this paper on the occassion of his 60th Birthday.

REFERENCES

1. Feynman R.P., *There is Plenty of Room at the Bottom*, in *Engineering and Science*, February (1960); see: www.zyvex.com/nanotech/feynman.html.
2. Yanson A.I. et al., *Nature* (London) **395**, 783 (1998); Ohnishi H., Kondo Y., and Takayanagi., *ibid.* p.780.
3. Frank S. et al., Science **280**, 1744 (199).
4. For review see e.g. Kouwenhoven L.P., Austing D.G., and Tarucha S., Rep. Prog. Phys. **64**, 701 (2001).
5. Bahn S.R. and Jacobsen W., Phys. Rev. Lett. **87**, 266101 (2001).
6. Tsukamoto S. and Kikuji H., Phys. Rev. B**66**, 161402(R) (2002); Ahn J.R. et al., Phys. Rev. B**66**, 153403 (2002).
7. Spałek J. et al., Phys. Rev. B**61**, 15676 (2000); Rycerz A. and Spałek J., *ibid.* **63**, 073101 (2001); *ibid.* **65**, 035110 (2002); *ibid.***64**, R161105 (2001).
8. Spałek J. et al., Acta Phys. Polonica B**31**, 2879 (2000); *ibid.***32**, 3189 (2001); Rycerz A., in *Lectures on the Physics of Highly Correlated Electron Systems VI*, edited by F. Mancini, AIP Conf. Proc., vol.629, pp.213–222 (2002); *ibid.* VII, to be published (2003); Spałek J. et al., in *Proc. of the NATO: Advanced Research Workshop: Concepts in Electron Correlations*, Kluwer Academic, Dordrecht, in press.
9. C.f. e.g. Shavitt R.A., in H. Schaeffer (editor), *Methods of Electronic Structure Theory*, Plenum Press, New York, 1977, pp.189–275.
10. Spałek J. et al., in preparation.
11. Zahorbeński R. and Spałek J., unpublished.
12. Rycerz A., Ph.D. Thesis, Jagiellonian University, Kraków–2003 (unpublished).
13. For review see: Voit J., Rep. Prog. Phys. **57**, 977 (1995); Solyom J., Adv. Phys. **28**, 201 (1979).
14. Rycerz A. and Spałek J., unpublished.
15. Kurzyk J., Spałek J., and Wójcik W., in preparation.

Spin ordering in the one-dimensional Kondo lattice and Double-exchange Models

D.J. Garcia*, K. Hallberg*, M. Avignon† and B. Alascio*

*Instituto Balseiro and Centro Atómico Bariloche, Comisión de Energia Atómica, 8400 San Carlos de Bariloche, Argentina.
†Laboratoire d'Etudes des Propriétés Electroniques des Solides, Associated with Université Joseph Fourier, C.N.R.S., BP 166, 38042 Grenoble Cedex 9, France.

Abstract. We study the spin ordering of localized quantum spins interacting with itinerant electrons via antiferromagnetic (Kondo model) or ferromagnetic (Hund coupling) exchange interactions in the one-dimensional lattice models. In the case of ferromagnetic coupling we also include a super-exchange interaction between the localized spins. Using the Density Matrix Renormalization Group (DMRG) we show, for both models, the existence of new magnetic phases with spin ordering consisting of *ferromagnetic islands* coupled antiferromagnetically.

INTRODUCTION

The study of a lattice of quantum localized moments interacting with conduction electrons is one of the very active field of condensed matter. An important problem remains concerning the ordering of the localized moments even in one dimension. Heavy-fermion systems and Kondo insulators are typical examples of systems in which these interactions are essential[1, 2]. Their physical properties result from an antiferromagnetic interaction J between the localized spins and itinerant electron spins, the so-called Kondo lattice model (KLM). The corresponding Hamiltonian has the well-known form:

$$H = -t \sum_{i,j,\sigma} c^{\dagger}_{i\sigma} c_{j\sigma} - J \sum_{i} \vec{S}_i \cdot \vec{\sigma}_i$$

The first term represents the conduction electron hopping between nearest-neighbor sites, $c^{\dagger}_{i\sigma}$ ($c_{i\sigma}$) being standard creation (anihilation) operators. In the second term the exchange interaction J is antiferromagnetic ($J < 0$), and $\vec{\sigma}_i = \frac{1}{2} \sum_{\sigma,\sigma'} c^{\dagger}_{i\sigma} \tau_{\sigma\sigma'} c_{i\sigma'}$, ($\tau_{\sigma\sigma'}$ are Pauli matrices).

It is interesting to note that, in recent years, the same model Hamiltonian with ferromagnetic coupling ($J > 0$) has been considered to contain the basic physics of manganites exhibiting "colossal" magnetoresistance effect[3, 4, 5, 6]. In this case, both localized spins and itinerant electrons originate from manganese d-states. The system is assumed to contain essentially Mn^{4+} ions with three localized t_{2g} orbitals represented as local spins $\vec{S}_i$ and additional itinerant electrons in the e_g orbital. Due to the strong Hund coupling the spin of the e_g electron is constrained to be parallel to the local spin on that site. Hund's rule together with the hopping term give rise to the "double-exchange"

(DE) interaction that favors ferromagnetic ordering of the local spins. In recent literature this model is often referred to as the ferromagnetic Kondo lattice (FKLM), however to avoid confusion with the Kondo model, we will call it Hund model (HM). Many authors have also considered the possibility of an antiferromagnetic interaction between local spins arising from super-exchange (SE) mechanism. Such interaction may be expressed as a Heisenberg interaction $K \sum_{i,j} \vec{S}_i \cdot \vec{S}_j$, $K > 0$ is the antiferromagnetic SE interaction between nearest neighbor local spins. The competition between DE and SE may lead to the formation of interesting magnetic superstructures. The resulting full Hamiltonian will be called the DE-SE model:

$$H = -t \sum_{i,j,\sigma} c_{i\sigma}^\dagger c_{j\sigma} - J \sum_i \vec{S}_i \cdot \vec{\sigma}_i + K \sum_{i,j} \vec{S}_i \cdot \vec{S}_j$$

where $J > 0$. We consider $S = 1/2$ localized quantum spins and present numerical results using the density-matrix renormalization group (DMRG)[7] with open boundary conditions for chains of different sizes. The different phases are characterized through the local spin-spin and itinerant charge-charge correlation functions and their Fourier transforms, the following spin and charge structure factors:

$$S(q) = \frac{1}{L} \sum_{i,j} e^{iq(R_j - R_i)} \langle \vec{S}_i \vec{S}_j \rangle$$

$$N(q) = \frac{1}{L} \sum_{i,j} e^{iq(R_j - R_i)} \langle (n_i - n)(n_j - n) \rangle$$

where L is the number of sites in the system, n the number of conduction electrons per site. The total spin can be obtained from $S(q=0)$.

THE KONDO-HUND MODELS ($K = 0$)

The phase diagrams of the one-dimensional KLM and HM (i.e. $K = 0$) present great similarities[2, 8, 9, 10]. The half-filled case $n = 1$ is pathological in both models whose ground state is very different from the $n \neq 1$ case: it is "spin liquid" in the Kondo case and "antiferromagnetic" in the Hund case. Both models have a Ferromagnetic phase at large J away from $n = 1$ in spite of the difference of their ground-states for $n = 1$. At lower J a phase qualified as "paramagnetic" in the KLM and "incommensurate" in the HM has been identified with exact diagonalization and DMRG, this phase is however much less understood than the ferromagnetic phase. The local spin-spin correlations are determined by the conduction electrons and $S(q)$ shows a peak at the wave number corresponding to $2k_F$ of the conduction electrons. The ground state undergoes a transition to the ferromagnetic phase with increasing J. Recently, the existence of a "spin dimerized" phase has been reported[11] for the Kondo model at quarter-filling ($n = 1/2$) for $J/t = -0.5$, through real space spin-spin correlations. The spin structure is of the island type ..↑↑↓↓.. similar to the one we have identified previously for the DE-SE model [12].

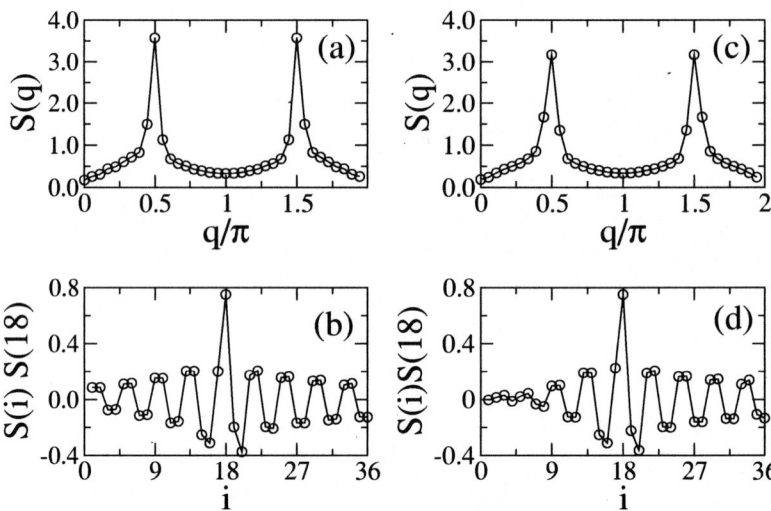

FIGURE 1. Spin-spin correlations S(i)S(L/2) and spin structure factor S(q) showing islands for n=1/2: (a)-(b) Hund coupling J/t=2, (c)-(d) Kondo coupling J/t=-0.5 ($K = 0$).

Our results confirm the existence of this phase and show that this "island" structure is not specific to the Kondo interaction but also exists in the ferromagnetic coupling case. We show that similar structures are present for other fillings $n = 1/3$ and $2/3$ as well. Besides the characterization of these phases, an interesting question is how the spin correlations evolve towards the ferromagnetic phase with increasing coupling J. Complete results will be published in detail elsewhere[13].

We report here new results for $n = 1/2, 1/3$ and $2/3$ and different values of J, both in the F and AF coupling cases.

Let us first present the quarter-filled case $n = 1/2$. For $J/t \gtrsim 14$, the F-phase is clearly identified with a peak of $S(q)$ at $q = 0$, and the total spin $S_T = N_s/2$, where N_s is the number of localized spins in the system. Each electron forms a triplet state $S = 1$ with the localized spins and all localized spins are ferromagnetically ordered. When J decreases ($J/t \lesssim 6$) the spin structure transforms into an "island" structure with a peak of $S(q)$ at $q = \pi/2$ as shown in Fig.1 (a) indicating a four sites periodicity. This phase remains stable down to very small values of J/t. The total spin $S_T = 0$ within the extrapolation error (the obtained values are ~ 0.1). The real space spin-spin correlations (Fig.2(b)) change sign every two sites indicating that the structure is effectively of the "island" type ↑↑↓↓ mentioned above with quasi-long range ordering. In the intermediate region it is difficult to identify the spin structure. Similar results are obtained in the Kondo case ($J < 0$). The island-type phase is conserved for small Kondo coupling ($J/t \gtrsim -1.2$) (Fig. 1 (c)-(d)) and the ferromagnetic phase is recovered for $J/t \lesssim -1.6$. In the strong Kondo coupling limit the ferromagnetic state differs from the Hund case in the sense that now, the conduction electrons form singlet states ($S = 0$) with the localized spins, so that the system contains $N_s/2$ singlets and $N_s/2$ unpaired localized spins. The unpaired spins are ferromagnetically coupled ($S_z = 1/2$) and the total spin is now $S_T = N_s(1-n)/2 = N_s/4$.

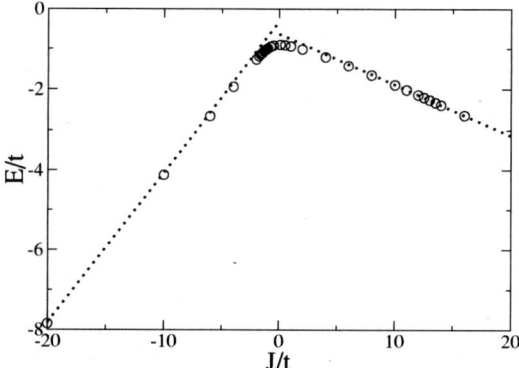

FIGURE 2. DMRG ground-state energy (circles) as function of J/t for the Kondo and Hund models (n=1/2 and $K=0$). Dotted lines represent the strong coupling limit.

We found that S_T increases with increasing the absolute value of the Kondo coupling $|J|$ and reaches the strong coupling maximum value $N_s/4$ for $J/t = -4$, while the spin-spin correlation $\langle S_i S_{i+1} \rangle$ (evaluated at the center of the chain for greater precision) continues to decrease smoothly towards $1/16$ for $J/t \leq -4$; this value is simply understood considering that the probability of nearest-neighbor localized spins will be $1/4$.

In Fig.2 we show the energy as function of J. The dotted line represents the energy of the ferromagnetic state in the strong coupling ($|J|/t \to \infty$) limit. This limit is more easily understood by transforming the Hamiltonian into a basis that diagonalizes the exchange interaction. The localized spins $S = 1/2$ are represented by fermion operators $f_{i\sigma}$, $\vec{S}_{i\sigma} = \frac{1}{2}\sum_{\sigma,\sigma'} f_{i\sigma}^\dagger \tau_{\sigma\sigma'} f_{i\sigma'}$, ($\tau_{\sigma\sigma'}$ are Pauli matrices) with the constraint that there is one f-fermion per site and boson operators representing singlet and triplet states between localized and itinerant fermions are introduced:

$$s_i^\dagger = \frac{1}{\sqrt{2}}\left(f_{i\uparrow}^\dagger c_{i\downarrow}^\dagger - f_{i\downarrow}^\dagger c_{i\uparrow}^\dagger\right) \quad t_{i,1}^\dagger = f_{i\uparrow}^\dagger c_{i\uparrow}^\dagger$$

$$t_{i,0}^\dagger = \frac{1}{\sqrt{2}}\left(f_{i\uparrow}^\dagger c_{i\downarrow}^\dagger + f_{i\downarrow}^\dagger c_{i\uparrow}^\dagger\right) \quad t_{i,-1}^\dagger = f_{i\downarrow}^\dagger c_{i\downarrow}^\dagger$$

Together with them the operators $\tilde{f}_{i\sigma}^\dagger = f_{i\sigma}^\dagger \left(1 - c_{i\uparrow}^\dagger c_{i\uparrow}\right)\left(1 - c_{i\downarrow}^\dagger c_{i\downarrow}\right)$ represent the unpaired localized spins.

These operators denote on-site states which are singly occupied or empty in terms of conduction electrons. Discarding doubly occupied sites (as justified below), only one of these states can be realized on a given site, so these operators have to satisfy the constraint $s_i^\dagger s_i + \sum_\mu t_{i,\mu}^\dagger t_{i,\mu} + \sum_\sigma \tilde{f}_{i\sigma}^\dagger \tilde{f}_{i\sigma} = 1$. Large exchange coupling $J \to \infty$ automatically prevents double occupation. It might be convenient to consider the limit of strong correlation among the conduction electrons ($U \to \infty$) to eliminate doubly occupied sites, however in this case, the physics might be different from that of an uncorrelated band for smaller coupling J. The conduction electron operators $c_{i\sigma}^\dagger$ ($c_{i\sigma}$) can be expressed in terms of the new operators, the kinetic energy term separates into a singlet, triplet and singlet-triplet parts. The full Hamiltonian then becomes:

$$H = H_{ex} + H_s + H_t + H_{st}$$

with

$$H_{ex} = -\frac{J}{4}\sum_i \sum_\mu t^\dagger_{i,\mu} t_{i,\mu} + \frac{3J}{4}\sum_i s^\dagger_i s_i \quad (1)$$

$$H_s = -\frac{t}{2}\sum_{i,\sigma} s^\dagger_{i+1} s_i \tilde{f}^\dagger_{i+1,\sigma} \tilde{f}_{i\sigma} + h.c \quad (2)$$

$$H_t = -t\sum_{i,\sigma}[(t^\dagger_{i+1,1} t_{i,1} + \frac{1}{2} t^\dagger_{i+1,0} t_{i,0})\tilde{f}_{i+1,\uparrow}\tilde{f}^\dagger_{i\uparrow} + (t^\dagger_{i+1,-1} t_{i,-1} + \frac{1}{2} t^\dagger_{i+1,0} t_{i,0})\tilde{f}_{i+1,\downarrow}\tilde{f}^\dagger_{i\downarrow}$$
$$+ \frac{1}{\sqrt{2}}\{(t^\dagger_{i+1,0} t_{i,-1} + t^\dagger_{i+1,1} t_{i,0})\tilde{f}_{i+1,\uparrow}\tilde{f}^\dagger_{i\downarrow} + (t^\dagger_{i+1,0} t_{i,1} + t^\dagger_{i+1,-1} t_{i,0})\tilde{f}_{i+1,\downarrow}\tilde{f}^\dagger_{i\uparrow}\} + h.c]$$

$$H_{st} = -\frac{t}{2}\sum_{i,\sigma}\sigma s^\dagger_{i+1} t_{i,0} \tilde{f}^\dagger_{i+1,\sigma}\tilde{f}_{i\sigma} - \frac{t}{\sqrt{2}}\{(s^\dagger_{i+1} t_{i,-1} - t^\dagger_{i+1,1} s_i)\tilde{f}_{i+1,\uparrow}\tilde{f}^\dagger_{i\downarrow}$$
$$+ (t^\dagger_{i+1,-1} s_i - s^\dagger_{i+1} t_{i,1})\tilde{f}_{i+1,\downarrow}\tilde{f}^\dagger_{i\uparrow}\}$$

As we have already mentioned, in the Kondo case one can retain, to a good approximation, only singlet states and f-fermions so the Hamiltonian reduces to $H = \frac{3J}{4}\sum_i s^\dagger_i s_i + H_s$, with hopping $t/2$ between the singlet states. In the Hund case, one can retain only $t_1(t_{-1})$ triplet states together with f-fermions and the Hamiltonian will reduce to $H = -\frac{J}{4}\sum_{i,\mu} t^\dagger_{i,\mu} t_{i,\mu} + t\sum_{i,\sigma}(t^\dagger_{i+1,1} t_{i,1} \tilde{f}^\dagger_{i\uparrow}\tilde{f}_{i+1,\uparrow} + h.c)$, with hopping t between $t_1(t_{-1})$ triplet states. Therefore the energy of the ferromagnetic state in the strong coupling limit becomes $-J/8 - 2t/\pi$ in the Hund case and $3J/8 - t/\pi$ in the Kondo case. Our numerical results compare well with these values for large J. However a detailed treatment of the "island" regime at small J has not yet been given.

Similar "island" phases are clearly evidenced for $n = 1/3$ for intermediate values of the coupling $-0.5 \lesssim J/t \lesssim 2$. Typical examples are for $J/t = 1$ and $J/t = -0.1$ (see Fig. 3 (a-d)). $S(q)$ shows a clear peak at $q = \pi/3$ and the spin correlation in real space presents an "island" structure of three ferromagnetic spins coupled antiferromagnetically between islands, basically ↑↑↑↓↓↓. The ferromagnetic phase as discussed above is clearly recovered for $J/t \lesssim -1$ and $J/t \gtrsim 5.5$. In the intermediate region (for values of J between these intervals) our numerical results are not so transparent and it is difficult to define precisely the structure of the ground state. The same occurs for $n = 2/3$. The island structure of the type ↑↑↓↑↑↓ is clearly identified for $-1 \lesssim J/t \lesssim 6$. In Fig.3 (e-h), we show results for $J/t = 0.5$ and -0.5.

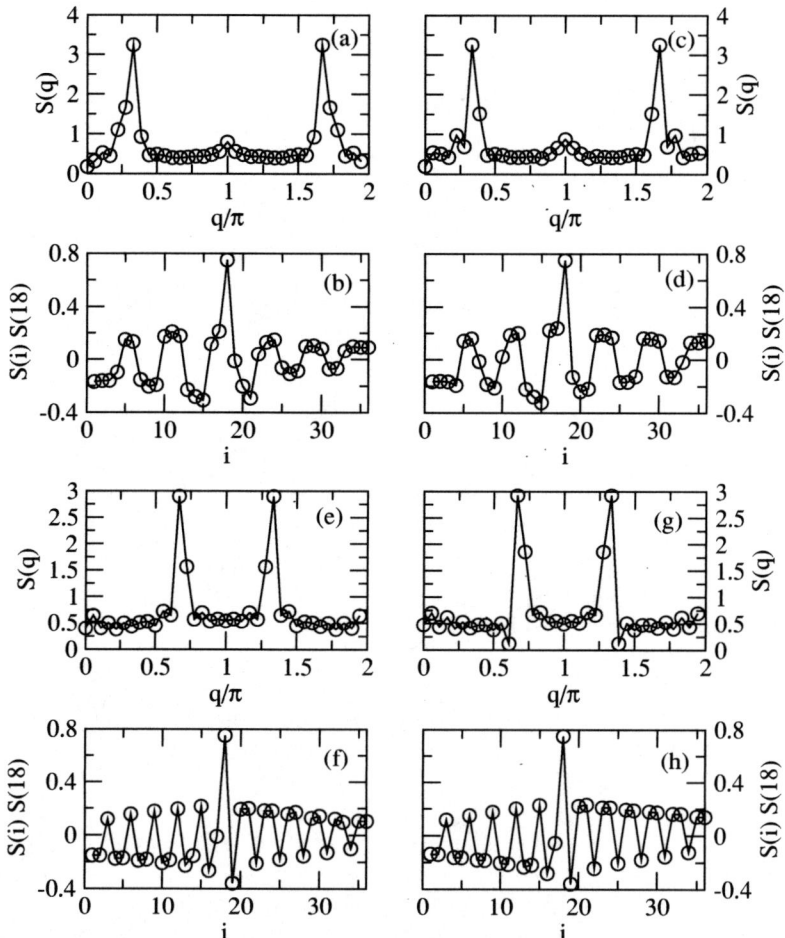

FIGURE 3. Same as fig.1 for: n=1/3 [J/t=1 (a)-(b) and J/t=-0.3 (c)-(d)] and n=2/3 [J/t=2.0 (e)-(f) and J/t=-0.5 (g)-(h).

THE DOUBLE-EXCHANGE-SUPER-EXCHANGE MODEL ($K \neq 0$)

We present our results obtained with competing double and super-exchange interactions[12]. We have considered the situation where the Hund coupling is large $J/t = 20$ so that, according to the above results, the ground state would be ferromagnetic in the absence of SE interaction ($K = 0$) except for close to $n = 1$, that is the case for the considered fillings $n = 1/3, 1/2$ and $2/3$. The antiferromagnetic SE interaction stabilizes an AF phase for $n = 0$ and will therefore compete with the DE term for intermediate fillings.

We describe below the different phases obtained:

(a) There is a fully polarized FM phase for small K/t and an AF phase for sufficiently

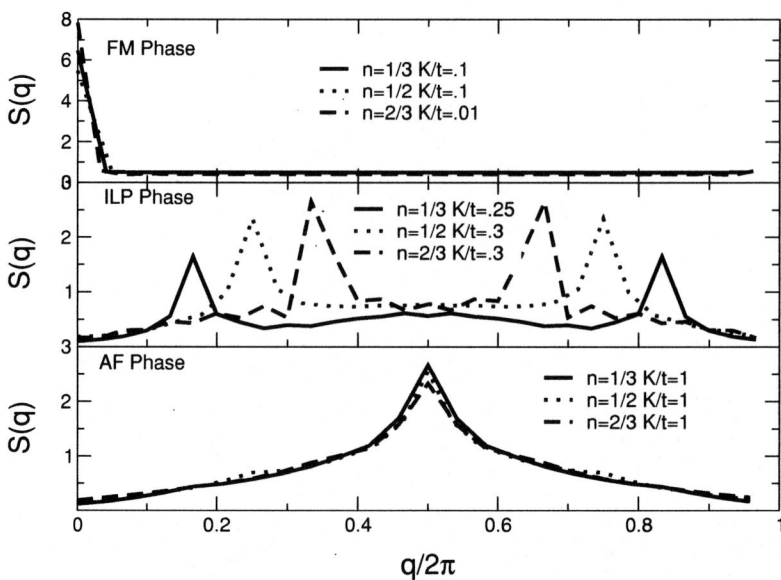

FIGURE 4. Spin correlations for the different phases found: a) Ferromagnetic phase (FM); b) Island-like Phase (ILP); c) Antiferromagnetic Phase (AF); $J/t = 20$

large K/t. This can be seen in Fig. 4 where we have plotted the spin structure factor $S(q)$ for some typical values of K. The corresponding real-space spin-spin correlations are also shown (Fig. 5 (a) and (f)).

(b) In the intermediate regime we get phases with clear peaks at $q = \pi/3, \pi/2$ and $2\pi/3$ for $n = 1/3, 1/2$ and $2/3$ respectively, correspoding to a 6, 4 and 3-sites periodicity for the localized spins. We found a nearly logarthmic increase of the peaks of $S(q)$ with the system size[12], which is indicative of a power-law decay with exponent one (however other values of this exponent cannot be excluded due to the system sizes considered) for the spin-spin correlation as a function of the distance (however other values of this exponent cannot be excluded due to the system sizes considered). In any case, the correlations are quasi-long-ranged. The absence of long-range order is a consequence of the one-dimension character of the model with SU(2) symmetry. This $q = \pi/2$ peak for $n = 1/2$ has been obtained with the Monte Carlo method for classical spins, however this has been interpreted as a spiral state[14]. From the real space spin-spin correlations (Fig. 5 (b-e)) we can rule out the spiral state since the "islands" are clearly distinguishable for the different fillings. For $n = 1/3$ and $n = 1/2$ the phase roughly consists of FM islands of three spins and two spins respectively which are aligned antiferromagnetically .(Fig. 5 (b) and (c)), schematically represented as .. ↑↑↑↓↓↓↑↑↑↓↓↓ . and ... ↑↑↓↓↑↑↓↓ ..For $n = 2/3$ we find two-site FM islands separated by antiparallel spins (see Fig. 5 (d)), the classical image being .. ↑↑↓↑↑↓↑↑↓ ..In Fig. 5 (e) we also report a result for the case $n = 1/4$ showing a structure of this type ..↑↑↑↓↑↑↑↓ ..In Fig. 5 we also show the charge distribution $n(i)$ for the different cases. Charge ordering (CO) is observed for $n = 1/3, 2/3$ and $1/4$ [Fig.5 (b)(d)(e)] due to the inequivalence of the sites within the islands. For $n = 1/2$ there is no CO, all sites being equivalent. However the effective hopping within a FM island is larger than the hopping between "antiparallel spins" therefore localizing the charges in the ferromagnetic bonds within the islands. As a consequence the charge structure factor $N(q)$ differs from the homogeneous FM case and the shape of $N(q)$ is closer to the charge structure factor of a completely dimerized state $(1-\cos q)/4$[12]. It is well known that this type of bond ordering also opens a gap that gives rise to an insulating behavior. Similar bond ordering also exists for other fillings in addition to the CO.

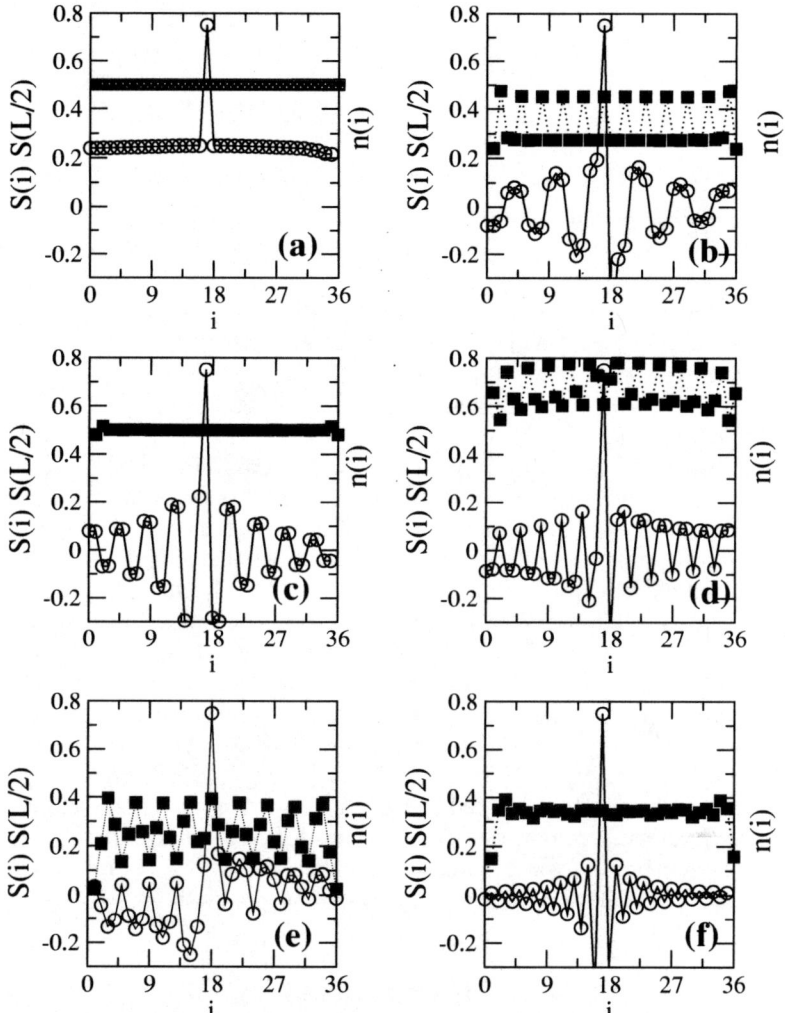

FIGURE 5. Real space spin-spin correlation $S(i)S(L/2)$ (left axis) and charge density $n(i)$ (right axis) in different phases for the DE-SE model for (a) F n=1/2, K/t=0.1 (b) Island (I) n=1/3 K/t=0.22 (c)I n=1/2 K/t=0.2 (d) I n=2/3 K/t=0.3 (e)I n=1/4 K/t=0.2 (f) AF n=1/3 K/t=1.

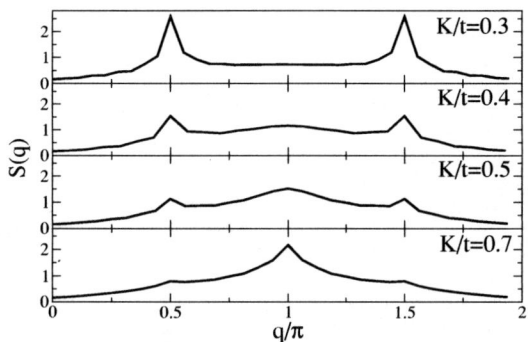

FIGURE 6. Spin structure factor S(q) for n=1/2 and different values of the super-exchange interaction K/t for the DE-SE model where the transition from the island to the AF phases is clearly seen.

It is interesting to examine in more detail how the island phase evolves towards the AF phase. As we have already mentioned it is quite clear that the phase is AF for K sufficiently large, however it is not so easy to determine precisely the boundary between the island phases and AF. We will consider the simplest case $n = 1/2$. In Fig. 6 we present the evolution of $S(q)$ with increasing K/t between 0.3 and 0.7. The $q = \pi$ peak increases at the expense of the $q = \pi/2$ peak. The amplitude of both peaks is plotted in Fig 7 (b) as function of K/t together with the ferromagnetic peak $S(0)$ clearly showing the transition between the F and island phases for $K/t \simeq 0.2$; for $K/t \gtrsim 1$, even though the $S(q)$ has still a finite amplitude at $q = \pi/2$, the peak disappears inside the broad $q = \pi$ peak and it becomes difficult to make a clear distinction between islands and antiferromagnetism. The same occurs for $n = 1/3$ and $2/3$. Larger system sizes would be necessary to clarify this point. The spin-spin correlations inside the island and between islands evaluated at the center of the chain shown in Fig. 8 illustrate the evolution of the islands. The correlation inside the island decreases continuously from completely ferromagnetic to antiferromagnetic and converges towards the value corresponding to a Heisenberg chain at large K/t (shown by the full line). For comparison, in this figure we also show the spin-spin correlations for classical spins represented by the dotted lines. A complete calculation with classical spins[15] shows that the structure within the island changes from ferromagnetic (↑↑↓↓↑↑↓↓) to canted (↖↗↙↘↖↗), the two spins inside the islands forming an angle θ, while between islands the spins remain antiparallel; the canting angle θ varies from $\theta = 0$ to $\theta = \pi$ when K/t increases. Strictly speaking the ground state is never antiferromagnetic for classical spins, however for quantum spins this issue is more difficult to settle from our numerical calculations.

CONCLUSIONS

Summarizing, our results show that the formation of commensurate spin and charge structures (which we call "island phases") appear naturally in the DE-SE model as a consequence of the competition between double-exchange and superexchange interac-

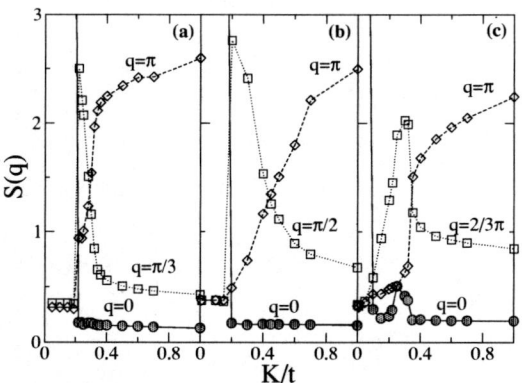

FIGURE 7. Amplitude of the peaks of S(q) as function of K for (a) n=1/3 (b) n=1/2 (c) n=2/3. For small K/t, the value of the FM peak at $q = 0$ is out of the scale of the figure.

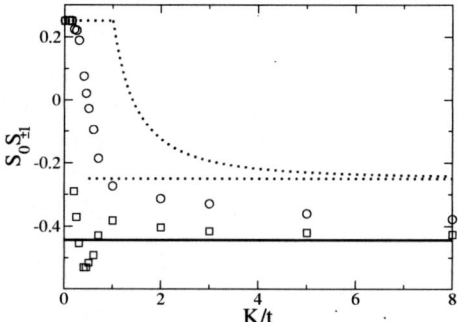

FIGURE 8. Spin-spin correlations inside an island and within islands as function of K for n=1/2. The dotted line corresponds to the classical calculation, where the upper curve is the correlation within the island and the lower one between islands. Circles (squares) are the numerically calculated correlations within (between) islands and the full line corresponds to the results for the Heisenberg chain.

tions. These phases are different from the spiral or canted phases proposed earlier. We find similar results for the Kondo and Hund models where the competition is between the hopping and the local magnetic interactions.

ACKNOWLEDGMENTS

We acknowledge partial support from the Collaboration Program between France and Argentina CNRS-PICS 1490.

REFERENCES

1. Hewson, A.C., *The Kondo Problem to Heavy Fermions*, Cambridge University Press, Cambridge, 1993.
2. Tsunetsugu, H., Sigrist, M., and Ueda, K., *Rev. Mod. Phys* **69**, 809-863 (1997).
3. Coey, J.M.D., Viret, M., and Von Molnar, S., *Advances in Physics* **48**, 167-293 (1999).
4. *Colossal magnetoresistive oxides*, edited by Y. Tokura, Advances in Condensed Matter Science, vol.**2**, Gordon and Breach Science Publishers, 2000.
5. *Physics of Manganites,* edited by T.A. Kaplan and S.D. Mahanti, Fundamental Materials Research, Kluwer Academic/Plenum Publishers, New York, 1999.
6. Dagotto, E., Hotta, T., and Moreo, A., Phys. Rep. 344,1 (2001).
7. *Density Matrix Renormalization, Lectures Notes in Physics*, Edited by I. Peschel, X. Wang, M. Kaulke, and K. Hallberg, Springer Verlag (1999).
8. Yunoki, S., et al., *Phys. Rev. Lett.* **80**, 845-848 (1998).
9. Dagotto, E., et al., *Phys. Rev. B* **58**, 6414-6427 (1998).
10. Caprara, S., and Rosengren, A., *Europhys. Lett.*, **39**, 55-60 (1997).
11. Xavier, J.C., Pereira, R.G., Miranda, E., and Affleck, I., cond-mat/0209623 (to be published).
12. Garcia, D.J., Hallberg, K., Batista, C.D., Avignon, M., and Alascio, B., *Phys. Rev. Lett.* **85**, 3720-3723 (2000). Garcia, D.J., et al., *Phys. Rev. B* **65**, 134444 (2002).
13. Garcia, D.J. et al, to be publish.
14. Yunoki, S., and Moreo, A., *Phys. Rev. B* **58**, 6403-6413 (1998).
15. Koshibae, W., Yamanaka, M., Oshikawa, M., and Maekawa, S., *Phys. Rev. Lett.* **82**, 2119-2122 (1999)

Low Temperature Transport Properties of Strongly Interacting Systems - Thermal Conductivity of Spin-1/2 Chains

N. Andrei[*], E. Shimshoni[†] and A. Rosch[**]

[*]*Center for Materials Theory, Rutgers University, Piscataway, NJ 08854–8019*
[†]*Department of Mathematics–Physics, University of Haifa at Oranim, Tivon 36006, Israel*
[**]*Institut für Theorie der Kondensierten Materie, Universität Karlsruhe, D-76128 Karlsruhe, Germany*

Abstract.
We outline a general approach to the computation of transport properties of interacting systems at low temperetures and frequencies. We show that if the fixed point and the irrelevant operators around it are known, then by studying the structure of the softly violated conserved currents chracterizing the fixed point one may set up an effective calculation in terms of a memory matrix formalism. We apply this approach to the computation of thermal conductivity of spin chains embedded in a matter matrix and interacting with its phonons. The results are found to be in very good agreement with experiment.

The study of transport properties of strongly interacting systems has been of great theortical and experimental interest for a long time. We shall concentrate in this contribution on developing an effective approach to the problem when the system is not far from its fixed point; in other words, when it is probed at low temperatures and at low frequencies. Subsequently we shall carry it out in detail for a system of spin 1/2 chains coupled to phonons.

Our approach consists of the following elements:

1. Identify the fixed point of the hamiltonian which describes the system, as well as the irrelevant operators around it.

The fixed point itself is typically insufficient to describe low energy transport properties; it is scale invariant and translationally invariant and thus unable to degrade a current, leading to infinite conductivity. To obtain finite conductivity one needs to take into account terms which break translational invariance and, more generally, violate the conservation laws associated with the fixed point.

2. Study the (weakly violated) conserved charges around the fixed point hamiltonian.

A fixed point H^* is scale invariant and often has several conserved quantities P, $[P, H^*] = 0$, associated with it. In 1-d, for example, if the fixed point is conformally invariant it has an infinite number of conserved quantities. When the irrelevant operators around the fixed point are taken into account, most of these quantities no longer commute with the low energy hamiltonian. The conservation of most of these quantites is strongly violated, but some may be only weakly violated and then significantly influence the low energy dynamics of the system.

Typically the current whose correlations determining the transport properties under considerations will be among those almost conserved charges, or "protected" by them in the following sense. When a system possesses some conserved quantities P, these may "protect" the current J from degrading (this occurs when the cross-susceptibility $\chi_{JP} \neq 0$) leading to a pure (i.e. $\delta(\omega)$) Drude peak and infinite d.c. conductivity. When the conservation of the pseudo-momenta P is softly violated they will, instead, lead to very long time tails in the decay of the current J. This occurs since states with a finite pseudo-momentum P typically carry also a finite current J since $\chi_{JP} \neq 0$. The component of the current "parallel" to P, $J_{\|P} = (\chi_{PJ}/\chi_{PP})P$ will therefore decay slowly. The presence of such approximately conserved quantities leads then to a natural hydrodynamic description of the system where a separation of fast and slowly decaying modes takes place and a consistent scheme of calculation of the slow mode conductivities can be carried out in terms of matrix of decay rates of these modes.

Let us explain and illustrate these ideas in more detail. We begin by arguing that given a conserved charge P, it will "protect" J if $\chi_{PJ} \neq 0$. Indeed, imagine preparing at $t=0$ a state carrying a current $\langle J \rangle$. Then necessarily that state will also have a non vanishing $\langle P \rangle$,

$$\langle P \rangle = \frac{\chi_{PJ}}{\chi_{JJ}} \langle J \rangle.$$

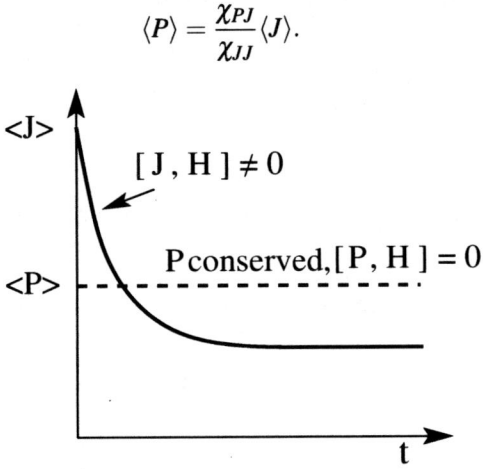

FIG. 1

In the limit $t \to \infty$ since P is conserved its expectation value will not have changed, and it will in its turn induce a non vanishing expectation value for J,

$$\lim_{t \to \infty} \langle J \rangle = \frac{\chi_{JP}}{\chi_{PP}} \langle P \rangle = \frac{\chi_{JP}^2}{\chi_{PP}\chi_{JJ}} \langle J(t=0) \rangle$$

(see Fig. 1). Since the current tends asymptotically to a constant value we find that the conductivity will have a $\delta(\omega)$-Drude peak containing a fraction $\frac{\chi_{JP}^2}{\chi_{PP}\chi_{JJ}}$ of the total weight; in other words, a Drude weight $D = \frac{1}{2}\frac{\chi_{JP}^2}{\chi_{PP}}$. An immediate consequence is that integrable models, having infinitely many conserved quantities, will typically also have an infinite dc conductivity.

If the charge P is not conserved but slowly decaying it will induce slow (long-time) decay in J (see Fig. 2).

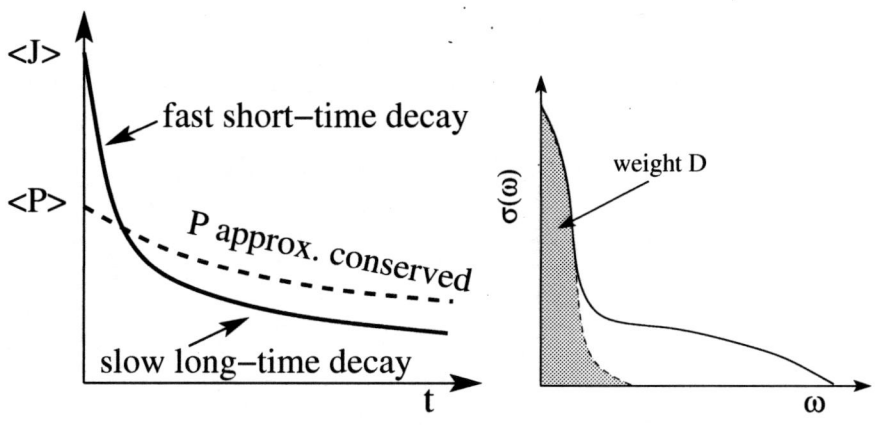

FIG. 2

This slow decay will show up as a peak in $\sigma(\omega)$ and the decay-rate Γ_P of P will determine the width of the peak and the dc value $\sigma(0) \approx D/\Gamma_P$ of the optical conductivity, the area under the peak being again, $D = \frac{1}{2}\frac{\chi_{JP}^2}{\chi_{PP}}$.

3. Having identified the slowly decaying charges of the effective model describing the low energy properties of the system we can compute the low temperature dc transport using a method that separates fast and slow modes, incorporating the former in the dynamics of the latter. Such a method is the memory matrix approach which can be used very efficiently and controllably when combined with the RG considerations outlined above.

Let us apply this approach to a system of spin chains embedded in some 3-d lattice and interacting with its phonons. Such systems (including, in particular, various compounds of SrCuO) have been recently studied in detail by, e.g., Sologubenko et al[1]. The authors have measured the heat conductivity along the three main axes of the sample. They observed that while conductivities along the a and c axes almost coincide, the conductivity along b, the axis along which the spin chains lie, has an enhancement which they interpret as being due to contribution of the spin degrees of freedom (Fig. 3, for example, presents data from a corresponding measurement in Sr_2CuO_3). In the temperature range $60\,\text{K} \leq T \leq 200\,\text{K}$, they gave the fit:

$$\kappa_s(T) \sim \exp(T^*/T),$$
$$T^* \approx 0.42\Theta_D$$

where κ_s is the spin contribution obtained after subtracting the phonon background from κ_b. The relevant energy scales of the system are: The Debye temperature characterizing the phonons - $\Theta_D \sim 400$ K and the spinon interaction scale $J/k_B \sim 2600$ K characterizing the spin chain.

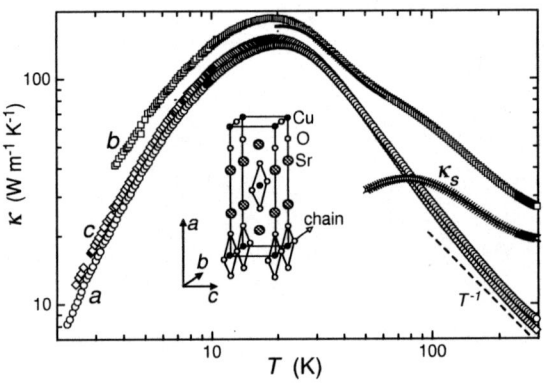

FIG. 3

One immediate question that arises is why is κ_s actually determined by the lower scale Θ_D. Furthermore, by plotting the above data of $\ln \kappa_s$ and of the pure phonon contribution $\ln \kappa_{a,c}$ vs. $1/T$, one observes that the slope of the latter is larger by a factor of 2. Again, why?

We shall find that a rather subtle interplay of (approximate) conservation laws and quantum dynamics underlies the experimentally observed heat conductivity, and the approach outlined above is necessary to fully account for it.

We begin by discussing the low energy effective hamiltonian. First consider a single spin chain,

$$H_s = \frac{1}{2}\sum_{i,j=1}^{N} J_{ij}\left(S_i^+ S_j^- + S_i^- S_j^+\right) + \sum_{i,j=1}^{N} J_{ij}^z S_i^z S_j^z$$

As is well known[2], for spin chains with short range interactions the fixed point hamiltonian is the Luttinger liquid,

$$H_{LL} = -i(Ja)\int dx(\psi_R^\dagger \partial_x \psi_R - \psi_L^\dagger \partial_x \psi_L) + J_z \int dx \rho(x)^2$$

with $\psi_{R/L}$ being right/left moving fermi field, and J and J_z are some average values of J_{ij} and J_{ij}^z respectively. As a reminder, the fixed point can be obtained by carrying out a Wigner-Jordan tranformation and then linearizing the resulting fermions around the

Fermi points $\pm k_F$ (see Fig. 4).

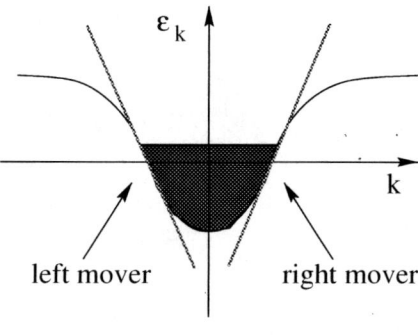

FIG. 4

It will be convenient to change to bosonic variables in terms of which the fixed point hamiltonian takes the form (to leading order in $|J_z|/J$),

$$H_{LL} = v\int \frac{dx}{2\pi}\left(K(\pi\Pi)^2 + \frac{1}{K}(\partial_x\phi)^2\right)$$

where

$$v \approx \left(J + \frac{J_z}{\pi}\right)a, \quad K \approx \frac{1}{1+\frac{2J_z}{\pi J}}.$$

We now consider the irrelevant operators around the fixed point. The irrelevant operators come from various sources. We shall divide them to Umklapp and non-umklapp operators. To the latter category belong local terms coming from band curvature around k_F: for example $\int \psi_R^\dagger \partial^2 \psi_R$. However, the operators that are important for transport are the Umklapp operators that reflect the underlying lattice structure. Only they break translation invariance and can degrade the currents. They have the following structure:

$$H^U = \sum_{nm} H^U_{nm}$$

$$H^U_{nm} = g^U_{nm}\int dx [e^{i\Delta k_{nm}x}\prod_{j=0}^{n}\psi_R^\dagger(x+ja)\psi_L(x+ja) + h.c.]$$

$$= \frac{g^U_{nm}}{(2\pi a)^n}\int dx[e^{i\Delta k_{nm}x}e^{i2n\phi(x)} + h.c.]$$

where $\Delta k_{nm} = n2k_F - mG$ is the momentum transfer associated with the Umklapp process where n particles are transferred from one Fermi point to another while giving up to the lattice m units of lattice momentum $G = 2\pi/a$. The particular values of the couplings g^U_{nm} depend on the couplings in the microscopic model. These terms are irrelevant perturbations around the fixed point not only by power counting but also because the x-dependent exponentials in them suppress their contribution exponentially. However they are the only source of dissipation.

We now consider the complete system consisting of an array of parallel spin chains interacting with 3-dimensional acoustic phonons. The 3-d phonon system projected along the axis describing deformations of the lattice *parallel* to the chains is described by

$$H_p^{3D} = \int \frac{d^3x}{2\pi} \left[(\pi P)^2 + \sum_\mu v_\mu^2 (\partial_\mu q)^2 \right],$$

with q the lattice deformation parallel to the spin chains direction, and P the conjugate momentum. Integrating the corresponding propagator over the perpendicular directions, we obtain the propagator along the chains:

$$\int d^2 k_\perp \frac{1}{(\omega^2 + \sum_\mu v_\mu^2 k_\mu^2)} \sim \ln[(\omega^2 + v_p^2 k^2)/\Theta_D^2].$$

In real space it takes the form $1/(x^2 + v_p^2 t^2)$ with v_p being the slowest phonon velocity. This again is the propagator of a $K = 1$ Luttinger liquid describing the phonons, $H_p = v_p \int \frac{dx}{2\pi} \left((\pi \Pi)^2 + (\partial_x q)^2 \right)$. Again we need to add to the combined phonon-spinon fixed point $H^* = \sum_\alpha H_p^\alpha + \sum_\alpha H_{LL}^\alpha$ (the summation α is over the spin chains) all the irrelevant operators. As before they fall into two categories, the Umklapp and non Umklapp operators. The most important ones of the former one have the form,

$$H_{nm}^{U,s-p} = \frac{g_{nm}^{U,p}}{(2\pi a)^n} \int dx [e^{i\Delta k_{nm} x} e^{i2n\phi} \partial_x q + h.c.]$$

(where ϕ is the bosonic field in a particular chain), while an example of the latter is $H_{s,p}^{nonU} = \int (\partial \phi)^2 \partial_x q$.

We now turn to our main interest, computing the thermal conductivity $\kappa_s(\omega, T)$ of the spin chains coupled to phonons. The thermal conductivity can be expressed in terms of a heat current correlation function,

$$\kappa(\omega, T) = \langle J_Q, J_Q \rangle(\omega, T)/\omega$$

where J_Q, the heat current expressed in bosonic variables, is $J_Q = -\sum_\alpha \int dx\, v^2 \Pi_\alpha \partial_x \phi_\alpha - \int d^3x\, v_p^2 P \partial_x q$. The correlation is to be computed with respect to the low-energy hamiltonian,

$$H_{low-E} = H^* + H^U + H^{nonU}.$$

As explained in the outline we need to identify the (approximately) conserved "charges" of the low-E Hamiltonian, to find out whether they induce a slow decay of the heat current. We now show that the quantities,

$$J_s = vK \sum_\alpha \int dx [\psi_{R\alpha}^\dagger \psi_{R\alpha} - \psi_{L\alpha}^\dagger \psi_{L\alpha}] = vK \sum_\alpha \int dx \Pi_\alpha$$

$$P_T = -\sum_\alpha \int dx \Pi_\alpha \partial_x \phi_\alpha - \int d^3x P \partial_x q$$

where, J_s is the spin current and P_T the momentum operator, are the "slow modes" which in turn protect J_Q rendering it slow too.

Indeed, J_s and P_T commute with H_{LL} and with H^{nonU} and their conservation is violated only through H^U with which they do not commute, thus inducing a slow current decay. More importantly, certain linear combinations of J_s and P_T, the "pseudo-momenta"

$$P_{nm} = \frac{1}{2n}\Delta k_{n,m} J_s + P_T$$

decay even slower as they commute with $H_{LL} + H^{nonU} + H^U_{nm} + H^{U,s-p}_{nm}$ and are therefore exactly conserved if only a *single* type of Umklapp with quantum numbers n and m is present. We note that the pseudo momenta can be written as $P_{nm} = P_{lat} + \frac{m}{2n}G(N_R - N_L)$ with $P_{lat} = \sum_k k c_k^\dagger c_k \approx k_F(N_R - N_L) + P_T$ being the lattice momentum. Unlike other (approximately) conserved quantities, they decay exponentially slowly with the temperature as their violation requires processes away from the Fermi energy[3].

The heat current on the other hand does not commute with both H^{nonU} and H^U and would therefore decay fast, but is protected by J_s, P_T and their linear combinations since $\chi_{J_Q,J_s}, \chi_{J_Q,P_T} \neq 0$. We proceed to discuss transport in the presence of several approximately conserved - "slow" - variables: $J_1, J_2...J_N$. We shall introduce the memory matrix formalism[4] which is very effective under the circumstances as it allows the separation of the slow modes (J_s, P_T, J_Q, in our particular case) from the fast modes.

To set up the formalism one introduces a scalar product in the space of operators of the theory

$$(A(t)|B) \equiv \frac{1}{\beta}\int_0^\beta d\lambda \left\langle A(t)^\dagger B(i\lambda)\right\rangle.$$

In terms of this scalar product one can express the dynamic correlation functions as follows,

$$\begin{aligned} C_{AB}(\omega) &= \int_0^\infty dt e^{i\omega t}(A(t)|B) \\ &= \left(A\left|\frac{i}{\omega - \mathcal{L}}\right|B\right) \\ &= \frac{iT}{\omega}\int_0^\infty dt e^{i\omega t}\langle [A(t),B]\rangle - \frac{(A|B)}{i\omega}\end{aligned}$$

where the Liouville operator $\mathcal{L}$ is defined as $\mathcal{L}A = [H,A]$.

The matrix of conductivities (Kubo formula) is then,

$$\hat{\sigma}_{pq}(\omega,T) = \frac{1}{TV}C_{J_p J_q}(\omega)$$

($p,q = 1\cdots N$). In our case the thermal conductivity is

$$\kappa_s(\omega,T) = \frac{1}{T}\sigma_{QQ}(\omega,T).$$

However the dc conductivity has no good perturbative expansion: $\sigma \sim 1/\Gamma$, with Γ the decay rate of the current is singular in perturbation theory. Put in other words, if we wish to compute the conductivity as a perturbative expansion of the irrelevant operators around the fixed point, this would be an arduous task in view of the fact that the conductivity computed from the fixed point is infinite. Furthermore, even the perturbative expansion for $1/\sigma(\omega=0)$ turns out to be ill behaved in the presence of slow modes. While the short-time decay rate (see Fig. 2) of the current is perturbative, the $\omega \to 0$ limit requires to take into account the presence of approximate conservation laws.

We seek therefore to compute a quantity which has a good perturbative expansion, and to this purpose introduce $\hat{M}(\omega, T)$ - the Memory Matrix, essentially the matrix of relaxation rates.

The matrix is defined as

$$\hat{M}_{pq}(\omega) = \frac{1}{T} \left(\partial_t J_p \left| \frac{i}{\omega - } \right| \partial_t J_q \right)$$

where is the projection away from slow modes

$$ = 1 - \sum_{pq} |J_p) \frac{1}{T} (\hat{\chi}^{-1})_{pq} (J_q|.$$

In terms of the memory matrix the conductivity matrix is,

$$\hat{\sigma}(\omega, T) = \hat{\chi}(T) \left(\hat{M}(\omega, T) - i\omega \hat{\chi}(T) \right)^{-1} \hat{\chi}(T)$$

with $\hat{\chi}$ the susceptibility matrix,

$$\hat{\chi}_{pq} = \frac{1}{TV} (J_p | J_q).$$

Applying the formalism in our case we find that the memory matrix is a sum over the Umklapp processes (nm), given - to leading order in g_{mn}^U - by,

$$\hat{M} = \frac{1}{T} \left[\sum_{nm} (\hat{M}_{nm} + \hat{M}_{nm, s-p}) \right]$$

where (the matrix indices p, q take the values s, T, Q)

$$M_{nm}^{pq} \equiv \frac{\langle F^p; F^q \rangle^0_\omega - \langle F^p; F^q \rangle^0_{\omega=0}}{i\omega},$$

$$M_{nm,s-p}^{pq} \equiv \frac{\langle F^p_{s-p}; F^q_{s-p} \rangle^0_\omega - \langle F^p_{s-p}; F^q_{s-p} \rangle^0_{\omega=0}}{i\omega};$$

here $F^p = i[J_p, H^U]$, $F^p_{s-p} = i[J_p, H^{U,s-p}]$ and $\langle F^p; F^q \rangle^0_\omega$ is the retarded correlation function calculated with respect to H^*. As the perturbative expansion is in irrelevant operators with respect to the fixed point the expansion is expected to be rapidly converging at low temperatures if slow and fast time scales are well seperated and all of the slowest modes have been taken into account.

Carrying out the computation of the various correlation functions (see Ref. [5] for details) we have:

$$\kappa_s(T) \approx v^2 T^3 \left[(\hat{M}^{-1})_{TT} + 2(\hat{M}^{-1})_{QT} + (\hat{M}^{-1})_{QQ} \right]$$

with the typical matrix elements,

$$M_{nm}^{pq} \sim (\Delta k_{nm})^{(n^2 K - 2)} e^{-v \Delta k_{nm}/2T}$$
$$M_{nm,s-p}^{pq} \sim T^{(2n^2 K - 1)} e^{-v_p \Delta k_{nm}/2T}.$$

Note that the spinon processes decay exponentially fast, with the exponent $-v\Delta k_{nm}/2T$, while the spinon-phonon exponents contain the much slower phonon velocity $v_p \ll v$, leading to a much slower decay with the exponent $-v_p \Delta k_{nm}/2T$. The latter will therefore clearly determine the thermal conductivity.

But which of the scattering processes (n,m) will dominate?

At low-T exponential factor prevails hence the smallest Δk_{nm}. At this point our discussion must distinguish between commensurate and incommensurate magnetization, or in fermionic language commensurate and incommensurate filling. The magnetization can be in principle tuned by varying an external magnetic field h.

Close to commensurate filling $k_F \approx G \frac{m_0}{2n_0}$, and then the dominant processes appear to be $H_{n_0 m_0}^U$ where $\Delta k_{n_0,m_0} \approx 0$. However, because of the conservation laws discussed earlier, it is the next leading term $H_{n_1 m_1}^U$ with $\Delta k_{n_1,m_1} = \pm G/n_0$ which determines the decay rate. Technically, this arises as one of the eigenvalues of the matrix $\hat{M}$ is not affected by $H_{n_0 m_0}^U$ (as $[P_{n_0 m_0}, H_{n_0 m_0}^U] = 0$) but determined by $H_{n_1 m_1}^U$. This smallest eigenvalue will then determine the size of $\hat{M}^{-1}$ and therefore the heat conductivity. The particular case of half filling is discussed in detail below.

On the other hand, at a typical incommensurate filling it will depend on the temperature which processes are dominant and subdominant and we need to sum over all terms (do saddle-point approximation with respect to n_1) and find,

$$\kappa_{typical} \sim \exp[c(\beta v G)^{2/3}].$$

with c a constant of order 1.

We now turn to the experiment by Sologubenko et al. discussed earlier. The experiment was carried out at $h = 0$ corresponding to half filling, $k_F = \pi/2a = G/4$. Therefore $\Delta k_{21} = 0$ (n=2, m=1). But we need at least two Umplapp terms and the next smallest is $\Delta k = G/2, (n=1, m=0)$. Recall also that as $v_p \ll v$, it follows that $\hat{M}_{nm,s-p} \gg \hat{M}_{nm}$. Hence, the dominant contribution to the thermal conductivity comes from, $(\hat{M}^{-1})_{TT} \approx 1/M_{n=1,s-p}^{TT}(G/2,T)$ and we have

$$\kappa(T) \approx \kappa_0 (T/T^*)^{2(1-K)} e^{T^*/T}$$

with

$$T^* = v_p G/4.$$

We find therefore that the second strongest rate wins (cf. the expression for $\hat{M}$): it is determined by v_p via a phonon process and is characterized by the momentum $G/2$. The $G/2$ transfer momentum characterizes the dominant spinon - phonon Umklapp process, and clearly distiguishes it from pure phonon Umklapp processes characterized by momentum transfer G. We expect therefore that in the pure phonon thermal conductivity (axes a and c in Fig. 3) the scale $2T^*$ would appear.

To compare our findings with the experimental data we need to express our expression in terms of Θ_D. Assuming an isotropic phonon dispersion one has: $\Theta_D \approx v_p(6\pi^2/a^3)^{1/3} \approx 0.6 v_p G$.

Therefore:

$$T^* \approx 0.4\Theta_D, (theory)$$
$$T^* \approx 0.42\Theta_D, (experiment).$$

Taking into account possible ambiguities in the fits to the experiments and that the phonon dispersion is probably not completely isotropic, part of this excellent agreement may by accidental. But further confirmation of our theory comes from the observation that the ratio of slopes of the spinon contribution compared to the pure phonon contribution (on a semilogarithmic graph of κ vs. $1/T$ - see Fig. 5[6]) is approximately 1:2 as discussed earlier.

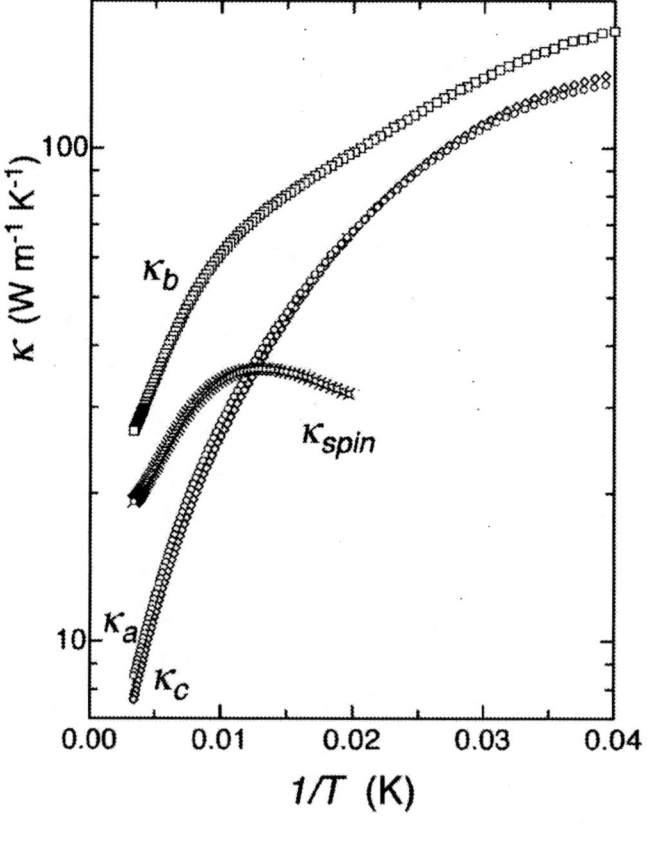

FIG. 5

We may also consider the effect of a magnetic field on the thermal conductivity. (To observe it experimentally one needs a material with much smaller spin energy scales J than SrCuO). The magnetic field modifies the value of k_F according to

$$k_F = \frac{\pi}{2a}(1+M) \approx \frac{\pi}{2a}(1+h/(\pi J)). \tag{1}$$

As the field h is varied the system passes through commensurate fillings, $\Delta k_{nm} = n2k_F - mG = 0$ and incommensurate fillings, $\Delta k_{nm} = n2k_F - mG \neq 0$. Thus different Umklapp operators become effective leading to a fractal-like dependence on $M \approx h/\pi J$ (see Fig. 6).

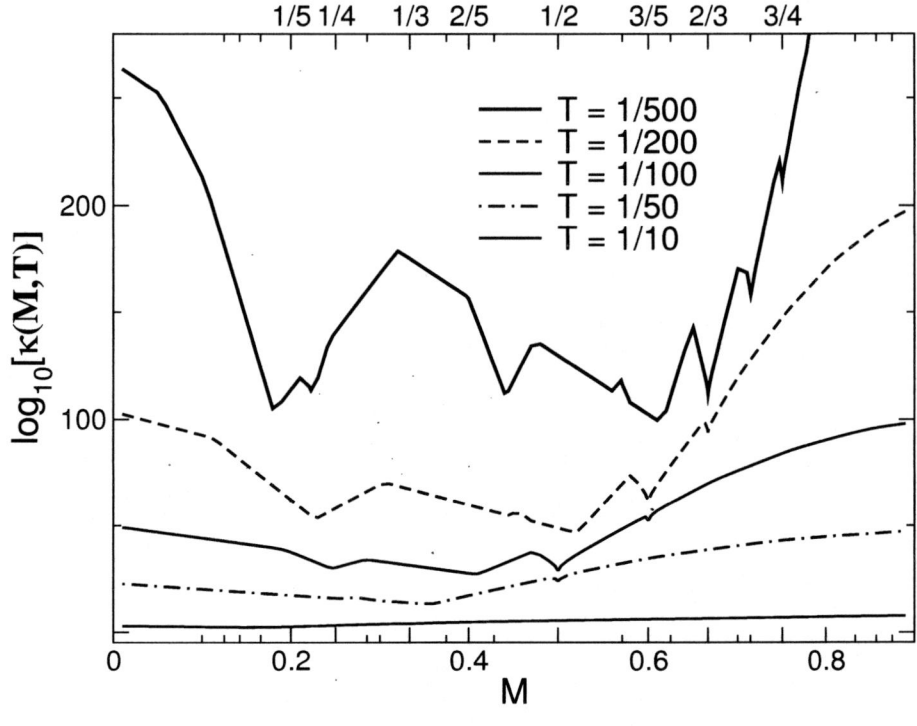

FIG. 6

We expect some dips to be experimentally observable. In conclusion,

- Transport is strongly influenced by conserved "charges": low energy processes cannot relax heat current.
- Exponents are determined by slowest mode in the system: typically phonons.
- Memory Matrix approach, separating slow and fast modes, allows controllable calculations.
- Calculations fit experiments
- Interesting predictions on the magnetization dependence of the heat transport have been made.

REFERENCES

1. A. V. Sologubenko et al., *Phys. Rev. B*, **62**, R6108 (2000).
2. I. Affleck, in *Fields, Strings and Critical Phenomena, Les Houches, Session XLIX*, edited by E. Brezin and J. Zinn-Justin (North-Holland, Amsterdam, 1988).
3. A. Rosch and N. Andrei, *Phys. Rev. Lett.*, **85**, 1092 (2000); A. Rosch and N. Andrei, *Journal Low Temp Phys.*, **126**, 1195 (2002).
4. D. Forster, *Hydrodynamic Fluctuations, Broken Symmetry, and Correlation Functions*, (Benjamin, Massachusetts, 1975).
5. E. Shimshoni, N. Andrei and A. Rosch, to be published in Phys. Rev. B (condmat/0304641).
6. We thank Alex Sologubenko for providing us with the replotted data.

Decoherence in dissipative systems

F. Guinea

Instituto de Ciencia de Materiales de Madrid. Cantoblanco. E-28049 Madrid. Spain.

Abstract. The quantum properties of a particle interacting with a dissipative environment are studied. We show the way in which different types of environments reduce quantum effects. The influence of certain environments, like that described by a Caldeira-Leggett bath of oscillators suppress very efficiently interference processes, while the coupling to other (ohmic) environments does not modify qualitatively the main features of a free particle. Examples of physical realizations of the different environments are also given.

INTRODUCTION

It is well known that the quantum properties of a particle are supressed by the interaction with an external system (defined in the following as the environment). In general, the enviroment acts as a measuring device, fixing the particle in one state and reducing the amount of entanglement between states which interact differently with the environment. We will consider here the case of a particle moving in free space, and a coupling to the external environment which depends on the position of the particle. Hence, the environment measures the position of the particle, which loses its ability to be in a wave-like superposition between states at different places. For charged particles, this is the usual case, as almost any external environment is coupled to the electrostatic potential induced by the particle.

In general, if the spectrum of the environment have a gap, the dynamics at temperatures lower than the gap, or time scales longer than the inverse of the gap, are not strongly modified, as there are no excitations in the environment capable to react sufficently slowly. The effect of an environment with a gap in the spectrum can be described in terms of a finite renormalization of the bare parameters which describe the particle.

Metallic systems, however, are, by definition, gapless, and are present in many common experimental setups. A particle in contact with a metal can lose its quantum properties in a non trivial way.

A gapless environment also modifies the motion, at low energies, of a *classical* particle. In this limit, the effects of many different environments can be described in terms of a single phenomenological parameter, the friction coefficient, η. While an environment has to be gapless in order to give rise to a finite friction coefficient at all velocities and zero temperature, not all gapless environments lead to friction. Those whose effects, in a classical system, can be described in terms of a finite friction coefficient are called ohmic. In most cases, clean or dirty metals behave as ohmic environments.

A combination of perturbative and selfconsistent analyses allowed a good understanding of dephasing in dirty metals[1, 2]. The interest in the decoherence effects of external environments revived when it become feasible the observation of quantum effects in macroscopic systems[3]. A. Caldeira and A. J. Leggett, in a seminal work, defined a model where a generic environment was described in terms of harmonic oscillators and which could be solved in a variety of situations[4, 5]. Recent experiments which suggest the existence of anomalous dephasing rates for quasiparticles in normal metals[6] have led to an increased interest in the role of dissipative environments[7, 8]. Quantum decoherence due to an external environment is obviously relevant in the operation of quantum information devices.

THE MODEL

We will use the path integral method to describe the properties of a quantum particle interacting with an external environment. The approach is a natural starting point for the study of properties in the quasiclassical limit, that is, when the coupling to the environment is strong. In addition, it provides a framework where the degrees of freedom from the environment can be integrated out, leading to a description where we have only to consider the possible paths of the particle.

Given a path, the environment leads to retarded interactions between the position of the particle at different times, as schematically shown in Fig.[1]. If the environment is bosonic, the only possible interactions are those connecting two positions, as shown in Fig.[1]. Generally, couplings involving more than two positions can also occur. If the coupling of the particle to each individual degree of freedom of the environment is weak, these contributions will be suppressed with respect to those in Fig.[1]. The existence of ohmic dissipation at low energies fixes the decay of the retarded interactions as function of the time difference, which is τ^{-2}. The generic form of the action to be studied is[9]:

$$S = \int d\tau \frac{M}{2} \left(\frac{\partial \vec{X}}{\partial \tau} \right)^2 + \int d\tau d\tau' \frac{\mathscr{F}\left[\left| \vec{X}(\tau) - \vec{X}(\tau') \right|^2 \right]}{|\tau - \tau'|^2} \quad (1)$$

where the function $\mathscr{F}$ depends on the local correlations in the environment. For the case of a metal electrostatically coupled to the particle, $\mathscr{F}$ can be written in terms of the charge-charge response function[9, 10]. Assuming that a single length scale, l^*, describes the structure of the environment, we can write:

$$\mathscr{F}(u) \equiv \mathscr{F}\left(\left| \vec{X}(\tau) - \vec{X}(\tau') \right|^2 \right)$$

$$u = \frac{\left| \vec{X}(\tau) - \vec{X}(\tau') \right|^2}{l^{*2}} \quad (2)$$

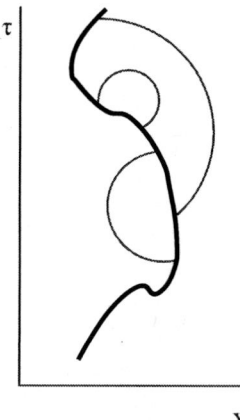

FIGURE 1. Retarded interactions along a given path of the particle (see text for details).

and the function $\mathscr{F}(u)$ usually satisfies:

$$\mathscr{F}(u) = \alpha f(u) = \begin{cases} \alpha u & u \ll 1 \\ \mathscr{F}_\infty & u \gg 1 \end{cases} \quad (3)$$

where α and $\mathscr{F}_\infty$ are dimensionless numbers. The asymptotic behavior of the function $f(u)$ is, in some interesting cases:

$$\begin{aligned}
f(u) &= u & &\text{Caldeira – Leggett model} \\
f(u) &= 2\sqrt{1+u} - 2 & &\text{1D dirty electron gas} \\
f(u) &= \log(1+u) & &\text{2D dirty electron gas} \\
f(u) &= 2 - \frac{2}{\sqrt{1+u}} & &\text{3D dirty electron gas}
\end{aligned} \quad (4)$$

The macroscopic friction coefficient is determined by the high temperature properties of the particle. In this semiclassical limit, the paths to be included in the action, eq.(1), are short, so that $u \ll 1$ in eq.(3). Then, one recovers an effective Caldeira-Leggett model, with a coupling which is quadratic in the particle coordinates. The friction coefficient is:

$$\eta \equiv \frac{\hbar \alpha}{\pi l^{*2}} \quad (5)$$

The length scale l^* is of order k_F^{-1} in a clean electron liquid, and of order of the mean free path, l, in a dirty metal.

RESULTS

A useful parameter, which describes well the main properties of the particle, is the expectation value

$$G(\tau) = \left\langle \left| \vec{X}(\tau) - \vec{X}(0) \right|^2 \right\rangle \tag{6}$$

The action, eq.(1), can be solved exactly in two opposite limits: i) In the absence of dissipation the particle is free, and, at long times, $\lim_{\tau \to \infty} G(\tau) \sim (\hbar \tau)/M$, where M is the mass of the particle. The opposite situation is described by the Caldeira-Leggett model, which gives $\lim_{\tau \to \infty} G(\tau) \sim \hbar/\eta \log[(\eta \tau)/M]$.

Most cases of interest lie between these limits. By expanding the function $f(u)$ in powers of u one can write a perturbative expansion around the exactly soluble limit $f(u) = u$ (the Caldeira-Leggett model). This expansion, however, contains an infinitely large number of dimensionless coupling constants, which makes unwieldly its analysis by Renormalization Group techniques.

There is an alternative scheme, which leads to the correct solution in the free particle and strong dissipation limit, a large N expansion[11, 12, 13], where N is the number of components of the vector $\vec{X}(\tau)$[14]. This calculation can be expressed as the sum of an infinite number of diagrams in the perturbative series mentioned earlier. It can be shown that, to lowest order, the vortex corrections which are not included are irrelevant, in the RG sense[14]. The calculation has to be performed numerically, and it amounts to the selfconsistent solution of the equations:

$$\begin{aligned} \Sigma(\tau) &= \frac{\alpha}{\tau^2} f'[G(\tau)] \\ G(\omega) &= \frac{l^{*2}}{\frac{Ml^{*2}\omega^2}{2} + \Sigma(\omega)} \end{aligned} \tag{7}$$

(we are setting, here and in the following equations, $\hbar = 1$).

At short times, the evolution of the particle is indistinguishable from that described by an effective Caldeira-Leggett model, as discussed earlier. Then:

$$G(\tau) \approx \frac{l^{*2}}{\alpha} \log\left(\frac{\alpha \tau}{Ml^{*2}}\right) \tag{8}$$

This expression ceases to be valid at times such that $G(\tau) \approx l^{*2}$, i. e. for times such that $\tau^* \sim (Ml^{*2} e^{\alpha})/\alpha$.

For frequencies much smaller than τ^{*-1}, the selfenergy expression in eq.(7) which describes the effect of the environment has a contribution quadratic in ω, and a second contribution coming from the propagator at long times:

$$\begin{aligned} \Sigma(\omega) &\sim \frac{\alpha}{l^{*2}} \omega^2 \tau^* + \Sigma_{\tau \gg \tau^*}(\omega) \\ \lim_{\tau \to \infty} \Sigma(\tau) &\propto \frac{\alpha}{\tau^2} \lim_{\tau \to \infty} f'[G(\tau)] \end{aligned} \tag{9}$$

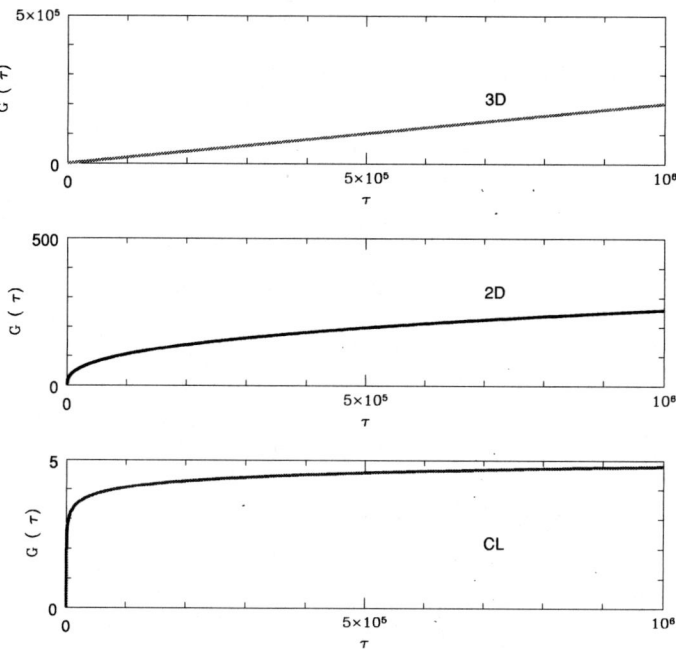

FIGURE 2. Time dependence of the propagator $G(\tau)$, for $\hbar^2/(Ml^{*2}) = 1$, $l^* = 1$ and $\alpha = 2$, and three different environments. Top: 3D dirty electron gas. Center: 2D dirty electron gas. Bottom: Caldeira-Leggett model. See eqs.(4). Note that the macroscopic friction coefficient is the same in the three cases.

Thus, the long time behavior of $\Sigma(\tau)$ is determined by $\lim_{u\to\infty} f'(u)$.

If $f'(u)$ decays faster than u^{-1}, $\lim_{\omega\to 0} \Sigma(\omega) = (\alpha\omega^2\tau^*)/l^{*2}$. This quadratic dependence on ω is the same behavior of a free particle, with an effective mass proportional to $M_{eff} \propto \alpha\tau^*/l^{*2}$. Using the estimation for τ^* in the previous paragraph, we find $M_{eff} \propto Me^{-c\alpha}$, where c is a constant of order unity.

If $f'(u)$ does not decay, or it decays more slowly than u^{-1} (as in the Caldeira-Leggett model, for example), the second term in eq.(9) dominates at low frequencies, and the particle cannot be described as having a renormalized mass. In this regime, the environment is sufficiently strong to suppress genuine quantum behavior at long times or low temperatures.

The main features of the two regimes mentioned above describe quite well the numerical solutions of eqs.(7) for different environmentss, as shown in Fig.[2].

CONCLUSIONS

We have analyzed the quantum properties of a particle coupled to dissipative environments using a method which intepolates correctly between the free and the strongly interacting limits.

Our results indicate that a variety of behaviors are possible as a function of the spatial dependence of the interactions mediated by the environment, even if the time dependence, which characterizes the ohmic behavior, is the same.

ACKNOWLEDGMENTS

This talk is a natural continuation of the lectures on dissipative quantum systems delivered, by invitation of Prof. F. Mancini, at Vietri sul Mare in 1997. That stay, and the scientific meeting for the ocassion of Prof. F. Mancini's 60th birthday held in Salerno in 2003, where this lecture was presented, were very fruitful and scientifically challenging, and very pleasant from a personal point of view. I send to Prof. Ferdinando Mancini my most sincere congratulations.

REFERENCES

1. Altshuler, B. L., Aronov, A. G., and Khmelnitskii, D. E., *J. Phys. C*, **15**, 7367 (1982).
2. Stern, A., Aharonov, Y., and Imry, Y., *Phys. Rev. A*, **41**, 3436 (1990).
3. Leggett, A. J., *J. Phys. (Paris), Colloq.*, **39**, C6–1264 (1978).
4. Caldeira, A., and Leggett, A. J., *Phys. Rev. Lett.*, **46**, 211 (1981).
5. Caldeira, A., and Leggett, A. J., *Ann. Phys. (N. Y.)*, **149**, 374 (1983).
6. Mohanty, P., Jariwala, E. M. Q., and Webb, R. A., *Phys. Rev. Lett.*, **78**, 3366 (1997).
7. Golubev, D. S., and Zaikin, A. D., *Phys. Rev. Lett.*, **81**, 1074 (1998).
8. Aleiner, I. L., Altshuler, B. L., and Gershenson, M. E., *Waves in Random Media*, **126**, 1377 (1999).
9. Guinea, F., *Phys. Rev. Lett.*, **53**, 1268 (1984).
10. Sols, F., and Guinea, F., *Phys. Rev. B*, **36**, 7775 (1987).
11. Fisher, M., k. Ma, S., and Nickel, B. G., *Phys. Rev. Lett.*, **29**, 917 (1972).
12. Stanley, H. E., *Phys. Rev.*, **176**, 718 (1973).
13. Coleman, S., *Aspects of symmetry*, Cambridge U. P., Cambridge, UK, 1985.
14. Guinea, F., *Phys. Rev. B*, **67**, 045103 (2003).

Structural Biology of Viruses: A Case of Synergy

Erika J. Mancini

Division of Structural Biology, Wellcome Trust Centre for Human Genetics, University of Oxford, Roosevelt Drive, Headington, Oxford OX3 7BN, U.K.

Abstract. In modern biology, there is an increasing emphasis on the relevance of the structure of a protein or other macromolecule to its biological functions. Similarly, the structure of a virus particle holds much information about its biology, and can be applied to such tasks as the interpretation of genetic data or the design of antiviral drugs. Recent technological advances in virus structure determination have allowed the structures of many viruses to be solved by x-ray crystallography and/or cryo-electron microscopy combined with image reconstruction.

INTRODUCTION

It is apparent that much interesting biology occurs due to the subtle interplay between molecular subunits in large macromolecular complexes and that some fundamental questions can be addressed by analysis of their structures. The folding and assembly of these complex structures are central to their function and analysis of assembly intermediates and final structures will, in turn, illuminate folding pathways.

In viruses a range of fundamental biological functions are achieved in a minimalist way. The limited viral genome size holds back increases in complexity, whereas rapid evolution produces diversity that, in the absence of structural information, can obscure profound similarities in assembly. It is clear that, in general, rather than seeking a solution to say, the protein folding problem, one has to address the appropriate biologically functional form. This may be a complex whose structure is intimately dependent not solely on the folding of its components, but on the pathway of assembly, to the extreme that some components may only fold in such a context. For this generalised folding problem, simplified systems such as viruses may still prove illuminating.

VIRUS ASSEMBLY

Viruses have served as useful models for studying the assembly of macromolecular complexes. The major reason for this has been the ability of the viral coat proteins to form a definite stable structure, the capsid. The capsid is formed from a large copy number of a small group of proteins, related by symmetry. As cells produce large numbers of viruses during infection, it was relatively easy to observe these populous proteins and structures even before the availability of recombinant DNA technologies. It has become evident that, in addition to the symmetrical capsids, virions contain elements that deviate from the symmetry and that there is always metastability built into the structure to allow the delivery of the nucleic acid upon stimuli derived from the interactions with the host cell. Bringing together many protein subunits, packaging the genome and maturing to an infective particle is a very complex process, even in the case of the smallest viruses. Understanding the principles governing this assembly process will surely lead ultimately to the design of new types of antiviral therapy.

Ideally, a viral assembly model system would use purified components and time-resolved methods to follow the assembly process, yielding infective particles. In addition, the structures of the initial components, intermediates and final virus should be determinable, preferably to high resolution [1].

TIME-RESOLVED STUDIES

Several approaches have been used to capture assembly intermediates and kinetic events: for instance, Lata et al. [2] triggered procapsid maturation of the double stranded RNA (dsDNA) phage HK97 by acidification and followed events by time-resolved low-angle X-ray diffraction and cryo-electron microscopy (cryo-EM). The former method monitors overall, average changes and the latter distinguishes changes in the ratios of identifiable particle classes. It is obvious that maturation moves from one local conformational free energy minimum to another, lower minimum. Thus, the assembly process parallels protein folding in having transitional states on the path to mature particles. These studies were done without DNA packaging and portal proteins, which would contribute to the in vivo energy landscape.

...The self-assembly of purified nucleocapsid coat proteins of phage Φ6 into empty shells is controlled by calcium ions. This process has been followed by time-resolved Raman spectroscopy and hydrogen isotope exchange [3]. Interestingly, the regions of the protein involved in subunit-subunit interactions in the intact shell undergo rapid exchange, which is indicative of low activation barriers, although one would expect less exchange upon capsid maturation.

These results suggest that this protein, which is involved with membrane interactions upon entry and maturation, is unusually dynamic.

Hepatitis B virus capsid protein assembly was followed by light scattering [4]. Empty capsids were produced at a rate that is a function of protein concentration and ionic strength. These results were incorporated into an evolving theoretical model in which assembly intermediates approach a low steady-state concentration. The work of Schwartz et al. [5] represents a further step towards in silico assembly. This work tests a local rules model for polyoma virus capsid assembly. The idea is that the information for capsid formation is in the subunits and that simplified (local) rules can be used to explain the formation of normal and aberrant structures.

These various approaches to understanding the dynamics of the assembly process are likely to be enriched in the future by combining them with high-resolution structural information on end-point structures and assembly-arrested intermediates.

ELECTRON MICROSCOPY OF VIRUS STRUCTURES

In addition to the X-ray-based atomic structures, cryo-EM-based virus structures at better than 15 Å resolution continue to multiply. The large and complex herpesvirus capsid, now seen at 8.5 Å resolution [6], possesses a heterotrimer at the local threefold axis. Intriguingly, this protein complex, with a unique quaternary structure, assembles by co-folding of its subunits. The Semliki Forest virus structure, resolved at 9 Å by cryo-EM [7], defines the spike-capsid interactions and the paired helical nature of the transmembrane segments, as well as the fusion peptide located in the cavity of the spike. These features help in the interpretation of interactions that take place during the assembly of an enveloped virus, a class of target for which we have, as yet, few model systems at reasonable levels of definition.

Krol et al. [8] showed how packaged RNA directs the capsid assembly of brome mosaic virus. The normal quasi-equivalent 180-subunit capsid is replaced by a 120-subunit capsid with two nonequivalent capsid protein environments when only a short RNA molecule is packaged. This implies a fundamental role for the RNA moiety in coat protein assembly. The arrangement of the 120-copy structure differs, however, from that found in the $T = 2$ capsids of double stranded RNA (dsRNA) viruses.

The cryo-EM-based structures of the dsDNA phage P22 procapsid and virion [9] reveal the structural transitions that occur upon maturation. A rationale for the existence of a metastable precursor particle is proposed, namely that this particle is able to distinguish between alternative coat protein conformations, ensuring correct assembly, and in addition provides a favourable template on its internal surface for DNA packaging.

It has proved very difficult to address the assembly and maturation of retroviruses. However, a significant advance has come from the analysis of helical tubes and cones that resemble native viral capsids [10]. Cryo-EM-based image reconstruction of six different helical families informed a molecular model for the HIV capsid in which the body of the cone is composed of curved hexagonal arrays of coat protein rings and the end is capped. The lack of regularity is a challenge for assembly. A range of different-volume capsids can be assembled; this may reflect the promiscuous capacity of retroviruses to package additional genetic material from the host cell. We can hope that further and more detailed cryo-EM analysis will allow the wealth of high-resolution structural information on the component domains of the gag protein to be integrated in a satisfyingly coherent model.

X-RAY CRYSTALLOGRAPHY OF VIRUS STRUCTURES

The human papillomavirus capsid protein forms pentamers that assemble in vitro into capsid-like structures [11]. The crystal structure of a 12-pentamer icosahedral particle made of N-terminally truncated (10 amino acid) coat proteins shows that the subunit closely resembles the corresponding structure in the 72-pentamer polyomavirus. However, no 72-pentamer particles are formed by the truncated protein, suggesting that the most N-terminal portion of the coat protein is responsible for switching the size of the assembly.

A wonderful new way of forming thin, but very robust, structures was revealed by the crystal structure of bacteriophage HK97, determined in the Johnson laboratory [12]. This capsid is some 660 Å across at its widest point and yet the protein shell is only 18 Å thick. Instead of being flimsy, this structure is made from protein 'chain mail'. Rings of five or six subunits are covalently linked together through lysine-asparagine isopeptide bonds and these rings are looped together. This is a remarkable final structure for a virus that undergoes some radical structural rearrangements during maturation. A full analysis of this morphogenesis will be fascinating.

The recent publication of the core of orthoreovirus shows the first structure of a representative of the 'turreted' members of the reoviridae [13]. In contrast to bluetongue virus (BTV) [14 and 15] and other 'smooth' members, such as rotavirus, the enzymatic activities involved in cap formation are found in turrets that protrude from the capsid surface. The orthoreovirus core is made up of 120 copies of the major structural protein 1 arranged on the so-called T=2 lattice seen in BTV. This arrangement of 120 protein subunits appears to be a common feature of dsRNA viruses. Whilst BTV strengthens this thin layer by cloaking it in 780 copies of a smaller structural protein (VP7), in reovirus the thin capsid is strengthened by a 'bolting' protein that sits at positions bracing joins between molecules of 1, as was proposed on the basis of EM analysis of another member of the reoviridae [16]. Despite these differences, the arrangement of the 120-copy layer is essentially identical for BTV and reovirus. Not only is there close similarity in the packing of equivalent proteins in this capsid layer, but there is also a striking similarity in the fold and architecture of the two proteins. Thus, structural alignment overlaps 568 residues with an root mean square deviation (rmsd) in C positions of 4.1 Å and an identical topology. This represents 63% of the residues for the BTV protein. Of the 568 residues, only 8.6% were chemically identical. The similarity of the structures of these two proteins from BTV and reovirus, and their identical arrangement in the viral capsids are further evidence that very many dsRNA viruses are linked via a common ancestor and suggest that the basic assembly mechanism is common to this large family of viruses.

CONCLUSIONS

Viruses provide ideal model systems for studying the related and sometimes intertwined issues of folding and assembly by using a combination of crystallography and electron microscopy to address structures of assembly intermediates and spectroscopic and light scattering techniques to follow the dynamics of assembly.

REFERENCES

1. J.N. Feng, P. Model and M. Russel, A trans-envelope protein complex needed for filamentous phage assembly and export. Mol. Microbiol. 34 (1999), pp. 745-755
2. R. Lata, J.F. Conway, N. Cheng, R.L. Duda, R.W. Hendrix, W.R. Wikoff, J.E. Johnson, H. Tsuruta and A.C. Steven, Maturation dynamics of a viral capsid: visualization of transitional intermediate states. Cell 100 (2000), pp.253-
3. R Tuma, JK Bamford, DH Bamford and GJ Thomas, Jr, Assembly dynamics of the nucleocapsid shell subunit (P8) of bacteriophage phi6. Biochemistry 38 (1999), pp. 15025-15033
4. A. Zlotnick, J.M. Johnson, P.W. Wingfield, S.J. Stahl and D. Endres, A theoretical model successfully identifies features of hepatitis B virus capsid assembly. Biochemistry 38 (1999), pp. 14644-14652
5. R. Schwartz, R.L. Garcea and B. Berger, 'Local rules' theory applied to polyomavirus polymorphic capsid assemblies. Virology 268 (2000), pp. 461-470
6. Z.H. Zhou, M. Dougherty, J. Jakana, J. He, F.J. Rixon and W. Chiu, Seeing the herpesvirus capsid at 8.5Å. Science 288 (2000), pp. 877-880
7. E.J. Mancini, M. Clarke, B.E. Gowen, T. Rutten and S.D. Fuller, Cryo-electron microscopy reveals the functional organization of an enveloped virus, Semliki Forest virus. Mol. Cell. 5 (2000), pp. 255-266
8. M.A. Krol, N.H. Olson, J. Tate, J.E. Johnson, T.S. Baker and P. Ahlquist, RNA- controlled polymorphism in the in vivo assembly of 180-subunit and 120-subunit virions from a single capsid protein. Proc. Natl. Acad. Sci. USA 96 (1999), pp. 13650-13655
9. Z Zhang, B Greene, PA Thuman Commike, J Jakana, PE Prevelige, Jr, J King and W Chiu, Visualization of the maturation transition in bacteriophage P22 by electron cryomicroscopy. J Mol Biol 297 (2000), pp. 615-62615

10. S. Li, C.P. Hill, W.I. Sundquist and J.T. Finch, Image reconstructions of helical assemblies of the HIV-1 CA protein. Nature 407 (2000), pp. 409-413

11. X.S. Chen, R.L. Garcea, I. Goldberg, G. Casini and S.C. Harrison, Structure of small virus-like particles assembled from the L1 protein of human papillomavirus 16. Mol. Cell. 5 (2000), pp. 557-567

12. W.R. Wikoff, L. Liljas, R.L. Duda, H. Tsuruta, R.W. Hendrix and J.E. Johnson, Topologically linked protein rings in the bacteriophage HK97 capsid. Science 289 (2000), pp. 2129-2133

13. K.M. Reinisch, M.L. Nibert and S.C. Harrison, Structure of the reovirus core at 3.6 Å resolution. Nature 404 (2000), pp. 960-967

14. J.M. Grimes, J.N. Burroughs, P. Gouet, J.M. Diprose, R. Malby, S. Zientara, P.P. Mertens and D.I. Stuart, The atomic structure of the bluetongue virus core. Nature 395 (1998), pp. 470-478.

15. P. Gouet, J.M. Diprose, J.M. Grimes, R. Malby, J.N. Burroughs, S. Zientara, D.I. Stuart and P.P. Mertens, The highly ordered double-stranded RNA genome of bluetongue virus revealed by crystallography. Cell 97 (1999), pp. 481-490.

16. C.L. Hill, T.F. Booth, B.V.V. Prasad, J.M. Grimes, P.P.C. Mertens, G.C. Sutton and D.I. Stuart, The structure of a cypovirus and the functional organization of dsRNA viruses. Nat. Struct. Biol. 6 (1999), pp. 565-568.

Compact Dark Objects and Gravitational Microlensing towards the Large Magellanic Cloud

L. Mancini*[†], Ph. Jetzer* and G. Scarpetta[†]**

*Institut für Theoretische Physik der Universität Zürich, CH-8057 Zürich, Switzerland
[†]Dipartimento di Fisica "E.R. Caianiello", Università di Salerno, I-84081 Baronissi (SA), Italy
**International Institute for Advanced Scientific Studies, Vietri sul Mare (SA), Italy

Abstract. Most of the matter in the galaxies is invisible to a direct observations. This dark matter is distributed in space differently than the stars and the gas, forming very vast and massive structures around galaxies, more spherical than disklike, called haloes. The composition of these haloes is unknown. It may comprise a mixture of exotic, hypothetical elementary particles and baryonic material, which can exist in several dark form, including planets, brown dwarfs, ancient degenerate dwarf stars, neutron stars and black holes, collectively known as massive compact halo objects (MACHOs). When a MACHO in the halo of our galaxy is sufficiently close to the line of sight between us and a more distant star, the light emitted from this source star suffers a gravitational deflection, according to the theory of General Relativity. The MACHO, acting as a *gravitational lens*, produces a *microlensing* effect, that is a potentially huge magnification of the source light. In this way, the MACHOs could be detected indirectly by their gravitational field. We analyzed the features of microlensing events found looking towards a Milky Way satellite, the Large Magellanic Cloud.

BARYONIC DARK MATTER

Today we know that in the universe there are different types of galaxies, each of them characterized by specific structures. Our galaxy, the Milky Way (MW), likewise the Andromeda galaxy, the nearest large neighbour, shows a spectacular rotating disk of stars and gas, together with a large central bulge. In addition to the visible structures, there is also an invisible halo of material that envelopes galaxies like the MW. The existence of dark haloes is directly related to the dark matter issue, namely the fact that more than 90% of the matter in the universe does not seem to emit any electromagnetic radiation. Its existence has been inferred from the gravitational effect that dark matter has on the observed motions of gas, stars, galaxies and cluster of galaxies.

The dark haloes are much more difficult to investigate than the stars and gas, because they are inaccessible to a direct observation. For our Galaxy, what we know for sure is that its halo is larger, ≈ 20 times more massive and differently shaped than the other visible galactic components.

The main unresolved problem of the dark haloes concerns our very poor knowledge of their physical composition, in particular if they are composed by ordinary matter (baryons) or by a new type of particles. A very important bound for the total amount of baryons in the universe comes from the agreement between the very recent observation of the cosmic background radiation by WMAP (Wilkinson Microwave Anisotropy Probe) [1] and the primordial post Big Bang nucleosynthesis theory. This requires that the density of the baryons ρ_b, in terms of the critical density of the universe ρ_c, $\Omega_b = \rho_b/\rho_c$, is ≈ 0.044. The critical density is the density leading to a flat universe. Since the luminous matter accounts for ≈ 0.006, the missing baryonic matter must be in form of dark matter. Part of this baryonic dark matter is probably distributed around galaxies, as a fraction of the total dark matter of their haloes.

There are many ways in which baryons can hide in dark forms: stellar remnants such as neutron stars or white dwarf, black holes, very small faint stars, brown dwarfs, planets, clouds of molecular H_2. These objects are generically called Massive Compact Halo Objects (MACHOs).

GRAVITATIONAL MICROLENSING

A good way to detect MACHOs was proposed by Paczyński [2]. He suggested that MACHOs could be revealed indirectly by their gravitational fields, which are able in principle to generate lensing effects by deflecting the light of background stars. This phenomenon, called gravitational lensing, was predicted by the Einstein theory of general relativity and was observed for the first time in 1979, when two very close quasars were clearly identified as being the lensed images of a single object.

For a cosmological situation, where the lens is a galaxy or even a cluster of galaxies and the source is a very distant quasar, one indeed sees two or more images which are typically separated by an angle of some arcseconds. Instead, in a galactic context, for a lens with a mass lower than one solar mass and located at a distance lower than 1 Mpc, the separation angle turns out to be of the order of some milliarcseconds. Thus, the images can not be seen separated. However, the measured brightnees of the source star varies with the time as the lens passes near the line of sight between the observer and the source. The name microlensing is associated precisely with this process. The corresponding light curve of the source is characterized by a symmetric peak as the microlens passes in front of it and the height and the width of the peak depends on how close it passes to the line of sight, as well on its velocity and mass. This time dependent particular variation of the light curve of the source is easily observable and it is a clear evidence of a microlensing event.

An important parameter of microlensing is the optical depth, that is the number of lenses inside the microlensing tube, with the axis along the line of view and transverse section radius equal to R_E, the so called Einstein radius, given by

$$R_E = \left(\frac{4GM}{c^2} \frac{D_{ol} D_{ls}}{D_{os}} \right)^{1/2}, \qquad (1)$$

where G is the Newton constant, c is the velocity of the light, M is the mass of the lens and D_{ol}, D_{ls} and D_{os} are the distance between observer and lens, lens and source, observer and source, respectively.

Assuming a spherical halo made entirely of MACHOs, one find an optical depth towards the nearest galaxy, the Large Magellanic Cloud (LMC), of $\tau \approx 5 \times 10^{-7}$. This means that in order to obtain a reasonable number of microlensing events, an experiment has to monitor several million of stars in the LMC or in other rich fields of stars like, for example, the Andromeda galaxy (M31), for a reasonable number of years.

Microlensing observations have now become an useful tool in searching for non-luminous astrophysical compact objects. Originally conceived to establish whether the halo of the Galaxy is composed of this type of objects, the ongoing search are also sensitive to the dark constituents of other galactic components of our galaxy (bulge, disk, spiral arms), as well the halo and the components of other target galaxies (essentially LMC, SMC and M31).

MICROLENSING TOWARDS THE LMC

The idea to use gravitational deflection of light to detect MACHOs in the halo of our galaxy by monitoring the light variability of millions of stars in LMC was first suggested by Paczyński in 1986 [2], and the theory was developed in a series of papers by De Rújula et al. [3] and Griest [4]. Since these first studies, the field has grown very rapidly, especially since the discovery of the first microlensing events at the end of 1993.

The main problem in microlensing is that the crucial observable, the event duration, suffers degeneracy in the three fundamental microlensing parameters: mass, distance and transverse velocity (to the line of sight) of the lens. In principle, we are not able to infer the location and the physical nature of the lenses. This makes it difficult to distinguish between the two principal geometric arrangements that may explain LMC microlensing:

i) MW microlensing, in which the lensing objects are part of the MW;

ii) self-lensing, in which both the lenses and the sources are part of the LMC.

While the MW is a well formed spiral galaxy, LMC is a irregular galaxy, which presents two main components: a disk and a central bar. LMC is tilted to the plane of the sky, with the north-east side closer to us than the south-west [5].

Until now, 16 microlensing events were found towards the LMC by the experimental team called MACHO collaboration [6], and 5 by the EROS team [7]. Among them, we have some information about the location and the physical nature of the lenses only for three events:

- One has been observed directly by the Hubble Space Telescope (HST), which was able to resolve the lensing object. Its mass was determined to be either $\simeq 0.04$ $M_\odot$ or in the range 0.095-0.13 $M_\odot$, so that it is a true brown dwarf or a M4-5V spectral type low mass star located in the disk of our Galaxy [8].
- The other two events were instead very particular, and probably generated by lenses in the LMC itself. In fact, one event was due to a binary lens, while the peculiarity of the other is that the source was a binary star. In the last case, the most realistic model gave an estimate of the mass of the lens equal to 0.24 $M_\odot$.

Presently, no other information about the lenses are available.

The measured optical depth by experimental team was $\tau = 1.2^{+0.4}_{-0.3} \times 10^{-7}$, significantly larger than allowed by known Galactic and LMC stellar populations, while the most likely MACHO mass is in the range 0.15 - 0.9 $M_\odot$. The better explanation of the experimental results given by MACHO collaboration was that the MW halo consists of about 20% white dwarfs. However, this test is not conclusive given the few events at disposal and the exact nature and location of the lenses is still a matter of controversy.

Optical Depth

By using specific density profiles both for the MW and LMC galactic components, obtained by a precise analysis of the geometrical structures and dynamics of these galaxies [9] [10] [11], we computed the optical depth for three cases:

- lenses located in the MW halo;
- lenses located in the LMC halo;
- self-lensing in the LMC.

In Fig. 1 we report the optical depth contour map for lenses in the Galactic halo, assuming that all the Galactic dark halo consists of compact lenses, together with the positions of the fields observed by the MACHO collaboration, of the microlensing events detected and of the line of nodes of the LMC disk. We observe that almost all the fields (except three of them) fall between the contour lines corresponding to $\tau = 46.3 \times 10^{-8}$ and $\tau = 49.3 \times 10^{-8}$; as expected the optical depth due to the Galactic halo is a slowly variable function, and presents a slight near-far asymmetry: moving from the nearer to the farer fields along a line passing through the center and perpendicular to the line of nodes, the increase of the optical depth is of the order of $\approx 6\%$.

In Fig. 2 and 3 are reported the optical depth contour maps for lenses belonging to the halo of LMC, respectively in the case of a spherical model and in the case of an ellipsoidal one, in the hypothesis that all the LMC dark halo consists of compact lenses. A striking feature of both maps is the strong near-far asymmetry.

For the spherical model, the maximum value of the optical depth, $\tau_{max,S} \simeq 8.05 \times 10^{-8}$, is assumed in a point falling in the field number 13, belonging to the fourth quadrant, at a distance of $\simeq 1.27$ kpc from the center of the LMC; the value assumed in the point, symmetrical with respect to the center, belonging to the second quadrant, falling about at the upward left corner of the field 82, is $\tau_S \simeq 4.30 \times 10^{-8}$. The increment of the optical depth is of the order of $\approx 87\%$.

The same happens if we use an elliptical model: the maximum value of the optical depth is higher than the previous, $\tau_{max,E} \simeq 9.88 \times 10^{-8}$, and is taken about in the same point, at the same distance from the center; in the symmetrical point with respect to the center, belonging to the field 82, the value of the optical depth is $\tau_E \simeq 5.05 \times 10^{-8}$. The ellipsoidal shape of the LMC halo gives rise to a further enhancement of the near-far asymmetry, placing the increment of the optical depth at level of $\approx 95\%$.

One can draw advantage from the different asymmetric behaviour of the optical depth in the two cases, both to confirm the existence of a proper LMC halo and to disantangle the microlensing events due to the Galactic halo from the ones due to the LMC halo. To this end, a good observation strategy could help; the goal beeing to allow the analysis of asymmetry of microlensing events belonging to two equivalent regions,

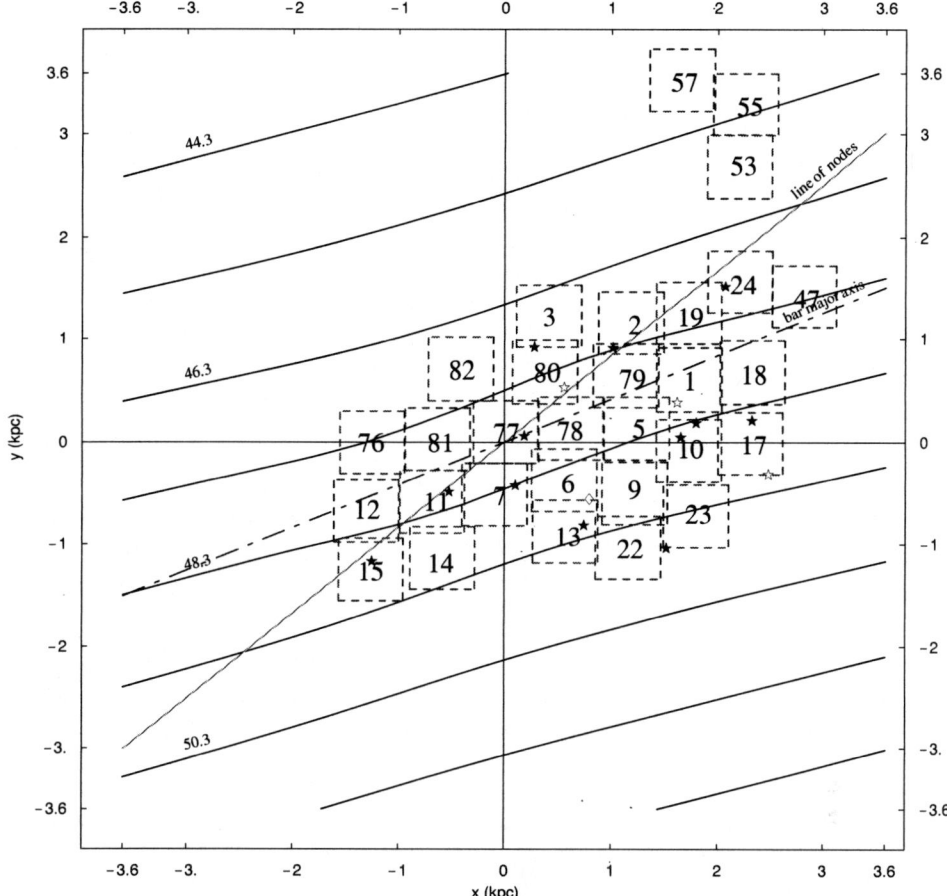

FIGURE 1. Contour map of the optical depth for lenses in the galactic halo. The locations and the nomenclatures of the MACHO fields (dashed boxes), and of the microlensing candidates (stars) are also shown, together with the position of the line of nodes and the bar major axis. The numerical values are in 10^{-8} units.

placed symmetrically with respect to the line of nodes. In figure 4 is reported the optical depth contour map for self-lensing, i.e. for events where both the sources and the lenses belong to the disk and to to the bulge of LMC. As expected, there is no evidence of a near-far asymmetry and the maximum value of the optical depth, $\tau = 4.8 \times 10^{-8}$, is assumed in the center of LMC. Starting from the center, for the central fields the optical depth would be rapidly decreasing along a line passing through the center and perpendicular to the minor axis of the elliptical disk. Instead, for the external fields, the optical depth would decrease more slowly.

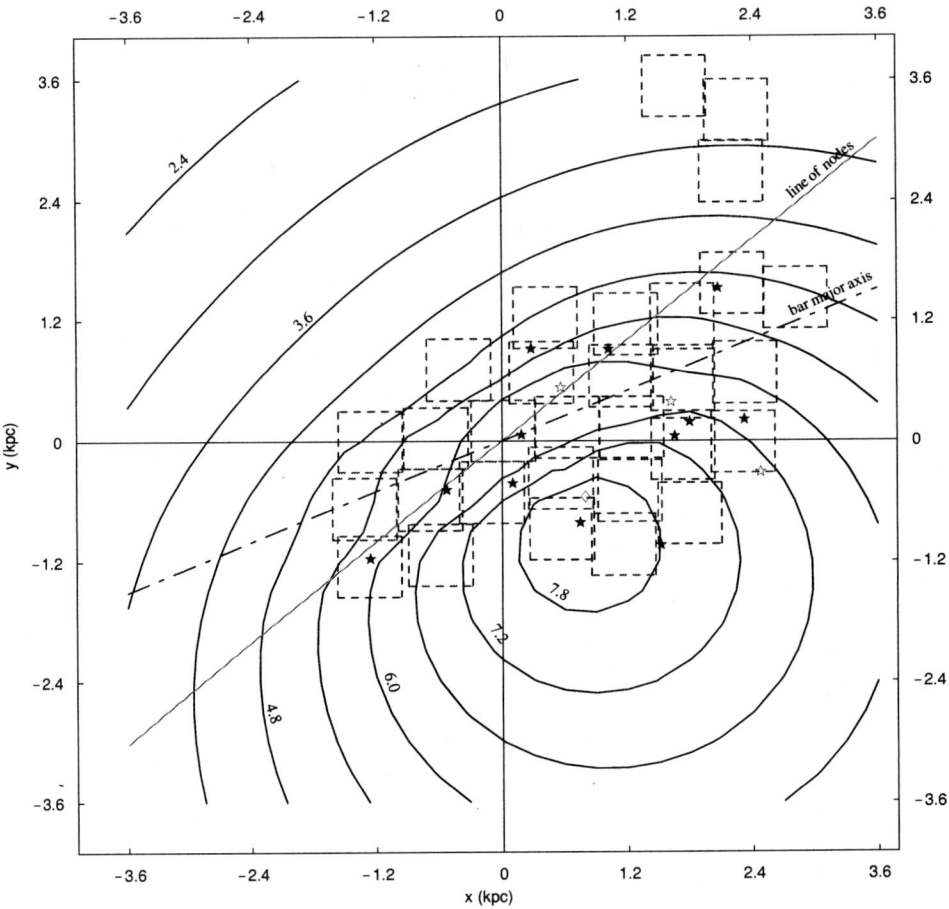

FIGURE 2. Contour map of the optical depth for lenses in the LMC halo, obtained by using a spherical model. The locations of the MACHO fields and of the microlensing candidates are also shown, together with the position of the line of nodes and the bar major axis. The numerical values are in 10^{-8} units.

MACHO mass fraction in the Galactic halo

Another important quantity to be determined is the fraction f of the local dark mass density detected in form of MACHOs in the Galactic halo.

Using the values given by the MACHO collaboration for their 5.7 years data, we estimated f for different models and also for the EROS2 data assuming that all their events are due to MACHOs in the halo. The fraction, obtained by assuming a standard spherical halo, varies between 5% and 13% [12].

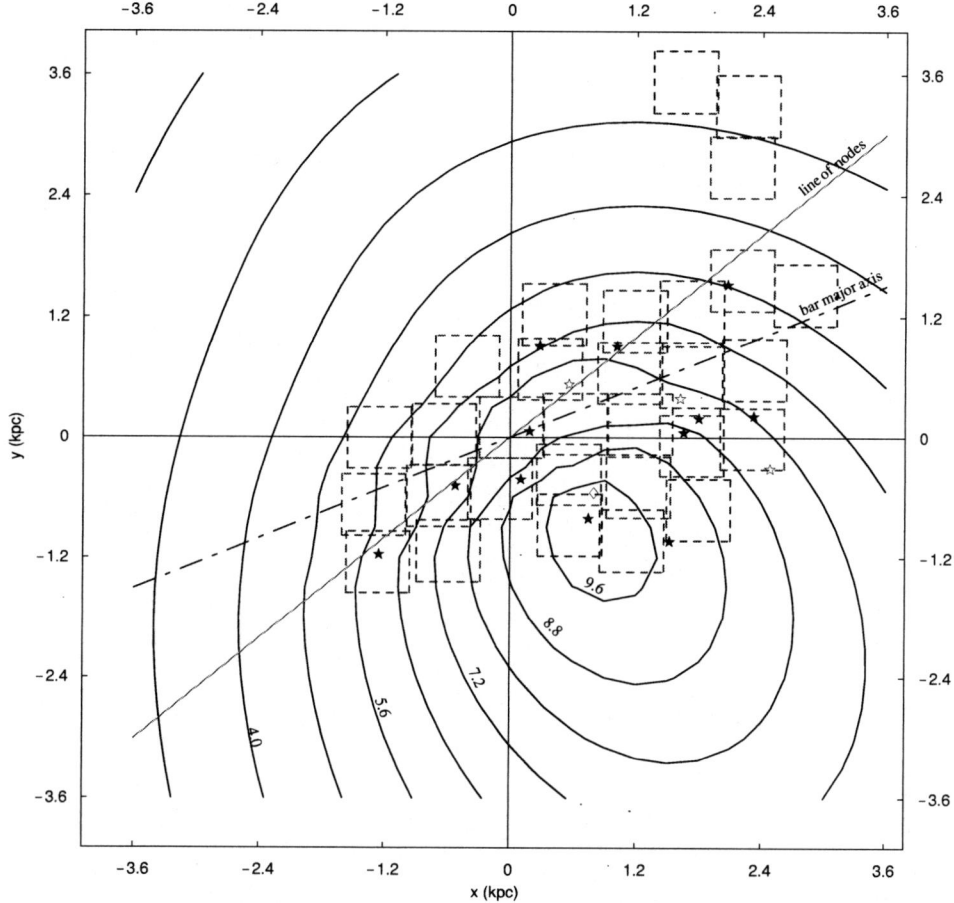

FIGURE 3. Contour map of the optical depth for lenses in the LMC halo, obtained by using an elliptical model. The locations of the MACHO fields and of the microlensing candidates are also shown, together with the position of the line of nodes and the bar major axis. The numerical values are in 10^{-8} units.

CONCLUSION

As already mentioned, the issue of the location and nature of the objects which act as lenses in the observed microlensing events is still an open problem, which can possibly be solved once more events will be available. The number of events found by the experimental collaborations are too few in comparison with that predicted for a halo composed entirely by MACHOs. In the last years several possible explanations of the experimental data have been proposed, but they are all not definitive and not completely satisfactory. One possibility might be that there are more MACHOs in the halo, but

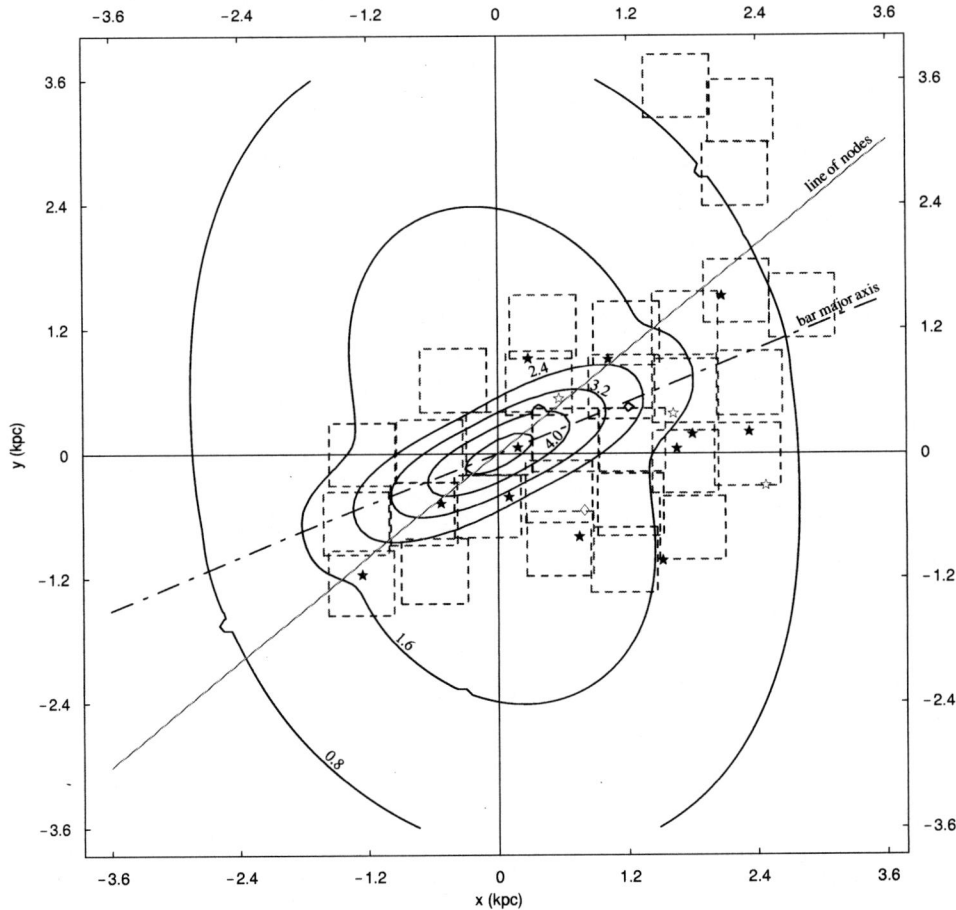

FIGURE 4. Contour map of the optical depth for the self-lensing case. The locations of the MACHO fields and of the microlensing candidates are also shown, together with the position of the line of nodes and the bar major axis. The inner contour line corresponds to an optical depth of 4.6. The numerical values are in 10^{-8} units.

that, for instance, they are associated with gas clouds [13] [14] [15], which would then produce non-achromatic events, that due to the present selection criteria have not been considered. On the other hand, one should also consider the possibility that some events might not be due to microlensing at all. This obviously underlines the fact that the present results have to be taken with care and that more observations are needed in order resolve this issue.

From the presently few available events and our phenomenological analysis it emerges clearly that, especially for the MACHO events, the lenses are due to different populations. Some are certainly due to LMC self-lensing, but this can hardly be the case for all the observed events especially for the few EROS events.

For the preferred LMC model we expect some 3 events among the MACHO ones (not including the possible self-lensing binary event) to be due to self-lensing. Moreover, with the mass moment method we find an average mass in the range $(0.1 - 0.5)\,M_\odot$ for the self-lensing events, which is consistent with the expectation that the lenses are low mass stars. The contribution of a LMC halo is, even if it exists, also very minor of at most $2 - 3$ event unless we are in the rather strange situation where the LMC halo is, contrary to our own, made almost entirely of MACHOs. From the galactic component, thin and thick disk, we expect roughly $1 - 2$ events on the MACHO data, but no contribution on the EROS data. This result seems to be in agreement with the fact that the event detected by the HST is a disk event. The inferred mass is small, though compatible with a low mass star, but clearly with only one event at disposal one has to take this value just as indicative.

As a result of our analysis we find that a plausible solution is that among the MACHO data some $2 - 3$ events are due to self-lensing, $2 - 3$ to the LMC halo, $1 - 2$ to the thick disk and the remaining ones due to the halo of MW.

Since the EROS events are less and given their larger spatial position, we do not expect that they get much contribution from self-lensing and the thick disk so that it is clearly not possible to draw more conclusions from them. On the other hand assuming that the EROS events are due to lenses in the halo, leads to a halo mass fraction which is in reasonable agreement with the corresponding MACHO value once the self-lensing and the disk events are subtracted. This way the MACHO and EROS results nicely fit together. Given our results it is also clear that one has to take with care values on the lens mass based on the assumption that all lenses belong to just one population. Clearly, once more data will be available, by using also the methods outlined in [12], it will be possible to draw more firm conclusions.

ACKNOWLEDGMENTS

This work is partially supported by the Swiss National Science Foundation.

REFERENCES

1. Bennet, C.L., Halpern M., Hinshaw G., et al., **astro-ph/0302207**
2. Paczyński B., Astroph. J. **301**, 503 (1986)
3. De Rújula, A., Jetzer, Ph., & Massó, E., Mont. Not. R. Astron. Soc. **250**, 348 (1991)
4. Griest, K., Astroph. J. **366**, 412 (1991)
5. Westerlund, B.E., *The Magellanic Clouds*, Camb. Astr. Se. **29** (1997)
6. Alcock, C., Allsman, R.A., Alves, D.R., et al., Astroph. J. **542**, 281 (2000)
7. Milsztajn, A., & Lasserre, A., Nucl. Phys. B Proc. Sup. **91**, 413 (2001)
8. Alcock, C., Allsman, R.A., Alves, D.R., et al., Nature **414**, 617 (2001)
9. van der Marel, R. P., & Cioni, M. R. 2001, Astron. J. **122**, 1807
10. van der Marel, R. P. 2001, Astron. J. **122**, 1827
11. van der Marel, R. P., Alves, D. R., Hardy, E., & Suntzeff, N. B., Astron. J. **124**, 2639 (2002)
12. Jetzer, Ph., Mancini, L., & Scarpetta, G., Astron. & Astroph. **393**, 129 (2002)
13. Bozza, V., Jetzer, Ph., Mancini, L., & Scarpetta, G., Astron. & Astroph. **382**, 6 (2002)
14. Bozza, V., & Mancini, L., Astron. & Astroph. **394**, L47 (2002)
15. Drake, A.J., & Cook, K.H., Astrophys. J. **589**, 281 (2003)

Summary of the Conference "Highlights in Condensed Matter Physics"

Andrzej M. Oleś

Department of Condensed Matter Theory, M. Smoluchowski Institute of Physics, Jagellonian University, Reymonta 4, PL-30059 Kraków, Poland

Abstract. The main ideas presented at the Conference "Highlights in Condensed Matter Physics" are shortly summarized. They might suggest important topics of future investigations in the theory of strongly correlated electrons including high T_c superconductivity, and in selected quantum and classical systems which are prominent examples of complex behavior in Nature. A special occasion celebrated by this Conference is also briefly emphasized.

The Conference "Highlights in Condensed Matter Physics", organized in honor of the 60th birthday of Prof. Ferdinando Mancini, took place at a spectacular site of Castello di Arechi (Arechi Castle) on the hill over Salerno and its bay on 9-11 May 2003, and brought together over 40 scientists. The scientific programme included 28 invited talks presented by the leading international experts in their fields. It was remarkable that we all enjoyed learning many new ideas, good scientific discussions, and friendly atmosphere, regrettably missing on some other occasions. I even heard somebody asking at a coffee break: *"Where are all these referees from Physical Review Letters?"*.

The majority of talks addressed the solutions of models describing the behavior of strongly correlated electrons. Among them, the discussion concentrated mainly on the possible mechanisms of high T_c superconductivity, which is an outstanding problem in this field. With over 15 years of efforts in theory and over 10^5 papers written, there is only little consensus at present concerning the mechanism of high T_c superconductivity. However, we definitely arrived at the stage when controlled analytical approaches or sophisticated numerical methods are used, frequently also a combination of the two.

Our understanding of the magnetic interactions in t-J model is now much improved by the numerical studies of *charge response*, with low energy excitations on the energy scale $\sim J$, suggesting polaron formation in a *spin liquid*. The charge response provides an explanation of doping dependence of breathing phonons, as shown by P. Horsch. We have heard a theory of anomalous ω/T scaling of the dynamic magnetic response in cuprates, with an overdamped collective mode at low doping being *inconsistent* with the Fermi liquid picture (P. Prelovšek). A better motivation for the relevant electronic structure comes from the charge transfer model. It suggests that all holes doped in the oxygen band are paired (N.M. Plakida), and explains in a natural way that the polaronic effects have to be strong in the cuprates at low doping, leading to *bipolaronic superconductivity* (A.S. Alexandrov). Finally, *phonons* also play a role and soften near the superconducting transition, as shown by recent experiments. Phonons can couple to charge fluctuations with low frequencies, and might thus induce pairing interactions only in some regions in **k**-space, leading to the observed $d_{x^2-y^2}$-wave symmetry of the superconducting parameter, as explained by M. Tachiki.

The existence of *two distinct energy scales*: T^* for the *pseudogap* and T_c for the collective (superconducting) behavior, is well established at present, and a better understanding of the transition at T^* would help us to resolve the origin of pairing and the mechanism of high T_c superconductivity by itself. We heard several different proposals of the origin of this remarkable behavior. First of all, within the Hubbard-Holstein model the pseudogap formation temperature T^* is governed by the proximity to an (incommensurate) charge ordering *quantum criticality* (stripe formation), and isotope effect occurs for T^*, but not for T_c (C. Di Castro). The two energy scales, and the high values of T_c, are well explained by bipolaronic superconductivity (A.S. Alexandrov). Yet, a richer two-component model of coexisting local two-electron pairs (bosons) and itinerant electrons, coupled by charge exchange mechanism, has very interesting properties and is able to explain the pseudogap, while a transition to a collective behavior happens at T_c simultaneously in both these subsystems. A closer look at Uemura plots, obtained both for s and $d_{x^2-y^2}$ pairing symmetries, allows to conclude that the *two-component scenario* is likely (R. Micnas). Finally, K. Maki suggested that the pseudogap phase in high T_c cuprates is a candidate for unconventional spin density wave, while unconventional charge density wave characterizes some other systems, such as organic conductors and heavy fermions. This discussion shows that several aspects are of importance, and various instabilities leading towards: antiferromagnetic, stripe, chiral, or charge order compete with superconductivity in the cuprates, as emphasized with variational wave functions by M. Randeria.

A fresh look at the problem of superconductivity in high T_c materials (and not only) was provided by J. Hirsch, who emphasized that: *Superconductors are giant atoms"*. After all, the clue to the understanding of the superconductivity might be the observation that there is a *fundamental difference* between electrons and holes. Electron and hole wave functions in the models of correlated electrons are *different*: while electron wave functions are *bonding* and smooth, hole functions are *antibonding*, have rather sharp structures, and depend on all other electrons in the system. Thus, the spectral functions of a hole moving in a correlated system consists of a weak quasiparticle peak and broad incoherent background, while it has just a single δ-function for low number of electrons. As a result, the apparent electron-hole symmetry of the models of correlated electrons (Hubbard model and the like) is only formal, and does not reflect a realistic situation in crystals. The superconductivity occurs as a collective behavior only in hole systems, and its mechanism is *undressing* by which antibonding electrons loose their incoherent part, and the kinetic energy decreases. The theory predicts that, as in atoms, the electric charge distribution in superconductors is nonhomogeneous, and could lead to currents which should be verified experimentally. Summarizing, many aspects are of importance to understand the puzzle of high T_c superconductivity. In order to solve it complementary approaches should be encouraged, and, as shown by the above discussion, certain extensions of the t-J model (by next-neighbor hopping t', intersite Coulomb and exchange interactions, or phonons) *are necessary*. Another reason that the t-J model might be oversimplified, and should in reality evolve gradually under doping, follows from the filling — it is relevant for holes which *have to know* about electrons in the underlying system.

Significant progress in the models of correlated electrons was reported in several other contributions. The physical properties of the Kondo lattice model are dominated by the quasiparticles which change with temperature T. Thus, one has to combine sophisticated band structure calculations at $T = 0$ with many-body approaches capable of describing finite T (W. Nolting). Inter alia, one finds a magnetic surface state for EuO films, which might give rise to a metal-insulator transition on the surface. The model of Kondo lattice plays an important role in manganites — it describes a competition between different phases and gives interesting phase diagrams bearing some similarity to the experimental ones (M. Avignon). The properties of *nanoscopic clusters* were discussed by J. Spałek within a new method which optimizes the one-particle wave functions by the correlation effects. This method seems promising, and it would be worthwhile to arrive at a combination of analytical and numerical approaches to describe correlated systems in the thermodynamic limit. A crossover from a band to a Mott insulator, which occurs in ionic Hubbard model, is not yet completely understood. While new topological excitations have been established, the mechanism of *spin and charge separation*, as well as the occurrence of second phase transition, are puzzling (G. Japaridze). Finally, I.V. Stasyuk analyzed the properties of pseudospin electron model, including the possibility of superconducting state.

When the on-site Coulomb interaction is large, electrons localize and spin models are typically more complex than just the familiar Heisenberg term for $S = 1/2$ spins in the t-J model. In spin-$1/2$ chains coupled to phonons, as in SrCuO compounds, the elementary excitations of the system are composite spinon-phonon objects (N. Andrei). When instead degenerate d orbitals are partly filled, the structure of superexchange in a Mott insulator ($n = 1$) is richer, and one finds *spin-orbital models* with intrinsic frustration of spin and orbital interactions, as I explained myself. In addition, only spin interactions have SU(2) symmetry, while this symmetry is removed for the *orbital* terms by multiplet splittings. In doped ferromagnetic e_g systems (with polarized and thus integrated out electron spins), one finds an *orbital t-J model*, in which the SU(2) symmetry is absent again, causing unconventional behavior, quite different from that encountered in spin systems (L.F. Feiner). We could appreciate that some familiar features of magnetic systems, for instance, the absence of superexchange (without orbital degeneracy) and quantum fluctuations in ferromagnetic states, or the Nagaoka's theorem, are simply consequences of the SU(2) symmetry. The gradual release of the kinetic energy in an *orbital liquid* state, when a Mott insulator at half-filling ($n = 1$) is doped by x holes ($n = 1 - x$) is a generic behavior characteristic of strongly correlated electron systems, and quantified by the Gutzwiller renormalization factor $q(x)$. Indeed, it was seen not only in the increasing values of T_c in the cuprates, but also in increasing ferromagnetic (double exchange) interactions $\propto q(x)$ in metallic manganites.

Other contributions addressed fundamental problems in condensed matter and selected topics in the behavior of complex systems. A criterion for the transition of a three dimensional Bravais lattice from bulk to molecular behavior (V. Shrinivasan) is an inverted question to that asked in the physics of mesoscopic systems, both being specific formulations of a more general question: *"How large is large enough?"*. Recent development in *Bose-Einstein condensation*, discovered in dilute alkaline atomic gases in magnetic traps, and in quantum coherence, was discussed by H. Matsumoto. We had a review of the polaronic behavior in various 3D/2D systems, and in quantum dots by J.T. Devreese. We heard as well about the complexity of quantum effects in quantum dots (F. Guinea). It was pointed out that the high T_c materials with their layered structure represent intrinsic Josephson junctions, which can be described within a phenomenological model explaining their optical properties (T. Koyama). The turbulence was presented as a universal phenomenon, common in such *complex systems* as: the tangles in superfluid He, the distribution of transverse momenta in electron-positron collisions, the distribution of relative velocities of stars and galaxies or the distribution of price changes. It is composed of singularities in physical space, and the origin of these singularities might be twofold — they might originate either from a vortex core itself, or from interactions between vortices (T. Arimitsu). Last but not least, the combinatorial background of parastatistics was presented in a fascinating lecture by S. Chaturvedi. The semion statistics of Haldane emerges here in a natural way, but only Fermi and Bose stastistics have *factorizable* partition functions.

On Saturday before lunch we had a very interesting and unconventional session with three talks by the young generation of family Mancini. This was really unique and showed both the complexity of related fields, and a constantly increasing impact of family Mancini in science. Erika J. Mancini gave an excellent talk about the role of physics in biology, where not only advanced experimental methods based on modern physics are used, but also the *way of thinking* used in physics helped to achieve a synthetic and focused advance on particular problems in structural virology. Francesco Paolo Mancini talked about the phase diagram of Josephson junction arrays with capacitive disorder. Finally, Luigi Mancini presented a talk about *dark matter*, which stimulated our imagination and allowed us to realize what a MACHO (massive compact halo object) is.

Let me finish with mentioning two talks on new methods for correlated electron systems. M. Imada presented the correlator projection method, which is by its nature non-perturbative and allows to control different energy scales, hence leading to a detailed and complete understanding of the phase diagram of the two dimensional frustrated Hubbard model. Our Jubillee (F. Mancini) presented an extensive report on the *composite operator method* (COM) with algebra constraints, which leads to self-consistent equations. The agreement between the COM and numerical methods demonstrated recently for static properties by F. Mancini and A. Avella is indeed impressive. It would be of interest to see as well how the COM works for the dynamic properties, in particular in multiband models, being so relevant for transition metal oxides. A combination of this (COM) approach with the state-of-the-art electronic structure calculations performed using local density approximation (LDA) might be a way towards a novel realistic approach for the spectral properties of correlated systems: LDA+COM. We hope that progress along this line is possible, and *"We wish that it would work!"*

Summarizing, we all agree that this successful Conference was a very appropriate way to celebrate 60th birthday of *our dear friend*, Professor Ferdinando Mancini.

"Dear Nando, we wish you all the best in your family and professional life!"

Our own roots

Ferdinando Mancini

Dipartimento di Fisica "E.R. Caianiello" - Unità di Ricerca INFM di Salerno
Università degli Studi di Salerno, I-84081 Baronissi (SA), Italy

Dear friends and colleagues,

The memory of this conference will accompany me for the coming years as one of the most memorable and enjoyable events of my life. It is very difficult to express the emotions and feelings of these days but certainly I wish to thank all the friends that organized this conference and the participants that greatly contributed to its success. In particular, my deepest gratitude goes to the chief organizer, Prof. G. Scarpetta, and to all the members of both the local and international organizing committee. A special thank goes to my colleagues and friendsCed Dr. A. Avella and Prof. V. Srinivasan. I would also like to thank the Rector of the University of Salerno, Prof. R. Pasquino; the Director of the Istituto Italiano per gli Studi Filosofici, Avv. G. Marotta; the President of the Istituto Internazionale per gli Alti Studi Scientifici, Prof. M. Marinaro; the Director of the Unità of the Istituto Nazionale di Struttura della Materia, Prof. G. Costabile; the President of the Provincia di Salerno, Dr. A. Andria.

On this occasion it is natural to draw a balance of what I call, being rather optimistic, the first 60 years of my life. I should say that I have been a very lucky person: in all these years I have always been surrounded by the love and friendship of so many people, and by the presence of brilliant and famous scientists. The first years of my life were marked, as naturally, by my family and, especially, by my mother. They taught me the importance of belonging to a solid structure where the roots and traditions are essential elements. To my mother, this year she has celebrated her 90th birthday, goes my deepest gratitude.

I should say that my approach to Physics was not systematic at all. Although in my boyhood I was reading books about Physics (I remember one book, the "Physics" by Carlton), I began my university career as a student in the Faculty of Law of the University of Naples. In one year I realized that most probably I would have never become a good lawyer, therefore I switched to the Faculty of Engineering, where I spent the following two years. In that period there was an important event at the Faculty of Science. Prof. Eduardo R. Caianiello returned to Naples from Princeton, founded the Institute of Theoretical Physics and started to build up the Neapolitan school of Theoretical Physics, gathering several famous physicists from all over Italy and abroad. These events attracted my attention (I must say that before Caianiello came back, there was, practically, no Physics in Naples) and I decided to go back to the dreams of my boyhood and I turned to Physics. Of course, this decision influenced the rest of my life. In those years the atmosphere at the Institute of Physics in Naples was very stimulating; before me many brilliant students had been attracted by the strong personality of Caianiello, had joined the Institute and contributed to its development. Among the large number of activities flourishing in the Institute I was particularly attracted by the group leaded by Maria Marinaro. In those years (early 60s) a new methodology within the Solid State Physics was rapidly developing. I am talking, of course, about the many-body theory of Condensed Matter Physics (CMP) based on the concepts and techniques of Quantum Field Theory (QFT). Caianiello was a QFT physicists and his presence was essential for the formation of a group in CMP. I graduated in 1966 under the supervision of Marinaro. After two years, spent at the Scuola di Perfezionamento in Fisica e Cibernetica, in 1968 I entered the Graduate School in Physics of the University of Wisconsin-Milwaukee (UWM), USA. This was the second fundamental decision I took in my life. Why Milwaukee? Between 1963 and 1966 the world-wide famous scientist, Hiroomi Umezawa, a very good friend of Caianiello's, was in Naples as a visiting professor. After Naples, Umezawa was called for a Chair at UWM in Milwaukee. Anyway, contacts were kept, and some young students from Naples moved to Milwaukee. After A. Aurilia, I was the second student involved in this collaboration. I must say that this choice was very lucky for me. At that time Umezawa was opening a new research program by applying methods of Quantum Field Theory to CMP and I naturally joined this program. Between 1968 and 1971 a strong group, leaded by Umezawa, constituted by Luc Leplae, V. Srinivasan and myself was established in Milwaukee. We managed to set up a formulation for superconductivity and apply it to real materials and well describe the vortex state. In those incredible and very fruitful years a new formulation of QFT, the Thermo Field Dynamics, was also developed. That period of my life also saw my marriage and the birth of my beloved daughter Erika.

In 1971 I completed the Ph.D. program at Milwaukee and decided to come back to Italy as professore incaricato at the University of Naples. With this choice I was just coming back to my original structure, remembering what I had learnt from my family. I took a third important decision just one year later. Together with Caianiello, Marinaro and Scarpetta I moved to the University of Salerno to found the Faculty of Science. With Marinaro, M. Fusco-Girard and other younger collaborators (too numerous to mention them all) we established a group in CMP. Since then this research line has been steadily expanding and, by now, it is one of the most relevant lines at the Department of Physics in Salerno. My collaboration with Umezawa continued and every year I spent some months in Milwaukee and Edmonton, where Umezawa had moved in 1975. On the occasion of one of these visits, actually in 1973, I met Hideki Matsumoto who joined Umezawa's group as a post-doc. Since then my collaboration with Matsumoto has continued uninterrupted up to the present days; particularly, between 1976 and 1978, when I spent two years as a visiting professor at the University of Alberta in Edmonton.

Since 1972 we have put much effort into the establishment and development of the Faculty of Science. The University of Salerno is rather young and before 1972 there was no scientific activity. I always like to recall how, from the original group of 4 people, we have been growing so much over the years. The Department of Physics in Salerno has now more than 120 members, 45 faculties, a large variety of experimental and theoretical groups, both in high and low energy physics, many international collaborations. I can certainly state that there is a Physical School in Salerno with a solid structure, known at international level. Between 1982 and 1992 my activity was mostly devoted to problems of organization and administration, not only of the Faculty of Science but of the entire University. During this period I was very little involved in research activity. Nevertheless, those years were quite important and fruitful. Sometime, we physicists live in our own world, forgetting that all around us there is the real one. The experience of those years somehow brought me back to real life. When, in 1992, I came back to research activity, I did it with a deeper motivation and a larger experience compared to how I had done it in the years of my youth.

I think that the last ten years have been the happiest and the most productive years of my professional life. In collaboration with Matsumoto we have been developing a formulation for highly interacting fields, where, as I just illustrated this morning, two main concepts, composite operators and algebra constraints, play a fundamental role. Matsumoto and I come from the same school, Umezawa's, and I believe that what we are doing now is the natural continuation and development of what we learnt from him. In these last ten years I have been very lucky to have in my group many brilliant students and post-docs who have brought very relevant contributions to the understanding and developing of our research project. Especially, I wish to mention Adolfo Avella; his presence in the group has been essential and thanks to him much progress has been made in the last five years.

Yesterday, I was very much gratified by my daughter Erika and my two sons Francesco Paolo and Luigi, who came here to participate to this event. They brought clear evidence that attention to the structures and traditions, the education and culture I received have been efficiently passed on. The existence of the Department of Physics in Salerno with so many young researchers and students is the clear evidence that the structure that Caianiello created almost fifty years ago in Naples has been transmitted. We can be happy of some personal success in physics, of writing some good papers and books, may-be of winning some prizes, but the most important of our contributions is, I believe, the one that we give to develop and to transmit what we received from our masters.

Let me close by recalling that in all these years I have always been accompanied, helped and assisted by my wife. The understanding and support has been constant all over the years, and I should like to close this recollection by expressing my deepest gratitude and love to my wife Fernanda.

AUTHOR INDEX

A

Alascio, B., 304
Alexandrov, A. S., 42
Andrei, N., 315
Arimitsu, N., 135
Arimitsu, T., 135
Avella, A., 258
Avignon, M., 304

B

Baizakov, B. B., 126
Bona, J., 101
Bussmann-Holder, A., 230

C

Chaturvedi, S., 145
Citro, R., 268
Cuoco, M., 215

D

Devreese, J. T., 47
Dóra, B., 10

F

Feiner, L. F., 188

G

Garcia, D. J., 304
Gentile, P., 215
Ghanashyam Krishna, M., 108
Görlich, E. M., 291
Guinea, F., 327

H

Hallberg, K., 304
Hirsh, J. E., 21
Horsh, P., 65

I

Imada, M., 75

J

Jetzer, P., 339

K

Khaliullin, G., 65
Konotop, V. V., 126
Koyama, T., 152

M

Maki, K., 10
Mancini, E. J., 333
Mancini, F., 240, 353
Mancini, F. P., 164
Mancini, L., 339
Marinaro, M., 268
Matsumoto, H., 114
Micnas, R., 230
Mizusaki, T., 75
Müller, W., 196

N

Noce, C., 215
Nolting, W., 196

O

Oleś, A. M., 176, 188, 348
Onoda, S., 75

P

Paramekanti, A., 34
Plakida, N. M., 92
Prelovek, P., 101

R

Randeria, M., 34
Robaszkiewicz, S., 230
Rosch, A., 315
Rycerz, A., 291

S

Salerno, M., 126
Santos, C., 196
Scarpetta, G., 339
Sega, I., 101
Shimshoni, E., 315

Sinjukow, P., 196
Sodano, P., 164
Spałek, J., 291
Srinivasan, V., 108
Stasyuk, I. V., 281

T

Tachiki, M., 1
Trivedi, N., 34
Trombettoni, A., 164

V

Virosztek, A., 10

W

Watanabe, S., 75

Z

Zahorbeński, R., 291